Edited by
Marc Descamps

Disordered Pharmaceutical Materials

Edited by Marc Descamps

Disordered Pharmaceutical Materials

WILEY-VCH
Verlag GmbH & Co. KGaA

Editor

Prof. Marc Descamps
UMET Batiment P5
University Lille1
59655 Villeneuve d'Ascq
France

Library of Congress Card No.: applied for

British Library Cataloguing-in-Publication Data
A catalogue record for this book is available from the British Library.

Bibliographic information published by the Deutsche Nationalbibliothek
The Deutsche Nationalbibliothek lists this publication in the Deutsche Nationalbibliografie; detailed bibliographic data are available on the Internet at <http://dnb.d-nb.de>.

Print ISBN: 978-3-527-33125-3
ePDF ISBN: 978-3-527-65272-3
ePub ISBN: 978-3-527-65271-6
Mobi ISBN: 978-3-527-65270-9
oBook ISBN: 978-3-527-65269-3

Cover Design Adam Design, Weinheim, Germany
Typesetting SPi Global, Chennai, India
Printing and Binding Markono Print Media Pte Ltd, Singapore

Printed on acid-free paper

Contents

List of Contributors

Karolina Adrjanowicz
University of Silesia
A. Chelkowski Institute of Physics
ul. Uniwersytecka 4
40-007 Katowice
Poland

Barbara Bellich
University of Trieste
Department of Chemical and Pharmaceutical Sciences
Trieste
Via Giorgieri 1
34127 Trieste
Italy

Pierre Bordet
Univ. Grenoble Alpes
Inst NEEL
F-38000 Grenoble
France

and

CNRS
Inst NEEL
F-38000 Grenoble
France

Silvia Borsacchi
CNR
Istituto di Chimica dei Composti OrganoMetallici (ICCOM)
via G. Moruzzi 1
56124 Pisa
Italy

and

Universitá di Pisa
Dipartimento di Chimica e Chimica Industriale
via G. Moruzzi 13
56124 Pisa
Italy

Kawthar Bouchemal
University Paris-Sud
School of Pharmacy
Institut Galien Paris Sud
UMR CNRS 861
5 rue Jean-Baptiste Clément
92296 Chatenay-Malabry Cedex
France

Elisa Carignani
Università di Pisa
Dipartimento di Chimica e Chimica Industriale
via G. Moruzzi 13
56124 Pisa
Italy

Attilio Cesàro
University of Trieste
Department of Chemical and Pharmaceutical Sciences
Via Giorgieri 1
34127 Trieste
Italy

and

Elettra-Sincrotrone Trieste S.C.p.A.
Strada Statale 14
34149 Trieste
Italy

Gerard Coquerel
Normandie Université
Crystal Genesis Unit
SMS EA 3233
Université de Rouen
Rue Tesniere
76821 Mont Saint Aignan Cedex
France

Marc Descamps
Université de Lille1
UMET (Unité Matériaux et Transformations)
UMR CNRS 8207
Bâtiment P5-215 59655
Villeneuve d'Ascq
France

Madeleine Djabourov
PSL Research University
Thermal Physics Laboratory
ESPCI-ParisTech
10 Rue Vauquelin
75231 Paris Cedex 5
France

Emeline Dudognon
Université de Lille1
UMET (Unité Matériaux et Transformations) UMR CNRS 8207
Bâtiment P5
59655 Villeneuve d'Ascq Cedex
France

Mark D. Eddleston
University of Cambridge
Department of Chemistry
Lensfield Road
CB2 1EW Cambridge
UK

Bertrand Gellie
Former Director Pharmacy & Pharmaceutical Chemistry
European Patent Office
Munich
Germany

Marco Geppi
Università di Pisa
Dipartimento di Chimica e Chimica Industriale
via G. Moruzzi 13
56124 Pisa
Italy

Giovanna Giannini
University of Trieste
Department of Chemical and Pharmaceutical Sciences
Via Giorgieri 1
34127 Trieste
Italy

Katarzyna Grzybowska
University of Silesia
A. Chelkowski Institute of Physics
ul. Uniwersytecka 4
40-007 Katowice
Poland

Alain Hedoux
Université de Lille 1
UMET (Unité Matériaux Et Transformations)
UMR CNRS 8207
Bâtiment P5-253
59655 Villeneuve d'Ascq
France

William Jones
University of Cambridge
Department of Chemistry
Lensfield Road
CB2 1EW Cambridge
UK

Khushboo Kothari
Scientist, Pharmaceutical Technology and Research and Development Laboratory
Takeda Oncology
35 Landsdowne Street
Cambridge MA 02139
USA

and

University of Minnesota
Department of Pharmaceutics
College of Pharmacy
308 Harvard St. SE
Minneapolis MN 55455
USA

Alessandro Maiocchi
Centro Ricerche Bracco
Bracco Imaging s.p.a.
via Ribes 5
10010 Colleretto Giacosa
Italy

Pauline Martinetto
Univ. Grenoble Alpes
Inst NEEL
F-38000 Grenoble
France

and

CNRS
Inst NEEL
F-38000 Grenoble
France

Marian Paluch
University of Silesia
A. Chelkowski Institute of Physics
ul. Uniwersytecka 4
40-007 Katowice
Poland

Sheng Qi
University of East Anglia
School of Pharmacy
Norwich Research Park
Earlham Road
Norwich NR4 7TJ
UK

Juraj Sibik
University of Cambridge
Department of Chemical Engineering and Biotechnology
New Museums Site Pembroke Street
Cambridge CB2 3RA
UK

Raj Suryanarayanan
University of Minnesota
Department of Pharmaceutics, College of Pharmacy
308 Harvard St. SE
MN 55455
Minneapolis
USA

Rui Tamura
Kyoto University
Graduate School of Human and Environmental Studies
606-8501 Kyoto
Japan

Jean-François Willart
Université de Lille1
UMET (Unité Matériaux et Transformations)
UMR CNRS 8207
Bâtiment P5-229 59655
Villeneuve d'Ascq
France

Zaneta Wojnarowska
University of Silesia
Institute of Physics
Universytecka 4
40-007 Katowice
Poland

J. Axel Zeitler
University of Cambridge
Department of Chemical Engineering and Biotechnology
New Museums Site Pembroke Street
Cambridge CB2 3RA
UK

1
Some Facets of Molecular Disorder in Crystalline and Amorphous Pharmaceuticals

Marc Descamps and Jean-François Willart

Most drugs, agrochemicals, and so on, are formulated in the solid state, which may be either crystalline or amorphous (i.e., glassy). It is well known that glassy compounds are very disordered solids. However, molecular crystals can also have varying degrees of disorder. Even perfect crystals are always disordered because of the thermal agitation of atoms and molecules. The intentional use of disordered solids and amorphous materials can be of great interest in pharmaceutical formulations because they may have favorable biopharmaceutical properties, for example, enhanced solubility and dissolution capabilities [1–3]. The drawback of this approach is that, often, the disordered solids can be metastable or unstable, physically or (and) chemically [4–6]. Glassy materials are in a nonequilibrium state and evolve upon aging. Formation of disordered solids may also be accidental, during the processing of crystalline materials. That can dramatically undermine the expected stability of the drug [7].

The differences in behavior between the different types of solids are fundamentally associated with the peculiarities of the molecular disorder. In the pharmaceutical literature related to amorphous solids, most interest has been focused on the molecular mobility and the way it may impact stability [8]. If mobility obviously plays an important role, many other aspects of the disorder also can determine the stability level of solids [9]. In this chapter we examine some of the facets of disorder that make it possible to differentiate between crystalline and amorphous states and also between various amorphous states of the same compound. That requires considering the structural, dynamic, thermodynamic, and even kinetic aspects of disorder.

Disordered Pharmaceutical Materials, First Edition. Edited by Marc Descamps.

1.1 The Crystal/Amorph Alternative

1.1.1 Crystal/Amorph Alternative: Terminology and Solidity Concept

Condensed matter can normally come in two forms: liquid and solid. The distinction between liquid and solid states lies in how they respond to the application of a shear stress.

A liquid deforms continuously when it is subjected to a shear stress: it is a fluid form of condensed matter.

A solid, on the contrary, can support a shear stress without flowing. If the shear is not too strong, it will deform elastically. This means that when the external forces are removed, an elastically deformed solid returns to its initial state: it is a rigid form of matter.

Solids can be either crystalline or amorphous: they differ in their structure and in the way they are solid.

Perfect crystals are those in which a "motif" – formed by a limited number of atoms or group of atoms – is repeated periodically in a three-dimensional array (lattice). This is reflected in the specific external shape that the crystals can adopt. The equilibrium state of a material at low temperature is expected to be crystalline.

The amorphous state of a material does not possess the long-range translational order (periodicity) of a crystal. It has no specific external shape. A liquid is always amorphous, but amorphous materials can be either solid or liquid. Noncrystalline solids formed as the result of the deep undercooling of a liquid are conventionally called glasses.

Crystalline and amorphous solids are not solids in the same way:

- *Crystalline solids* are "real solids" in that the elastic reversibility does not depend on the length of time that the shear stress acts. The elastic behavior is characterized by the shear modulus G, which is the constant of proportionality between the strain and the applied stress.
- *Noncrystalline (amorphous) solidity* needs considering the viscoelastic property of a real liquid. Generally, a real liquid is capable of responding at first in an elastic way (instantaneous shear modulus G_∞). But, in the words of Maxwell [10], the elastic behavior is fugitive. After a given time τ (τ = relaxation time, which is a function of the temperature T), the behavior is that of a liquid. The behavior is then characterized by the viscosity $\eta(T)$. According to Maxwell's model [11], the value of τ is linked to that of η and G_∞ by

$$\tau = \eta / G_\infty \tag{1.1}$$

For liquid water at room temperature, $\tau \cong 10^{-3}\ \text{Pa·s}/10^{9}\ \text{Pa} = 10^{-12}\ \text{s}$. This very low value of τ gives rise to a high effective fluidity. The designation of an amorphous compound as a liquid or solid depends on the value of τ relative to the time of observation t_0. If $t_0 < \tau$, the material behaves as a solid. If τ becomes on the order of a few hours or days, a viscoelastic liquid can be considered operationally as a solid: it is a glass. G_∞ depends on the material and temperature, but its order of magnitude is always some gigapascals (GPa) or tens of GPa. When a liquid can be undercooled (i.e., cooled without crystallizing), the value of G_∞ is observed to vary with temperature. However, its temperature dependence is insignificant compared to the large temperature dependence of η. As a consequence, τ and η are roughly proportional. Values of $10^3 - 10^4$ s for τ correspond to viscosity values of about 10^{12} Pa·s ($\equiv 10^{13}$ poises). When the viscosity of an undercooled liquid – which fails to crystallize – reaches such values, we start observing a solid behavior. This marks the entry into the glassy domain. The glass transition temperature T_g has often been defined as that at which the viscosity of an undercooled liquid reaches a value of 10^{13} poise. We will later give another definition of T_g based on calorimetric observations.

1.1.2
Crystal/Amorph Alternative: Structural Order and Disorder

In this subsection we focus on the structural aspects of order and disorder. We are concerned about the relative positions and orientations of the molecules, ignoring the possible dynamic aspects of the disorder (molecular mobility). In a simple manner, an amorphous solid is sometimes defined as a "disordered solid"; however, crystals, even perfect, are always disordered at some level, and sometimes very disordered. On the other hand, the structure of many amorphous solids is, in fact, non-random at certain length scales. We will briefly discuss the various forms of disorders that will help us to specify the boundary between crystalline and amorphous disorders.

X-ray (and neutron) diffraction by a sample are probably the best techniques to provide direct information about the structural organization of condensed matter. The effects of the different types of disorders on X-ray diffractograms will be presented to help identifying them.

1.1.2.1 Perfect Crystals

A perfect crystal is the periodic repetition in three dimensions (in principle to an infinite extent) of unit cells containing an atomic or molecular motif composed of a few atoms only. The unit cell is built on three noncoplanar vectors $\mathbf{a}_1$, $\mathbf{a}_2$, and $\mathbf{a}_3$ (Figure 1.1). The structural situation at some point in space is exactly reproduced at every other point obtained by adding a lattice translation vector:

$$\mathbf{r}_m = m_1\mathbf{a}_1 + m_2\mathbf{a}_2 + m_3\mathbf{a}_3 \cdot (m_1, m_2, m_3 \text{ are integers})$$

In brief, a crystal $=$ a motif $+$ a lattice.

The translational invariance defines the long-range order (LRO) of a crystalline state.

X-Ray diffraction by perfect crystals (for details, see [12–14])

The geometry of an X-ray diffraction experiment is shown in Figure 1.2.

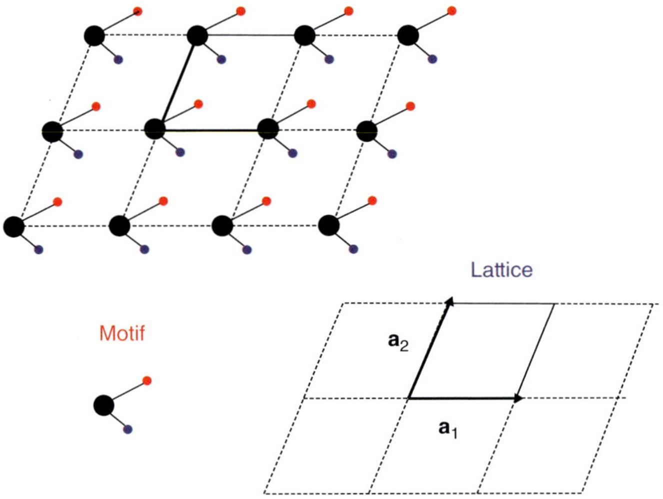

Figure 1.1 Two-dimensional representation of the periodic property of a molecular crystal.

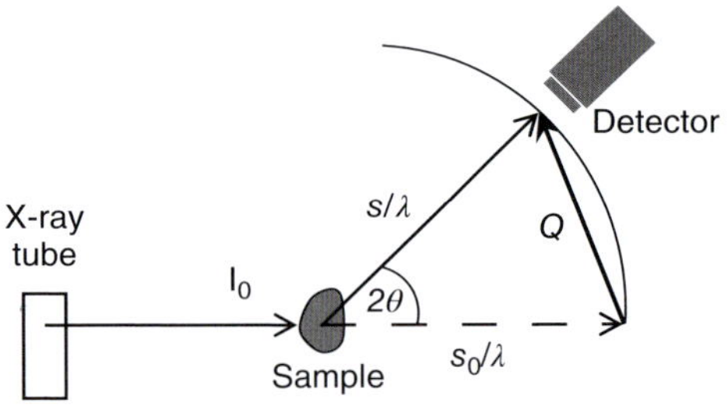

Figure 1.2 Usual setup for X-ray diffraction experiments. λ is the wavelength. **Q** is the scattering vector. $|\mathbf{Q}| = |(\mathbf{s}-\mathbf{s}_0)/\lambda| = 2\sin(\theta)/\lambda$ $|\mathbf{s}|=|\mathbf{s}_0|=1$.

The general expression of the intensity diffracted by the atoms of a sample – whatever its structure, crystalline or not – is given by

$$I(\mathbf{Q}) = \Sigma_i\Sigma_j f_i f_j \exp[i\, 2\pi\, \mathbf{Q}\cdot (\mathbf{r}_i - \mathbf{r}_j)] \tag{1.2}$$

where **Q** is the scattering vector, which is a vector in the reciprocal space as defined in Figure 1.2. f_n is the atomic scattering factor of the nth atom of the sample situated at $\mathbf{r}_n$ from the origin. Summations are taken over the full sample. This expression simply shows that X-ray diffraction provides an image, in the reciprocal space, of the structure of the sample.

In the case of a perfect crystal, the translational repetition of the motif allows simplifying the expression of $\boldsymbol{I}(\mathbf{Q})$ in the form of a product:

$$I(\mathbf{Q}) = |F(\mathbf{Q})|^2 \cdot \Im(\mathbf{Q}) \tag{1.3}$$

$F(\mathbf{Q})$ is the *structure factor* of the unit cell. It reflects the distribution of positions of the molecules (the *motif*) within the unit cell relative to the lattice points.

$$F(\mathbf{Q}) = \Sigma_j f_j \exp[i2\pi\,\mathbf{Q}\cdot\mathbf{r}_j] \tag{1.4}$$

where the summation is taken on the atoms j of the unit cell only.

$\Im(\mathbf{Q})$ is called the *interference function*. It reflects the geometry of the *lattice.*

$$\Im(\mathbf{Q}) = \Sigma_m \Sigma_{m'} \exp[i2\pi\,\mathbf{Q}\cdot(\mathbf{r}_m - \mathbf{r}_{m'})] \tag{1.5}$$

where $\mathbf{r}_m$ is the vector specifying the origin of the mth unit cell. Summations are taken on the full sample.

- Because of the translational invariance, if the crystalline sample is very large, $\Im(\mathbf{Q})$ has nonzero values only for $\mathbf{Q}$ corresponding to nodes of the reciprocal lattice (RL) of the crystal [12–14]. This determines the positions of the Bragg peaks. If the size of the crystal is large enough (more than a few micrometers), the width of the $\Im(\mathbf{Q})$ function around each RL node is negligible.
- The intensity integrated over one diffraction line is proportional to the number of unit cells.
- The positions of the diffraction peaks are given in the θ scale by the Bragg law:

$$\sin(\theta) = n\lambda/2d_{hkl} \tag{1.6}$$

where n is an integer and d_{hkl} is the spacing of a given set of crystallographic planes. Equation 1.6 also shows that Bragg peaks are observed for $|\mathbf{Q}| = n/d_{hkl}$. The position of the Bragg peaks depends only on the parameters of the unit cell. The intensity of the diffraction measured at one Bragg peak depends on the spatial organization of the motif in the unit cell, via the value of the structure factor $F(\mathbf{Q})$.

1.1.2.2 Crystal Size Effect

No crystal is fully perfect because of its finite size. Very often, pharmaceuticals are formulated as powders. Processes used to obtain the powder (milling, drying, etc.) may either give rise to micro/nanocrystalline or amorphous grains with very important consequences on the stability and functionality of the final drug. It is thus important to have a method to identify the structural and microstructural nature of the grains. Reduction of size can also be the result of accumulation of dislocations, which subdivide the crystal into small crystalline domains that diffract independently. It must be, however, noted that crystallite size is most often not the same thing as the particle size. A particle can be composed of several small crystallites. Particle sizes are measured, for example, by light scattering rather than by X-ray scattering.

Crystal size reduction leads to a specific type of X-ray peak broadening. In practice, this broadening can be observed only for crystallite sizes lower than 0.1 – 1 μm. Such peak broadening is only linked to a broadening of $\Im(\mathbf{Q})$ around the RL node while $F(\mathbf{Q})$ is not modified. As a consequence, the integrated intensity is still proportional to the number of unit cells composing the nanocrystal.

The width of a diffracted line, in Q units, is inversely proportional to the number of unit cells (N) along that direction.

$$\delta Q \propto 1/L_{hkl} \quad \text{where } L_{hkl} = Nd_{hkl} \tag{1.7}$$

The important point is that δQ does not change with the order of the line (i.e., with Q). If the line profiles are plotted on the 2θ scale, the equation relating the crystallite size L to the broadening $\delta(2\theta)$ of the line observed at Bragg angle θ_0 is given by the Scherrer equation

$$\delta(2\theta) \cong 0.9l\lambda/L \cos\,\theta_0 \tag{1.8}$$

For this type of plot, a specific cosine dependence of the width is observed. When the crystal sizes are very small, and the Bragg peaks very broad, a considerable overlapping of the peak wings can give rise to an apparent diffuse background, which should not be interpreted as such.

Figure 1.3 shows the diffractograms of the same crystalline compound for various crystal sizes. For the smallest size (~30 Å = 3 nm), the diffractogram can become very similar to that of the amorphous form. The interpretation in

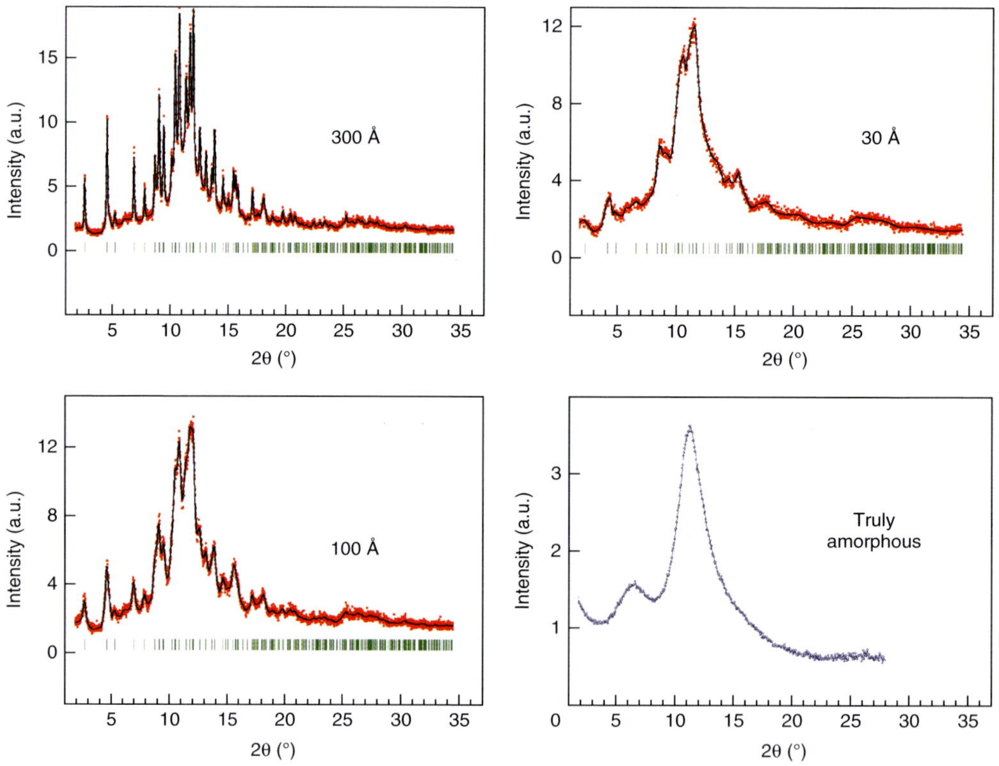

Figure 1.3 X-ray powder diagrams of the same molecular crystal for three different average sizes of the crystallites compared with the X-ray diagram of the fully amorphous form.

that case is challenging. Recent PDF (pair distribution function) analyses have made good progress in differentiating nanocrystalline and really amorphous compounds (for details see [15, 17] and the chapter by Bordet). When it is difficult to discriminate between an amorphous and a nanocrystalline situation, from the mere inspection of the X-ray diffraction pattern the presence or not of a calorimetric glass transition (existing only for the amorphous state) also helps in interpreting the structure.

1.1.2.3 Imperfect Crystals: How Disordered Can a Crystal Be?

The translational periodicity determines the LRO of crystals. However, some disorder or randomness is not necessarily incompatible with the crystalline nature. Such a disorder can arise from a local displacement of the structural elements (such as atoms, monomer units, motif) or a chemical substitution. A disorder induces fluctuations of the distance between homologous atoms. The characteristics of these fluctuations allow us to catalog crystalline imperfections and clarify the limit of crystallinity. In this respect, it is necessary to distinguish two kinds of crystalline imperfections [13, 14].

1) Imperfections of the first kind: They are such that the fluctuations of the interatomic (inter-motif) distances do not increase with the distance between motifs. Such imperfections preserve the long-range positional order of the lattice, on average. The motif that repeats translationally is an average one. This average is taken over either time or space.

$I(\mathbf{Q})$ can still be expressed in the form of the product shown in Eq. (1.3). The presence of $\mathfrak{J}(\mathbf{Q})$ still ensures the existence of Bragg peaks. Imperfections of the first kind do not induce broadening of the Bragg peaks. The structure factor is now an effective one: $\langle F(\mathbf{Q}) \rangle$.

It corresponds to the average motif, and takes into account all molecular positions inside the unit cell with an appropriate statistical weight. This type of disorder produces only a reduction in the intensity of the diffraction peaks at high Q values. The diffracted intensity removed from the Bragg peaks is spread throughout the reciprocal space in the form of a low-intensity, diffuse scattering.

Example 1.1. Thermal Agitation

The typical example of imperfection of the first kind is the thermal agitation of atoms and molecules that exist in every crystal. The regular positions of molecules in a crystal are only average positions around which they are continuously vibrating and librating. The centers of mass of the vibrating molecules are perfectly ordered because the vibrations of the different molecules are statistically similar. Figure 1.4 outlines a simplified example of a crystalline chain of atoms. It simply demonstrates how the fluctuations induced by vibrations are independent of the distance between the atoms. The amplitude of the fluctuation can be as large as 1/10 of the lattice parameter. It is highly temperature dependent. For X-ray diffraction, the atomic form factors have to be replaced by the temperature-dependent expression

$$f_j(T) = f_j \exp(-K\mathbf{Q}^2 \mathbf{u}_j^2) = f_j \exp(-M) \tag{1.9}$$

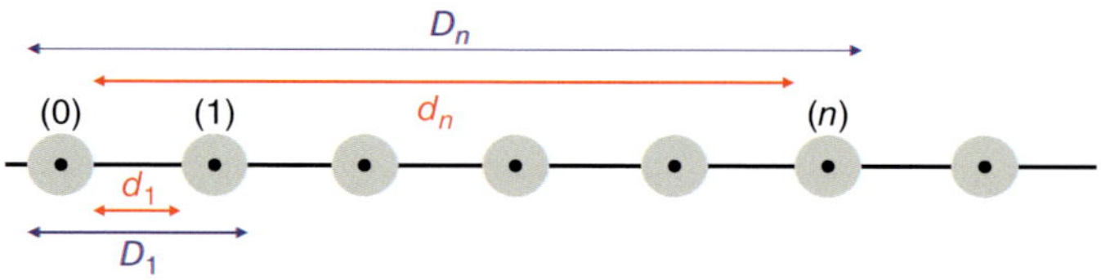

Figure 1.4 One-dimensional monoatomic model of imperfection of the first kind: thermal agitation. The gray zones show the spatial extension of atomic vibrations around the average position (black point). The fluctuation of the distances is independent of the interatomic distance: $D_1 - d_1 = D_n - d_n \, \forall n$.

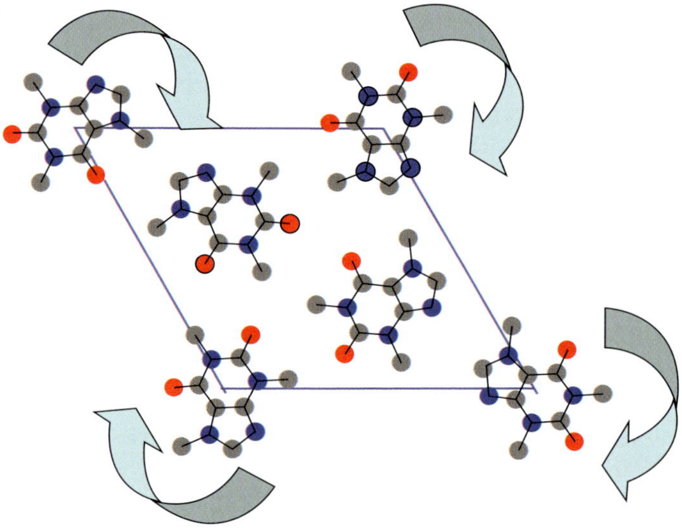

Figure 1.5 High-temperature hexagonal crystalline phase of caffeine. Molecules are rotating around the *C*-axis. The structure is perfectly crystalline but only on average.

where K is a constant and $\mathbf{u}_j$ is the average displacement of the atom.

M is usually called the Debye–Waller factor.

As a result of expression (1.9), $|F(\mathbf{Q})|^2$ and the Bragg scattering intensity are reduced. The reduction is more pronounced at high Q values (i.e., high Bragg angles) and for high temperatures where $\mathbf{u}_j^2$ is large.

Example 1.2. Rotationally Disordered Crystals ("plastic crystals") [18–20]

It is the case of the high-temperature crystalline phase of anhydrous caffeine [21]. Figure 1.5 shows the average hexagonal array of molecules. The high crystalline symmetry can be understood only if the caffeine molecules – which have a low symmetry – can rotate around their centers of mass. The dynamic aspect of this rotational disorder has been confirmed by dielectric relaxation experiments [21, 22]. Figure 1.6 shows the X-ray powder diffraction pattern of this phase [22]. It is characterized by very sharp peaks. The intensity of these peaks is, however,

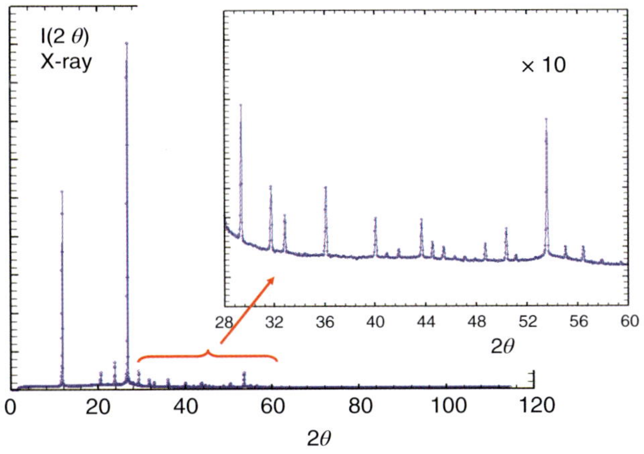

Figure 1.6 X-ray powder diagram of the high-temperature, hexagonal, disordered phase of caffeine. Notice the considerable decrease of peak intensities at high Bragg angle, which is due to the disorder. However, the Bragg peaks remains extremely sharp, which shows that the disorder is fully compatible with the existence of a perfect (average) crystalline periodicity.

decreasing fast with the Bragg angle. We can check that the width of the peaks does not increase with the diffraction angle. The situation is fully illustrative of the case of an imperfection of the first kind. The disorder is very important, but the crystal quality is very high.

We can notice the presence of a conspicuous diffuse scattering, which is spread in the angular range 10–30° (in 2θ). The diffuse scattering intensity corresponds to the intensity removed from the Bragg peaks due to the rotational disorder. This disordered crystalline phase is an example of mesophase. Other cases of mesophases are the liquid crystalline phases [23]. These latter types of mesophases are, however, no more crystalline. They are rather structured liquids (nematic, smectic, etc.).

Example 1.3. Ice Disorder

The crystalline polymorphism of ice is very complex, but the ordinary phase at atmospheric pressure (ice I_h) has a hexagonal structure, as shown in Figure 1.7a. The oxygen atoms form a regular hexagonal lattice, and each atom has four oxygen neighbors arranged tetrahedrally around it. One hydrogen atom resides along each bond connecting two neighboring oxygens. But each hydrogen lies close to one of the two oxygens and binds them. Each oxygen has two hydrogens close to it, which creates a local configuration similar to that of the H_2O molecule. However, the arrangement of hydrogens is not the same, but varies from one cell to another. The distribution of hydrogens is disordered: on each bond, the hydrogen atom can be closely bound to one or the other of the two oxygens. For the H_2O formula to be satisfied permanently, the jump of a hydrogen on one bond must entail a jump of another hydrogen on an adjacent bond. That is equivalent to saying that a water molecule can take six orientations around each oxygen site

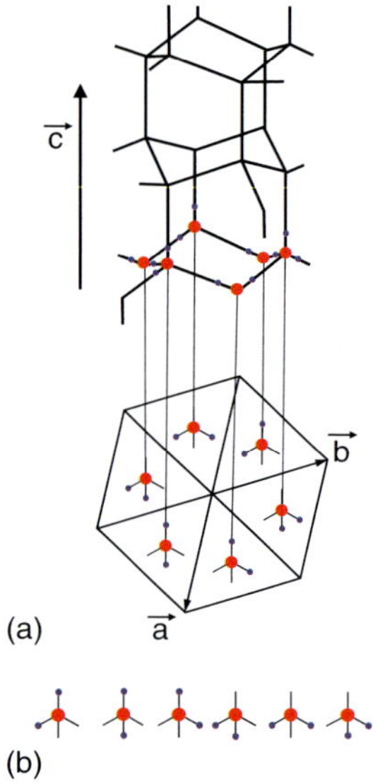

Figure 1.7 (a) Structure of hexagonal ice (I_h), showing the proton disorder. (b) The six possible configurations of a water molecule around each oxygen atom.

(Figure. 1.7b). That constitutes the Pauling "ice rules" [24, 25]. It is easy to imagine that the rotational motions of the water molecules are strongly coupled. The disorder of hydrogens and thus of water molecules is also manifested in the fact that the entropy of ice does not obey the third law of thermodynamics ($S(T = 0\,\text{K}) \neq 0$: the well-known residual entropy of ice). As for other disordered crystals, the intensity of the diffraction peaks decreases at high Bragg angles. The neutron intensity away from the Bragg peaks is spread across the reciprocal space of the crystal in the form of a diffuse scattering. Because of the specificity of the ice disorder, as well as the strong intermolecular correlations resulting from the ice rules, the diffuse scattering takes the form of very specific patterns in the reciprocal space [25].

Example 1.4. Substitutional Disorder

The disorders outlined above are all related to shifts in atomic or molecular positions. There are also imperfections of the first kind, which result from chemical substitutions. These are molecular alloys in which two kinds of molecules can coexist in a same perfect average crystalline structure (Figure 1.8). The different molecular species that are present within the crystal, in a certain ratio, are mixed without any LRO. Note that the different molecular species can be the result of a conformational disorder of a given molecule (see chapters by Cesaro *et al.* and Coquerel *et al.*). Each unit cell can be described by an "average molecule." The atomic positions are determined, but the description of the unit cell – of the motif – needs introducing a statistical weight corresponding to the molecular ratio. Such a system is a crystalline solid solution. An important example is the chiral compounds that exhibit a so-called pseudo-racemic crystalline phase (sometimes called pseudo-racemate or racemic solid solution). Contrary to the racemic compounds or the conglomerates, the two enantiomers coexist in the crystal lattice but not in an ordered manner [26].

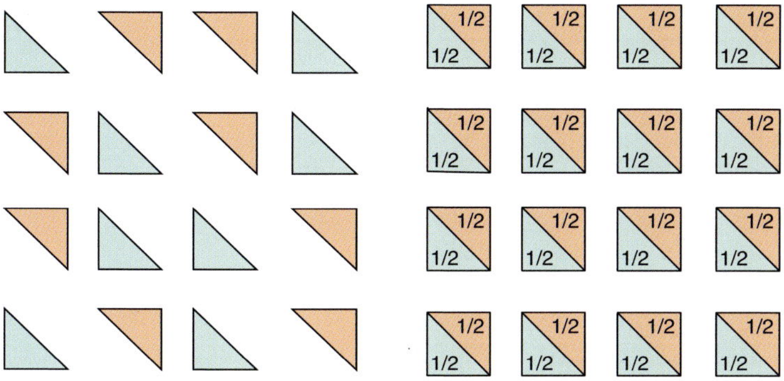

Figure 1.8 Two-dimensional illustration of a crystalline solution. Real local structure (a) and average structure (b) as determined by X-ray diffraction analysis.

2) Imperfections of the second kind: They are such that the fluctuations of the inter-motif distances increase with the distance between motifs. In that case, there is no more strict long-range crystalline periodicity or average motif.

When the fluctuations increase only very slightly, the molecular positioning is still rather well defined, though not perfect. The solid is then often designated as "badly crystallized". Sometimes, a "paracrystalline" description [16] can be adopted.

The X-ray diffraction of such a disordered system still shows a sort of diffraction peaks, which are however broadened. Unlike size broadening, the peak width δQ *increases with the order of the line* (i.e., with Q). The peak intensity also decreases rather fast as $|Q|$ increases.

Figure 1.9a gives a one-dimensional illustration of an imperfection of the second kind [13, 14]. Two kinds of atoms, with slightly different diameters, are stacked randomly. It is clear that in this case the fluctuation of the interatomic distance increases with the distance: $D_n - d_n > D_1 - d_1 (n > 1)$.

Figure 1.9b shows the X-ray picture corresponding to this system (in Q units). Pseudo-diffraction peaks are clearly visible. However, their width increases very rapidly with the order of the peak (as the square of the scattering vector $\mathbf{Q}^2$). At the same time, the maximum intensity decreases as $1/\mathbf{Q}^2$, so that only the first orders are observable. If the atomic scattering factors of the two types of atoms are different, the peaks are dissymmetric and their maxima are displaced from ideal reciprocal lattice nodes. By comparison, the peak width due to crystal size is independent of the peak index.

Another interesting situation corresponds to that of a collection of crystals with parameters that fluctuate around an average value. In that case, the peak broadening is proportional to $|\mathbf{Q}|$.

For solids of that type, which are often designated as "defective" crystals, line broadening analysis allows characterizing the microstructure and the type of

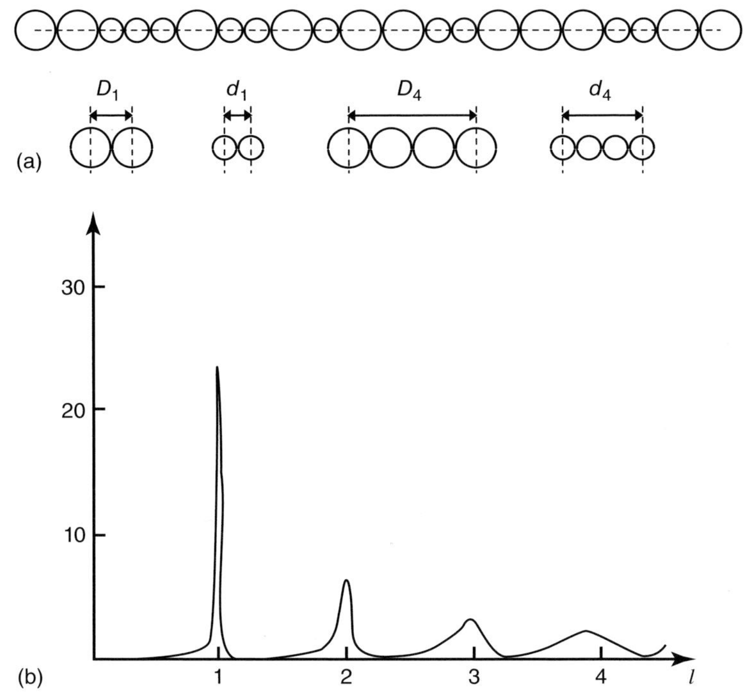

Figure 1.9 (a) One-dimensional illustration of an imperfection of the second kind [16]. Two types of atoms, with slightly different diameters, are stacked randomly. The fluctuation of the interatomic distance increases with the atomic separation [13]. (b) Corresponding calculated X-ray diagram [13, 14, 27].

disorder and identifying possible crystal size reductions. Many sophisticated software are available for this purpose (see the chapter by Bordet).

1.1.2.4 **Structure of Amorphous – Liquid or Glassy – Materials**

In these materials, the fluctuations increase so fast that any reference to a crystalline organization is difficult to make. It is the extreme form of imperfections of the second kind. The diffractograms have smooth variations because of a lack of strong interference effect. Figure 1.10 (bottom) shows the neutron diffraction pattern of amorphous ice. The profile has the particulars of what has been just described, but driven to the extreme: (i) considerable line broadening and (ii) fast increase in broadening with the order of the line. If an attempt is made to interpret the neutron diffractogram in terms of a nanocrystalline picture, the Bragg law (1.6) and Scherrer equation (1.8), respectively, give a repeat distance ($d \sim$ inverse of the first peak position) and a correlation length ($D \sim$ inverse of the peak width). This kind of brief analysis would predict a ratio $D/d \leq 3$. This value is much smaller than expected to consider the notion itself of crystallite being applicable. A statistical description of the structure is, therefore, preferred. In this frame, the structure is specified by the space variation of the radial pair distribution function (PDF) $g(r)$,

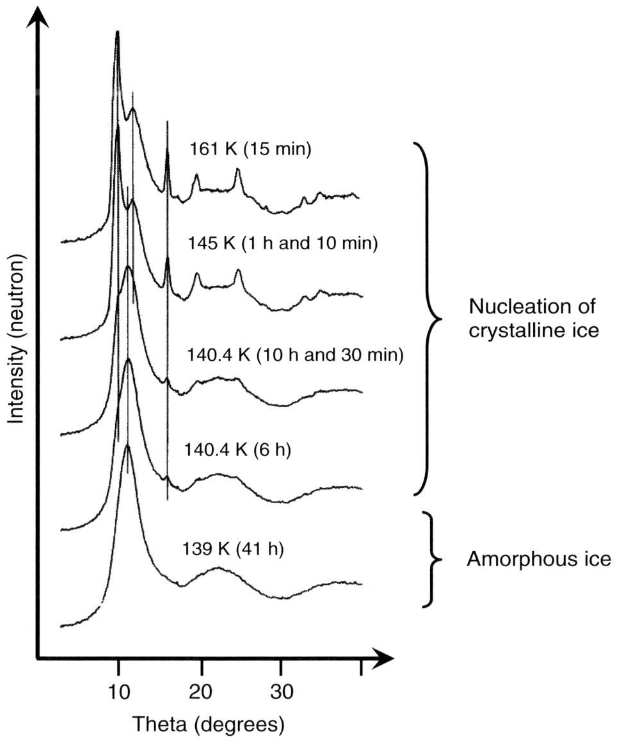

Figure 1.10 (Bottom) Neutron intensity measurement for low-temperature amorphous ice. The top diagrams, obtained upon heating, show the appearance of Bragg peaks corresponding to the progressive nucleation of crystalline ice (From Elarby *et al.* [28], Figure 1.10. Reproduced with permission).

which gives the probability of finding an atom at the distance r from another one. Experimental determination of $g(r)$ allows describing the short-range order (SRO) that possibly develops between the molecules.

Figure 1.11 illustrates the meaning of $g(r)$ in the case of a very simple monoatomic amorphous system.

The formalism to calculate the diffracted intensity $I(\mathbf{Q})$ is similar to that previously given (Eq. (1.2)). The difference is that the decomposition into the unit cell and lattice no longer makes sense, nor is the expression of $I(\mathbf{Q})$ in the form of a product. The average intensity $I(Q)$ can be expressed by introducing the average PDF $g(r)$. If we assume that all atoms in the sample are similar,

$$< I(Q) > \propto 1 + \rho_0 \int (g(r) - 1) 4\pi r^2 [\sin(2\pi Qr)/2\pi Qr]\, dr \tag{1.10}$$

where ρ_0 is the average value of density (for details see [14, 15] and the chapter by Bordet).

The formalism can be extended to polyatomic and molecular amorphous compounds in a rather straightforward way. The notation is, however, heavier. The

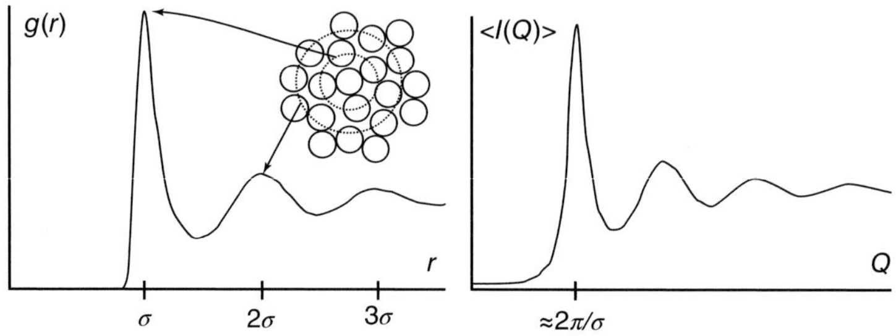

Figure 1.11 (a) Radial pair distribution function (PDF) $g(r)$ for a simple atomic liquid (atomic diameter σ). (b) The corresponding diffracted intensity $<I(Q)>$. A sample structure is also depicted where the solvation shells are indicated by the dotted lines. The exclusion radius can be seen in the absence of amplitude of $g(r)$ for $\sigma < r < 2\sigma$.

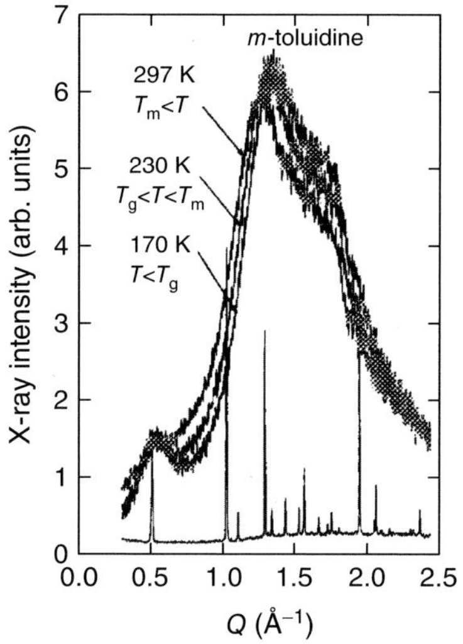

Figure 1.12 X-ray diffraction patterns of *meta*-toluidine. Liquid state: 297 K, undercooled liquid 230 K, glassy state 170 K. Crystalline phase: 230 K. (From Descamps *et al.* [32], Figure 1.12. Reproduced with permission of Progress of Theoretical Physics.)

PDF analysis is based on this formalism. It is widely used for amorphous compounds. It can also be used to investigate nanocrystalline materials. It is the case when the crystallite size becomes so small that distinction with a real amorphous state becomes dubious (cf. Figure 1.3). For details, see [28–31] and the chapter by Bordet. There are some cases where the simple observation of the diagram can give useful information. It is the case, for example, when a "pre-peak" is observed at a Bragg angle smaller than that of the main halo. In the real space, it is an indication of the formation of an intermediate range order that develops in the amorphous system. Such pre-peaks have been observed for several amorphous molecular compounds such as *meta*-toluidine and *meta*-cresol (Figure 1.12) [32]. Their presence suggests the formation of small clusters of chemically associated molecules [32, 33].

1.1.3
Crystal/Amorph Alternative: Metastability and Interconversion

1.1.3.1 Thermodynamic Measure of Physical Stability, Driving Force, Disorder

At given pressure, a crystal melts at temperature T_m. The melted compound – which is amorphous – can be undercooled below T_m with more or less ease. The possible existence below T_m of two states, namely crystalline or amorphous, poses the question of the relative stability of these states and of the rates at which the conversion between the two can occur. For transformations that occur at a given T and P, the stability of a state is measured by the Gibbs free energy (G), also named free enthalpy. G is defined by

$$G = H - TS \tag{1.11}$$

The differential form of $G(T,P)$ when T and P vary is given by

$$dG = -SdT + VdP \tag{1.12}$$

Here, H is the enthalpy, which measures the heat content of the system, and is reflective of its mechanical energy. T is the absolute temperature. S is the entropy of the system, which measures the level of molecular disorder. The value of G can be calculated using the measured values of the specific heat $C_p(T)$. The most stable state is that for which the Gibbs function is the lowest. The equilibrium value of G is a compromise between a low mechanical energy and a high disorder. Figure 1.13 shows a schematic variation of G with the molecular configurations at a given temperature T. Stable states correspond to minima. The lowest minimum corresponds to a stable configuration, and the highest to a metastable one.

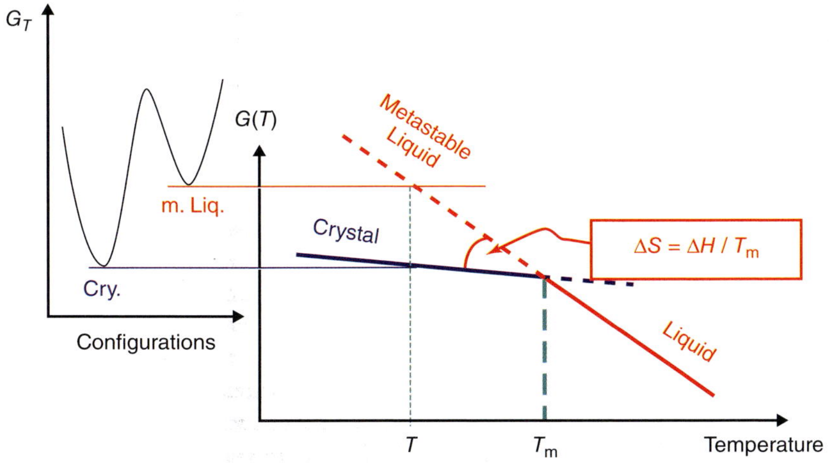

Figure 1.13 Schematic variations of the isothermal Gibbs curve as a function of molecular configurations for $T < T_m$. Stable states correspond to the minima. The figure also shows the evolution with temperature, in isobaric condition, of the free enthalpy curves of crystal and liquid phases.

Figure 1.13 also shows the evolution with temperature, under isobaric condition, of the free enthalpy curves of crystal and liquid phases. At the "equilibrium" melting temperature T_m, the values of G of the two phases are equal. The derivatives of $G(T)$ (which is $-S$) has a step change due to the increase in entropy (i.e., in disorder) ΔS_m when passing from the low-temperature crystal phase to the high-temperature liquid phase. Equality of the values of G for the two phases also means the existence of a latent heat ($\Delta H_m = T_m \cdot \Delta S_m$) for the transformation. Below T_m, the liquid state, which has a higher value, is metastable with respect to the crystal. In dealing with crystallization, we are concerned with the difference in G values between the two phases. Crystallization will be accompanied by a decrease in free enthalpy ΔG (J mol^{-1}). This ΔG decrease is the driving force for crystallization. At T_m, $\Delta G_m = 0$. Consequently crystallization cannot occur exactly at T_m. ΔG increases with the undercooling ($\Delta T = T_m - T$). A nonzero value of ΔG and, consequently, a minimum degree of undercooling ($\Delta T = T_m - T$) are necessary for a crystallization from the melt to occur. For small undercooling, and ignoring the difference in the specific heats of the liquid and crystal, a useful approximate expression of ΔG can be obtained:

$$\Delta G(T) \cong \Delta S_m \cdot \Delta T \cong \Delta H_m \, \Delta T / \, T_m \tag{1.13}$$

An exact expression should take into account the difference in heat capacity of the liquid and crystal, which is nonzero and especially large for the molecular compounds with which we are concerned. This difference is of the order of the magnitude of the heat capacity jump observed at the glass transition temperature, as will be shown later.

If a thermodynamic driving force is a prerequisite for a crystallization to occur, it is not its value alone that determines the occurrence and rate of the transformation itself. Several other factors play significant roles. In what follows, we discuss the interplay between the three main factors that determine the kinetics of recrystallization of an undercooled liquid and therefore determine the lifetime of metastability. In addition to the crystallization thermodynamic driving force, these factors are the liquid molecular mobility and the similarity in structure between the liquid and crystal (in particular near the interface).

1.1.3.2 Stability of the Amorphous State, Kinetics of Crystallization

The physical stability of the amorphous state is governed by the crystallization [34]. Crystallization occurs in the domain of metastability of the undercooled liquid ($T < T_m$). An amorphous, undercooled compound can be maintained in a metastable state for a length of time that depends strongly on the temperature. An understanding of the lifetime of metastability requires the investigation of the kinetics of recrystallization at each temperature. The progress of an isothermal crystallization as a function of time and temperature can be conveniently represented by the TTT (time, temperature, transformation rate) diagram, as shown in Figure 1.14 for the specific example of L-arabitol [35]. Specific isothermal experiments, which are sensitive to crystallization (DSC (differential scanning calorimetry), X-ray, dielectric measurement, etc.), allow us to plot such curves. A TTT

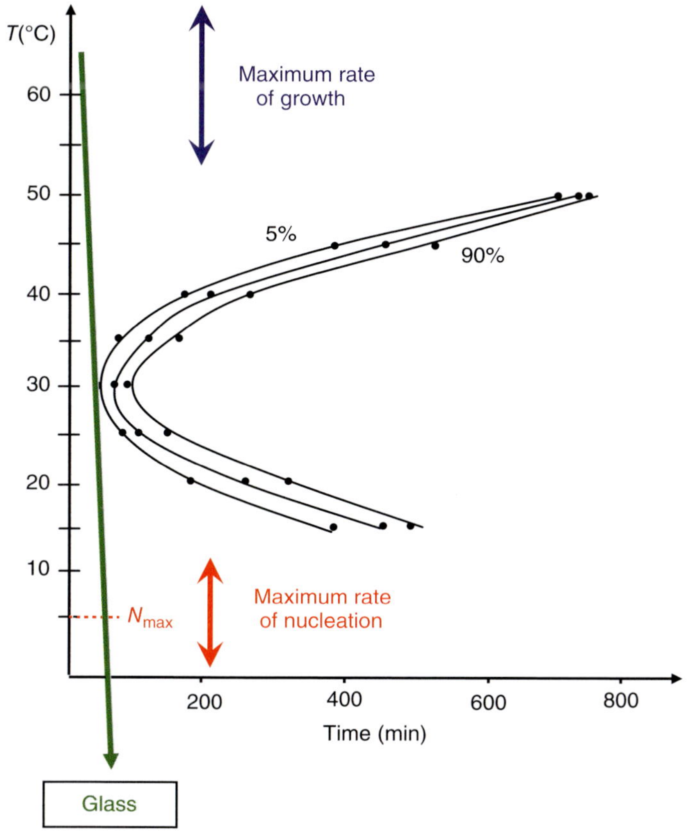

Figure 1.14 TTT (time, temperature, transformation rate) diagram for the specific example of L-arabitol. Also shown are the zones of maximum rate of nucleation (*N*) and growth (*V*). (From Descamps and Dudognon [35], Figure 1.14. Reproduced with permission of Wiley.)

diagram usually shows a nose-shaped feature of most rapid crystallization at some temperature. The overall behavior can be explained by the fact that the crystallization rate decreases when approaching the equilibrium melting temperature T_m because the thermodynamic driving force also decreases. At temperatures lower than that of the nose, the crystallization time increases as the molecular mobility decreases when approaching T_g. Going into detail, this description is too simplistic. Crystallization does not operate by a progressive homogeneous ordering process involving the entire sample at the same time. Crystallization of an undercooled amorphous melt results from the complex interplay between nucleation and growth processes. Nucleation is the process by which crystalline clusters having a minimum critical size appear randomly, with a given rate N (with dimension $t^{-1}L^{-3}$ in the usual 3D space), within the amorphous matrix. Once formed, the crystalline grains grow with a domain wall velocity V (with dimension $t^{-1}L$).

The combination of these two processes appears, for example, in the expression of the simplest JMAK (Johnson–Mehl–Avrami–Kolmogorov) model for the isothermal time (t) evolution of the fraction of crystallized material, $X(t)$ [36]:

For homogeneous nucleation and isotropic 3D growth

$$X(t) = 1 - \exp -[K(t/t_0)^4] \quad K = 4\pi/3 \tag{1.14}$$

This is a universal sigmoïdal function of the timescale

$$t_0 = (NV^3)^{-1/4} \tag{1.15}$$

where 4 ($=d+1$ where d is the dimensionality of the space in which the nucleation and growth transformation is proceeding) is the Avrami exponent. The temperature variation of t_0, which determines the shape of the TTT curve (Figure 1.14), depends on the rates of the nucleation and growth processes that have their proper temperature variations. Both have a maximum at some degree of undercooling. However, the two maxima do not occur at the same temperature, as shown in Figure 1.14. The maximum of $N(T)$ usually occurs slightly above T_g [37, 38], whereas the maximum of $V(T)$ occurs at a higher temperature often not far below T_m [39]. The width of the temperature gap between the nucleation and growth zones is important in determining whether a compound is a good glass former or not. It is also a determining factor in the physical stability of the amorphous compound.

Figure 1.15 shows the DSC curves recorded upon reheating an L-arabitol sample previously undercooled to temperatures $T_a < (T_m \cong 100\,°C)$ ranging from -83 to $82\,°C$. No calorimetric events can be observed on the DSC heating curves for samples annealed at $T_a > 35\,°C$. The absence of any crystallization exotherm or melting endotherm shows that no crystallization process occurred during the cooling/heating cycle. Similarly, for $T_a < -10\,°C$, crystallization exotherms are not detectable. On these curves, we can see the localization of the glass transition temperature at $T_g \approx -12\,°C$. These results show that the sample was totally amorphized during cooling. On the contrary, for annealing temperatures T_a ranging between -5 and $30\,°C$, pronounced crystallization exotherms, followed by melting endotherms, are observed. The exotherms have their maxima located not far below the melting point, in the range 60–95 °C, that is, at temperatures clearly separated from the annealing T_a domain. The interpretation of these results is that the exotherms correspond to the growth of crystals nucleated at a significantly lower temperature. Figures 1.14 and 1.15, respectively, show the estimated localization of the maxima of $N(T)$ (N_{max}) and $V(T)$ (V_{max}) and their positions with respect to the nose of the TTT curve. That underlines the relative influence of the factors that contribute to the recrystallization.

Both $N(T)$ and $V(T)$ have a maximum, which results from the antagonistic effects of

1) The thermodynamic driving force, which increases when the degree of undercooling increases;
2) The molecular mobility, which decreases when the temperature decreases.

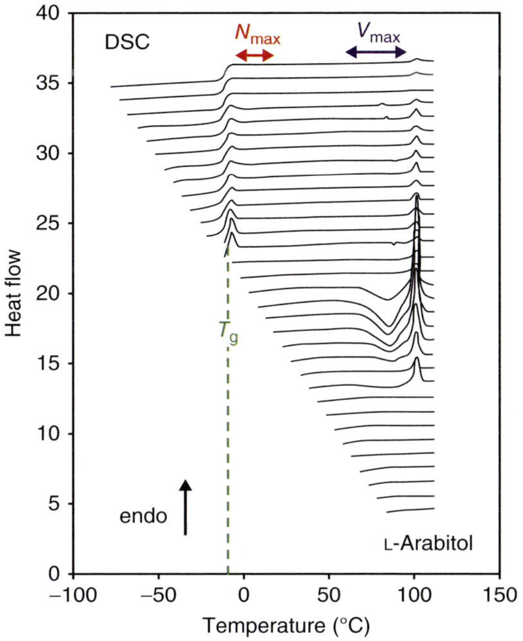

Figure 1.15 (L-Arabitol) DSC curves recorded upon heating (heating rate of $2\,K\,min^{-1}$) for samples undercooled at temperatures T_a ranging from −83 to 82 °C and annealed for 60 min at this temperature. (From Descamps and Dudognon [35], Figure 1.15. Reproduced with permission of Wiley.)

The latter prevails at low temperatures and becomes the limiting factor which permits to avoid the crystallization when quenching a liquid rapidly enough. The correlation between the decrease of $N(T)$ and $V(T)$ and the decrease of molecular mobility at low temperature has been shown in [35, 37] for indomethacin. The temperature positions of N_{max} and V_{max}, however, differ, mainly because of the effect of the surface tension γ between the crystallite starting to form and the amorphous metastable matrix. This is reflected in the expressions of $N(T)$ and $V(T)$ resulting from conventional theories [36, 40–44]:

- The classical "capillary" nucleation theory for $N(T)$ can be written in the form

$$N(T) = f_0(T)\exp(-g*_{3-d}/RT) \quad (1.16)$$

The mobility term f_0 is an "attempt frequency" for the addition of molecules from the metastable melt to the stable crystal across the interface. The thermodynamic effect is expressed in the exponential factor. $g^*_{3-d} = 16\pi\gamma^3/3\Delta G^2$ is the nucleation barrier. This expression results from the competition between an unfavorable, positive interfacial free enthalpy (γ) and the negative bulk driving force (ΔG).

- The semiphenomenological expression of Turnbull [45] for $V(T)$ can be written in the form

$$V(T) \propto V_0(T) \cdot \Omega \cdot [1 - \exp(-\Delta G/RT)] \tag{1.17}$$

The mobility term $V_0(T)$ plays a role similar to that of f_0. The term inside the brackets expresses the influence of the thermodynamic driving force ΔG on the net flux from the amorphous matrix to crystal. Ω is a function of ΔG and depends on the specific mechanism of growth. γ has a strong impact on $N(T)$ via g^*_{3-d} but has no (or limited for lateral growth) explicit influence on $V(T)$. The temperature separation of the preferential areas of nucleation and growth is mainly influenced by the value of the interfacial free enthalpy (γ). A high value of γ contributes to an increase in the distance between the two maxima N_{max} and V_{max} [35]. Because of this distance, and since N and V combine to give the characteristic time of the global kinetics t_0, the temperature variation of the latter is often connected to more than one cause. At low temperature, the decrease of molecular mobility has certainly a dramatic influence. However, due to the separation of $N(T)$ from $V(T)$, the effect of mobility on t_0 will result of a combination of factors, which have different temperature evolutions.

1.1.3.3 The Interfacial Free Enthalpy γ: Structure Dependence and Disorder Effect

It is useful to briefly address the molecular origin of the surface tension γ in order to better identify what can make a compound a good glass former. The value of γ is strongly influenced by the difference in the degree of order (disorder) between the amorphous mother phase and the crystallites, especially in the interfacial region. The surface tension γ is a free enthalpy (Gibbs free energy), and as such has an energetic and entropic contribution. Turnbull [45], Spaepen [46], and Oxtoby [47] pointed out that molecules of the liquid have to increase their ordering near the crystal boundary in order to optimize fitting with the translational order of the crystal and eventually allow an ordered embryo to grow. The entropy loss corresponding to this interfacial ordering is mainly the origin of the crystal/melt interfacial free enthalpy. Spaepen justified the fact (at least for atomic compounds) that entropy rises more slowly than enthalpy when going from a crystal to the bulk liquid. That is taken into account in the negentropic model [46, 48] of the interface. It provides an interesting guide to understanding the important and potential role of the relative disorder in the crystal embryo and relative order in the surrounding liquid.

Figure 1.16 summarizes schematically the situation. It shows the evolution of molecular ordering in the liquid at the interface and of the corresponding physical quantities, namely the enthalpy H and entropy S. At T_m, for example, there is a balance between enthalpic and entropic effects such as $\Delta H_m = T_m \Delta S_m$. At the interface, the balance is no longer satisfied, which gives rise to the excess free enthalpy γ. The maximum value of γ is obviously $\Delta H_m = T_m \Delta S_m$(bulk). At the interface, the molecular organization locally decreases the configurational entropy jump, which becomes ΔS(interface) < ΔS(bulk).

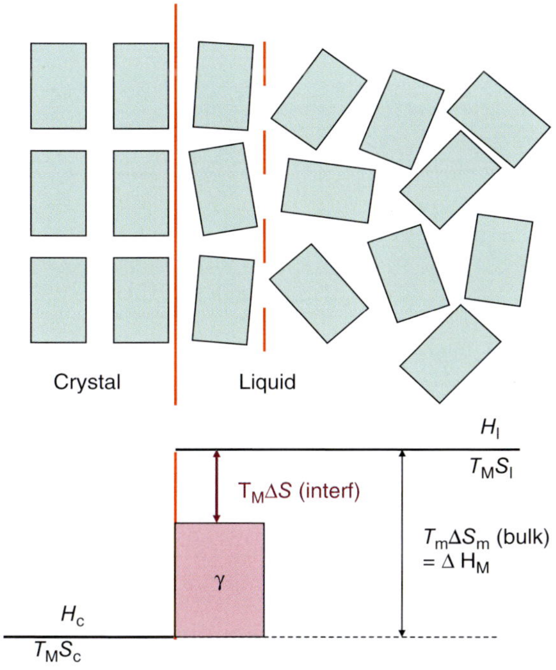

Figure 1.16 Schematic representation of the crystal/liquid interface showing short-range molecular ordering. The corresponding evolution of entropy that illustrates the negentropic approach of Spaepen (see Ref. [46, 48]).

The difference in free enthalpy (per molecule at the interface) between the liquid, which develops an interface, and that of the reference bulk is: T_m $[\Delta S(\text{bulk}) - \Delta S(\text{interface})]$.

Spaepen [46, 48] proposed the following formal expression for γ:

$$\gamma \propto \alpha T \Delta S_m \quad \text{with } \alpha = [\Delta S(\text{bulk}) - \Delta S(\text{interface})]/\Delta S_m (< 1) \tag{1.18}$$

The above considerations allow us to understand some of the main factors that make a compound easy to undercool, or not, from the liquid state.

If the melting entropy is low, the maximum value of the crystal/melt surface tension will also be low, and the undercooling ability will be often very poor. This is the case for crystals that are very disordered, such as plastic crystals [18–20]. It is also the case recently found for the high-temperature phase of caffeine [21, 22, 49], and the reason why metastable polymorphic varieties (in a monotropic situation), which have lower melting enthalpies than the stable phase and thus higher crystalline entropies, are often the first to nucleate. The first crystalline phase to appear is not that of highest thermodynamic driving force but that of highest entropy. It is a way to justify the observations made by Oswald, as set out in his "rule of stage" [50, 51].

Poor undercooling ability is also found in the case of molecular liquids that are locally organized in a way mimicking crystalline order. In this case, much further molecular organization is not needed at the interface. That leads to a low value of α. Turnbull and Spaepen used this argument to explain the small undercooling ability of n-alkane liquids [45]: The occurrence of linear configurations in the melt minimizes the entropy loss when adjusting to the crystal plane.

On the contrary, in the case of racemic compounds for example, the melt is a muti-component mixture of the R and S species. The entropy of melting is thus rather high. The ordered racemic compound is very difficult to nucleate from the melt, because a specific molecular organization is to be built at the interface that is costly in entropy. The first crystalline phase to appear is often a disordered solid solution of the R and S species. That seems to be the case recently found for racemic ibuprofen [35, 52, 53].

1.1.3.4 Concluding Remarks

Several factors may influence the recrystallization from the melt and thus play a role in the stability of the amorphous systems. These factors often compete in a delicate way. These are as follows:

- The thermodynamics that drives the process.
- The molecular mobility that facilitates (or not) the transformation.
- The interface energy that modulates the splitting of nucleation and growth processes. The interface energy is basically related to the structural similarity between bulk liquid, crystal, and the interface region between the two.
- The heterogeneities and cracks that may amplify considerably the rate of transformation. The role of these heterogeneities results from modifications of interface energy [36] and surface mobility [54, 55].

It should always be borne in mind that these different factors play a role when trying to interpret or to predict the stability behaviors of amorphous compounds.

Molecular compounds may exhibit a rich crystalline polymorphism with different structures and levels of disorder. The structural similarity with the liquid and the condition of recrystallization are modified accordingly. That plays a significant role in the phase selection.

1.1.4 Crystal/Amorph Alternative in the Context of Solubility

The low bioavailability of the active pharmaceutical ingredient (API), which is due to its poor solubility in water, is a major issue in pharmaceutical science. The use of the amorphous rather than crystalline form of API is considered one of the best formulation strategies to enhance the oral bioavailability. The principal reason that has long been recognized is that the amorphous state offers a way to extraordinarily increase the apparent aqueous solubility of poorly soluble pharmaceuticals [1, 5, 56–58]. Amorphous solid dispersion technologies have been emerging in the past years [59] in order to overcome the intrinsic instability of the amorphous state, both in the dosage form and during supersaturation in the intestinal environment. More recently, it has been shown that the use of the amorphous forms, which allow increasing the apparent solubility without the addition of a solubilizer (co-solvents, surfactants, cyclodextrines, etc.), can concomitantly increase the drug flux through the intestinal membrane and thus provide higher overall absorption [60–63]. This could have a significant impact on oral drug delivery of lipophilic compounds.

In this subsection, we identify the reasons for the higher apparent solubility of the amorphous solid form of a pharmaceutical compound compared with that of a stable crystal form. Traditionally, the term "solubility" refers to the equilibrium limit of solubility of the stable crystalline form in a solvent (in practice, water for pharmaceutical applications). It is the maximum concentration of the most stable crystalline form of the compound for which only a single phase solution occurs under equilibrium condition. The term *apparent solubility* refers to the more loosely defined limit of solubility experimentally reached when starting with an amorphous – unstable – solid form of the compound.

Take, for example, the case of indomethacine, whose solubility of the different forms has been reported in several publications [1, 5]. Figure 1.17 allows us to compare the *in vitro* experimental aqueous solubility profiles (measured at room temperature) of the most stable crystalline form γ and of the amorphous form obtained by quench-cooling of the melt. The enhanced solubility of the amorphous form is clearly demonstrated by the shape of the concentration versus time curve. The latter shows a peak of high concentration ("spring effect"), which occurs within the first 20 min. It corresponds to an apparent solubility enhancement ratio of ~5 with respect to that of the crystalline form. The peak then declines to reach a regime of very slow decrease ("parachute effect"), during

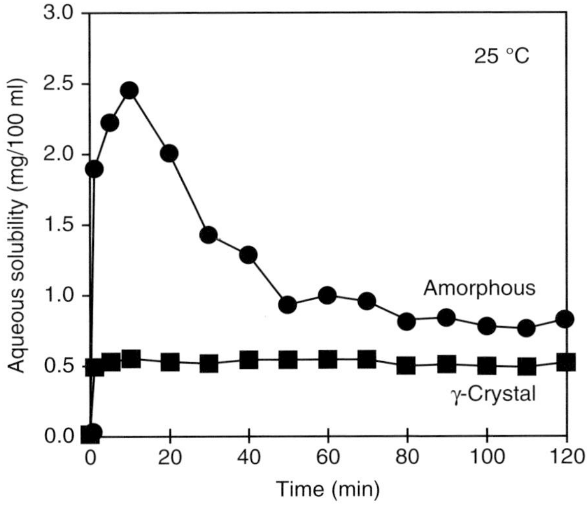

Figure 1.17 *In vitro* experimental aqueous solubility profiles of indomethacin (measured at room temperature). Comparison of the behaviors of the most stable crystalline form γ and of the amorphous form obtained by quench-cooling of the melt (Adapted from Hancock *et al.* [5]. Reproduced with permission of Springer). Similar results are reported in reference [1] figure 4.

which the concentration remains significantly higher than while the dissolution of the crystal form. The ratio of concentration values is still ~1.6 after 2 h. The "spring and parachute" effect is typical of the dissolution profiles found for the amorphous forms of many APIs. The significance of these results to the behavior of pharmaceutical dosage forms is clear. Even if the *in vivo* dissolution kinetics is more complex, it is likely that this type of behavior is able to deliver higher drug concentration for several hours.

It was mentioned in [1] that the presence of crystals could not be detected during the early "spring" stage. However, X-ray diffraction revealed a partial conversion to the crystalline form after 2 h of the "parachute" regime. That is a factor that contributes to the reduction of the ratio of concentration values. These results thus reveal the potential interest in using the amorphous forms, but at the same time helps us to imagine the complexity of the dissolution process – more properly the dilution process – of the nonequilibrium amorphous solid forms. This complexity results from the combination of modified supersaturations, kinetics of dilution, and recrystallization.

The concept of solubility limit itself is different for crystalline and amorphous solid states. This is because the dissolution of a crystal, unlike an amorphous solid, needs to provide the energy, which destroys the crystal lattice. To realize that, it has to be remembered that melting of a crystal is a first-order transition, which also involves a latent heat of melting to destroy the lattice symmetry. On the contrary, "glass melting" is a continuous process in which the viscosity decreases gradually when the temperature increases without requiring the important intake of heat. An amorphous solid is a liquid of extremely high viscosity. There is thus no symmetry-breaking between the amorphous state and the liquid solution state. Therefore, the solubility limit of the amorphous API solute arises from a possible liquid–liquid phase separation. It occurs only if molecules of the solvent and solute "dislike" each other enough ("homo interactions"). For an ideal solution (with supposedly no energy of mixing), there is a limit of solubility for the crystal, since energy is needed to destroy the lattice. On the contrary, there is no solubility limit for the amorphous solid state since a random mixing of the two types of molecules is always favored for entropy reasons. This is also the situation experienced in the case of "hetero interactions" between molecules.

In order to clarify this difference and to show the origin of the apparent solubility enhancement of amorphous API, we adopt a thermodynamic approach. We use a graphical description of the binary system API–solvent, which allows us to intuitively visualize the situation and gives a tangible meaning to the equations.

The stability condition specific to multicomponent systems needs evaluating the right Gibbs free energy, which now includes the mixing enthalpy ΔH_{mix} and the mixing entropy ΔS_{mix} as

$$\Delta G_{\text{mix}}(T) = \Delta H_{\text{mix}} - T\Delta S_{\text{mix}} \tag{1.19}$$

where ΔH_{mix} and ΔS_{mix} are, respectively, the differences between the enthalpy and entropy between the mixed and unmixed states. $\Delta S_{\text{mix}} > 0$, while ΔH_{mix} can be of either sign according to the type of interaction.

It also implies the evaluation of the partial molar Gibbs free energy (the chemical potential μ_i) of each component as

$$G = \mu_A X_A + \mu_B X_B \tag{1.20}$$

for a binary system A (solvant) – B (API), where X_A and X_B are the mole fractions of A and B. Obviously, $X_A = (1 - X_B)$.

It is the equality between the chemical potentials of a given component in different phases that determines the chemical equilibrium conditions between these phases. When the free enthalpy curve for a solution, G, is known as a function of X_B, the chemical potentials are obtained by extrapolating the tangent to the $G(X_B)$ curve to the vertical axis. The equilibrium states of a binary system can be obtained from the $G(X_B)$ curves at a given temperature and using the common tangent construction [36, 64]. Figure 1.18a,b shows examples of free energy diagrams and common tangent constructions. The heat of mixing ΔH_{mix} is due to the interaction energies between neighboring molecules; its value also depends on the composition X_B.

We may distinguish two cases:

1. $\Delta H_{mix} \leq 0$. No miscibility gap in the liquid state. It is the case where A (solvent) and B (solute) molecules "like" each other ($\Delta H_{mix} < 0$; exothermic solutions) or are indifferent to each other ($\Delta H_{mix} = 0$; ideal solution). Figure 1.18a shows a schematic free energy diagram for the liquid solution and crystal B phases at a given temperature T_0. In this case, the $G_{sol}(X_B)$ curve for the liquid solution is U-shaped. At low enough temperature, its value for $X_B = 1$ is that for the amorphous state B^{am}. Figure 1.18a also shows the Gibbs free energy curve of the crystalline API (B^{cryst}). It is supposed that the solvent does not dissolve B^{cryst} so that $G_{cryst}(X_B)$ rises rapidly from its minimum at $X_B = 1$. The two different free enthalpy curves are necessary to describe the equilibrium between the crystal form of B and the liquid solution of B in A. The composition of the liquid solution in equilibrium with the crystal B phase is defined by the common tangent to the $G_{sol}(X_B)$ and $G_{cryst}(X_B)$ curves. X_c is the limit of solubility of B^{cryst}. At the opposite, there is no limitation of the mutual solubility between A and B^{am} in the liquid states, since it is not possible to draw a tangent to $G(X_B)$ starting from $G_{Am}(X_B = 1)$. As mentioned above, in this case there is no thermodynamic limit of solubility for the amorphous state of B (B^{am}). Any limitation is only kinetic.

Mathematically, the limit of solubility of B^{cryst} (X_C) is given by

$$\mu_B^{cryst} = \mu_B^{sol}(X_C) \tag{1.21}$$

with

$$\mu_B^{cryst} \cong G_B^{cryst} \text{ and } \mu_B^{sol}(X_C) = G_B^{am} + RT \ln a_B(X_C) \tag{1.22}$$

where a_B is the activity of the component B in the solution. It is also written as $a_B(X_C) = \gamma_C X_C$, where γ_C is the solution activity coefficient, which takes into account the interaction energies; $\gamma = 1$ for an ideal solution.

Equations 1.21 and 1.22 give the limit of solubility of B^{cryst} as

$$\ln X_C = -\left[G_B^{am} - G_B^{cryst}\right] / RT - \ln \gamma_C$$

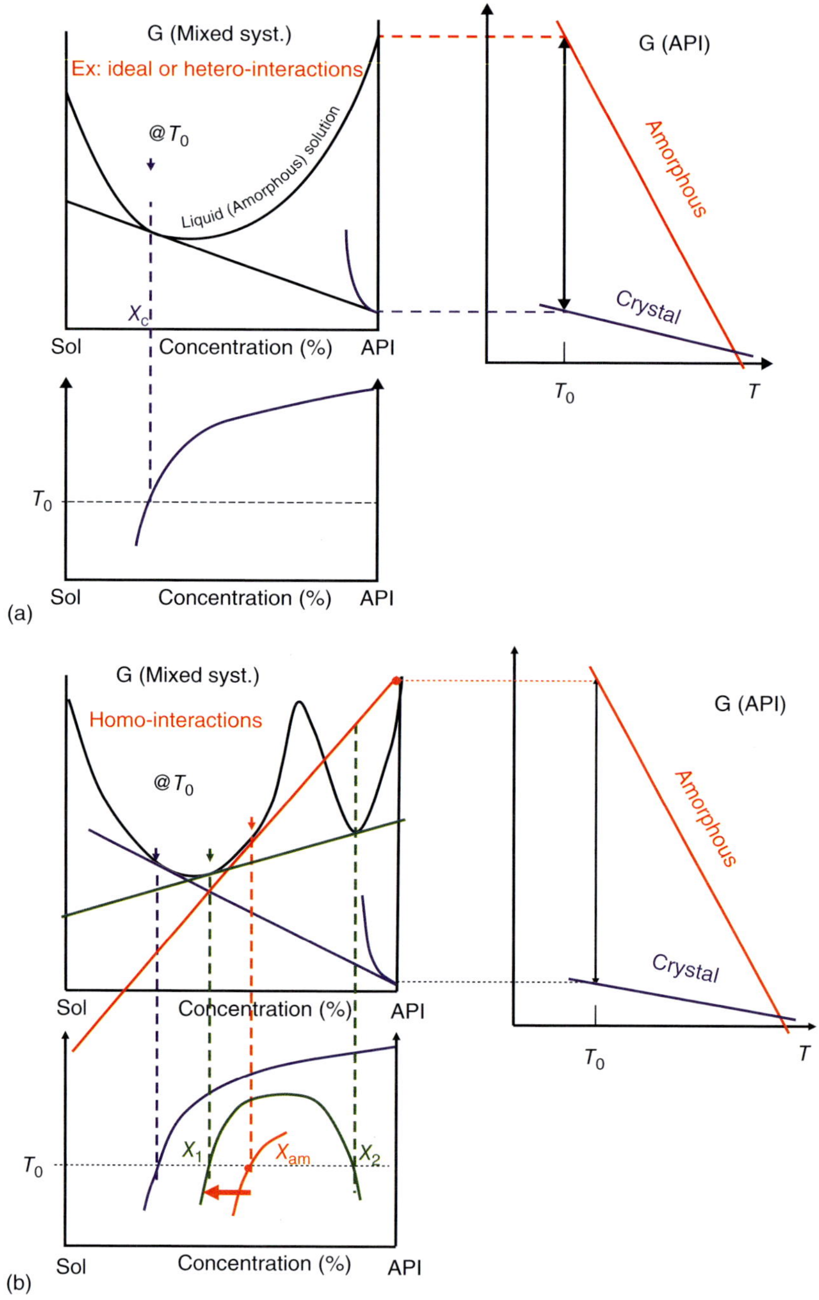
G (Mixed syst.)
Ex: ideal or hetero-interactions
@T_0
Liquid (Amorphous) solution
X_c
Sol
Concentration (%)
API
G (API)
Amorphous
Crystal
T_0
T
T_0
Sol
Concentration (%)
API
(a)
G (Mixed syst.)
Homo-interactions
@T_0
Sol
Concentration (%)
API
G (API)
Amorphous
Crystal
T_0
T
T_0
X_1
X_{am}
X_2
Sol
Concentration (%)
API
(b)

Figure 1.18 (a) $\Delta H_{mix} \leq 0$. No miscibility gap in the liquid state. Phase diagram and common tangent construction. There is a solubility limit for the crystal state of the API, but not for its amorphous state. (b) $\Delta H_{mix} > 0$. Existence of a miscibility gap in the liquid state (green line). Phase diagram and common tangent construction. The red tangent (and the red curve in the bottom diagram) corresponds to the nonequilibrium situation in which the glassy amorphous state of the API is put in contact with the solvent. The green tangent corresponds to the equilibrated situation: X_1 corresponds to the liquid solution in which the amorphous API dissolves in the liquid solvent. X_2 corresponds to the amorphous solution in which the solvent penetrates into the amorphous API. X_{am} is the initial apparent limit of solubility of the glassy API before penetration of the solvent into it **N.B.** The A and B components in the text correspond respectively to solvent (SOL) and API in the figure.

For an ideal solution, it gives

$$\ln X_C = -[G_B^{am} - G_B^{cryst}]/RT$$

That gives the solubility limit of the crystal, while there is no miscibility gap in the liquid (amorphous) state.

$[G_B^{am} - G_B^{cryst}] = \Delta G(T)$ can be approximately estimated from the heat of melting and melting temperature, as

$$\Delta G(T) \cong \Delta H_m \Delta T/T_m = \Delta H_m (T_m - T)/T_m$$

2. $\Delta H_{mix} > 0$. Existence of a miscibility gap in the liquid state. It is the situation where A and B molecules "dislike" each other. Figure 1.18b shows a schematic free energy diagram for the liquid solution and crystal B phases at a given temperature. Since the amorphous state B^{am} and the liquid solution have the same liquid structure, they lie on the same free energy curve $G_{sol}(X_B)$. At low enough temperature, the $G_{sol}(X_B)$ curve assumes a W-shape with a negative curvature in the middle. This is due to the fact that the enthalpy term is much higher than the entropy term. In that case, the most stable liquid solution is a mixture of two liquid-like phases of different compositions. The compositions of the two phases are given by the common tangent rule, which ensures equilibrium between the chemical entities. These two phases are, respectively, a liquid solution where the amorphous compound B^{am} dissolves in the liquid solvent (concentration X_1) and an amorphous solution where the solvent penetrates into B^{am} (concentration X_2). X_1 is the limit of solubility for B^{am} in the amorphous solid state. As in the previous case, the composition of the liquid solid solution in equilibrium with B^{cryst} phase is defined by the common tangent to the $G_{sol}(X_B)$ and $G_{cryst}(X_B)$ curves. We suppose again that the solvent is insoluble in B^{cryst} The limit of solubility of B^{cryst} is designated by X_C.

The above consideration regarding the solubility limit of the amorphous API is valid for an equilibrium situation. If the penetration of the solvent in the glassy API is slow, the initial value of the solubility limit of the amorphous API (X_{am}) is higher than X_1.

Mathematically, the phase equilibrium between the two liquid phases is such that

$$\mu_B(X_1) = \mu_B(X_2) \quad (1.23)$$

that is,

$$G_B^{am} + RT \ln \gamma_1 X_1 = G_B^{am} + RT \ln \gamma_2 X_2 \quad (1.24)$$

which gives the expression of the limit of solubility X_1 of B^{am}:

$$\ln X_1 = -\ln \gamma_1 + \gamma_2 X_2$$

As above, the limit of solubility X_C of B^{cryst} is given by

$$\ln X_C = -[G_B^{am} - G_B^{cryst}]/RT - \ln \gamma_C$$

If we assume that the activity coefficients at X_C and X_1 are equal, the amorphous and crystal solubility limits are linked by

$$\ln X_1 = \ln X_C + [G_B^{am} - G_B^{cryst}]/RT + \ln \gamma_2 X_2$$

In the above equation, the last term on the right-hand side is the correction associated with the possible absorption of the solvent by the amorphous material.

For the case where $\Delta H_{mix} > 0$, it thus appears that a stable and a metastable limit of solubility are conceivable. The first is related to solubility limit of crystalline B. The metastable limits of solubility X_1 and X_2 are those corresponding to the miscibility gap of the amorphous states when the liquid–liquid phase separation has reached equilibrium. Since the amorphous glassy state is out of equilibrium, the limit of the solubility X_1 depends on its effective level of free energy. It varies with the way of preparing the amorphous state, aging conditions, penetration of water in the glass, and so on.

The considerations given above are to be taken into account in the case of the dissolution in water of a crystalline or amorphous API. They are also to be considered when looking at the dissolution of an API in a polymer excipient [65].

It is important to note that the noncrystalline phase separation associated with a liquid miscibility gap may well be the first stage of a crystallization–devitrification [66]. It is thus important to acquire, as far as possible, a good knowledge of the stable and metastable limits of solubility at all the temperature ranges corresponding to the manufacturing process and drug storage.

1.2 Characteristics of the Disorder in Glass Formers

1.2.1 Glass Formation by Supercooling: Calorimetric Phenomenology

Glass formation (vitrification) is a generic behavior of matter [42, 67–69]. It concerns metals as well as polymers, oxides, or salts. The conventional way to form a glass is by continuous undercooling a liquid below T_m such that crystallization is avoided. Figure 1.14 shows the route that leads to glass formation in a TTT diagram. It needs to pass quickly the nose of most rapid crystallization at T_n. In so doing, the undercooled liquid is retained in a metastable state with regard to the crystal. Many other amorphization routes can be used for pharmaceutical formulations [70].

A remark, however, is necessary to be made. As shown above, the TTT curve is obtained from experimental determination of the kinetics of recrystallization. It thus results from a combination of nucleation and growth. Passing quickly the TTT nose may certainly avoid experimentally detectable crystallization but it does not necessarily avoid catching some crystalline nuclei that are unable to grow within the amorphous matrix. This is due to the fact that the maxima of nucleation and of growth are separated from each other by temperature. The maximum nucleation rate is situated at a temperature lower than that of the TTT nose. Therefore, it is often observed that recrystallization occurs upon reheating as a result of crystal growth from these nuclei.

Figure 1.19a shows the evolution of C_p for indomethacin upon melting and undercooling. The increase of C_p above melting is due to the dynamic release of the molecular degrees of freedom of large amplitude that characterize the

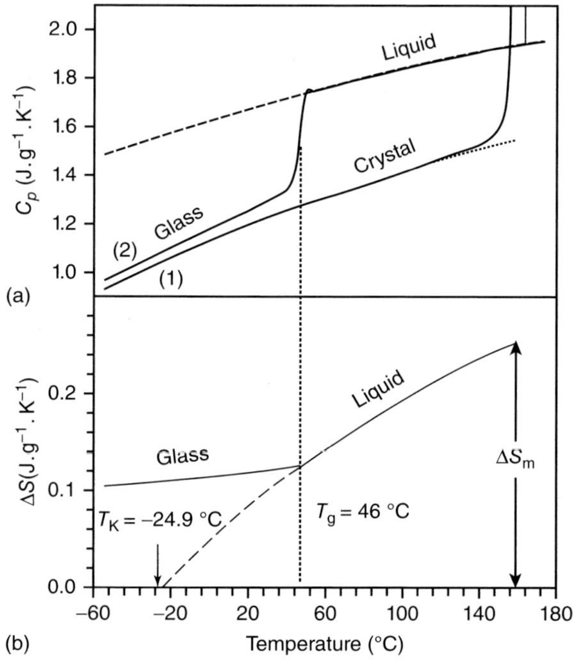

Figure 1.19 (a) Temperature evolution of the specific heat (C_p) of indomethacin in the crystal, liquid, undercooled, and glassy state. The positions of T_m (melting point) and glass transition T_g are shown. (b) Evolution of the excess entropy ($\Delta S = S_{liq} - S_{cry}$) of the supercooled liquid indomethacin in metastable equilibrium relative to the stable crystalline material. $\Delta S(T)$ is obtained by integration of $C_p(T)/T$. ($C_p(T) = T\partial S/\partial T$). It extrapolates to zero at the temperature T_K which is rather close to T_g.

liquid state. By undercooling the liquid rapidly enough (typically 10 °C min^{-1}), crystallization can be avoided. Upon undercooling, C_p keeps a value higher than that of the crystal down to a temperature T_g, at which it decreases continuously (but rather abruptly) to a value comparable (while slightly higher) to that of the crystalline solid. This event defines the *calorimetric glass transition*. It occurs at a temperature (T_g) that is typically around $2T_m/3$ (T_m in kelvin). A dilatometric investigation would also show a drop of the expansion coefficient α_p. The drop of C_p and α_p correspond to a change in the slopes of enthalpy $H(T)$ ($C_p = \partial H/\partial T$) (Figure 1.20), entropy $S(T)$ ($C_p = T\partial S/\partial T$) (Figure 1.19b), and volume $V(T)$ ($\alpha = -[1/V]\ \partial V/\partial T$). T_g marks the transition from the amorphous undercooled liquid to the glassy amorphous state. The value of C_p for a crystal is linked to the vibrational degrees of freedom alone. That suggests that this type of fast motion is also the primary factor contributing to C_p in the glassy state. The much slower and larger relaxational motions, which characterize the liquid state, only contribute for $T > T_g$.

1.2.1.1 The Glass Transition is not a Phase Transition

The glass transition It does not involve any discontinuous structural change, as shown in Figure 1.12. Below T_g, the aspect of the structure factor $S(Q)$, typical of a liquid, is unchanged. The structure essentially remains that of the liquid starting material. The slight shift in the position of the amorphous X-ray bump, which occurs continuously upon cooling, is linked to a progressive contraction.

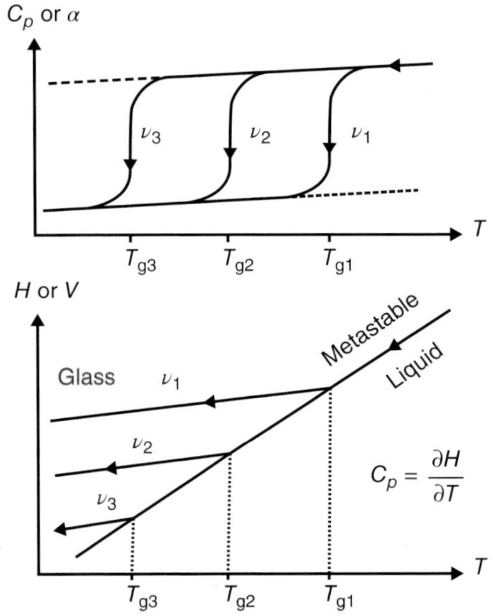

Figure 1.20 Temperature evolution of (C_p, α) for the liquid undercooled at different rates. The corresponding evolution of enthalpy (H) and volume (V). The greater the cooling rate, the higher the value of T_g, and the higher the level of "frozen" H and V.

The change at T_g that can be detected on the structure factor is only the rate of change of this shift with respect to temperature, which slows down below T_g.

1.2.1.2 The Glass Transition is a Kinetic Phenomenon

The vitrification process, in fact, occurs over a temperature interval. Furthermore, the value of T_g depends on the cooling rate. The higher the cooling rate, the higher the value of T_g. Typically, T_g changes by 3–5 °C when the rate changes by one order of magnitude. Figure 1.20 illustrates the evolution of the enthalpy (H) and of C_p as a function of temperature for different cooling rates. Above T_g, the rate of change of enthalpy with respect to temperature is independent of the cooling rate. However, at T_g the slope of $H(T)$ decreases suddenly to a value similar to that of a crystalline solid. Furthermore, enthalpy seems to be frozen at a value which increases with the cooling rate. Similar behavior is found for the entropy $S(T)$ (Figure 1.19b), volume $V(T)$, and so on. Below T_g, the thermodynamic quantities H, S, V, and so on, have values that depend on the way the glass has been formed: they are especially higher as the cooling rate becomes high. Not too far below T_g, slow relaxations of H, V, S, and so on, and the related properties are observed: this is the aging phenomenon by which the glass tries to evolve – on a timescale higher than 100–1000 s – toward the metastable liquid state in internal equilibrium at the temperature of aging. Aging is the manifestation of the glass being in an out-of-equilibrium state. A simple evidence of the nonequilibrium nature of the glassy state is linked to the violation of the Nernst law (third law of thermodynamics, which states that for an equilibrium crystal state at 0 K the entropy should be zero). Because of the break in the evolution of $S(T)$ at T_g, the experimental value of the entropy of a glass at 0 K is positive [71].

In a warm liquid – ordinary liquid – several dynamic processes occur on different timescales. Molecular motions can be roughly divided into two categories: (i) very fast (frequency on the order of terahertz) vibrational and librational motions localized inside the cage formed by neighboring molecules and (ii) the less frequent relaxation processes of diffusion from the cage and molecular rotations of high angles. In glass physics, these processes are named α, β, and so on. The corresponding motions are of a much larger amplitude than that involved in vibrations. They effectively involve a jump over an activation energy barrier. The corresponding frequencies are typically in the gigahertz range at the melting temperature. The two categories of motions contribute to the C_p of an ordinary liquid, while only vibrational motions contribute to the C_p of a crystalline solid. As the temperature of an undercooled liquid decreases, the relaxation time, which is the timescale that characterizes the diffusion processes, becomes longer and longer. T_g is the temperature at which the relaxation time becomes on the order of 100–1000 s, which is the laboratory timescale. At these low temperatures, molecules rearrange so slowly that they cannot explore the entire possible sample configurations in the available time allowed by the cooling rate. The consequence is that slow relaxational motions (the so-called main or α relaxations) no more contribute to the specific heat of the glass below T_g. Only the remaining fast motions may contribute to C_p. For the main part, these are vibrations of molecules.

The slight excess of the specific heat of the glass relative to the crystal seen below T_g (cf. Figure 1.19a) is, in part, due to the fact that vibrations do not take place exactly in the same molecular environment and in part to remaining fast secondary relaxations (β relaxations) [72].

The amplitude of the C_p jump (ΔC_p) at T_g can vary widely according to the compound. The tendency is that liquids with very strong intermolecular (or atomic) bonding generally exhibit small $\Delta C_p(T_g)$ values. Liquids with weak intermolecular bonding (fragile bonding) generally exhibit larger $\Delta C_p(T_g)$ values. As shown for indomethacine in Figure 1.19a, ΔC_p is generally large for molecular materials. According to a terminology proposed by Angell, corresponding glass-forming liquids are called *strong* or *fragile* [4, 73, 74]. It will be shown in the following that these different types of liquids also show differences in the manifestation of their molecular mobility.

As mentioned above, according to the Maxwell relation, the shear viscosity η is proportional to the structural relaxation time (also named the α or the main relaxation process). It is thus equivalent to associating T_g with the temperature at which the viscosity has the value corresponding to a structural relaxation time of 100–1000 s. All glass formers have thus a viscosity of about 10^{13} poises at T_g. Upon cooling, the glass transition occurs when the viscosity of the supercooled liquid becomes so large that molecular motions involved in the viscosity mechanism are arrested on the timescale of the experiment.

1.2.2
T_g as a Transition from an Ergodic to a Non-Ergodic Situation

The differentiation of the states of a system above and below T_g can be depicted schematically in the form of a Gibbs free energy plot (Figure 1.21). Between T_m

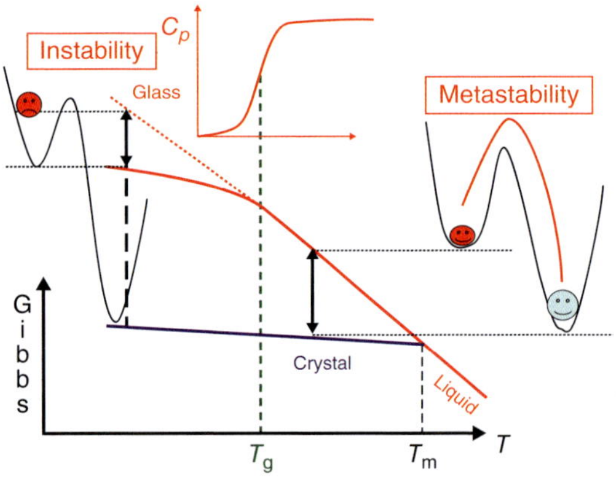

Figure 1.21 Schematic representation of the evolution of the Gibbs function as a function of temperature. Also represented are schematic aspects of the Gibbs function as a function of a configurational variable for two temperatures, respectively, in the domain of the metastable undercooled liquid ($T_g < T < T_m$) and that of the unstable glass ($T < T_g$).

and T_g, the system is metastable but in an equilibrium state. The situation corresponds to the bottom of the free energy well corresponding to the undercooled liquid. Its level is, however, higher than that of the bottom of the well corresponding to the crystal. Below T_g, the system is not even metastable, but it is unstable. Schematically, we can assume its energy position as being frozen somewhere on the side of a potential well, which is not a stable situation. This position depends on the way the glass has been formed. Also, the nonequilibrium system is prone to slow relaxation toward the bottom of a free enthalpy potential well: it is the aging phenomenon [69]. It should, however, be remarked that such a description in term of a well is an oversimplified one. It has never been demonstrated that a unique free energy curve exists out of equilibrium. At equilibrium, what is known is only the existence and position of a potential energy minimum, as well as the curvature of the potential well at that point that determines the responses to a slight perturbation.

Above T_g, the properties can be described with a very limited number of variables (state variables or state functions) energy, specific volume, and so on. This is a consequence of the fact that the motions are extremely fast. The system has enough time to explore all possible molecular configurations. Measured values of the system variables are the results of a good "statistical evaluation" performed by the system itself on all its configurational states. In the language of statistical physics, the system is ergodic because it is equivalent to taking an average over time or over the phase space of the system. On the contrary, in the glassy state it is impossible to describe the properties with such a limited number of variables. Strictly speaking, it would be necessary to know the frozen configurational state of each molecule to describe the glass and its properties. This led to the development of the concept of "potential energy hypersurface" (also known as *energy landscape* [69, 75–77]) on which there are innumerable potential energy minima corresponding to specific molecular configurations of the entire system. For a glass, the experimentally measured physical quantities are the result of an averaging performed on a very limited number of domains of the phase space, that is, of the energy landscape, which is accessible to the system during the measurement. Depending on the way the glass has been formed, the duration on the measurements, the time of the measurement, and so on, different values will be found for the quantities characterizing the system, such as density, optical index, and so on; and this cannot be avoided. It is the reason why patenting the glassy form of an amorphous sample is really a challenging question. In practice, we often try to find at least one additional internal parameter that is able, operationally, to describe the main specific properties. For example, that gave rise to the introduction of the fictive temperature (T_f) concept, to be defined in the following.

1.2.3
The Entropy Below T_g: The Kauzmann Paradox

At the melting temperature, the entropy of a liquid is higher than that of the corresponding crystal. Since C_p ($C_p = T\partial S/\partial T$) of a liquid is higher than that of the crystal, the entropy (S) of the liquid decreases faster than that of the crystal. For fragile liquids, the entropy difference between the liquid and the crystal decreases by a factor of about 3 between T_m and T_g. For most glass formers, especially for molecular compounds, extrapolating the data obtained above T_g leads to a crossing of the entropy of the liquid with that of the crystal at a positive temperature T_K (the Kauzmann temperature) [78]. Figure 1.19b shows, for instance, the temperature evolution of the excess entropy $\Delta S = S_{liq} - S_{cry}$ for indomethacin. In this case, $T_K \cong -24.9\,°C$ (it is located $\approx 173\,°C$ below T_m and $\approx 71\,°C$ below T_g). If the liquid could be supercooled under equilibrium conditions down to T_K, we would have arrived at the astonishing situation that the liquid (disordered) and the crystal (ordered) would have the same entropy. Figure 1.22 shows some available entropy data plotted by means of a T_m-scaled Kauzmann representation. T_K is all the more close to T_g, so that the rate of change of liquid entropy with respect to temperature is large (more specifically, if the slope at T_m of $\Delta S/\Delta S_m$ with respect to T/T_m (i.e., $\Delta C_p/\Delta S_m$) is larger, and, in any case, larger than 1). From that, it results that most fragile compounds, with high ΔC_p, have their T_K rather close to T_g. The change in slope of S that occurs at T_g, for purely dynamic reasons, permits avoiding this paradoxical situation. If realized, S would become negative when approaching 0 K. That is impossible in the light of the statistical definition of entropy:

$$S(N, E, V) = k_B \ln(\Omega) \tag{1.25}$$

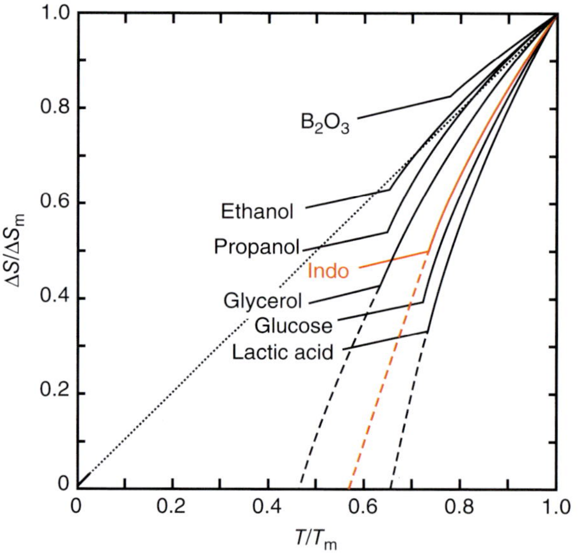

Figure 1.22 Temperature dependence (schematic) of the excess entropy for several supercooled liquids. In red: undercooled liquid indomethacin (data from figure 1.19b) Normalization of ΔS and T have been done against melting values. The temperature T_K, where $\Delta S/\Delta S_m$ extrapolate to zero, depends on the "fragility" of the compound. (Adapted from Kauzmann [78]. Reproduced with permission of American Chemical Society.)

where k_B is the Boltzmann constant and Ω is the number of states accessible to the system of N molecules with fixed energy and volume. Obviously, the minimum value of Ω is 1.

Something should prevent this situation if the equilibrated liquid could be tracked below T_g. Obviously, that is impossible experimentally. The resolution of the Kauzmann paradox is still a matter of theoretical debate. It is, in fact, related to the true thermodynamics of the deeply undercooled liquid state. Several proposals have been made to solve the Kauzmann paradox. It has been suggested that the vanishing of the entropy difference between the liquid and crystal would reveal an ideal glass state only reachable near T_K at infinitely slow cooling. This is the essence of the Adam–Gibbs (AG) approach [79]. The existence of “such a state of high order for the liquid” [78], however, seemed difficult to conceive for Kauzmann himself. He suggested, instead, the existence of an intrinsic limit of metastability of the liquid situated between T_g and T_K. Such a metastability limit (pseudo-spinodal) means that an equilibrated liquid could never be cooled down to T_K because the nucleation barrier opposing crystallization would decrease critically before reaching this temperature. If the glass transition did not occur, the crystallization of the metastable liquid could not then be avoided before reaching T_K. Other works suggest that the extrapolation is no more valid below some temperature higher than T_K and, therefore, the entropy catastrophe at T_K is unphysical [80]. The fact remains that the fast decrease of the entropy of the liquid relative to that of the crystal is one of the most notable aspects of the phenomenology of glass formers. In view of the statistical definition of entropy, it indicates that correlations of a certain type are developing fast in the liquid. However, there is no significant experimental manifestation of ordering in X-ray structural investigations. It is one of the puzzles of glass formation.

1.2.4
Dynamic Features of the Disorder in Glass Formers: The Three Nons. Fragile *versus* Strong Classification

The features of the glass formation and glassy state are not related to the structural (“static”) aspects of the compounds but to the distinctive peculiarities of their dynamics. If the nature of the thermodynamic manifestation of the glass transition is well understood, understanding the nature and the temperature evolutions of the relaxation processes while approaching T_g is one of the main challenges in the physics of glasses. Specificities of the dynamics are sometimes referred to as the “three nons” of glass formers [81–83].

- The non-Arrhenius temperature dependence of the relaxation time above T_g
- The non-exponential decay in the response to perturbations from the equilibrium state
- The nonlinearity of the response to thermodynamic perturbations below T_g (The latter is equivalent to saying that the relaxation time is a function of time.).

1.2.4.1 Above Tg: The Dramatic Non-Arrhenius Temperature Dependence of Viscosity and Relaxation Times

Of interest here is the temperature dependence of the viscosity η and of the average relaxation time τ_α. τ_α is the time associated with the so-called main or α relaxation process. It is the characteristic time for the relaxation of some properties (strain, enthalpy, electric polarization, etc.) toward equilibrium after the application of a small perturbation of the intensive conjugate variable (respectively, stress, temperature, electric field, etc.). τ_α can be calculated from the viscosity by means of Maxwell's relation 1.1. It can also be obtained as the inverse mechanical, specific heat, or dielectric loss peak frequency. These definitions of τ_α are not strictly identical, but the differences can be neglected for our purpose. τ_α measures the rate of molecular rearrangements, which involve relatively large rotational and translational molecular displacements over potential energy barriers.

For all normal liquids, under atmospheric pressure, the viscosity η and relaxation time τ_α do not evolve too much between the boiling point and the melting point: η is of order 10^{-2} poise and τ_α is of order of 10^{-12} s. At T_g, after supercooling, η is of the order of 10^{13} poises and τ_α is of the order of 10^3 s. At the glass transition, when the temperature decreased by just about 50% from the boiling point, and about 33% from T_m, η, and τ_α have increased by more than 15 orders of magnitude. It is this spectacular increase at the approach of T_g and the temperature dependence of this increase according to the nature of the glass formers that are challenging.

For oxide glasses (e.g., pure silica) and a few other *strong* liquids (covalent bonds), the temperature evolution of the viscosity η and τ is well represented by an Arrhenius law:

$$\tau_\alpha(T) = \tau_0 \exp E/k_B T \tag{1.26}$$

where the activation energy E is almost constant. It is roughly the energy that is necessary for breaking intermolecular bonds. The prefactor τ_0 is a relaxation time at very high temperature ($\tau_0 \cong 10^{-14}$ s). This expression "predicts" a divergence at 0 K. However, for most liquids – it is the case with most molecular compounds that are rather fragile (weak dispersive chemical bonds) – the evolution of the relaxation while approaching T_g is even more rapid (super-Arrhenius behavior). An attempt to use an Arrhenius description thus requires an activation energy E that is temperature dependent, $E = E(T)$, and strongly increases upon cooling. The effective activation energy $E(T)$ is of the order of 40 $k_B T$ at T_g and increases by a factor of 5–6 between T_m and T_g. In these cases, the activation energy for the structural relaxation is much higher than the bonding energy. Over a rather large temperature range above T_g, (which covers 4–5 order of magnitude in relaxation time), the considerable variation of the viscosity and relaxation times is well described by the Vogel–Fulcher–Tamman (VFT) equation [84]:

$$\tau_\alpha = \tau_0 \exp[DT_0/(T - T_0)] \tag{1.27}$$

where τ_0, D (the "strength" parameter [A]), and $0 < T_0 < T_g$ are adjustable parameters. The VFT equation predicts a virtual divergence at the temperature

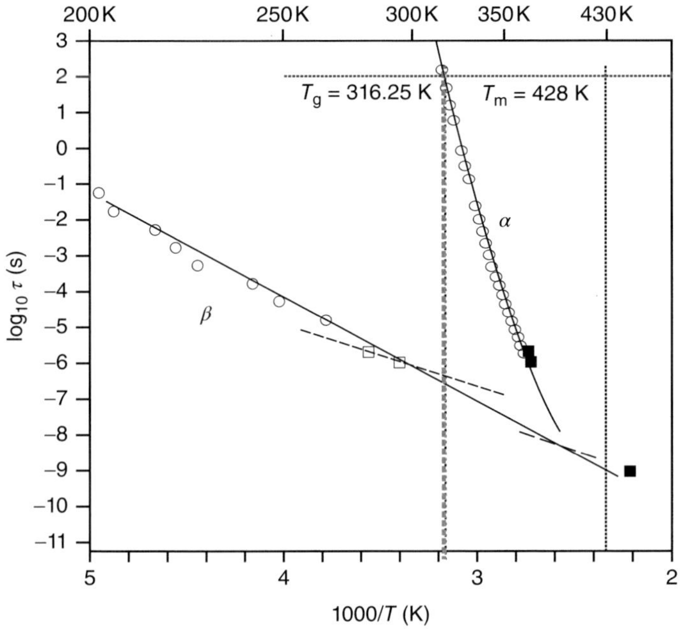

Figure 1.23 Evolution of the main (τ_α) and secondary (τ_β) in the case of amorphous indomethacin. (Adapted from Carpentier *et al.* [85]. Reproduced with permission of American Chemical Society.)

$0\,\mathrm{K} < T_0 < T_g$. For molecular compounds, T_0 is typically a few tens of degrees below T_g. For many systems, T_0 is found to be very close to T_K: typically $0.9 < T_K/T_0 < 1.1$ [69]. Figure 1.23 shows the evolution of τ_α (and an other sub-T_g relaxation to be discussed latter) in the case of amorphous indomethacin [85]. The best fit to the VFT equation is obtained for $\tau_0 = 2.6 \times 10^{-20}$ s, $T_0 = 230.5$ K, and $D = 18.5$. (Similar fitting parameters were found by Paluch *et al.* [86].)

Figure 1.24 shows a T_g-scaled Arrhenius representation, proposed by Angell [67], of viscosity data for a wide variety of glass formers from strong to fragile. This figure illustrates clearly the corresponding change from Arrhenius to super-Arrhenius behavior of relaxational mobilities. In terms of VFT representation, the "strength" parameter D falls between ~3 (fragile) and ∞ (extremely strong). In this plot, the slop at T_g is also a possible measure of fragility. The fragility (or steepness) index m is defined as

$$m = d(\log_{10}\tau_\alpha)/d(T_g/T)(\text{taken at } T = T_g) \tag{1.28}$$

The experimentally observed extreme values are $m = 16$ (strong) and $m = 200$ (very fragile behavior). As has been proposed by Angell [67, 73], glass formers are now classified, in practice, by their more or less non-Arrhenius behavior, that is, by their degree of fragility. For indomethacin, the value of the fragility index was estimated to be $m \approx 79-83$ [85, 86]. This a typical value for a complex "small molecule" compound as found, for example, for salol ($m = 73$) and for nifedipine

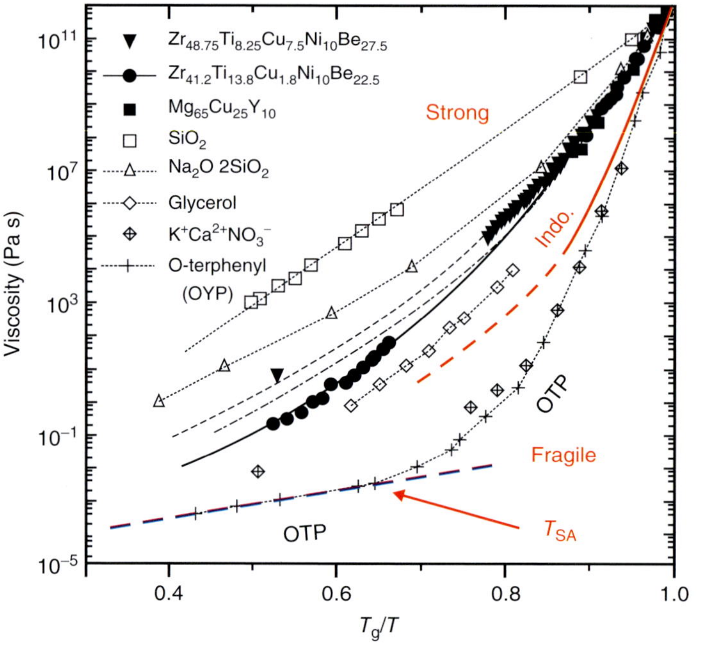

Figure 1.24 T_g-scaled Arrhenius representation, proposed by Angell [71, 83], of viscosity for a wide variety of glass formers from strong to fragile. (Adapted from Busch [87]. Reproduced with permission of Springer.)

($m = 83$). For the rather fragile sorbitol, $m = 93$, while glycerol is found to have an intermediate fragility index (between strong and fragile) $m = 53$. For polymers, the fragility index is generally larger: for polystyrene $m = 139$ and for poly(vinyl chloride) $m = 191$.

Considering that $m = 16$ corresponds to the Arrhenius behavior, D and m are related by [74]

$$m \cong 16 + 590/D \tag{1.29}$$

Another empirical equation was established by Williams, Landel, and Ferry (WLF) [83] to describe the super-Arrhenius relaxation mechanisms in glass-forming liquids. The WLF equation is equivalent to the VFT equation and gives the viscosity as

$$\log \eta/\eta_S = -C_1(T - T_S)/[C_2 + (T - T_S)] \tag{1.30}$$

where η_S is the viscosity at the reference temperature T_S, and C_1 and C_2 are constants. Generally, $T_S = T_g$ is chosen as the reference temperature for convenience. C_1 does not vary so much with the material and has a value of ~16–17, which corresponds to the number of decades of viscosity decrease between T_g and an infinitely high temperature (where $\eta_\infty \approx 10^{-4}$ poise) [88]. On the contrary, C_2 depends on the fragility and thus on the material. The relation between C_2

and the fragility index m is given by

$$m = C_1 \cdot T_g / C_2 \tag{1.31}$$

C_2 is linked to T_g and the VFT temperature T_0 by

$$T_0 = T_g - C_2 \tag{1.32}$$

It is generally observed that both VFT and WLF models are able to describe only the behavior in temperature domain close to T_g.

1.2.4.2 Possible Connection Between Dynamics and Thermodynamics

From what we have seen, a connection between dynamics an thermodynamics seems to be emerging:

Strong glass formers have a small ΔC_p at T_g and an Arrhenius behavior of relaxations.
Fragile glass formers generally have large ΔC_p at T_g and a pronounced non-Arrhenius behavior of the relaxations.

Furthermore, the Kauzmann temperature T_K is usually found to be very close to the VFT temperature T_0. This suggests that a structural arrest of configurational rearrangements would occur at T_K if the system could be cooled under equilibrium condition down to that temperature.

Adam and Gibbs [79] developed a theory that rationalizes the relationship between dynamics and thermodynamics. It assumes a phase transition to a state of zero configurational entropy and infinite relaxation time at T_K. This leads to expressing the temperature evolution of the viscosity, or equivalently of the relaxation times, as a function of the configurational entropy S_c:

$$\eta \propto \tau \propto A \exp C/TS_c(T) \tag{1.33}$$

where C is a constant containing an energy barrier term. If it is assumed that structural equilibrium is maintained at all temperatures, the configurational entropy varies with temperature according to the following expression:

$$S_c(T) = \int_{T_K}^{T} (\Delta C_p / T') dT' \tag{1.34}$$

where ΔC_p is the configurational specific heat [$\Delta C_p \approx C_p$(liquid) $- C_p$(crystal)], and T_K is the temperature at which the configurational entropy is supposed to fall to zero. If we use Eq. (1.33) and the approximation $\Delta C_p = A/T$ (this approximation is usually valid for non-polymeric systems but the result is not very sensitive to the form of ΔC_p), the empirical VFT equation is recovered with $T_0 = T_K$:

$$\eta \propto \tau \propto A \exp[DT/(T - T_K)] \tag{1.35}$$

In the formulation of the AG equation, it is assumed that structural equilibrium is maintained at all temperatures, even below T_g, which is experimentally impossible. The equation is thus strictly applicable only for $T > T_g$. For $T < T_g$, the AG equation is usable only to provide a formal expression that links the virtual critical slowing down to the thermodynamic pseudo-singularity at T_K. It would make no sense to use the argument that for $T < T_K$ there is no molecular mobility anymore and that amorphous compounds are stable at these low temperatures.

Even if the possible existence of an ideal glass state, having an entropy equal to that of the crystal, remains speculative, the AG expression provides a very good correlation between dynamic and thermodynamic data (for $T > T_g$) for a large majority of compounds. This analogy between the VFT law and AG prediction seems to show that the temperature at which relaxation times tend toward infinite values corresponds to the temperature at which the configurational entropy reduces to zero. It should be noted, however, that the experimentally measured excess heat capacity may include some vibrational contributions in addition to purely configurational ones. That may induce some uncertainty in the evaluation of T_K and be a source of slight discrepancy between T_0 and T_K.

AG envisaged the idea of cooperatively rearranging regions (CRRs), whose size increases as the temperature decreases [89]. As a consequence, their number decreases, which reduces the number of effective configurational degrees of freedom and therefore the configurational entropy. The AG theory predicts that the minimum size of CRRs of molecules is inversely proportional to S_c. This size would be in the nanometer scale near T_g. One picture is that the CRR blocks would virtually become of infinite size at T_K as well as the characteristic time required to cause a transition from one configuration to another. However, the nature of these CRRs, their possible interactions, and so on, remain poorly understood [90] as well as their real involvement in the dynamics of glass formers [68].

1.2.4.3 Above Tg: Non-Exponential Relaxations and Dynamic Heterogeneity

For ordinary "warm" liquids, the relaxation toward equilibrium of any microscopic or macroscopic property $f(t)$, after a small applied perturbation, can be usually described by an exponential time decrease. The relaxation function $f(t)$ measures, for example, the instantaneous polarization following a step change of the electric field. On the other hand, for a supercooled liquid, the response $f(t)$ is no longer exponential. It has a sluggish behavior, which contrasts with the behavior above T_m. At long enough times, which correspond to that of the α relaxation process, it may be generally expressed with the Kohlrausch–Williams–Watts function [91] or the "stretched exponential":

$$f(t) \propto \exp[-(t/\tau)^{\beta}] \tag{1.36}$$

where the value of the parameter β is between 0 and 1. The exponential, the so-called Debye behavior, is recovered for $\beta = 1$. The smaller the value of β, the more stretched the relaxation. The stretching generally increases when temperature decreases and reaches values on the order of 0.3–0.8 at T_g. It was observed [92, 93] that low values of β are generally correlated with a high fragility index m.

Non-exponential relaxation described by Eq. (1.36) can be interpreted in terms of a superposition of exponential function with a distribution of relaxation times. It is (was) a matter of considerable interest to know whether this non-exponential nature is intrinsic or whether it is the result of a spatial heterogeneity [94]. A number of experiments have shown the existence of kinetic heterogeneities in glass-forming systems with regions of slower and faster molecular mobility. The size of these heterogeneities is estimated to be of some nanometers at T_g. It must be mentioned that these heterogeneities are purely dynamic in nature. As to whether these dynamic heterogeneities may have some structural connection remains a challenging issue.

Ngai proposed a coupling model [92, 95] according to which the primitive relaxation is exponential – $\exp(t/\tau_0)$ – for $t < t_c$. τ_0 is the "uncoupled" relaxation time. For $t > t_c$, molecules move cooperatively and the relaxation becomes stretched and fittable with $\exp[-(t/\tau)^\beta]$. For usual glass formers, the crossover time $t_c \approx 2$ ps. The fractional exponent coupling parameter $n = 1 - \beta$ is a measure of the strength of the intermolecular cooperativity. This formalism allows us to derive a simple relation between $\tau(T)$, $\tau_0(T)$, β, and t_c. This relation is operationally useful to identify the secondary β relaxation since $\tau_0(T)$ is supposed to be "close" to $\tau_\beta(T)$ (see below).

Some experimental methods can characterize directly the relaxation in the time domain [96]. However, the dynamic response of a system can also be analyzed in the frequency (ω) domain by applying a sinusoidal perturbation. What is measured in that case is a frequency-dependent complex susceptibility $\chi^*(\omega)$. The imaginary part (loss spectrum) $\chi''(\omega)$ is the Fourier transform of $df(t)/dt$, where $f(t)$ is the corresponding relaxation function. A relaxation is highlighted by a peak in the $\chi''(\omega)$ spectrum, with a maximum for the characteristic frequency $\omega_c \propto 1/\tau$. Figure 1.25 shows the $\chi''(\omega)$ spectra for the α relaxation of amorphous indomethacin measured at different temperatures and plotted as a function of $\log_{10}(\omega)$. When the temperature decreases, the peak shifts toward low frequencies (i.e., large relaxation times). The width of the peak gives information on the departure from pure Debye behavior. The broadening of the peak observed at low temperature reflects the stretching of the relaxation function mentioned above.

1.2.4.4 *Below T_g: Aging, Nonlinearity, Secondary Relaxations*

Aging and Variability of Glass Structure

When a liquid is undercooled and crystallization is avoided, a temperature ($\approx T_g$) is finally reached where cooling is too fast for the molecules to arrange into an equilibrated state. The liquid "falls out of thermal equilibrium." The liquid structure nearly freezes. As a consequence, below T_g, the sample is in a structural state corresponding to a higher temperature. The faster the cooling rate, the higher will be this temperature. The glassy state thus depends markedly on the cooling conditions in relation to the relaxational properties of the liquid. Practically all physical properties of a glass are modified to some extent if the conditions of the cooling process are varied. Other consequences of the nonequilibrium state of the glass are a variety of phenomena such as aging, memory effect, or rejuvenation [97]. Indeed, we must understand that, even though slow, the molecular mobility still remains. It provides the mechanism by which the glassy compound slowly evolves

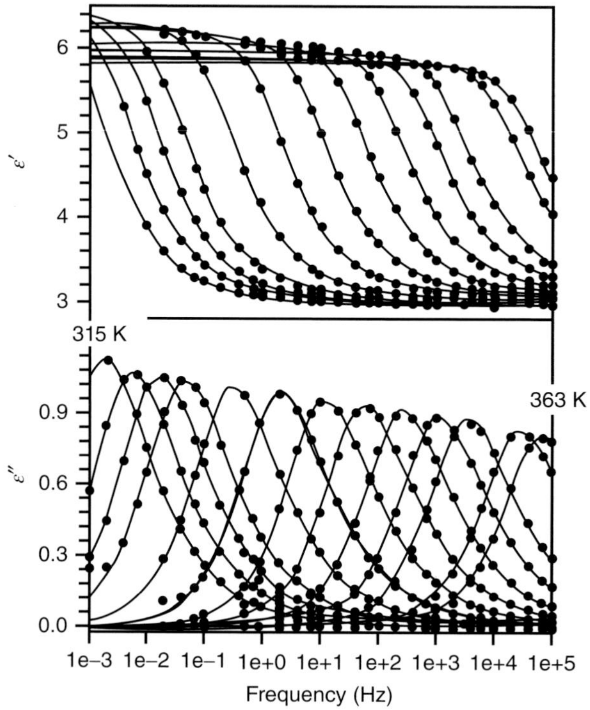

Figure 1.25 Real and imaginary parts of the dielectric susceptibility of amorphous indomethacin at different temperatures between 315 and 363 K. Solid lines are a fit with a Havriliak–Negami function. (Adapted from Carpentier *et al.* [85]. Reproduced with permission of American Chemical Society.)

with time or ages. Therefore, for identically prepared glasses, the larger the deviation from T_g, the slower the aging.

Upon aging, the volume of the glassy material changes with time. Similarly, an evolution of enthalpy will also take place. DSC is a common method to characterize this enthalpy evolution. The result of such an experiment is depicted in Figure 1.26 for amorphous indomethacin in which two different aging temperatures T_a (20 and 35 °C) were imposed. In these experiments, the sample is first melted ($T > T_m \approx 161$ °C) and then cooled (at 10 K min^{-1}) to a temperature $T_a < T_g$, where it is aged for a time t_a. The heat flow corresponding to each value pair (T_a, t_a) is then measured upon reheating. For each T_a, the experiment is repeated for different values of t_a. The main pieces of information provided by these curves are the following: (i) An overshoot appears and grows as t_a increases; (ii) The rate of increase of the overshoot is faster for higher aging temperature. For $T_a = 35$ °C, a saturation is nearly reached at $t_a = 1$ h ($\approx T_g - 8$ °C); (iii) For $T_a = 20$ °C, the overshoot grows more slowly but reaches higher values; (iv) The stronger growth of the overshoot for $T_a = 20$ °C is correlated to a shift toward higher temperatures, which gives rise to an apparent increase of T_g.

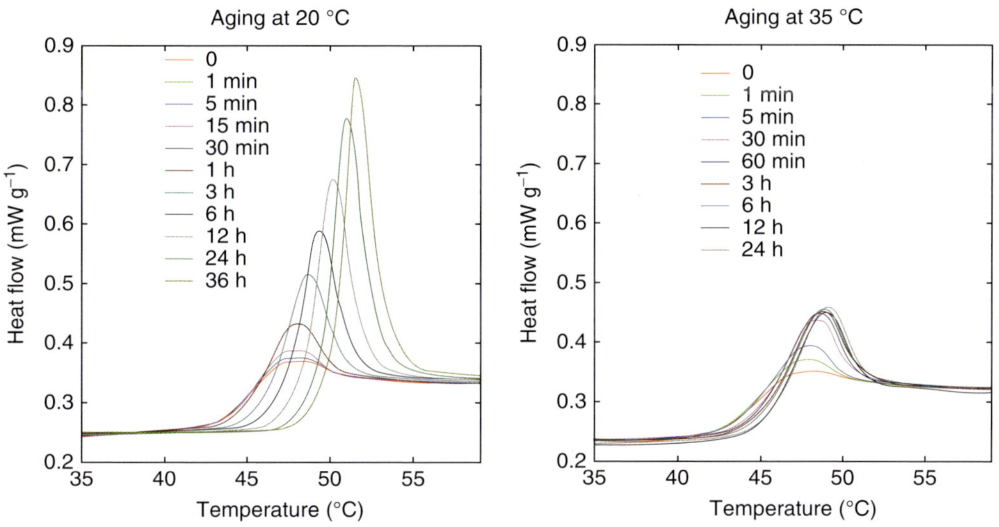

Figure 1.26 DSC heating scans for amorphous indomethacin aged during various times t_a at two different temperatures (20 and 35 °C).

The heat flow response is related to the heat capacity, which is the temperature derivative of enthalpy. Consequently, an overshoot reflects a fast transient evolution of enthalpy as illustrated in figure 1.27. During annealing, the glass relaxes slowly toward metastable equilibrium and loses. When heated through the glass transition temperature, the lost enthalpy is recovered rapidly, and gives the overshoot. Enthalpy recovers the supercooled metastable liquid value upon heating at the temperature where molecular mobility is high enough to allow this. At low aging temperatures, molecular mobility is very slow and the structural relaxation process is kinetically limited. However, the enthalpy difference between the nonequilibrium glass and the equilibrium supercooled metastable liquid is large (see for example figure 1.20). The glass relaxes more slowly but may reach low enthalpy values. As a result the recovery may give rise to a large overshoot after long enough aging. When T_a is close to T_g, the glass nearly reaches equilibrium during a rather short annealing, but the energy loss is small and so is the size of the overshoot. The progressive shift of the overshoot toward higher temperatures is an indication that annealed glasses have less molecular mobility compared to unannealed glasses. This is due to the fact that part of the residual enthalpy and free volume are lost upon annealing. The resulting lesser mobility delays the recovery upon heating. That the molecular mobility, which allows structural relaxation, is decreasing as the structural relaxation progress expresses the nonlinearity of the phenomenon.

Fictive Temperature

While a glass is a non-ergodic state that cannot be described by state functions, its properties and structural relaxation are often tentatively described in terms

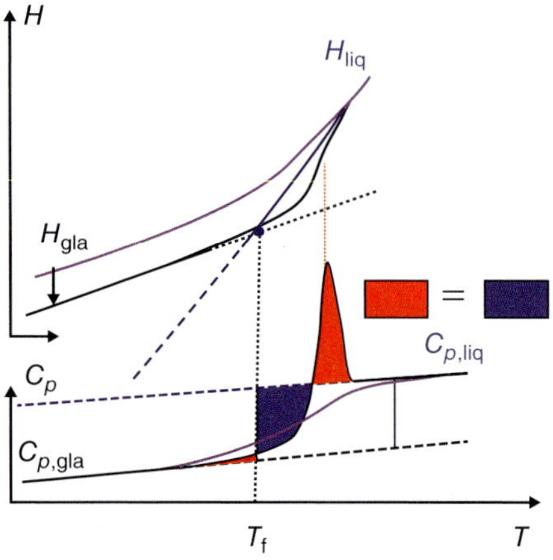

Figure 1.27 Fictive temperature. (Top) Definition on the $H(T)$ curve. It shows how T_f decreases when the glass is annealed. (Bottom) Experimental determination of T_f by application of an "equal area rule" on $C_p(T)$ curve.

of the change in the fictive temperature T_f. The notion of a fictive temperature T_f was formulated by Tool [98]. T_f of a nonequilibrium glassy system is defined as the actual temperature of the same compound in the equilibrium, metastable, undercooled state whose structure is expected to be similar to that of the nonequilibrium compound. The definition is illustrated in Figure 1.27, which also shows a method ("equal area rule") to determine T_f for a glassy compound using experimental values of a property of this compound (e.g., C_p) when the temperature evolutions of this property for the liquid ($C_p^{\,l}$) and glassy (C_p^g) states are known. From this picture, we can see that there is a one-to-one correspondence between the enthalpy level of the glass and its fictive temperature. Upon aging, H and T_f decrease together. H tends to reach the equilibrium metastable level at the same time as T_f tends to reach the actual aging temperature T (at equilibrium $T_f = T$, which is always the case for $T > T_g$).

Nonlinearity

It has long been experimentally observed that below T_g, the shear viscosity depends on the degree of structural relaxation [68, 99]. As a result of the Maxwell relation 1.1 between the shear viscosity η_s and the shear stress relaxation time τ_s ($\eta_s = G_\infty\ \tau_s$), another way of saying the same thing is that the relaxation time, that is, the rate of structural relaxation, is itself dependent on the progression of the relaxation process. This is basically due to the fact that the relaxation time τ_s depends on the enthalpy and free volume values, which decrease upon annealing at a rate that is determined by τ_s. If the nonequilibrium state of the glass is approximately characterized by a fictive temperature alone, this is reflected in the fact that τ depends on the fictive temperature T_f in addition to the ambient temperature T [81].

To summarize, at a given temperature T below T_g, the instantaneous values of H and V (depicted by T_f) determine the mobility (τ), while τ determines the rate dT_f/dt at which T_f (and therefore H, V, etc.) changes. Aging can thus be understood via following the closed-loop scheme [100]:

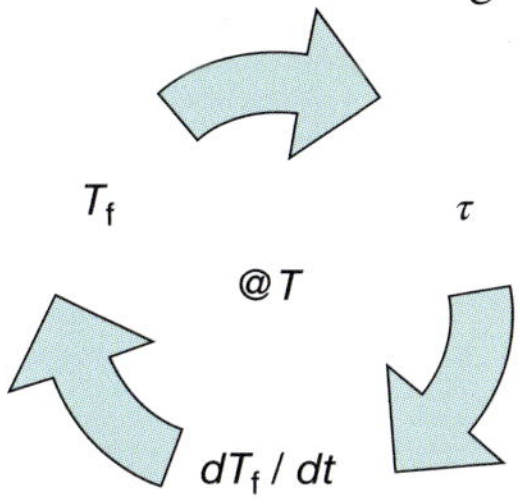

The following TNM (Tool–Narayaswamy–Moynihan) expression [101] is commonly used to describe the dependence of τ on both T and T_f:

$$\tau(T, T_f) = \tau_0 \exp[\, xA/(k_B T) + (1 - x)A/(k_B T_f)] \tag{1.37}$$

where x is the nonlinearity parameter ($0 \leq x \leq 1$). The maximum nonlinearity is for $x = 0$. The first term in the bracket expresses the effect of temperature, and the second term expresses that of the fictive temperature [102]. This description is highly empirical.

Another equation adapted from the more physical AG model and the entropy evolution (referred to as AGF: Adam–Gibbs–Fulcher [103–105]) has been proposed to express the time dependence of τ in terms of T_f evolution. The generalization of the AG equation allows us to describe the mobility both above and below T_g for a real system. It is obtained by expressing the configurational entropy S_c as a function of T_f rather than T.

$$\tau(T, T_f) = A \exp[C/TS_c(T_f)] \tag{1.38}$$

For $T > T_g$, the system is in a state of equilibrium and $T_f = T$, and Eq. (1.33) is recovered.

For $T < T_g$, in the fictive temperature formalism, the configurational entropy of the nonequilibrium system (the glass) at temperature T ($S_c^g(T)$) is taken to be identical to the configurational entropy of the equilibrium supercooled system at T_f ($S_c^{eq}(T_f)$).

$$S_c^g(T) = S_c^{eq}(T_f) = \int_{T_K}^{T_f} (\Delta C_p/T')\,dT' \tag{1.39}$$

where ΔC_p is the extrapolated value of the configurational heat capacity of the equilibrium supercooled system. Substitution of Eq. (1.39) (rather than Eq. (1.33)) in Eq. (1.38) gives the AGF equation for the relaxation time:

$$\tau(T, T_f) = \tau_0 \exp[\, E/T(1 - T_0/T_f)] \tag{1.40}$$

and thus allows us to define an isostructural activation energy

$$E(T_f) = E/(1 - T_0/T_f) \tag{1.41}$$

Since $T \leq T_f$, $E(T_f)$ is lower than the equilibrated value of the activation energy $E(T)$. Aging induces a slow decrease of T_f, which moves toward the aging temperature T. Aging thus induces an increase of $E(T_f)$, which slowly tends to be closer

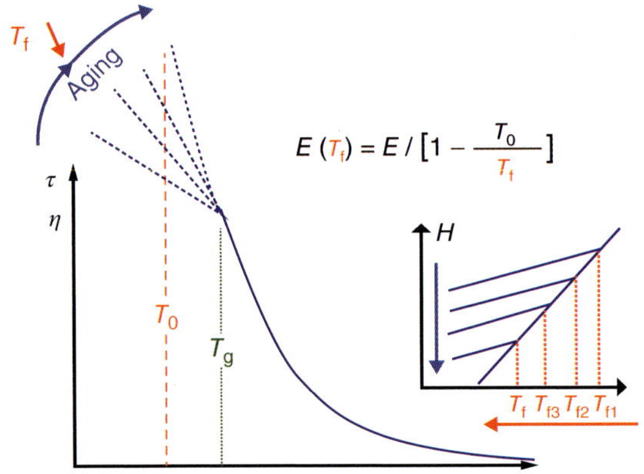

Figure 1.28 Evolution of the apparent activation energy upon aging the glass. Correlation with the evolution of the fictive temperature T_f is underlined.

to the equilibrated value $E(T)$. The evolution of the relaxation times corresponding to the forgoing description is depicted in Figure 1.28. The change of slope at T_g is related to the evolution of the measured configurational entropy.

Combination of the AGF and NMN equations allows us to express the nonlinearity factor $x \approx 1 - T_K/T_g$. This relation and that linking m to β [92] show that the levels of fragility, non-exponentiality, and nonlinearity are connected and rise together. When comparing different glass formers, the tendency is as follows:

x decreases when m increases, β decreases, and T_K increases toward T_g.

To obtain an accurate – albeit phenomenological – description of relaxation features and evolution of the specific heat as a function of the temperature upon cooling and heating through the glass transition, it is necessary to combine a stretched exponential expression of the relaxation function with the T_f-dependent expression of the relaxation time $\tau(T, T_f)$ [103, 106].

Secondary Relaxations

Below Tg, not all molecular mobility ceases to occur. In addition to the fast vibrational motions of atoms or molecules, and the very slow relaxations associated with the irreversible aging of glasses (main or α process), there are small-scale molecular motions that can be detected generally by dielectric (see the chapter by Paluch *et al.* in this book) and mechanical relaxation studies. The timescale of these secondary relaxation processes is generally several orders of magnitude lower than that of the main relaxation motions. Some of these motions can be attributed to the internal degrees of freedom of the molecules when the latter are flexible. However, there can exist other types of secondary relaxation processes, which have been identified for the first time by Johari and Goldstein [107]. These relaxations can be observed even for rigid molecular substances and are thus an intrinsic property of amorphous systems. They are named Johari–Goldstein (JG) β processes (β_{JG}), or simply β processes. The other types of secondary processes are often designated by other greek letters γ, δ, and so on.

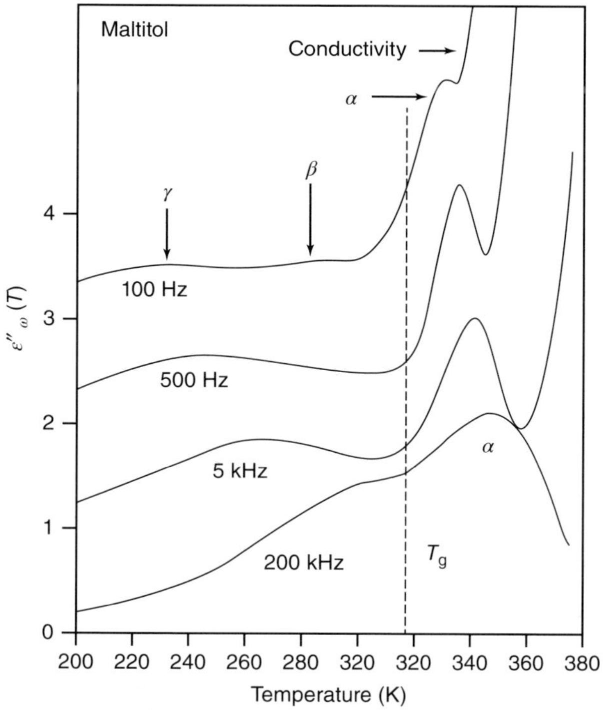

Figure 1.29 Isochronal imaginary part of the dielectric response of amorphous maltitol plotted against the temperature for four typical frequencies. (Adapted from Carpentier and Descamps [108]. Reproduced with permission of American Chemical Society.)

Figure 1.29 shows the dielectric loss factor of amorphous maltitol as a function of temperature for several frequencies [108]. The main (α) and secondary relaxations are seen for temperatures, respectively, above and below T_g. Below T_g, for a frequency = 100 Hz, two secondary processes are visible. It is sometimes difficult to distinguish JG β processes from other secondary processes. This aspect is discussed in [109]. In the case of maltitol, the JG β process is most probably that observed at the highest temperature. The other one (γ) is associated with faster intramolecular motions. This latter mode is also observed in sorbitol [108]. The relaxation map shown in Figure 1.30 summarizes the temperature evolution of the mean relaxation times of maltitol corresponding to the different modes of relaxation. The secondary relaxation processes may persist above T_g. α and β processes show a tendency to merge above T_g (see Figures 1.23 and 1.30). Secondary relaxations are more easily observed below T_g, where they have an Arrhenius type evolution with activation energies much lower than that of the α process. For all these aspects, see the chapter by Paluch *et al.*

There is general agreement that the β_{JG} relaxation involves localized molecular motions as opposed to α-type motions that are linked to cooperative rearrangements. However, the nature of theses motions and there spatial distribution are

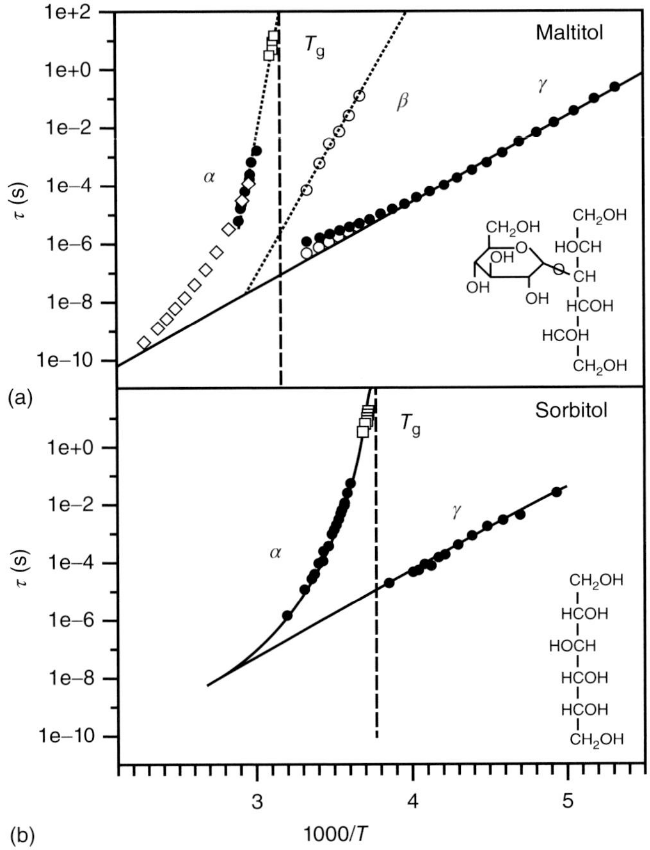

Figure 1.30 Arrhenius plot summarizing the temperature evolutions of the relaxation times of (a) amorphous maltitol and (b) amorphous sorbitol. (Adapted fromCarpentier and Descamps [108]. Reproduced with permission of American Chemical Society.)

still matters of debate. There are two main types of interpretation. The first is that of Johari [110], which considers that the glass is structurally nonuniform. β relaxation would correspond to motions of molecules situated at local regions of low density, the so-called *islands of mobility*. Another interpretation [111] considers that the main and secondary relaxations are continuous and represent the evolution with time of a single process whose long-time behavior corresponds to the main process. The model of Ngai has some connection with the latter model. Ngai proposed that the primitive relaxation of his coupling model is approximately located near the most probable relaxation time τ_β of the JG relaxation since both relaxations are noncooperative in nature [95]. The long-time behavior (α process) would correspond to the motions of molecules moving cooperatively.

The irreversible decrease in enthalpy and volume due to physical aging brings about a decrease of the height of the JG β relaxation peak. That is interpreted by Johari as being the result of a decrease in the number of molecules involved in the islands of mobility. It is easy to imagine that aged glasses require more cooperativity for structural relaxation because the packing is denser.

1.2.5
Fragility and Polyamorphism in the Energy Landscape View Point

1.2.5.1 Fragility and Landscape Topology View Point

Goldstein [112], more than 45 years ago, already pointed out the existence of the crossover temperature (such as T_{SA} in Figure 1.24) that occurs well above T_g in the moderately undercooled liquid state. He proposed that for $T < T_{SA}$, the dynamics corresponding to the slow α process is dominated by thermally activated relaxation processes over potential energy barriers that are larger than the typical thermal energy ($k_B T$). This relates to the energy landscape view point and its topology (Figure 1.31). From this view point, the dramatic slowing down of relaxations

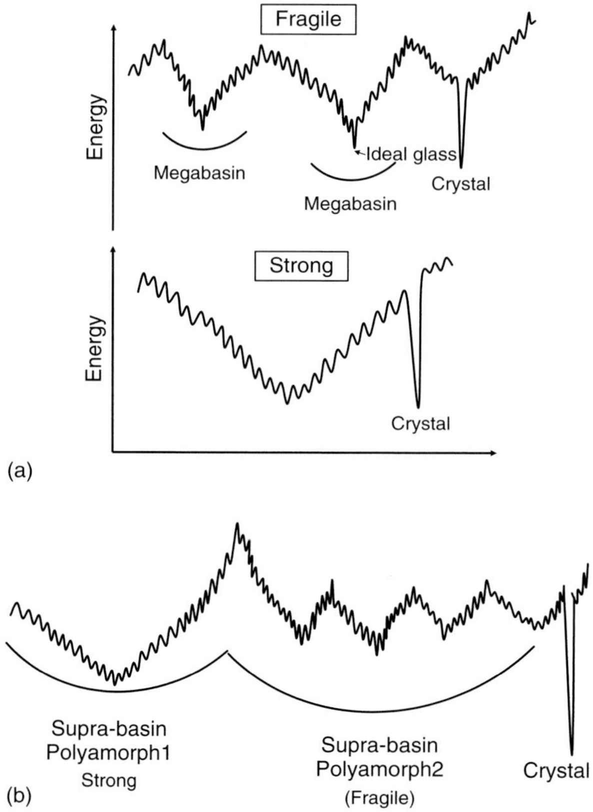

Figure 1.31 Examples of energy landscape topologies. (a) Difference between strong and fragile glass formers. In the latter case, "megabasins" are separated by relatively high barriers. Smaller basins inside "megabasins" could be linked to β relaxations. (b) Example of polyamorphism. Each polyamorph has its proper "supra-bassin".

at low T – which is typical of fragile glass formers – is linked to the increasing difficulty for the amorphous system to find new minima in the landscape. Fragile behavior is thus linked to a specific landscape topology with "megabasins" linked by relatively high energy barriers. Moving from one "megabasin" to another one involves a high energy activation (much larger than the energy involved in individual bonding energy between molecules) and cooperative rearrangement of many molecules (similar idea as that of CRR). In this case, the slowing down of α relaxations is associated with the fast decrease of the configurational entropy, which measures the number of minima that are available at any given temperature. These minima become increasingly hard to find for the amorphous system when the temperature decreases. It has been suggested that the faster, well-separated β relaxations would be linked to the elementary relaxations between small, contiguous basins that decorate the interior of a "megabasin" [69, 113]. On the contrary, strong formers could be associated to a single "megabasin" with a multitude of "small" barriers inside, which have roughly the activation energy corresponding to the breaking of the bonds between individual molecular entities. Figure 1.31a shows a schematic representation of the different types of landscape topologies.

1.2.5.2 Polyamorphism and Landscape Topology

The concept of polyamorphism has become very popular for a while both in the field of fundamental physics and pharmaceutical science. In the latter case, it is because it is well established that the properties of amorphous glasses, which can impact pharmaceutical behavior, can be different according to the previous history of the formulation and (or) the time of storage. As mentioned above, the primary reason for such a variability is linked to the non-ergodicity of the glassy state for the same compound. There is thus some kind of unavoidable "multiamorphicity" of a glass connected, for example, to a variable fictive temperature. It reflects the landscape localization of the frozen molecular configuration and the limited extension of this landscape that is accessible during experimental times. In contrast to this behavior, a truly polyamorphic situation for a compound implies the existence of two different liquid structures for the same material. This may lead to a possible liquid/liquid phase transition and existence of an interface between these phases. The polyamorphism of water is certainly the most widely studied example [71, 114, 115]. There are a few documented examples of suspected polyamorphism for organic and pharmaceutical compounds (for a discussion see [116]). The experimental identification of a true polyamorphism in such compounds is very difficult. From this perspective, the case of the "glacial phase" of triphenyl phosphite (TPP) is particularly illustrative. Indeed, there is a danger of confusion with the existence of a mesophase [117], or a situation of a fully nanocrystallized liquid [118, 119]. A situation of polyamorphism can be described by a potential energy hypersurface with "supra-basins" corresponding to the specific physical properties of the different polyamorphic varieties. Figure 1.31b gives a schematic representation of this landscape. One "polyamorph" may be strong, while the other is fragile. In connection with the situation of amorphous ice, Angell [71] has suggested that "polyamorphs" may indeed differ in their strength: a low-density form acting as a strong glass former, while a high-density form as a fragile glass former.

Acknowledgments

The authors thank their colleagues F. Affouard, E. Dudognon, N. Correia, L. Carpentier, F. Danède, P. Derollez, A. Hedoux, Y. Guinet, and L. Paccou for their kind cooperation and interaction. This work was funded by the EU INTERREG IV A 2 Mers-Seas-Zeeën Crossborder Cooperation Programme (IDEA and AMPTECH).

References

1. Murdande, S.B., Pikal, M.J., Shanker, R.M., and Bogner, R.H. (2010) Solubility advantage of amorphous pharmaceuticals: I. A thermodynamic analysis. *J. Pharm. Sci.*, **99** (3), 1254–1264.
2. Craig, D.Q.M., Royall, P.G., Kett, V.L., and Hopton, M.L. (1999) The relevance of the amorphous state to pharmaceutical dosage forms: glassy drugs and freeze dried systems. *Int. J. Pharm.*, **179**, 179.
3. Yu, L. (2001) Amorphous pharmaceutical solids: preparation, characterization and stabilization. *Adv. Drug Delivery Rev.*, **48**, 27.
4. Hancock, B.C. and Zografi, G. (1997) Characteristics and significance of the amorphous state in pharmaceutical systems. *J. Pharm. Sci.*, **86**, 1.
5. Hancock, B.C. and Parks, M. (2000) What is the true solubility advantage for amorphous pharmaceuticals? *Pharm. Res.*, **17**, 397.
6. Baird, J.A. and Taylor, L.S. (2012) Evaluation of amorphous solid dispersion properties using thermal analysis techniques. *Adv. Drug Delivery Rev.*, **64**, 396.
7. Graeser, K.A., Patterson, J.E., and Rades, T. (2008) Physical stability of amorphous drugs: evaluation of thermodynamic and kinetic parameters. *J. Pharm. Pharmacol.*, **60**, 116.
8. Bhattacharya, S. and Suryanarayanan, R. (2009) Local mobility in amorphous pharmaceuticals-characterization and implications on stability. *J. Pharm. Sci.*, **98**, 2935.
9. Paudel, A., Geppi, M., and Van Den Mooter, G. (2014) Structural and dynamic properties of amorphous solid dispersions: the role of solid-state nuclear magnetic resonance spectroscopy and relaxometry. *J. Pharm. Sci.*, **103**, 2635.
10. Massey, B.S. and Ward-Smith, J. (1998) *Mechanics of Fluids*, 7th edn, Taylor & Francis.
11. Dyre, J.C., Christensen, T., and Olsen, N.B. (2006) Elastic models for the non-Arrhenius viscosity of glass-forming liquids. *J. Non-Cryst. Solids*, **352**, 4635.
12. Warren, B.E. (1970) *X-Ray Diffraction*, Addison-Wesley, New York.
13. Guinier, A. (1994) *X-Ray Diffraction in Crystals, Imperfect Crystals, and Amorphous Bodies*, Dover Publications.
14. Guinier, A. (1964) *Théorie et technique de la radiocristallographie*, Dunod, Paris.
15. Billinge, S. and Thorpe, F. (1998) *Local Structure from Diffraction*, Springer, New York: Plenum Press.
16. Hosemann, R. (1950) Der ideale Parakristall und die von ihm gestreute kohärente Röntgenstrahlung. *Z. Angew. Phys.*, **128**, 465.
17. Elliott, S.R. (1990) *Physics of Amorphous Materials*, Longman Scientific & Technical.
18. Sherwood, J.N. (1979) *The Plastically Crystalline State*, John Wiley & Sons,

Ltd., Chichester, New York, Brisban, Toronto.

19. Suga, H. and Seki, S. (1974) Thermodynamic investigation on glassy states of pure simple compounds. *J. Non Cryst. Solids*, **16**, 171.
20. Descamps, M., Caucheteux, C., Odou, G., and Sauvajol, J.L. (1984) Local molecular order in the glassy crystalline phase of cyanoadamantane: diffuse X-ray scattering analysis. *J. Phys. Lett.*, **45**, 719.
21. Descamps, M., Correia, N.T., Derollez, P., Danede, F., and Capet, F. (2005) Plastic and glassy crystal states of caffeine. *J. Phys. Chem. B*, **109**, 16092.
22. Derollez, P., Correia, N.T., Danède, F., Capet, F., Affouard, F., Lefebvre, J., and Descamps, M. (2005) Ab initio structure determination of the high-temperature phase of anhydrous caffeine by X-ray powder diffraction. *Acta Crystallogr., Sect. B: Struct. Sci.*, **61**, 329.
23. Stevenson, C.L., Bennett, D.B., and Lechuga-Ballesteros, D. (2005) Pharmaceutical liquid crystals: the relevance of partially ordered systems. *J. Pharm. Sci.*, **94**, 1861.
24. Ziman, J. (1979) *Model of Disorder: The Theoretical Physics of Homogeneously Disordered Systems*, Cambridge University Press, New York.
25. Descamps, M. and Coulon, G. (1977) Series expansion calculation of the elastic neutron diffuse scattering: ice Ih. *Chem. Phys.*, **25**, 117.
26. Suchod, B. and Lajzerowicz-Bonneteau, J. (1991) Interactions between substitutional and orientational orders – the phase-transitions in enantiomeric and racemic crystals of tmhp. 1. X-ray structural studies of the orientational order-disorder transition in enantiomeric tmhp. *J. Phys. I*, **1**, 553.
27. Elarby, A., Jal, J.F., Dupuy, J., Chieux, P., Wright, A., and Parreins, R. (1982) Nucléation homogène et cristallisation delaglace cubique (Ic) dans les verres d'électrolytes LiCl.D2O, *Journal de Physique Lettres*, **43**, 355.
28. Proffen, T., Billinge, S.J.L., Egami, T., and Louca, D. (2003) Structural analysis of complex materials using the atomic pair distribution function – a practical guide. *Z. Kristallogr.*, **218**, 132.
29. Dmowski, W. and Swider-Lyons, K.E. (2004) PDF analysis – from atomic displacements to nanocrystals. *Z. Kristallogr.*, **219**, 136.
30. Kodama, K., Iikubo, S., Taguchi, T., and Shamoto, S.I. (2006) Finite size effects of nanoparticles on the atomic pair distribution functions. *Acta Crystallogr., Sect. A: Found. Crystallogr.*, **62**, 444.
31. Pauchet, M., Gervais, C., Courvoisier, L., and Coquerel, G. (2004) Successful application of the derived crystal packing (DCP) model in resolving the crystal structure of a metastable polymorph of (±) modafinil†. *Cryst. Growth Des.*, **4**, 1143.
32. Descamps, M., Legrand, V., Guinet, Y., Amazzal, A., Alba, C., and Dore, J. (1997) "Pre-peak" in the structure factor of simple molecular glass formers. *Prog. Theor. Phys. Suppl.*, **126**, 207.
33. Morineau, D. and Alba-Simionesco, C. (1998) Hydrogen-bond-induced clustering in the fragile glass-forming liquid m-toluidine: experiments and simulations. *J. Chem. Phys.*, **109**, 8494.
34. Qi, S., Weuts, I., De Cort, S., Stokbroekx, S., Leemans, R., Reading, M., Belton, P., and Craig, D.Q.M. (2010) An investigation into the crystallisation behaviour of an amorphous cryomilled pharmaceutical material above and below the glass transition temperature. *J. Pharm. Sci.*, **99**, 196.
35. Descamps, M. and Dudognon, E. (2014) Crystallization from the amorphous state: nucleation-growth decoupling, polymorphism interplay, and the role of interfaces. *J. Pharm. Sci.*, **103**, 2615.
36. Porter, D.A., Easterling, K.E., and Sherif, M. (2009) *Phase Transformations in Metals and Alloys*, 3rd (Revised Reprint) edn, Taylor & Francis.
37. Andronis, V. and Zografi, G. (2000) Crystal nucleation and growth of indomethacin polymorphs from the amorphous state. *J. Non-Cryst.-Solids*, **271**, 236.
38. Carpentier, L., Desprez, S., and Descamps, M. (2003) Crystallization and glass properties of pentitols: xylitol,

adonitol arabitol. *J. Therm. Anal.*, **73**, 577.
39. Legrand, V., Descamps, M., and Alba-Simionesco, C. (1997) Glass-forming meta-toluidine: a thermal and structural analysis of its crystalline polymorphism and devitrification. *Thermochim. Acta*, **307**, 77.
40. Kelton, K.F. (1991) Crystal nucleation in liquids and glasses. *Solid State Phys.*, **45**, 75.
41. Christian, J. (1975) *The Theory of Transformations in Metals and Alloys*, Pergamon Press, Oxford.
42. Debenedetti, P.G. (1996) *Metastable Liquids – Concepts and Principles*, Princeton University Press.
43. Gutzow, I. (1977) The mechanism of crystal growth in glass forming systems. *J. Cryst. Growth*, **42**, 15.
44. Gutzow, I. (1980) Kinetics of crystallization processes in glass forming melts. *J. Cryst. Growth*, **48**, 589.
45. Turnbull, D. (1956) *Solid State Physics – Advances in Research and Applications*, vol. **3**, Academic Press, p. 225.
46. Spaepen, F. (1975) A structural model for the solid-liquid interface in monatomic systems. *Acta Metall.*, **23**, 729.
47. Oxtoby, D.W. and Haymet, A.D.J. (1981) A molecular theory of the solid-liquid interface. II. Study of bcc crystal-melt interfaces. *J. Chem. Phys.*, **76**, 6262.
48. Spaepen, F. and Meyer, R.B. (1976) The surface tension in a structural model for the solid-liquid interface. *Scr. Metall.*, **10**, 257.
49. Descamps, M. and Decroix, A.A. (2014) Polymorphism and disorder in caffeine: dielectric investigation of molecular mobilities. *J. Mol. Struct.*, **1078**, 165.
50. Ostwald, W. (1897) Studien über die Bildung und Umwandlung fester Körper. *Z. Angew. Phys.*, **22**, 289.
51. Threlfall, T. (2003) Structural and thermodynamic explanations of ostwald's rule. *Org. Process Res. Dev.*, **7**, 1017.
52. Dudognon, E., Correia, N.T., Danède, F., and Descamps, M. (2013) Solid-solid transformation in racemic Ibuprofen. *Pharm. Res.*, **30**, 81.
53. Dudognon, E., Danède, F., Descamps, M., and Correia, N.T. (2008) Evidence for a new crystalline phase of racemic Ibuprofen. *Pharm. Res.*, **25**, 2853.
54. Zhu, L., Brian, C.W., Swallen, S.F., Straus, P.T., Ediger, M.D., and Yu, L. (2011) Surface self-diffusion of an organic glass. *Phys. Rev. Lett.*, **106**.
55. Capaccioli, S., Ngai, K.L., Paluch, M., and Prevosto, D. (2012) Mechanism of fast surface self-diffusion of an organic glass. *Phys. Rev. E: Stat. Nonlinear Soft Matter Phys.*, **86**.
56. Alonzo, D.E., Gao, Y., Zhou, D., Mo, H., Zhang, G.G., and Taylor, L.S. (2011) Dissolution and precipitation behavior of amorphous solid dispersions. *J. Pharm. Sci.*, **100**, 3316.
57. Alonzo, D.E., Zhang, G.G.Z., Zhou, D., Gao, Y., and Taylor, L.S. (2010) Understanding the behavior of amorphous pharmaceutical systems during dissolution. *Pharm. Res.*, **27**, 608.
58. Newman, A., Knipp, G., and Zografi, G. (2012) Assessing the performance of amorphous solid dispersions. *J. Pharm. Sci.*, **101**, 1355.
59. Qi, S., McAuley, W.J., Yang, Z., and Tipduangta, P. (2014) Physical stabilization of low-molecular-weight amorphous drugs in the solid state: a material science approach. *Ther. Delivery*, **5**, 817.
60. Dahan, A., Beig, A., Ioffe-Dahan, V., Agbaria, R., and Miller, J.M. (2013) The twofold advantage of the amorphous form as an oral drug delivery practice for lipophilic compounds: Increased apparent solubility and drug flux through the intestinal membrane. *AAPS J.*, **15**, 347.
61. Frank, K.J., Rosenblatt, K.M., Westedt, U., Hölig, P., Rosenberg, J., Mägerlein, M., Fricker, G., and Brandl, M. (2012) Amorphous solid dispersion enhances permeation of poorly soluble ABT-102: true supersaturation vs. apparent solubility enhancement. *Int. J. Pharm.*, **437**, 288.
62. Gardner, C.R., Walsh, C.T., and Almarsson, Ö. (2004) Drugs as materials: valuing physical form in drug discovery. *Nat. Rev. Drug Discovery*, **3**, 926.

63. Miller, J.M., Beig, A., Carr, R.A., Spence, J.K., and Dahan, A. (2012) A win-win solution in oral delivery of lipophilic drugs: supersaturation via amorphous solid dispersions increases apparent solubility without sacrifice of intestinal membrane permeability. *Mol. Pharm.*, **9**, 2009.
64. Kittel, C. and Kroemer, H. (1980) *Thermal Physics*, W. H. Freeman.
65. Amharar, Y., Curtin, V., Gallagher, K.H., and Healy, A.M. (2014) Solubility of crystalline organic compounds in high and low molecular weight amorphous matrices above and below the glass transition by zero enthalpy extrapolation. *Int. J. Pharm.*, **472**, 241.
66. Ilevbare, G.A. and Taylor, L.S. (2013) Liquid-liquid phase separation in highly supersaturated aqueous solutions of poorly water-soluble drugs: implications for solubility enhancing formulations. *Cryst. Growth Des.*, **13**, 1497.
67. Angell, C.A. (1988) Perspective on the glass transition. *J. Phys. Chem.*, **49**, 863.
68. Jäckle, J. (1986) Models of the glass transition. *Rep. Prog. Phys.*, **49** (2), 171.
69. Debenedetti, P.G. and Stillinger, F.H. (2001) Supercooled liquids and the glass transition. *Nature*, **410**, 259.
70. Willart, J.F. and Descamps, M. (2008) Solid state amorphization of pharmaceuticals. *Mol. Pharm.*, **5**, 905.
71. Angell, C.A. (1995) Formation of glasses from liquids and biopolymers. *Science*, **267**, 1924.
72. Goldstein, M. (1976) Viscous liquids and the glass transition. V. Sources of the excess specific heat of the liquid. *J. Chem. Phys.*, **64**, 4767.
73. Angell, C.A. (1995) The old problems of glass and the glass transition, and the many new twists. *Proc. Natl. Acad. Sci. U.S.A.*, **92**, 6675.
74. Hodge, I.M. (1996) Strong and fragile liquids – a brief critique. *J. Non-Cryst. Solids*, **202**, 164.
75. Angell, C.A., McMillan, P.F., Oganov, A.R., and Wolynes, P.G. (2005) Energy landscapes for cooperative processes: nearly ideal glass transitions, liquid-liquid transitions and folding transitions. *Philos. Trans. R. Soc. A: Math. Phys. Eng. Sci.*, **363**, 415.
76. Angell, C.A. (2008) Glass-formers and viscous liquid slowdown since David turnbull: enduring puzzles and new twists. *MRS Bull.*, **33**, 544.
77. Descamps, M., Moura Ramos, J.J., and Correia, N.T. (2002) Ageing exploration of the energy landscape of a glass by the TSDC technique. *Mol. Phys.*, **100**, 2669.
78. Kauzmann, W. (1948) The nature of the glassy state and the behavior of liquids at low temperatures. *Chem. Rev.*, **43**, 219.
79. Adam, G. and Gibbs, J.H. (1965) On the temperature dependence of cooperative relaxation properties in glass-forming liquids. *J. Chem. Phys.*, **43**, 139.
80. Stillinger, F.H. (1988) Supercooled liquids, glass transitions, and the Kauzmann paradox. *J. Chem. Phys.*, **88**, 7818.
81. Dyre, J.C. (2006) Colloquium: the glass transition and elastic models of glass-forming liquids. *Rev. Mod. Phys.*, **78**, 953.
82. Ediger, M.D., Angell, C.A., and Nagel, S.R. (1996) Supercooled liquids and glasses. *J. Phys. Chem.*, **100**, 13200.
83. Angell, C.A., Ngai, K.L., McKenna, G.B., McMillan, P.F., and Martin, S.W. (2000) Relaxation in glassforming liquids and amorphous solids. *J. Appl. Phys.*, **88**, 3113.
84. Mauro, J.C. (2014) Grand challenges in glass science. *Front. Mater.*, **1**, 20.
85. Carpentier, L., Decressain, R., Desprez, S., and Descamps, M. (2006) Dynamics of the amorphous and crystalline α-, β-phases of indomethacin. *J. Phys. Chem. B*, **110**, 457.
86. Wojnarowska, Z., Adrjanowicz, K., Wlodarczyk, P., Kaminska, E., Kaminski, K., Grzybowska, K., Wrzalik, R., Paluch, M., and Ngai, K.L. (2009) Broadband dielectric relaxation study at ambient and elevated pressure of molecular dynamics of pharmaceutical: indomethacin. *J. Phys. Chem. B*, **113**, 12536.

87. Busch, R. (2000) Thermophysical properties of bulk metallic glass-forming liquids. *JOM*, **52**, 39.
88. Angell, C.A. (1997) Why C1 = 16-17 in the WLF equation is physical – and the fragility of polymers. *Polymer*, **38**, 6261.
89. Donth, E. (1996) Characteristic length of the glass transition. *J. Polym. Sci., Part B: Polym. Phys.*, **34**, 2881.
90. Stevenson, J.D., Schmalian, J., and Wolynes, P.G. (2006) The shapes of cooperatively rearranging regions in glass forming liquids. *Nat. Phys.*, **21**, 268.
91. Williams, G. and Watts, D.C. (1970) Non-symmetrical dielectric relaxation behaviour arising from a simple empirical decay function. *Trans. Faraday Soc.*, **66**, 80.
92. Böhmer, R., Ngai, K.L., Angell, C.A., and Plazek, D.J. (1993) Nonexponential relaxations in strong and fragile glass formers. *J. Chem. Phys.*, **99**, 4201.
93. Böhmer, R. and Angell, C.A. (1994) in *Disorder Effects on Relaxational Processes* (eds R. Richert and A. Blumen), Springer, Berlin, Heidelberg, p. 11.
94. Richert, R. (2002) Heterogeneous dynamics in liquids: fluctuations in space and time. *J. Phys.: Condens. Matter*, **14**, R703.
95. Ngai, K.L. (2011) *Relaxation and Diffusion in Complex Systems*, Springer, New York.
96. Correia, N.T., Moura Ramos, J.J., Descamps, M., and Collins, G. (2001) Molecular mobility and fragility in indomethacin: a thermally stimulated depolarization current study. *Pharm. Res.*, **18**, 1767.
97. Lunkenheimer, P., Wehn, R., Schneider, U., and Loidl, A. (2005) Glassy aging dynamics. *Phys. Rev. Lett.*, **95**.
98. Tool, A.Q. and Eicitlin, C.G. (1931) Variations caused in the heating curves of glass by heat treatment. *J. Am. Ceram. Soc.*, **14**, 276.
99. de Bast, J. and Gilard, P. (1963) Variation of the viscosity of glass and relaxation of stresses during stabilization. *Phys. Chem. Glasses*, **4**, 117.
100. Struik, L.C.E. (1977) Physical aging in plastics and other glassy materials. *Polym. Eng. Sci.*, **17**, 165.
101. Moynihan, C.T., Easteal, A.J., DeBolt, M.A., and Tucker, J. (1976) Dependence of the fictive temperature of glass on cooling rate. *J. Am. Ceram. Soc.*, **59**, 12.
102. Pappin, A.J., Hutchinson, J.M., and Ingram, M.D. (1992) Enthalpy relaxation in polymer glasses: evaluation and interpretation of the tool-narayanaswamy parameter x for poly(vinyl chloride). *Macromolecules*, **25**, 1084.
103. Hodge, I.M. (1994) Enthalpy relaxation and recovery in amorphous materials. *J. Non-Cryst. Solids*, **169**, 211.
104. Hodge, I.M. (1997) Adam-Gibbs formulation of enthalpy relaxation near the glass transition. *J. Res. Nat. Inst. Stand. Technol.*, **102**, 195.
105. Hodge, I.M. (2013) Parameterization of annealing kinetics in pharmaceutical glasses. *J. Pharm. Sci.*, **102**, 2235.
106. Luthra, S.A., Hodge, I.M., and Pikal, M.J. (2008) Effects of annealing on enthalpy relaxation in lyophilized disaccharide formulations: mathematical modeling of DSC curves. *J. Pharm. Sci.*, **97**, 3084.
107. Johari, G.P. and Goldstein, M. (1970) Viscous liquids and the glass transition. II. Secondary relaxations in glasses of rigid molecules. *J. Chem. Phys.*, **53**, 2372.
108. Carpentier, L. and Descamps, M. (2003) Dynamic decoupling and molecular complexity of flass-forming maltitol. *J. Phys. Chem. B*, **107**, 271.
109. Ngai, K.L. and Paluch, M. (2004) Classification of secondary relaxation in glass-formers based on dynamic properties. *J. Chem. Phys.*, **120**, 857.
110. Johari, G.P. (1973) Intrinsic mobility of molecular glasses. *J. Chem. Phys.*, **58**, 1766.
111. Reid, C.J. and Evans, M.W. (1979) Zero-THz absorption profiles in glassy solutions. High frequency γ process and its characterisation. *J. Chem. Soc., Faraday Trans. 2*, **75**, 1218.
112. Goldstein, M. (1969) Viscous liquids and the glass transition: a potential

energy barrier picture. *J. Chem. Phys.*, **51**, 3728.

113. Stillinger, F.H. (1995) A topographic view of supercooled liquids and glass formation. *Science*, **267**, 1935.

114. Stanley, H.E., Kumar, P., Franzese, G., Xu, L., Yan, Z., Mazza, M.G., Buldyrev, S.V., Chen, S.H., and Mallamace, F. (2008) Liquid polyamorphism: possible relation to the anomalous behaviour of water. *Eur. Phys. J.: Spec. Top.*, **161**, 1.

115. Mishima, O. and Stanley, H.E. (1998) The relationship between liquid, supercooled and glassy water. *Nature*, **396**, 329.

116. Shalaev, E. and Zografi, G. (2002) in *Progress in Amorphous Food and Pharmaceutical Systems*, vol. **23** (ed H. Levine), The Royal Chemistry Society, p. 11.

117. Cohen, I., Ha, A., Zhao, X., Lee, M., Fischer, T., Strouse, M.J., and Kivelson, D. (1996) A low-temperature amorphous phase in a fragile glass-forming substance. *J. Phys. Chem.*, **100**, 8518.

118. Hédoux, A., Guinet, Y., Derollez, P., Hernandez, O., Lefort, R., and Descamps, M. (2004) A contribution to the understanding of the polyamorphism situation in triphenyl phosphite. *Phys. Chem. Chem. Phys.*, **6**, 3192.

119. Hédoux, A., Guinet, Y., Derollez, P., Hernandez, O., Paccou, L., and Descamps, M. (2006) Micro-structural investigations in the glacial state of triphenyl phosphite. *J. Non-Cryst. Solids*, **352**, 4994.

2 Influence of Disorder on Dissolution

Khushboo Kothari and Raj Suryanarayanan

2.1 Introduction

New drug discovery methods such as high-throughput screening and combinatorial chemistry have led to the identification of a large number of highly target-selective and potent new drug candidates [1, 2]. Although these techniques enable rapid identification of new lead molecules, a majority of these compounds (80–90%) are practically insoluble in water [3]. According to the U.S. Pharmacopeia (USP), a compound is considered insoluble if one part of solute dissolves in 10 000 or more parts of a solvent [4]. This poses a major challenge in the effective oral delivery of these compounds.

For a drug to be systemically effective following oral administration, drug absorption into the systemic circulation is a prerequisite [5, 6]. There are two main barriers to oral drug absorption – *dissolution* in the gastrointestinal (GI) fluid, followed by *permeation* across the GI membrane [7–9]. Figure 2.1 shows the sequence of events following the oral administration of a solid oral dosage form, as a tablet. The first step is disintegration of the tablet, leading to granules or directly to primary particles, followed by dissolution of the drug in the GI fluid. Thus, solubility in the GI fluid is a potential barrier to drug absorption. Solubility can be enhanced transiently by the use of a metastable physical form of the drug. However, the formation of a supersaturated solution can be followed by the precipitation of the stable form. The permeability across the membrane is the second barrier to drug absorption. These two barriers, namely solubility and permeability, form the basis of the Biopharmaceutical Classification System (BCS) [7]. Table 2.1 summarizes the distribution of both the drugs in the market and the new chemical entities in the pipeline, based on the BCS [3].

Membrane permeability is an intrinsic property of a compound and is dependent on the molecular structure of the compound (molecular weight, functional groups, etc.).There are numerous approaches to enhance membrane permeability, including chemical modification of the drug [10]. However, these issues are outside the scope of this chapter. Our discussion will focus on some of the approaches

Disordered Pharmaceutical Materials, First Edition. Edited by Marc Descamps.

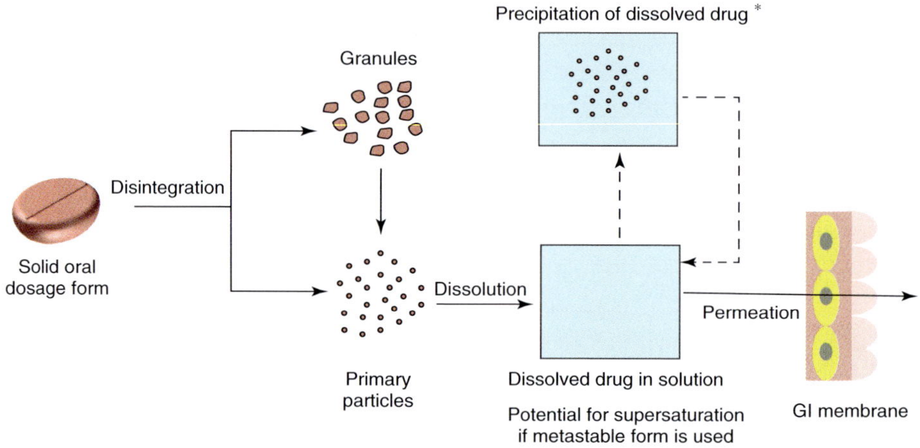

Figure 2.1 Schematic of the sequence of events following the oral administration of a tablet. *In spite of supersaturation, precipitation is not universal. Only the dissolved drug permeates across the membrane, though it is recognized that there are exceptions.

Table 2.1 Biopharmaceutics Classification System of drugs in the market and in the pipeline.

BCS class	Solubility	Permeability	Marketed drugs (%)	NCEs (pipeline) (%)
I	High	High	35	5–10
II	Low	High	30	60–70
III	High	Low	25	5–10
IV	Low	Low	10	10–20

Adapted from Babu and Nangia [3]. Reproduced with permission of American Chemical Society.

to enhance drug solubility, with the ultimate goal of enhancing oral bioavailability of poorly water soluble drugs.

"Solubility" at a given temperature can be defined as the amount of a substance (of a specific physical form) that can be completely dissolved in a specified quantity of a solvent [11]. Of much greater practical interest is the dissolution rate, a kinetic process (dM/dt) defined as the amount (M) of a compound dissolved in its own solution in a given time (t). It is proportional to the surface area of the solid (S) and the difference between its saturation solubility (C_s) and bulk concentration in solution (C_b). In compounds exhibiting diffusion-controlled dissolution (a large fraction of pharmaceuticals), the dissolution rate constant $k = D/h$, where D is the diffusion coefficient and h is the thickness of the boundary layer [12, 13].

$$dM/dt = kS(C_s - C_b) \ \ldots \tag{2.1}$$

2.2 Approaches to Enhance Solubility

Drugs belonging to BCS class II are characterized by poor aqueous solubility – believed to be the major oral absorption barrier. Recently, according to the developability classification system (DCS), the BCS class II drugs were further subdivided into IIa and IIb based on whether the drugs show dissolution-rate-limited or solubility-limited absorption, respectively [14]. Drugs are classified under IIa if the dissolution rate is slower than the transit time of the drug through the absorption site. In class IIb drugs, there is insufficient GI fluid to dissolve the administered dose.

Current efforts to enhance the drug dissolution rate focus on approaches to increase both the surface area and the solubility [15–17]. Particle size reduction, leading to an increase in the surface area, is potentially a simple approach. Various size reduction techniques, including milling, have resulted in enhancement in the *in vitro* dissolution rate and consequently the oral bioavailability [18–20]. Wu *et al.* studied the influence of particle size on the oral absorption of MK-0869 in beagle dogs [21]. A decrease in the drug particle size led to an increased *in vivo* exposure (Figure 2.2). Conversion of a conventional suspension into a nanocrystalline material caused a 42-fold increase in surface area and a 4-fold increase in

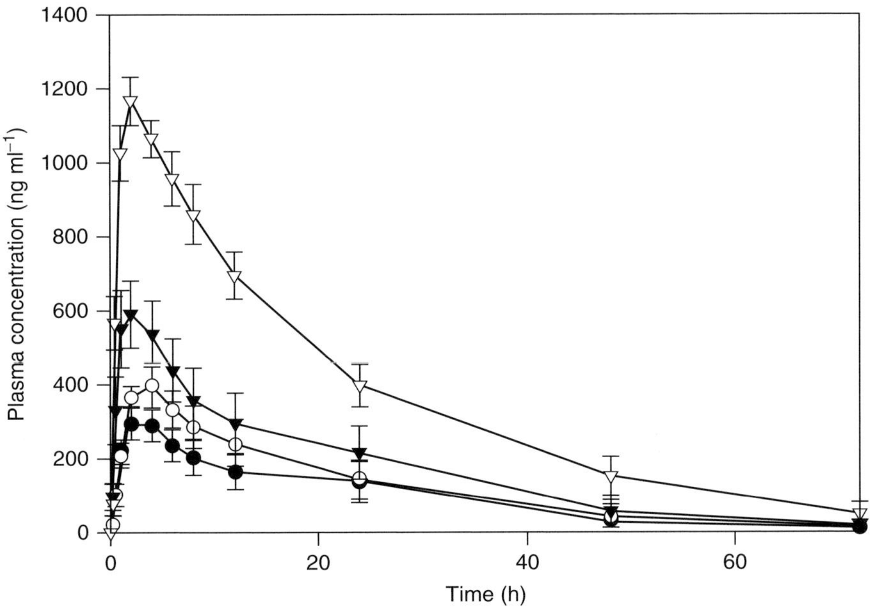

Figure 2.2 Plasma concentration versus time profiles following oral administration of different suspensions (conventional (● 5.5 μm); wet-milled (○ 1.8 μm); jet-milled (▼ 0.48 μm); and nano-milled (▽ 0.12 μm)) of MK-0869 to beagle dogs. The mean drug particle size in each suspension is given in parentheses [21]. (Adapted from Wu *et al.* [21]. Reproduced with permission of Elsevier.)

the area under the plasma drug concentration–time curve (area under the curve, AUC). Because of the potential for agglomeration, the milling-induced increase in surface area may not always translate to enhancement in the *in vivo* dissolution rate [22]. While there are several examples of enhancement in oral bioavailability brought about by particle size reduction, the effect can be modest to pronounced (typically 2–20-fold) [21–27]. While particle size reduction techniques may work well for dissolution-rate-limited drugs belonging to class IIa of the DCS, for practically insoluble drugs belonging to class IIb, this approach may not provide the required bioavailability enhancement. For these compounds, effective oral delivery may necessitate increased aqueous solubility [14]. In addition, there are numerous practical problems with milled particles. Alteration in particle size and surface characteristics brought about by milling can adversely affect both the blending and flow properties [28]. These issues warrant particular consideration in large-scale manufacturing. Milling can also lead to disorder of the crystal lattice and, in extreme cases, cause amorphization [29–31]. While this can enhance the dissolution rate, accidental and uncontrolled lattice disruption can potentially lead to erratic dissolution behavior.

Several other approaches, such as the use of metastable polymorphs, salts, and co-crystals, have also been utilized to improve the solubility and consequently the dissolution rate of poorly soluble drugs [3, 32–34]. The use of a metastable polymorph, much like particle size reduction, usually provides a limited enhancement in the dissolution rate and bioavailability [35, 36]. For example, Stagner and Guillory compared the dissolution rates of two polymorphs of iopanoic acid and found only a small difference [37]. Pudipeddi and Serajuddin compiled the solubility ratios of 65 pharmaceutical polymorphs [38]. They reported a typical solubility enhancement of only nearly fourfold, which is unlikely to translate into significant increase in bioavailability. Weak acids and bases can be converted to the corresponding salts, resulting in enhanced solubility and bioavailability [33, 39]. However, this approach has several limitations. Neutral molecules without ionizable functional groups are not amenable to salt formation. Secondly, compounds with very poor intrinsic solubility may not be able to provide the required dissolution advantage. There is also the potential for the salt to convert back to the free acid (or base), thereby negating the solubility advantage – a problem encountered with salts of weak acids in the acidic stomach pH [40]. Finally, the mechanical properties and the hygroscopic nature of some salts may pose processing problems [41–43]. Co-crystal formation of the active pharmaceutical ingredient (API) with a second pharmaceutically acceptable component is another potential means to overcome the problem of poor aqueous solubility [34, 44]. This approach has resulted in bioavailability enhancement, sometimes substantial, for several drugs [45–50]. This approach is usually limited to APIs that can hydrogen-bond with the co-former, the most common interaction for co-crystal formation.

Amorphization is increasingly emerging as an effective approach to increase the solubility of drugs. An amorphous compound, unlike its crystalline counterpart, lacks the three-dimensional long-range lattice periodicity. As a result, it has a higher free energy and is thermodynamically unstable. A pharmaceutically desirable consequence is the higher apparent solubility than its crystalline counterpart [51–54]. Gupta *et al.* documented the solubility difference between amorphous (estimated as peak solubility) and crystalline celecoxib at several temperatures (Figure 2.3). The authors determined that the favorable entropy and free energy of the solution was responsible for the enhanced solubility of the amorphous form [55]. The enhancement in apparent solubility has been noted for many drug compounds, and in several cases has resulted in an enhanced dissolution rate [3, 56]. For example, Six *et al.* have shown nearly an order of magnitude increase in the dissolution rate of amorphous itraconazole as compared to its crystalline form (Figure 2.4) [57]. Stagner and Guillory

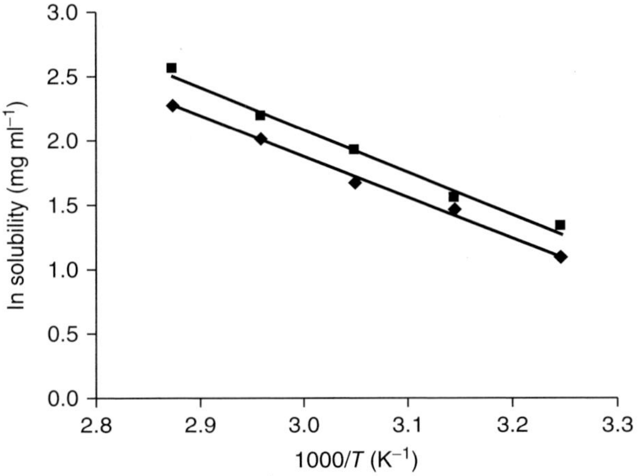

Figure 2.3 van't Hoff type plots revealing the higher solubility of amorphous celecoxib (■) over the crystalline form (♦), in the temperature range studied (Adapted from Gupta *et al.* [55]. Reproduced with permission of American Chemical Society.)

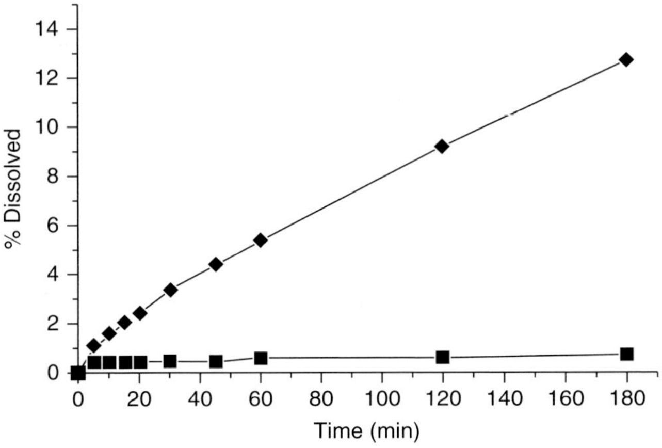

Figure 2.4 Dissolution profiles of crystalline (■) and amorphous itraconazole (♦) in simulated gastric fluid [57]. (Adapted from Six *et al.* [57]. Reproduced with permission of Wiley.)

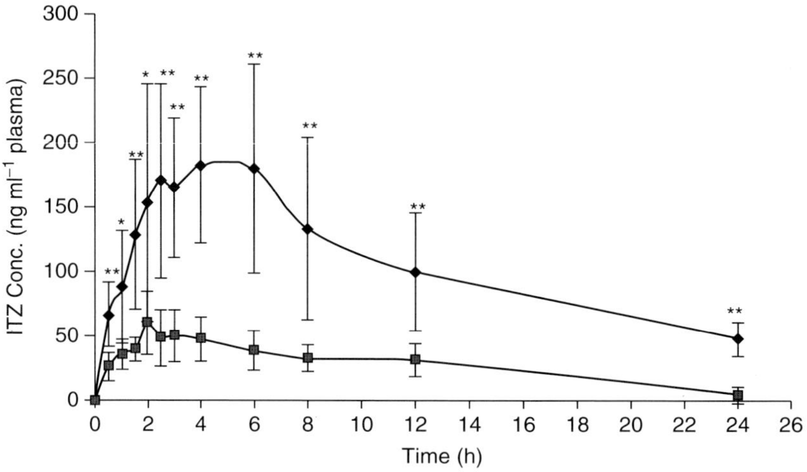

Figure 2.5 Plasma concentration versus time profiles of ITZ in rats following inhalation of nebulized aqueous wet-milled (■) and amorphous ITZ dispersion (•). (*$p < 0.05$ and **$p < 0.01$) [59]. (Yang *et al.* [59]. Reproduced with permission of Elsevier.)

demonstrated that the enhancement in dissolution rate brought about by using the amorphous form was much greater than with the metastable polymorph of iopanoic acid. A pronounced enhancement in the solubility and consequent *in vitro* dissolution rate following the conversion of crystalline compounds into amorphous solid dispersions have been reported for several compounds (solid dispersions will be discussed in detail later) [58]. Formulation of nanocrystalline itraconazole as an amorphous solid dispersion resulted in a substantial (nearly fourfold) enhancement in the bioavailability (Figure 2.5) [59]. Solid dispersion of a poorly water soluble VR1 antagonist (AMG517), when compared with micronized powder, revealed superior drug dissolution behavior, translating to a substantial enhancement in oral bioavailability [60]. Fakes and coworkers also compared the dissolution rate and oral bioavailability enhancement obtained by two approaches – particle size reduction and amorphization. They used a BCS class II compound (BMS-488043), characterized by very poor aqueous solubility ($0.04\,\mathrm{mg\,ml^{-1}}$) but an acceptable permeability ($178\,\mathrm{nm\,s^{-1}}$ in Caco-2 cell line model). Particle size reduction caused a modest (~80%) increase in bioavailability, while there was almost an order of magnitude increase in bioavailability following amorphization (Table 2.2) [61].

While it is clear from the above examples that amorphization can be successfully utilized as a strategy to improve the solubility and consequently the dissolution rate of poorly soluble drug candidates, the physical instability associated with the amorphous form is a major challenge, both in the solid as well as in the solution state. Several studies have attributed the role of molecular mobility in the physical instability of amorphous pharmaceuticals [62, 63]. Bhardwaj *et al.* established a correlation between molecular mobility (measured as structural relaxation time) in the supercooled state and the physical stability of itraconazole and trehalose [64,

Table 2.2 Relative exposure ratios obtained upon dosing 200 mg equivalents of a single oral dose of BMS-488043 in dogs [61].

Formulation	Treatment	Relative exposure ratios	
		C_{max} (ng ml^{-1})	AUC (0–24) (ng-h ml^{-1})
Capsule	Wet-milled	1	1
Tablet	Micronization	1.8	1.8
Nanosuspension	Nano-sizing	4.7	4.6
Solid dispersion	Amorphization	18.2	7.0
Solid dispersion	Amorphization	15.7	8.7

Adapted from Fakes *et al.* [61]. Reproduced with permission of Elsevier.

65]. This correlation has been extended to the glassy state [66]. Preventing drug crystallization in the dosage form right until use is necessary to realize the potential solubility advantage of the amorphous form. Assuming oral administration, the next critical issue for potentially enhanced bioavailability is the maintenance of *in vivo* supersaturation. The solubility advantage offered by the amorphous form can be short lived, an effect attributed to rapid crystallization into the more stable form of the drug in solution [37, 56, 67, 68]. Figure 2.6 is a schematic describing the dissolution behavior of an amorphous API. The higher apparent solubility (spring effect) of the amorphous form, with respect to its crystalline counterpart, can lead to elevated initial API concentration in solution (curve 2) [69]. However, since the

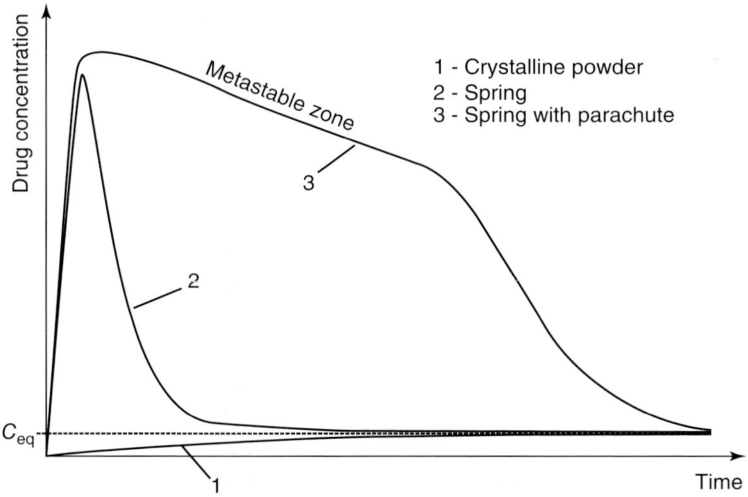

Figure 2.6 Schematic drug concentration versus time profile illustrating the dissolution profiles of the crystalline (curve 1) and amorphous forms (curve 2) of a hypothetical drug. The higher apparent solubility of the amorphous form results in the "spring" effect (curve 2). The polymer in a solid dispersion, by inhibiting drug crystallization, can yield the "parachute" effect (curve 3) [69]. (Adapted from Brouwers *et al.* [69]. Reproduced with permission of Wiley.)

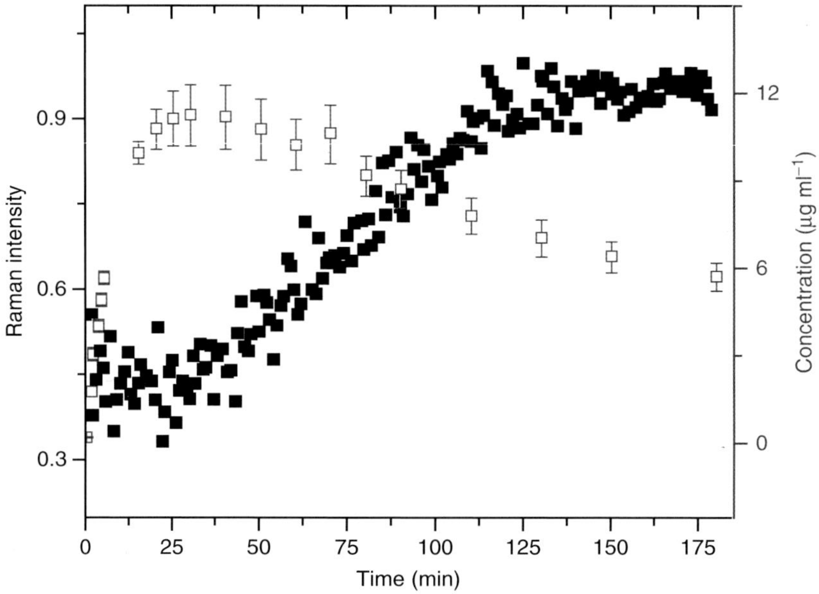

Figure 2.7 Dissolution profile of amorphous indomethacin at 37 °C. The indomethacin concentration in solution (□) and the normalized Raman peak intensity (■) of alpha indomethacin at 1648 cm^{-1} were simultaneously monitored [70]. (Adapted from Alonzo *et al.* [70]. Reproduced with permission of Springer.)

system is supersaturated with respect to the thermodynamically stable crystalline form (the form under the test conditions), there is potential for solute crystallization. Figure 2.7 shows the dissolution behavior of amorphous indomethacin, wherein the amorphous form yielded a supersaturated solution following which solute crystallization resulted in rapid decrease in solution concentration [70]. Raman spectroscopy enabled direct confirmation of α-polymorph crystallization.

2.3 Measuring the Solubility Advantage of Amorphous Compounds

As we evaluate the use of amorphous pharmaceuticals, it is instructive to consider the potential advantage (enhancement in solubility) as well as liability (risk for crystallization). In order to estimate the dissolution advantage offered by the amorphous form, the solubility advantage should be first assessed. This information will enable the formulator to make an informed decision of the suitability of the amorphous route. Since the solution-mediated crystallization of the amorphous form can be very rapid, determining the solubility of the amorphous form can be intrinsically challenging. Hancock and Parks were the first to report a method to estimate the solubility of the amorphous form [56]. A simple thermodynamic approach was used to estimate the apparent solubility of the amorphous form, which was compared with the experimentally determined values. Very broadly, the solubility ratio of the amorphous to the crystalline form (R_s) was related to the free energy difference ($\Delta G_{x,a}$) between the two forms (Eq. (2.2)).

$$R_s = \exp^{(\Delta G_{x,a}/T)} \tag{2.2}$$

To experimentally obtain the solubility ratio, the authors determined the true (equilibrium) solubility of the thermodynamically stable form, and obtained the apparent solubility of the amorphous form as the "peak concentration" during the dissolution experiment.

Murdande *et al.*, using a rigorous thermodynamic approach, obtained the accurate free energy difference ($\Delta G_{x,a}$) between the crystalline and the glassy states (Eq. (2.3)) [67].

$$\Delta G_{x,a}(T) = \frac{\Delta H_f}{T_m}(T_m - T) + T\Delta C_p^{xl} \left[\ln\left(\frac{T_m}{T}\right) - \gamma \ln\left(\frac{T_g}{T}\right) - \left(\frac{T_m}{T}\right)(1-\gamma) - \left(\gamma\frac{T_m}{T_g} - 1\right)\right] \tag{2.3}$$

where ΔH_f and T_m are, respectively, the enthalpy of fusion and melting temperature of the crystalline material, T is the temperature of interest, $\Delta C_p{}^{x,l}$ is the difference in heat capacity between the crystal and the supercooled liquid, $\Delta C_p{}^{x,g}$ is the difference in heat capacity between crystal and glass, and $\gamma = 1 - (\Delta C_p^{x,g}/\Delta C_p^{x,l})$.

In addition to relating the free energy differences to the solubility advantage offered by the amorphous form, the authors also took into consideration the reduction in the thermodynamic activity of the amorphous solute in contact with water. They applied the Gibbs–Duhem equation to calculate the activity of the hydrated amorphous solid. This correction is necessary because the reduction in the thermodynamic activity of the amorphous solid will ultimately reduce its solubility advantage. They also calculated the reduction in the fraction ionized for saturated solutions of an amorphous solid relative to its crystalline counterpart, which moderates the solubility advantage of the amorphous phase. The refined thermodynamic approach predicted with reasonable accuracy, the potential solubility advantage of the amorphous form of a large number of compounds differing widely in their physicochemical properties (Table 2.3).

In compounds that crystallized rapidly from solution, the theoretical prediction of the solubility advantage offered by the amorphous form was not practically realized. In such cases, the ease of crystallization of compounds with different glass transition temperatures was predicted using the reduced temperature approach [71]. The experimental results were in good agreement with the theoretical predictions. Finally, in an effort to accurately determine the solubility of these "problem" compounds (spironolactone is one example), rapid crystallization in the dissolution medium was inhibited using polymers. This provided a much improved estimate of the solubility advantage of these fast crystallizing systems [72].

Table 2.3 Comparison between experimentally and theoretically estimated solubility enhancement ratio [68].

Drug	Experimental solubility enhancement ratio (R_s Expt.)	Theoretically estimated solubility enhancement ratio (R_s Theory)			
		Contributing factors considered			
		Ionization, water sorption, and Gibbs free energy	Gibbs free energy	Water sorption and Gibbs free energy	Gibbs free energy and ionization only
Indomethacin	4.9	7.0	29.0	20.8	9.9
Iopanoic acid	1.5	5.7	12.5	10.9	5
Glipzide	9.2	11.1	50.6	23.8	23.5
Glibenclamide	22.6	17.1	130.6	86.3	25.9
Hydrochlorthiazide	4.9	33.9	69.1	33.9	69.1
Terfenadine	10	13.0	43.6	29.0	19.4
Griseofulvin	1.4	29.1	32.1	29.1	32.1
Spironolactone	2.1	110.3	195.9	110.3	195.9
Danazol	3.0	26.5	34.4	26.5	34.4

Adapted from Murdande *et al.* [68]. Reproduced with permission of Springer.

2.4 Solid Dispersions

From the above discussion, it is clear that the amorphous form can provide significantly higher solubility than its crystalline counterpart, albeit transiently. This can translate into increased dissolution rate and enhanced bioavailability, provided supersaturation is maintained during absorption. However, as noted earlier, while supersaturation can be achieved using the amorphous form, rapid recrystallization can negate the solubility advantage. Solid dispersions are a promising strategy to overcome the physical instability associated with amorphous compounds. They were first defined by Chiou and Riegelman as the "dispersion of one or more active ingredients in an inert carrier" [73]. Solid dispersions are of great interest to the pharmaceutical community, and several excellent reviews have comprehensively summarized their preparation, characterization, and the challenges associated with their use [58, 74–77]. Using this approach, dissolution enhancement and the consequent improvement in bioavailability has been realized in numerous compounds [58, 76, 78–81]. For example, when crystalline vemurafenib was formulated as a solid dispersion, a 30-fold increase in the apparent solubility was observed. This translated into a nearly fivefold enhancement in human bioavailability [80]. Several marketed drugs, encompassing numerous therapeutic classes (antifungals, antivirals, anti-inflammatory, etc.), are formulated as solid dispersions (Table 2.4) [77, 82].

Our emphasis will be on the dissolution behavior of amorphous solid dispersions. Figure 2.6 schematically represents the "spring and parachute effect"

Table 2.4 Current drug products marketed as solid dispersions [58, 82].

Brand name	Drug	Major excipient	Manufacturer
Certican	Everolimus	HPMC	Novartis
Cesamet	Nabilone	PVP	Valeant
Isoptin SR-E	Verapamil	HPC/HPMC	Abbott
Intelence	Etravirine	HPMC	Tibotec
Kaletra	Lopinavir, ritonavir	PVPVA	Abbott
Gris-PEG	Griseofulvin	PEG6000	Pedinol Pharmacal Inc.
Sporanox	Itraconazole	HPMC	Janssen Pharmaceutica
Nivadil	Nilvadipine	HPMC	Fujisawa Pharmaceutical Co., Ltd
Prograf	Tacrolimus	HPMC	Fujisawa Pharmaceutical Co., Ltd
Rezulin	Troglitazone	PVP	Developed by Sankyo, manufactured by Parke-Davis division of Warner-Lambert
Zelboraf	Vemurafenib	HPMCAS	Roche
Incivek	Telaprevir	HPMCAS-M	Vertex pharmaceuticals

(curve 3), commonly observed during the dissolution of solid dispersions [69]. The initial rapid rise in concentration ("spring effect") is attributed to the higher apparent solubility of the amorphous form. In the second stage, a crystallization inhibitor (typically a polymer) acts as a "parachute" and delays drug crystallization. By influencing the extent of supersaturation, the polymer used to prepare the solid dispersion had a pronounced influence on the dissolution behavior of a model drug (Figure 2.8) [83]. The degree of supersaturation brought about by the polymer depends on its structural properties, concentration, the nature of drug–polymer interactions, other components in the formulation, as well as the formulation variables. In order to maintain supersaturation, it is necessary to inhibit crystallization, and the potential mechanisms of crystallization inhibition have been elucidated [84]. The polymer, by increasing viscosity, can inhibit nucleation. Even if there is nucleation (or seed crystals), by adsorbing to the growing crystal surface a polymer can hinder crystal growth. Thus it can act as a sequential barrier, both to nucleation and growth. Since these topics have been covered elsewhere, our discussion will be restricted to the recent literature, with a focus on rational polymer selection [85, 86].

2.5 Polymer Properties

Polymer structure, including the nature and number of hydrophilic and hydrophobic functional groups, backbone rigidity, and molecular weight have been found to

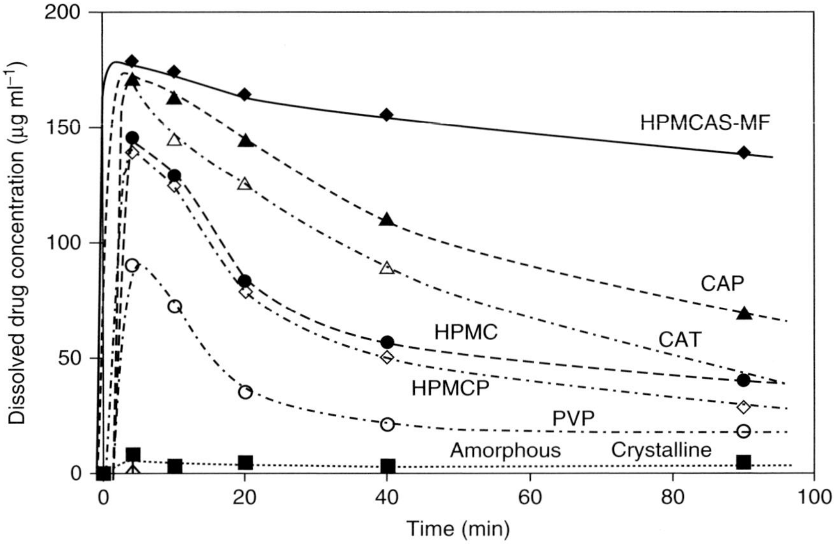

Figure 2.8 Comparison of the dissolution behavior of solid dispersions of a model drug compound. The polymers used were hydroxypropyl methylcellulose acetate succinate (HPMCAS) polyvinlypyrrolidone (PVP), hydroxypropyl methylcellulose pthalate (HPMCP), cellulose acetate phthalate (CAP), and cellulose acetate trimellitate (CAT). The drug loading was 10% w/w. Dissolution behavior of various solid dispersions of a model compound at 10% with PBS as the dissolution medium (37 °C) [83]. (Adapted from Curatolo *et al.* [83]. Reproduced with permission of Springer.)

influence the extent and duration of drug supersaturation. Recently, an extensive study by Warren *et al.* evaluated the effectiveness of polymers as precipitation inhibitors, based on their structural properties [87]. Cellulose-derived polymers, in general, were the most effective in maintaining drug supersaturation. The influence of structure, of both the model drug (neutral, weak acid/base) and the polymer, on crystallization inhibition is summarized in Table 2.5. However, the results in the table are specific to the model drugs, and several exceptions can exist. Nevertheless, such studies can serve as an excellent starting point for rational polymer selection. Other reports in the literature have also provided a mechanistic understanding of the precipitation inhibition behavior of several polymers. Curtalo *et al.* have shown the effectiveness of hydroxypropylmethylcellulose acetate succinate (HPMCAS) in maintaining supersaturation for a large number of amorphous drugs [83]. The superior stabilization properties of this polymer, when ionized, have been attributed to the rapid formation of stable amorphous drug–polymer nanostructures during dissolution. The hydrophobic substituents in the polymer are reported to interact with insoluble drug molecules, leading to the formation of amorphous drug–polymer nanostructures. The ionized succinate groups, by preventing aggregation, are believed to stabilize the nanostructures. The drug can dissolve rapidly from these nanostructures, leading to high levels of supersaturation [88]. Recently, the effect of systematic alteration

Table 2.5 Correlation between the precipitation inhibition of drugs and the molecular properties of polymers based on principal component analysis.

Drug	Correlation		
	None	**Positive**	**Negative**
Danazol – non-electrolyte	Molecular weight Hydroxyls Ethers Hydrogen-bond acceptors	None	Carboxylic acids Hydrogen-bond donors
Halofantrine – weak base	Hydroxyls Hydrogen-bond acceptors	Ethers	Molecular weight Carboxylic acids Hydrogen-bond donors
Meclofenamic acid – weak acid	Hydroxyls	Ethers	Molecular weight Carboxylic acids Hydrogen-bond donors Hydrogen-bond acceptors Amides
Danazol, Halofantrine, Meclofenamic acid			1° amines Carboxylic acids Hydroxyls Hydrogen-bond acceptors

Adapted from Warren *et al.* [87]. Reproduced with permission of Elsevier.

of the acetate to succinate ratio on drug dissolution behavior was studied. The authors attributed the crystallization inhibition capacity of the polymer mainly to drug–polymer hydrophobic interactions, with a decrease in succinoyl content favoring drug supersaturation. For one of the model compounds studied, 1D ^1H NMR measurements revealed a significant decrease in the drug mobility in the HPMCAS solution. The drug–polymer interactions responsible for this decrease in mobility also led to crystallization inhibition [89]. The work of Ilevbare *et al.* also revealed the importance of polymer hydrophobicity in preventing nucleation from supersaturated drug solutions [90]. They postulated that the hydrophobic nature of the polymer helped in forming stronger polymer–solute interactions as compared to polymer–solvent or polymer–polymer interactions. In another study, the effectiveness of cellulose derivatives in preventing crystallization from solution was attributed to their moderate hydrophobicity, semirigid structure, and amphiphilic nature [91]. These properties favored adsorption of the polymer onto the growing crystal surface, thereby preventing further crystal growth. The functional groups on the polymer backbone dictate its state of ionization, which will in turn influence its effectiveness as a precipitation inhibitor. In alkaline pH, cellulosic polymers were effective in preventing crystallization, an effect attributed to the existence of the adipate groups (pK_a of 4.43) in the ionized state, yielding a favorable conformation [92]. Recently, Yang *et al.* have shown the

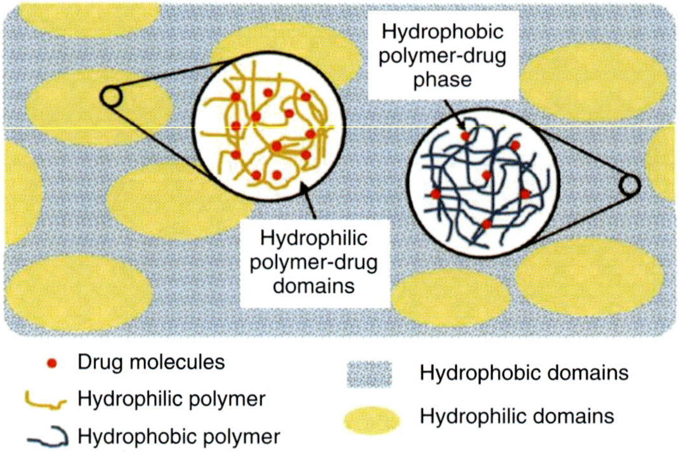

Figure 2.9 Schematic illustration of the microstructure of immiscible polymer blend formulation loaded with model drug [93]. (Adapted from Yang *et al.* [93]. Reproduced with permission of American Chemical Society.)

contrasting properties of a mixture of a hydrophobic ionic polymer (Eudragit® E PO) and a hydrophilic polymer (PVP-VA, copolymer of polyvinylpyrrolidone (PVP) and vinyl acetate (VA)) in the stabilization of amorphous felodipine. In solid dispersions, the hydrophilic polymer enabled high drug loading, whereas the hydrophobic polymer protected the drug from moisture-induced recrystallization (Figure 2.9) [93]. The molecular weight of the polymer is another parameter that can influence the dissolution behavior of drugs from dispersions. The enhanced dissolution of piroxicam observed with low molecular weight PVP was attributed to a thin diffusion layer around the drug particles. On the other hand, the high molecular weight PVP retarded dissolution due to the thicker diffusion layer [94]. Sulfathiazole, indomethacin, and prednisolone solid dispersions also exhibited similar dependence on the molecular weight of the polymer [95–97]. On the other hand, an increase in molecular weight, by increasing the viscosity of the dissolution medium, helped in maintaining supersaturation of itraconazole [98]. Interestingly, this effect was not universal, and in halofantrine and meclofenamic acid the polymer molecular weight had a negative or marginal effect [87]. Hence, based on current reports, it seems that the use of molecular weight as a parameter to modify dissolution behavior will be highly system-specific.

2.6 Drug–Polymer Interactions

Drug–polymer interaction in solution is one of the mechanisms for facilitating the maintenance of drug supersaturation. However, a majority of literature reports probe drug–polymer interactions in the solid dispersions with an implicit assumption of the interaction continuing in solution leading to supersaturation. A recent study revealed the correlation between drug–polymer interactions, calculated using the Flory–Huggins theory, and the dissolution behavior of telmisartan solid dispersions. An increase in the strength of interaction enabled longer duration of supersaturation with an exception noted for the HPMC (hydroxypropylmethylcellulose) solid dispersion [99]. The delayed precipitation from solution of several compounds including celecoxib, aceclofenac, propranolol HCl, diphenhydramine HCl, and ER-34122 (a drug under development) was attributed to drug–polymer hydrogen bonding in the dispersion [100–103]. The *in vivo* implications of this approach are evident from the work of Kojima *et al.* using mefenamic acid and a methacrylate polymer (Eudragit® EPO) solid dispersions. The enhanced bioavailability was attributed to strong drug–polymer interactions in the solid dispersions [104]. Ionic interactions, which are stronger than hydrogen bonding, are believed to further enhance the effectiveness of the polymer in maintaining supersaturation for a prolonged period and also enhance the dissolution rate. This was observed with the weakly acidic drug indomethacin, which interacted with a methacrylate polymer (Eudragit® EPO) (Figure 2.10a) [105]. In such cases, when drug ionization is inhibited ($pH \ll pK_a$), the observed solubility is governed by the intrinsic solubility (solubility of the un-ionized form). Therefore, the selection of a polymer that can ionically interact with the drug can be particularly valuable. Similar principles will typically govern the behavior of weak bases at intestinal pH values. Use of anionic polymers (HPMCAS-LF, HPMCAS-MF, Eudragit® – L-100-55) enhanced the dissolution of itraconazole, a weakly basic drug, in simulated intestinal fluid (Figure 2.10b). In general, halofantrine (a weak base) in the ionized state was better stabilized by cationic polymers, whereas anionic polymers were more effective in inhibiting the precipitation of meclofenamic acid [87]. As mentioned earlier, drug–polymer interactions were investigated only in solid dispersions. Any potential interactions in solution, leading to sustained supersaturation, were not directly probed. A recent study investigated the solution-state polymer assemblies in solid dispersions of probucol by dynamic light scattering. The sustained supersaturation was attributed to nanoaggregates of the drug and the polymer. The drug and the polymer were postulated to exist as micellar structures, where the drug is in the core of the polymeric micelle interacting with the hydrophobic part of the polymer [106, 107].

2.7 Polymer Concentration

For a given drug–polymer system, with an increase in polymer concentration, drug supersaturation in solution can be maintained for a longer duration. The stabilization is attributed to an increase in solution viscosity [84]. Another potential mechanism of stabilization is the increase in the extent of drug–polymer interaction as a function of polymer concentration. The utility of this approach

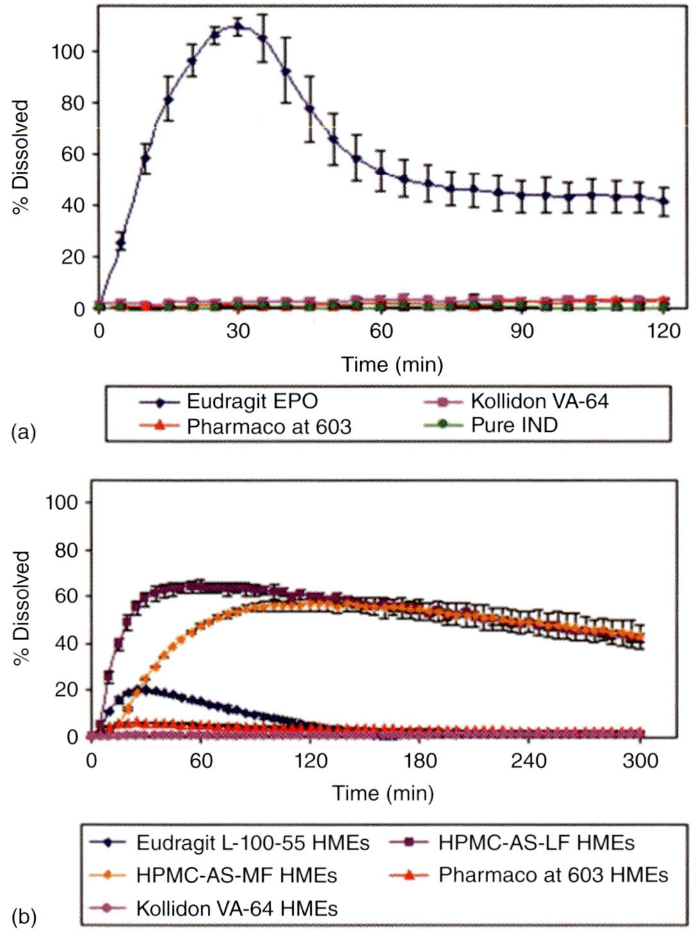

Figure 2.10 (a) Dissolution behavior of indomethacin solid dispersions (30% w/w drug loading) in simulated gastric fluid [105]. (b) Dissolution behavior of itraconazole solid dispersions (30% w/w drug loading) in simulated intestinal fluid [105]. (Adapted from Sarode *et al.* [105]. Reproduced with permission of Elsevier.)

has been shown in several compounds including felodipine and celecoxib [86, 100]. The role of the strength of the drug–polymer interaction was evident in indomethacin solution in the presence of each PVP (K90), HPMC, and methacrylate polymer (Eudragit® E100). Although all polymers delayed drug crystallization in a concentration-dependent manner, the effect was most pronounced with PVP due to its strong interaction with the drug [108]. While increasing the polymer concentration is a simple and effective approach to maintain supersaturation, identifying the lowest effective polymer concentration is not trivial. This issue is particularly important and challenging when dealing with high-dose drugs. In these systems, because of the high drug loading in the formulation, the amount

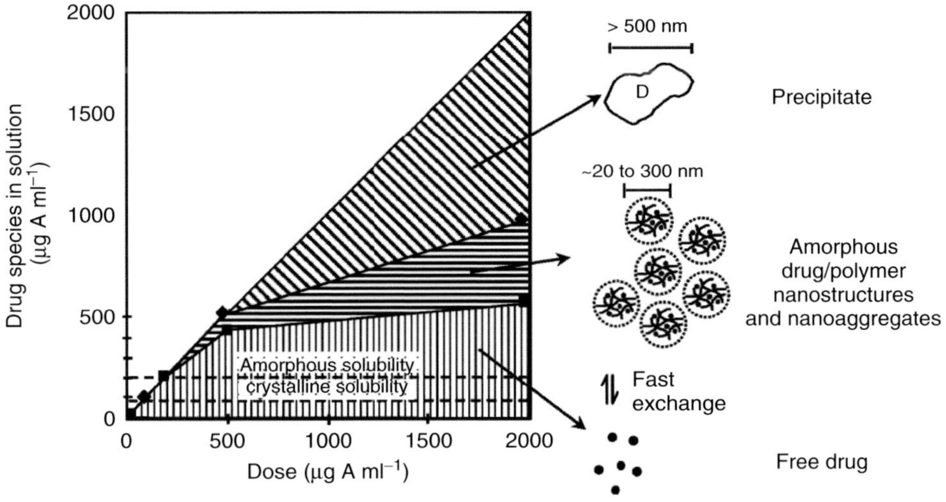

Figure 2.11 Drug species formed in solution as a function of drug dose from amorphous solid dispersions [88].

of polymer available to inhibit drug crystallization *in situ* (the GI fluid) may be insufficient with potential bioavailability implications. Recently, Friesen *et al.* listed the possible species that can exist in solution during the dissolution of a solid dispersion (Figure 2.11) [88]. As expected, at low drug concentration, below the solubility limit of the drug, no precipitation was observed. As the drug loading was increased and the drug concentration exceeded the saturation solubility, amorphous drug–polymer nanostructures were formed. Drug crystallization occurred only at very high drug loading. The drug release behavior will also change as a function of the polymer concentration. Therefore, at high drug loadings, the high initial supersaturation can cause drug phase separation and precipitation. The "properties" of this precipitate (physical form, particle size distribution, composition) will then dictate the observed dissolution behavior. When the drug concentration in the dispersion is low (high polymer loading), since the drug is expected to be completely dissolved in the polymer, the observed dissolution behavior will mimic that of the polymer [109].

2.8 Other Formulation Components

Solid dispersions can be complex dosage forms, and may contain several excipients including plasticizers and surfactants. The influence of these excipients has been elaborated in several recent publications [110, 111]. Since our focus is on solubility enhancement by modifying the physical form of the drug, we will not address the role of these excipients.

2.9 Formulation Variables

The rate of supersaturation generation, which can be modulated by the polymer, can be an important determinant of product performance. Rapid attainment of very high levels of supersaturation may not be desirable, since it can lead to rapid recrystallization from solution [112, 113]. Recently, in indomethacin–PVP solid dispersions, for a fixed degree of supersaturation, the interplay between the rate of supersaturation buildup and the overall time of its maintenance was investigated [113]. A higher rate of supersaturation generation led to a higher maximum kinetic solubility, but faster recrystallization from solution. On the other hand, lower rates of supersaturation generation delayed drug precipitation for longer duration (Figure 2.12). The authors built a mechanistic model based on the classical nucleation theory taking into account the role of supersaturation in both the nucleation and the crystallization processes and predicted that a modest *in vitro* dissolution rate will maximize the AUC of the concentration–time profile. The experimental results were in excellent agreement with theoretical predictions. The supersaturation behavior could be modulated with water-insoluble, cross-linked hydrogels acting as drug dispersion carriers. These hydrogels can help in prolonging supersaturation via a feedback-controlled mechanism, thereby also avoiding the initial rapid drug dissolution, as has been observed in indomethacin [114, 115]. Most studies in the literature deal with dissolution from powdered solid dispersions. However, these powders are usually formulated as tablets or capsules for use by patients. It will be instructive to evaluate the drug dissolution behavior in the final formulation. When felodipine solid dispersions

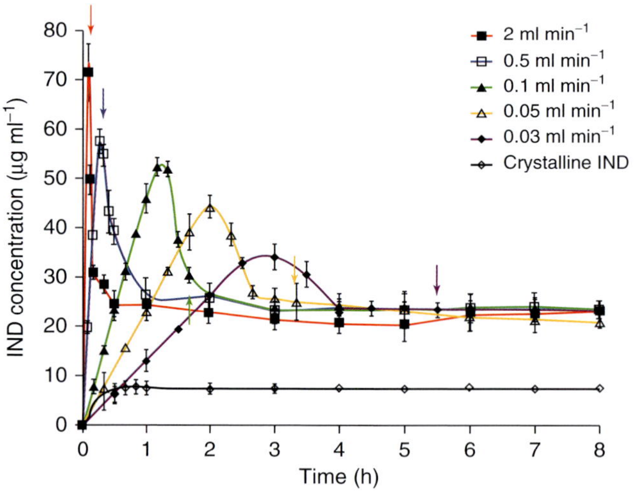

Figure 2.12 Experimental kinetic solubility profile of indomethacin as a function of supersaturation rate generated with various drug solution infusion rates [113]. (Adapted from Sun and Lee [113]. Reproduced with permission of American Chemical Society.)

were compacted, there was a pronounced decrease in rate and extent of drug dissolution. When compared with powders, the altered dissolution behavior was attributed to changes in porosity, surface area, water uptake kinetics, and drug recrystallization on the tablet surface [107]. A similar observation was made following the administration of a capsule formulation wherein a nondispersible plug was formed in the dissolution medium [116].

2.10
Reliable Measurement of Supersaturation

As mentioned earlier, solid dispersions are predominantly designed with the goal of enhancing the bioavailability of poorly water soluble drugs. The ability of the polymer to maintain supersaturation in the GI fluids is crucial for the successful formulation of solid dispersions. However, quantifying supersaturation as well as detecting precipitation of drugs in solution can be challenging. The *in vitro* dissolution tests should therefore be designed to provide a reliable measure of the solubility enhancement and maintenance of supersaturation. A well-designed *in vitro* dissolution test should at least enable us to rank order a series of formulations based both on the extent and duration of supersaturation. Ideally, these studies should also predict *in vivo* performance, though the challenges of *in vitro–in vivo* correlation are widely recognized. USP dissolution tests (types I and II; immediate-release dosage forms) are predominantly designed to ensure drug product quality and batch-to-batch consistency. These dissolution tests are conducted under sink conditions and do not adequately simulate the complex GI environment. Hence, these methods are not designed to assess the dissolution behavior of formulations that are intended to yield supersaturated solutions. The issue is further complicated in APIs exhibiting pH-dependent solubility, that is, weak acids and bases. For example, the traditional USP method is unsuitable for evaluating the supersaturation of itraconazole from solid dispersions [98]. When the dissolution behavior of experimental itraconazole solid dispersions (prepared with cellulose acetate phthalate) was compared with that of a commercial (Sporanox®) solid dispersion formulation, in the acidic medium (pH 1.2), the degree of supersaturation was much higher with the latter. However, when the dissolution medium was changed to phosphate buffer (pH 7.2), the experimental formulation exhibited a longer duration of supersaturation. The higher bioavailability of this solid dispersion was attributed to the enhanced intestinal targeting and longer duration of supersaturation. The use of an acidic dissolution medium is recommended by both the USP and the Food and Drug Administration (FDA). If the goal is to use *in vitro* dissolution tests to gage *in vivo* performance, the compendial dissolution test (specifically the dissolution medium) may be unsuitable. The unsuitability of the USP type II dissolution apparatus to predict the *in vivo* behavior was also demonstrated for BMS-A solid dispersions [117]. In vitro, the PVP–VA64 solid dispersion dissolved faster than the HPMCAS system, while the *in vivo* studies revealed more enhanced bioavailability with the latter, attributed to its effectiveness in prolonging supersaturation. Newman *et al.* recently reported that USP Apparatus 3 (reciprocating cylinder) and Apparatus 4 (flow-through cell), by allowing for controlled pH and volume changes, more closely mimicked the GI environment. The use of biorelevant media is a step toward simulation of the GI fluids, though this will not necessarily lead to *in vitro–in vivo* correlation [118–121]. A better prediction of *in vivo* behavior from the *in vitro* data will be enabled by more realistic simulation of the GI fluid [122, 123]. In some model dispersions, the simulated intestinal fluid overpredicted the precipitation inhibition capacity of some polymers. This conclusion was based on a comparison with the supersaturation achieved in human intestinal fluid [124]. Recently, the composition of human GI fluid has been evaluated in detail [125–127]. In addition, physiologic factors and food effects have been incorporated in an effort to make the *in vitro* evaluations more realistic. These studies should enable a rational modification in the composition of the simulated GI fluid. Once the appropriate dissolution testing conditions and medium are identified, the role of the formulation components in maintaining supersaturation can also be assessed. For example, Bevernage *et al.* have calculated the "excipient gain factor," which enables a comparison of the precipitation inhibition capacity of different polymers. This can be used as a quick screening tool to identify the formulation with the best *in vivo* potential.

While evaluating supersaturation, it is instructive to also simultaneously consider drug permeation through the GI membrane. Surfactants and inclusion compounds, while enhancing solubility, can have a detrimental effect on permeability, an effect not seen with HPMC-AS solid dispersion [128]. This observation, coupled with *in vitro* dissolution tests designed to study the influence of precipitation as well as permeation, may improve the prediction of *in vivo* performance [129]. Thus, in an "absorptive" environment (Caco-2 Transwell®inserts), a much higher degree of lovirdie supersaturation was maintained when compared to a regular supersaturation assay.

2.11 Conclusion

Amorphization is a very attractive approach to enhance aqueous solubility. Solid dispersions, with the potential to physically stabilize the drug in the amorphous state, may enable effective oral delivery of compounds with poor aqueous solubility. In order to formulate a robust solid dispersion, we not only need to consider the intrinsic properties of drug and polymer but also exploit the potential interactions between the drug and polymer to physically stabilize the drug and prevent its crystallization. The success of the formulation will be gaged by the stabilization during processing, shelf storage, and *in vivo* (i.e., maintenance of supersaturation). The recent literature summarized in this chapter can serve as a reference point to select a polymer that can afford the best stabilization from precipitation in the GI fluid.

The ultimate success of the solid dispersion can be predicted using appropriate *in vitro* dissolution conditions, and keeping in mind the complex GI environment.

References

1. Teague, S.J., Davis, A.M., Leeson, P.D., and Oprea, T. (1999) The design of leadlike combinatorial libraries. *Angew. Chem. Int. Ed.*, **38** (24), 3743–3748.
2. Carnero, A. (2006) High throughput screening in drug discovery. *Clin. Transl. Oncol.*, **8** (7), 482–490.
3. Babu, N.J. and Nangia, A. (2011) Solubility advantage of amorphous drugs and pharmaceutical cocrystals. *Cryst. Growth Des.*, **11** (7), 2662–2679.
4. General Chapters. (2013) *"Description and Relative Solubility" The United States Pharmacopeia 36–National Formulary-31*, United States Pharmacopeial Convention, Inc, Rockville, MD.
5. Dressman, J., Amidon, G., Reppas, C., and Shah, V. (1998) Dissolution testing as a prognostic tool for oral drug absorption: immediate release dosage forms. *Pharm. Res.*, **15** (1), 11–22.
6. Yu, L. (1999) An integrated model for determining causes of poor oral drug absorption. *Pharm. Res.*, **16** (12), 1883–1887.
7. Amidon, G., Lennernäs, H., Shah, V., and Crison, J. (1995) A theoretical basis for a biopharmaceutic drug classification: the correlation of in vitro drug product dissolution and in vivo bioavailability. *Pharm. Res.*, **12** (3), 413–420.
8. Zhang, Y. and Benet, L. (2001) The gut as a barrier to drug absorption. *Clin. Pharmacokinet.*, **40** (3), 159–168.
9. Martinez, M. and Amidon, G. (2002) A mechanistic approach to understanding the factors affecting drug absorption: a review of fundamentals. *J. Clin. Pharmacol.*, **42** (6), 620–643.
10. Aungst, B.J. (1993) Novel formulation strategies for improving oral bioavailability of drugs with poor membrane permeation or presystemic metabolism. *J. Pharm. Sci.*, **82** (10), 979–987.
11. Gong, Y., Grant, D.J.W., and Brittain, H. (2007) in *Solvent Systems and Their Selection in Pharmaceutics and Biopharmaceutics* (eds P. Augustijns and M. Brewster), Springer, New York, pp. 1–27.
12. Bircumshaw, L.L. and Riddiford, A.C. (1952) Transport control in heterogeneous reactions. *Q. Rev. Chem. Soc.*, **6** (2), 157–185.
13. Noyes, A.A. and Whitney, W.R. (1897) The rate of solution of solid substances in their own solutions. *J. Am. Chem. Soc.*, **19** (12), 930–934.
14. Butler, J.M. and Dressman, J.B. (2010) The developability classification system: application of biopharmaceutics concepts to formulation development. *J. Pharm. Sci.*, **99** (12), 4940–4954.
15. Stegemann, S., Leveiller, F., Franchi, D., de Jong, H., and Lindén, H. (2007) When poor solubility becomes an issue: from early stage to proof of concept. *Eur. J. Pharm. Sci.*, **31** (5), 249–261.
16. Williams, H.D., Trevaskis, N.L., Charman, S.A., Shanker, R.M., Charman, W.N., Pouton, C.W. *et al.* (2013) Strategies to address low drug solubility in discovery and development. *Pharmacol. Rev.*, **65** (1), 315–499.
17. van Hoogevest, P., Liu, X., and Fahr, A. (2011) Drug delivery strategies for poorly water-soluble drugs: the industrial perspective. *Expert Opin. Drug Delivery*, **8** (11), 1481–1500.
18. Müller, R.H. and Peters, K. (1998) Nanosuspensions for the formulation of poorly soluble drugs: I. Preparation by a size-reduction technique. *Int. J. Pharm.*, **160** (2), 229–237.
19. Hecq, J., Deleers, M., Fanara, D., Vranckx, H., and Amighi, K. (2005) Preparation and characterization of nanocrystals for solubility and dissolution rate enhancement of nifedipine. *Int. J. Pharm.*, **299** (1–2), 167–177.
20. Vogt, M., Kunath, K., and Dressman, J.B. (2008) Dissolution improvement of four poorly water soluble drugs by

cogrinding with commonly used excipients. *Eur. J. Pharm. Biopharm.*, **68** (2), 330–337.

21. Wu, Y., Loper, A., Landis, E., Hettrick, L., Novak, L., Lynn, K. *et al.* (2004) The role of biopharmaceutics in the development of a clinical nanoparticle formulation of MK-0869: a beagle dog model predicts improved bioavailability and diminished food effect on absorption in human. *Int. J. Pharm.*, **285** (1–2), 135–146.
22. Perrut, M., Jung, J., and Leboeuf, F. (2005) Enhancement of dissolution rate of poorly-soluble active ingredients by supercritical fluid processes: part I: micronization of neat particles. *Int. J. Pharm.*, **288** (1), 3–10.
23. Hu, J., Johnston, K.P., and Williams, R.O. (2004) Nanoparticle engineering processes for enhancing the dissolution rates of poorly water soluble drugs. *Drug Dev. Ind. Pharm.*, **30** (3), 233–245.
24. Kesisoglou, F., Panmai, S., and Wu, Y. (2007) Nanosizing — oral formulation development and biopharmaceutical evaluation. *Adv. Drug Delivery Rev.*, **59** (7), 631–644.
25. Liversidge, G.G. and Cundy, K.C. (1995) Particle size reduction for improvement of oral bioavailability of hydrophobic drugs: I. Absolute oral bioavailability of nanocrystalline danazol in beagle dogs. *Int. J. Pharm.*, **125** (1), 91–97.
26. Jinno, J.-I., Kamada, N., Miyake, M., Yamada, K., Mukai, T., Odomi, M. *et al.* (2006) Effect of particle size reduction on dissolution and oral absorption of a poorly water-soluble drug, cilostazol, in beagle dogs. *J. Controlled Release*, **111** (1–2), 56–64.
27. Sigfridsson, K., Lundqvist, A.J., and Strimfors, M. (2009) Particle size reduction for improvement of oral absorption of the poorly soluble drug UG558 in rats during early development. *Drug Dev. Ind. Pharm.*, **35** (12), 1479–1486.
28. Rasenack, N. and Müller, B. (2002) Dissolution rate enhancement by in situ micronization of poorly water-soluble drugs. *Pharm. Res.*, **19** (12), 1894–1900.
29. Sharma, P., Denny, W.A., and Garg, S. (2009) Effect of wet milling process on the solid state of indomethacin and simvastatin. *Int. J. Pharm.*, **380** (1–2), 40–48.
30. Govindarajan, R. and Suryanarayanan, R. (2006) in *Polymorphism: in the Pharmaceutical Industry* (ed R. Hilfiker), Wiley-VCH Verlag GmbH, New York, pp. 333–364.
31. Mah, P.T., Laaksonen, T., Rades, T., Aaltonen, J., Peltonen, L., and Strachan, C.J. (2013) Unravelling the relationship between degree of disorder and the dissolution behavior of milled glibenclamide. *Mol. Pharm.*, **11**, 234–242.
32. Brittain, H.G., Grant, D.J.R., and Myrdal, P.B. (2009) in *Polymorphism in Pharmaceutical Solids*, 2nd edn (ed H.G. Brittain), Informa Healthcare, London, pp. 436–480.
33. Serajuddin, A.T.M. (2007) Salt formation to improve drug solubility. *Adv. Drug Delivery Rev.*, **59** (7), 603–616.
34. Shan, N. and Zaworotko, M.J. (2008) The role of cocrystals in pharmaceutical science. *Drug Discovery Today*, **13** (9–10), 440–446.
35. Aguiar, A.J., Krc, J., Kinkel, A.W., and Samyn, J.C. (1967) Effect of polymorphism on the absorption of chloramphenicol from chloramphenicol palmitate. *J. Pharm. Sci.*, **56** (7), 847–853.
36. Khalafallah, N., Khalil, S.A., and Moustafa, M.A. (1974) Bioavailability determination of two crystal forms of sulfameter in humans from urinary excretion data. *J. Pharm. Sci.*, **63** (6), 861–864.
37. Stagner, W.C. and Guillory, J.K. (1979) Physical characterization of solid iopanoic acid forms. *J. Pharm. Sci.*, **68** (8), 1005–1009.
38. Pudipeddi, M. and Serajuddin, A.T.M. (2005) Trends in solubility of polymorphs. *J. Pharm. Sci.*, **94** (5), 929–939.

39. Berge, S.M., Bighley, L.D., and Monkhouse, D.C. (1977) Pharmaceutical salts. *J. Pharm. Sci.*, **66** (1), 1–19.
40. Hörter, D. and Dressman, J. (2001) Influence of physicochemical properties on dissolution of drugs in the gastrointestinal tract. *Adv. Drug Delivery Rev.*, **46** (1), 75–87.
41. Sun, C. and Grant, D.J.W. (2001) Compaction properties of L-lysine salts. *Pharm. Res.*, **18** (3), 281–286.
42. Newman, A.W., Childs, S.L., and Cowans, B.A. (2008) in *Preclinical Development Handbook* (ed S.C. Gad), John Wiley & Sons, Inc., Hoboken, NJ, pp. 455–481.
43. Bastin, R.J., Bowker, M.J., and Slater, B.J. (2000) Salt selection and optimisation procedures for pharmaceutical new chemical entities. *Org. Process Res. Dev.*, **4** (5), 427–435.
44. Schultheiss, N. and Newman, A. (2009) Pharmaceutical cocrystals and their physicochemical properties. *Cryst. Growth Des.*, **9** (6), 2950–2967.
45. Stanton, M.K. and Bak, A. (2008) Physicochemical properties of pharmaceutical co-crystals: a case study of ten AMG 517 co-crystals. *Cryst. Growth Des.*, **8** (10), 3856–3862.
46. Hickey, M.B., Peterson, M.L., Scoppettuolo, L.A., Morrisette, S.L., Vetter, A., Guzmán, H. *et al.* (2007) Performance comparison of a co-crystal of carbamazepine with marketed product. *Eur. J. Pharm. Biopharm.*, **67** (1), 112–119.
47. Cheney, M.L., Shan, N., Healey, E.R., Hanna, M., Wojtas, L., Zaworotko, M.J. *et al.* (2009) Effects of crystal form on solubility and pharmacokinetics: a crystal engineering case study of lamotrigine. *Cryst. Growth Des.*, **10** (1), 394–405.
48. Remenar, J.F., Morissette, S.L., Peterson, M.L., Moulton, B., MacPhee, J.M., Guzmán, H.R. *et al.* (2003) Crystal engineering of novel cocrystals of a triazole drug with 1,4-dicarboxylic acids. *J. Am. Chem. Soc.*, **125** (28), 8456–8457.
49. Smith, A.J., Kavuru, P., Wojtas, L., Zaworotko, M.J., and Shytle, R.D. (2011) Cocrystals of quercetin with improved solubility and oral bioavailability. *Mol. Pharm.*, **8** (5), 1867–1876.
50. Jung, M.-S., Kim, J.-S., Kim, M.-S., Alhalaweh, A., Cho, W., Hwang, S.-J. *et al.* (2010) Bioavailability of indomethacin-saccharin cocrystals. *J. Pharm. Pharmacol.*, **62** (11), 1560–1568.
51. Zallen, R. (1998) in *The Physics of Amorphous Solids* (ed R. Zallen), Wiley-VCH Verlag GmbH, New York, pp. 1–32.
52. Hancock, B.C. and Zografi, G. (1997) Characteristics and significance of the amorphous state in pharmaceutical systems. *J. Pharm. Sci.*, **86** (1), 1–12.
53. Petit, S. and Coquerel, G. (2006) in *Polymorphism: in the Pharmaceutical Industry* (ed R. Hilfiker), Wiley-VCH Verlag GmbH, New York, pp. 259–285.
54. Yu, L. (2001) Amorphous pharmaceutical solids: preparation, characterization and stabilization. *Adv. Drug Delivery Rev.*, **48** (1), 27–42.
55. Gupta, P., Chawla, G., and Bansal, A.K. (2004) Physical stability and solubility advantage from amorphous celecoxib: the role of thermodynamic quantities and molecular mobility. *Mol. Pharm.*, **1** (6), 406–413.
56. Hancock, B. and Parks, M. (2000) What is the true solubility advantage for amorphous pharmaceuticals? *Pharm. Res.*, **17** (4), 397–404.
57. Six, K., Verreck, G., Peeters, J., Brewster, M., and Mooter, G.V.D. (2003) Increased physical stability and improved dissolution properties of itraconazole, a class II drug, by solid dispersions that combine fast- and slow-dissolving polymers. *J. Pharm. Sci.*, **93** (1), 124–131.
58. Vasconcelos, T., Sarmento, B., and Costa, P. (2007) Solid dispersions as strategy to improve oral bioavailability of poor water soluble drugs. *Drug Discovery Today*, **12** (23–24), 1068–1075.
59. Yang, W., Johnston, K.P., and Williams, R.O. III, (2010) Comparison of bioavailability of amorphous versus crystalline

itraconazole nanoparticles via pulmonary administration in rats. *Eur. J. Pharm. Biopharm.*, **75** (1), 33–41.
60. Kennedy, M., Hu, J., Gao, P., Li, L., Ali-Reynolds, A., Chal, B. *et al.* (2008) Enhanced bioavailability of a poorly soluble VR1 antagonist using an amorphous solid dispersion approach: a case study. *Mol. Pharm.*, **5** (6), 981–993.
61. Fakes, M.G., Vakkalagadda, B.J., Qian, F., Desikan, S., Gandhi, R.B., Lai, C. *et al.* (2009) Enhancement of oral bioavailability of an HIV-attachment inhibitor by nanosizing and amorphous formulation approaches. *Int. J. Pharm.*, **370** (1–2), 167–174.
62. Bhattacharya, S. and Suryanarayanan, R. (2009) Local mobility in amorphous pharmaceuticals—characterization and implications on stability. *J. Pharm. Sci.*, **98** (9), 2935–2953.
63. Bhugra, C. and Pikal, M.J. (2008) Role of thermodynamic, molecular, and kinetic factors in crystallization from the amorphous state. *J. Pharm. Sci.*, **97** (4), 1329–1349.
64. Bhardwaj, S.P. and Suryanarayanan, R. (2012) Molecular mobility as an effective predictor of the physical stability of amorphous trehalose. *Mol. Pharm.*, **9** (11), 3209–3217.
65. Bhardwaj, S.P., Arora, K.K., Kwong, E., Templeton, A., Clas, S.-D., and Suryanarayanan, R. (2013) Correlation between molecular mobility and physical stability of amorphous itraconazole. *Mol. Pharm.*, **10** (2), 694–700.
66. Kothari, K., Ragoonanan, V., and Suryanarayanan, R. (2013) Hydrogen bonding and molecular mobility: predictors of drug crystallization in solid dispersions. Poster presentation at the 2013 Annual Meeting and Exposition, San Antonio, TX, November 10–13, 2013, Poster W5140.
67. Murdande, S.B., Pikal, M.J., Shanker, R.M., and Bogner, R.H. (2010) Solubility advantage of amorphous pharmaceuticals: I. A thermodynamic analysis. *J. Pharm. Sci.*, **99** (3), 1254–1264.
68. Murdande, S., Pikal, M., Shanker, R., and Bogner, R. (2010) Solubility advantage of amorphous pharmaceuticals: II. Application of quantitative thermodynamic relationships for prediction of solubility enhancement in structurally diverse insoluble pharmaceuticals. *Pharm. Res.*, **27** (12), 2704–2714.
69. Brouwers, J., Brewster, M.E., and Augustijns, P. (2009) Supersaturating drug delivery systems: the answer to solubility-limited oral bioavailability? *J. Pharm. Sci.*, **98** (8), 2549–2572.
70. Alonzo, D., Zhang, G.Z., Zhou, D., Gao, Y., and Taylor, L. (2010) Understanding the behavior of amorphous pharmaceutical systems during dissolution. *Pharm. Res.*, **27** (4), 608–618.
71. Zhou, D., Zhang, G.G.Z., Law, D., Grant, D.J.W., and Schmitt, E.A. (2002) Physical stability of amorphous pharmaceuticals: importance of configurational thermodynamic quantities and molecular mobility. *J. Pharm. Sci.*, **91** (8), 1863–1872.
72. Murdande, S.B., Pikal, M.J., Shanker, R.M., and Bogner, R.H. (2011) Solubility advantage of amorphous pharmaceuticals, part 3: is maximum solubility advantage experimentally attainable and sustainable? *J. Pharm. Sci.*, **100** (10), 4349–4356.
73. Chiou, W.L. and Riegelman, S. (1971) Pharmaceutical applications of solid dispersion systems. *J. Pharm. Sci.*, **60** (9), 1281–1302.
74. Leuner, C. and Dressman, J. (2000) Improving drug solubility for oral delivery using solid dispersions. *Eur. J. Pharm. Biopharm.*, **50** (1), 47–60.
75. Serajuddin, A.T.M. (1999) Solid dispersion of poorly water-soluble drugs: early promises, subsequent problems, and recent breakthroughs. *J. Pharm. Sci.*, **88** (10), 1058–1066.
76. Newman, A., Knipp, G., and Zografi, G. (2012) Assessing the performance of amorphous solid dispersions. *J. Pharm. Sci.*, **101** (4), 1355–1377.
77. Janssens, S. and Van den Mooter, G. (2009) Review: physical chemistry of solid dispersions. *J. Pharm. Pharmacol.*, **61** (12), 1571–1586.
78. Othman, A.A., Cheskin, H., Locke, C., Nothaft, W., and Dutta, S. (2012) A phase 1 study to evaluate the bioavailability and food effect of

2 solid-dispersion formulations of the TRPV1 antagonist ABT-102, relative to the oral solution formulation, in healthy human volunteers. *Clin. Pharmacol. Drug Dev.*, **1** (1), 24–31.

79. Piao, Z.-Z., Choe, J.-S., Oh, K.T., Rhee, Y.-S., and Lee, B.-J. (2014) Formulation and in vivo human bioavailability of dissolving tablets containing a self-nanoemulsifying itraconazole solid dispersion without precipitation in simulated gastrointestinal fluid. *Eur. J. Pharm. Sci.*, **51**, 67–74.
80. Shah, N., Iyer, R.M., Mair, H.-J., Choi, D.S., Tian, H., Diodone, R. *et al.* (2013) Improved human bioavailability of vemurafenib, a practically insoluble drug, using an amorphous polymer-stabilized solid dispersion prepared by a solvent-controlled coprecipitation process. *J. Pharm. Sci.*, **102** (3), 967–981.
81. Tran, P.H.-L., Tran, T.T.-D., Piao, Z.Z., Van Vo, T., Park, J.B., Lim, J. *et al.* (2013) Physical properties and in vivo bioavailability in human volunteers of isradipine using controlled release matrix tablet containing self-emulsifying solid dispersion. *Int. J. Pharm.*, **450** (1–2), 79–86.
82. Surikutchi, B.T., Shete, G., Patel, S., and Bansal, A.K. (2013) Drug-excipient behavior in polymeric amorphous solid dispersions. *J. Excipients Food Chem.*, **4** (3), 25.
83. Curatolo, W., Nightingale, J., and Herbig, S. (2009) Utility of hydroxypropylmethylcellulose acetate succinate (HPMCAS) for initiation and maintenance of drug supersaturation in the GI milieu. *Pharm. Res.*, **26** (6), 1419–1431.
84. Warren, D.B., Benameur, H., Porter, C.J.H., and Pouton, C.W. (2010) Using polymeric precipitation inhibitors to improve the absorption of poorly water-soluble drugs: a mechanistic basis for utility. *J. Drug Targeting*, **18** (10), 704–731.
85. Raghavan, S., Trividic, A., Davis, A., and Hadgraft, J. (2001) Crystallization of hydrocortisone acetate: influence of polymers. *Int. J. Pharm.*, **212** (2), 213–221.
86. Konno, H., Handa, T., Alonzo, D.E., and Taylor, L.S. (2008) Effect of polymer type on the dissolution profile of amorphous solid dispersions containing felodipine. *Eur. J. Pharm. Biopharm.*, **70** (2), 493–499.
87. Warren, D.B., Bergström, C.A.S., Benameur, H., Porter, C.J.H., and Pouton, C.W. (2013) Evaluation of the structural determinants of polymeric precipitation inhibitors using solvent shift methods and principle component analysis. *Mol. Pharm.*, **10** (8), 2823–2848.
88. Friesen, D.T., Shanker, R., Crew, M., Smithey, D.T., Curatolo, W.J., and Nightingale, J.A.S. (2008) Hydroxypropyl methylcellulose acetate succinate-based spray-dried dispersions: an overview. *Mol. Pharm.*, **5** (6), 1003–1019.
89. Ueda, K., Higashi, K., Yamamoto, K., and Moribe, K. (2014) The effect of HPMCAS functional groups on drug crystallization from the supersaturated state and dissolution improvement. *Int. J. Pharm.*, **464** (1–2), 205–213.
90. Ilevbare, G.A., Liu, H., Edgar, K.J., and Taylor, L.S. (2012) Maintaining supersaturation in aqueous drug solutions: impact of different polymers on induction times. *Cryst. Growth Des.*, **13** (2), 740–751.
91. Ilevbare, G.A., Liu, H., Edgar, K.J., and Taylor, L.S. (2012) Understanding polymer properties important for crystal growth inhibition—impact of chemically diverse polymers on solution crystal growth of ritonavir. *Cryst. Growth Des.*, **12** (6), 3133–3143.
92. Ilevbare, G.A., Liu, H., Edgar, K.J., and Taylor, L.S. (2012) Inhibition of solution crystal growth of ritonavir by cellulose polymers - factors influencing polymer effectiveness. *CrystEngComm*, **14** (20), 6503–6514.
93. Yang, Z., Nollenberger, K., Albers, J., Craig, D., and Qi, S. (2013) Microstructure of an immiscible polymer blend and its stabilization effect on amorphous solid dispersions. *Mol. Pharm.*, **10** (7), 2767–2780.
94. Kogermann, K., Penkina, A., Predbannikova, K., Jeeger, K., Veski,

P., Rantanen, J. *et al.* (2013) Dissolution testing of amorphous solid dispersions. *Int. J. Pharm.*, **444** (1–2), 40–46.
95. Simonelli, A.P., Mehta, S.C., and Higuchi, W.I. (1969) Dissolution rates of high energy polyvinylpyrrolidone (PVP)-sulfathiazole coprecipitates. *J. Pharm. Sci.*, **58** (5), 538–549.
96. Hilton, J.E. and Summers, M.P. (1986) The effect of wetting agents on the dissolution of indomethacin solid dispersion systems. *Int. J. Pharm.*, **31** (1–2), 157–164.
97. Jachowicz, R. (1987) Dissolution rates of partially water-soluble drugs from solid dispersion systems. I. Prednisolone. *Int. J. Pharm.*, **35** (1–2), 1–5.
98. DiNunzio, J.C., Miller, D.A., Yang, W., McGinity, J.W., and Williams, R.O. (2008) Amorphous compositions using concentration enhancing polymers for improved bioavailability of itraconazole. *Mol. Pharm.*, **5** (6), 968–980.
99. Dukeck, R., Sieger, P., and Karmwar, P. (2013) Investigation and correlation of physical stability, dissolution behaviour and interaction parameter of amorphous solid dispersions of telmisartan: a drug development perspective. *Eur. J. Pharm. Sci.*, **49** (4), 723–731.
100. Abu-Diak, O.A., Jones, D.S., and Andrews, G.P. (2011) An investigation into the dissolution properties of celecoxib melt extrudates: understanding the role of polymer type and concentration in stabilizing supersaturated drug concentrations. *Mol. Pharm.*, **8** (4), 1362–1371.
101. Maulvi, F.A., Dalwadi, S.J., Thakkar, V.T., Soni, T.G., Gohel, M.C., and Gandhi, T.R. (2011) Improvement of dissolution rate of aceclofenac by solid dispersion technique. *Powder Technol.*, **207** (1–3), 47–54.
102. Kushida, I., Ichikawa, M., and Asakawa, N. (2002) Improvement of dissolution and oral absorption of ER-34122, a poorly water-soluble dual 5-lipoxygenase/cyclooxygenase inhibitor with anti-inflammatory activity by preparing solid dispersion. *J. Pharm. Sci.*, **91** (1), 258–266.
103. Maniruzzaman, M., Morgan, D.J., Mendham, A.P., Pang, J., Snowden, M.J., and Douroumis, D. (2013) Drug–polymer intermolecular interactions in hot-melt extruded solid dispersions. *Int. J. Pharm.*, **443** (1–2), 199–208.
104. Kojima, T., Higashi, K., Suzuki, T., Tomono, K., Moribe, K., and Yamamoto, K. (2012) Stabilization of a supersaturated solution of mefenamic acid from a solid dispersion with Eudragit® EPO. *Pharm. Res.*, **29** (10), 2777–2791.
105. Sarode, A.L., Sandhu, H., Shah, N., Malick, W., and Zia, H. (2013) Hot melt extrusion (HME) for amorphous solid dispersions: predictive tools for processing and impact of drug–polymer interactions on supersaturation. *Eur. J. Pharm. Sci.*, **48** (3), 371–384.
106. Dalsin, M.C., Tale, S., and Reineke, T.M. (2013) Solution-state polymer assemblies influence BCS class II drug dissolution and supersaturation maintenance. *Biomacromolecules*, **15** (2), 11.
107. Langham, Z.A., Booth, J., Hughes, L.P., Reynolds, G.K., and Wren, S.A.C. (2012) Mechanistic insights into the dissolution of spray-dried amorphous solid dispersions. *J. Pharm. Sci.*, **101** (8), 2798–2810.
108. Chauhan, H., Kuldipkumar, A., Barder, T., Medek, A., Gu, C.-H., and Atef, E. (2014) Correlation of inhibitory effects of polymers on indomethacin precipitation in solution and amorphous solid crystallization based on molecular interaction. *Pharm. Res.*, **31** (2), 500–515.
109. Craig, D.Q.M. (2002) The mechanisms of drug release from solid dispersions in water-soluble polymers. *Int. J. Pharm.*, **231** (2), 131–144.
110. Ghebremeskel, A.N., Vemavarapu, C., and Lodaya, M. (2007) Use of surfactants as plasticizers in preparing solid dispersions of poorly soluble API: selection of polymer–surfactant combinations using solubility parameters and testing the processability. *Int. J. Pharm.*, **328** (2), 119–129.

111. Ghebremeskel, A., Vemavarapu, C., and Lodaya, M. (2006) Use of surfactants as plasticizers in preparing solid dispersions of poorly soluble API: stability testing of selected solid dispersions. *Pharm. Res.*, **23** (8), 1928–1936.
112. Augustijns, P. and Brewster, M.E. (2012) Supersaturating drug delivery systems: fast is not necessarily good enough. *J. Pharm. Sci.*, **101** (1), 7–9.
113. Sun, D.D. and Lee, P.I. (2013) Evolution of supersaturation of amorphous pharmaceuticals: the effect of rate of supersaturation generation. *Mol. Pharm.*, **10** (11), 4330–4346.
114. Sun, D.D., Ju, T.-c., and Lee, P.I. (2012) Enhanced kinetic solubility profiles of indomethacin amorphous solid dispersions in poly(2-hydroxyethyl methacrylate) hydrogels. *Eur. J. Pharm. Biopharm.*, **81** (1), 149–158.
115. Sun, D.D. and Lee, P.I. (2014) Crosslinked hydrogels – a promising class of insoluble solid molecular dispersion carriers for enhancing the delivery of poorly soluble drugs. *Acta Pharm. Sin. B*, **4** (1), 10.
116. Puri, V., Dantuluri, A.K., and Bansal, A.K. (2011) Investigation of atypical dissolution behavior of an encapsulated amorphous solid dispersion. *J. Pharm. Sci.*, **100** (6), 2460–2468.
117. Qian, F., Wang, J., Hartley, R., Tao, J., Haddadin, R., Mathias, N. *et al.* (2012) Solution behavior of PVP-VA and HPMC-AS-based amorphous solid dispersions and their bioavailability implications. *Pharm. Res.*, **29** (10), 2766–2776.
118. MüLlertz, A. (2007) in *Solvent Systems and Their Selection in Pharmaceutics and Biopharmaceutics* (eds P. Augustijns and M. Brewster), Springer, New York, pp. 151–177.
119. Dressman, J.B. and Reppas, C. (2000) In vitro–in vivo correlations for lipophilic, poorly water-soluble drugs. *Eur. J. Pharm. Sci.*, **11** (S2), S73–S80.
120. Tang, L., Khan, S.U., and Muhammad, N.A. (2001) Evaluation and selection of bio-relevant dissolution media for a poorly water-soluble new chemical entity. *Pharm. Dev. Technol.*, **6** (4), 531–540.
121. Nicolaides, E., Galia, E., Efthymiopoulos, C., Dressman, J., and Reppas, C. (1999) Forecasting the in vivo performance of four low solubility drugs from their in vitro dissolution data. *Pharm. Res.*, **16** (12), 1876–1882.
122. Klein, S. (2010) The use of biorelevant dissolution media to forecast the in vivo performance of a drug. *AAPS J.*, **12** (3), 397–406.
123. Kleberg, K., Jacobsen, J., and Müllertz, A. (2010) Characterising the behaviour of poorly water soluble drugs in the intestine: application of biorelevant media for solubility, dissolution and transport studies. *J. Pharm. Pharmacol.*, **62** (11), 1656–1668.
124. Bevernage, J., Forier, T., Brouwers, J., Tack, J., Annaert, P., and Augustijns, P. (2011) Excipient-mediated supersaturation stabilization in human intestinal fluids. *Mol. Pharm.*, **8** (2), 564–570.
125. Jantratid, E., Janssen, N., Reppas, C., and Dressman, J. (2008) Dissolution media simulating conditions in the proximal human gastrointestinal tract: an update. *Pharm. Res.*, **25** (7), 1663–1676.
126. Clarysse, S., Tack, J., Lammert, F., Duchateau, G., Reppas, C., and Augustijns, P. (2009) Postprandial evolution in composition and characteristics of human duodenal fluids in different nutritional states. *J. Pharm. Sci.*, **98** (3), 1177–1192.
127. Marques, M.R., Loebenberg, R., and Almukainzi, M. (2011) Simulated biological fluids with possible application in dissolution testing. *Dissolution Technol.*, **18** (3), 15–28.
128. Miller, J.M., Beig, A., Carr, R.A., Spence, J.K., and Dahan, A. (2012) A win–win solution in oral delivery of lipophilic drugs: supersaturation via amorphous solid dispersions increases apparent solubility without sacrifice of intestinal membrane permeability. *Mol. Pharm.*, **9** (7), 2009–2016.
129. Bevernage, J., Brouwers, J., Annaert, P., and Augustijns, P. (2012) Drug precipitation–permeation interplay: supersaturation in an absorptive environment. *Eur. J. Pharm. Biopharm.*, **82** (2), 424–428.

3
Crystal Imperfections in Molecular Crystals: Physical and Chemical Consequences

William Jones and Mark D. Eddleston

3.1
Introduction

The existence of defects in pharmaceutical materials is well recognized [1]. The nature and characterization of such defects is, however, complex and extremely poorly understood. Frequently, defects are referred to in a very general sense without any clear description of what the actual nature of the defect is. In instances involving milling or micronization of a pharmaceutical, for example, the introduction of "defects" is closely associated with a transition state between crystalline and amorphous [2–6]. The material might be considered as a highly defective crystal or a partially ordered (crystalline) amorphous phase, whose evidence comes from a broadening of the lines in a powder X-ray diffraction pattern or the development of a glass transition temperature [7–12]. In this chapter, we aim to give an overview of defects that may be present even when a crystal diffracts perfectly: in other words, where the defects can be crystallographically well defined in terms of unit cell dimensions, lattice planes, and crystallographic directions [13–16]. Chapter 4 will then describe how such defects may be characterized.

Here we describe crystal defects more fully in terms of their nature and their potential physical and chemical consequences. While long studied in the metallurgical field (where their important role, for instance, in influencing mechanical properties is incontrovertible), such a direct link between the number and type of imperfections in organic molecular materials and their influence on the properties of individual crystals or drug products have not been established.

Various reasons can be given for why this is the case. Models of dislocations in metals – stacking faults and twins, for example (as discussed later) – invariably depict the lattice as a series of organized spheres (usually in a cubic lattice) [17]. For the vast majority of molecular crystals, however, we deal with very asymmetric species with conformational flexibility. While models of "atomic species" on lattice sites can illustrate, in a general sense, the ideas behind crystalline imperfections, issues that cannot be readily addressed by these simple models arise. For example, the idea of a lattice vacancy is simple and readily understood in, for example, ionic crystals, and their role has been extensively studied [13].

Disordered Pharmaceutical Materials, First Edition. Edited by Marc Descamps.

In contrast, the existence of a vacancy in a molecular crystal raises questions about how molecules close to the vacancy may adjust conformationally – perhaps to a geometry where the (unwanted) formation of an intramolecular degradation product is facilitated. The existence of edge dislocations – again readily appreciated in atomic models – provides further opportunities for an "incorrect" conformation to be adopted near the core of the dislocation. The existence of stacking faults may bring molecules into an alignment ideal for unexpected reaction to occur, that is, unexpected in terms of the perfect crystal. Simple microtwins similarly also require an acknowledgment that, across the whole of the twin boundary within the crystal, molecules will be brought together in an intermolecular arrangement different from that in the bulk crystal, again with the potential for the formation of an intermolecular degradation product not expected from a knowledge of the perfect lattice. While the idealized pictures that emerge from single-crystal X-ray diffraction give us a general idea of the arrangement of molecules in a perfect lattice, the microstructure associated with imperfections is far from clear [18].

The crystal surface itself can also be considered as an "imperfection," with the precise arrangement at the surface unlikely to replicate exactly the three-dimensional structure below. For example, our understanding of the surface catalysis of metals has long drawn on concepts such as surface reconstruction [19, 20]. In a molecular crystal, where the intermolecular forces are relatively weak, some molecular rearrangement at the surface would seem inevitable. If we then factor in issues such as solvent inclusion [21] (saturated solution/crystal interface) and, even more importantly, amorphous regions and the interface between amorphous and crystalline domains, the issue becomes increasingly complex [1].

All this, perhaps, would not matter if the number of dislocations, microtwins, vacancies, and interstitials and so on, was small and their influence negligible. But a question arises as to whether even small numbers of defects could act as sites of initial physical or chemical transformation and promote further reactivity via autocatalytic processes. Indeed, the heterogeneous nucleation of reactions at dislocations in photoinduced reactions (referred to as *non-topochemical* – see later) was heavily investigated in the 1970s and 1980s. In recent times, however, the issue has been less studied [22]. When dealing with rather small compositional changes (perhaps on the order of a few percent chemical degradation), such lattice imperfections may play an important role.

A challenge is how to relate some unexpected solid-state phenomenon to the existence of defects. In metals, the movement of dislocations has a significant impact on the deformation properties [15, 16]. The number and types of such dislocations have been long analyzed. In pharmaceutical crystals, the mechanical properties are predicted, nevertheless, on the basis of the perfect crystal, but perhaps this will not always be so. If the crystallization conditions can influence the concentration and types of defects in a crystal, there may be a point when the defects dominate the compression properties – perhaps in an advantageous manner – allowing plastic deformation to occur more readily than might be anticipated.

3.2
General Aspects of Defects in Crystals

The presence of crystalline imperfections in a solid may lead both to local changes in topology at the imperfection and to changes in energy due to any strain set up or relieved by the imperfection [23].The chemical potential of a molecule associated with an imperfection will be different from that of a molecule in a perfect lattice. In order to anticipate the role that imperfections will play in the properties of an organic solid, it will be necessary to establish completely their character. Some salient properties of imperfections in general, and point defects, linear (dislocations), and planar defects (e.g., stacking faults), in particular are now given. We also note that the concept of a volume defect might apply where solvent occlusion is also present.

Structural imperfections can be classified into three general types based on their geometry: point, line, and planar.

1) *Point Imperfections*: In a pure material, there are several types of point defects. Vacancies exist where atoms or molecules are missing from their normal positions in the crystal lattice. Molecules may also occupy non-lattice (interstitial) sites (Figure 3.1). At temperatures above 0 K, there is always an equilibrium number of vacancies present in a crystal. This number N is given by $N_o = N_s$ $\exp(-\Delta G/RT)$, where N is the number of lattice sites, ΔG is the free energy for formation of a mole of vacancies, R is the gas constant, and T is the absolute temperature. As the temperature increases, the equilibrium number of vacancies also increases. During crystal growth, a number of these defects in excess of the equilibrium number may be introduced through rapidly quenching from high temperatures, through plastic deformation, or through radiation damage.
Plastic deformation or irradiation processes also serve to displace species from their normal lattice sites to interstitial sites. Irradiation, for example, produces equal numbers of vacancies and interstitials. Rapid quenching may generate a large number of nonequilibrium point defects, which may not be able to equilibrate because of low mobility. These point imperfections are

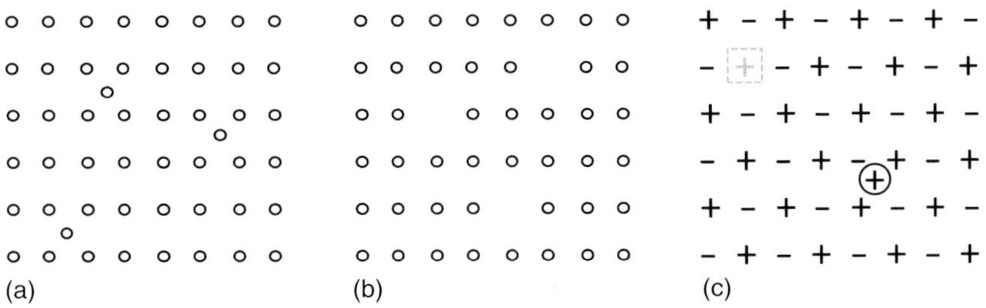

Figure 3.1 Schematic of point defects in a simple square lattice of spheres. (a) Interstitial, (b) vacancy, and (c) simultaneous formation of vacancy and interstitial for a salt-like structure with charge balance preserved.

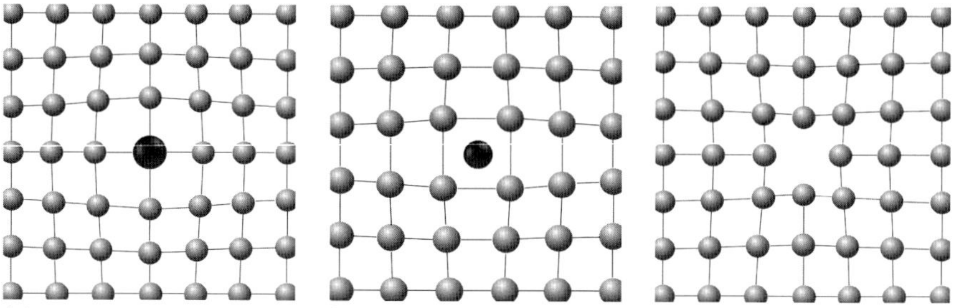

Figure 3.2 Schematic representation of the lattice strain associated with (a) an impurity, (b) an interstitial, and (c) a lattice vacancy. Such simple "spherical atom" diagrams fail to address possible changes in orientation for molecular crystals when inter and intramolecular contacts may be significantly modified, resulting in possible centers of reaction.

centers of relatively small elastic strain, but the resulting local dilations and contractions of the crystal lattice may still be sufficient to allow chemical change to occur by bringing molecules close together or by providing the space necessary for molecular motion to occur (Figure 3.2). In salts, it is possible to have vacancies on both sublattices, but interstitials will most likely involve the displacement of the smaller ion.

Impurity molecules in a material may also be considered point defects. These impurity molecules play an important part in the physical and mechanical properties of the material. The impurities may take up either substitutional or interstitial sites, but for molecular crystals they are likely to be substitutional because of the close packing within the crystal [24]. In the case of anisotropic molecules, point defects may exist where a molecule takes up a position out of orientational registry with those around it. In such a case, X-ray diffraction would point to a large amount of scattering associated with lattice disorder – as is frequently observed during single-crystal structure refinement of organic solids [25]. In anthrone, for example, molecules appear to stack randomly in terms of molecular orientation although still centered on lattice sites [26].

The formation of vacancies and interstitials in molecular crystals will require a fairly large amount of energy – at least as large as the enthalpy of sublimation. As a consequence, the relative abundance of point defects in molecular crystals will be low compared to the numbers formed in metals and ionic salts. Sherwood [18] has measured point defect properties in organic crystals by radio tracer studies of self-diffusion coefficients. He found a factor of 10 difference in the absolute diffusion coefficients when he compared plastic crystals (camphone, adamantane, hexamethyl ethane) and brittle crystals (benzene, anthracene, biphenyl, and naphthalene). This agrees with Thomas and Williams' [27] statement that point defect concentrations are low in organic materials, except those that enter a rotator phase prior to melting.

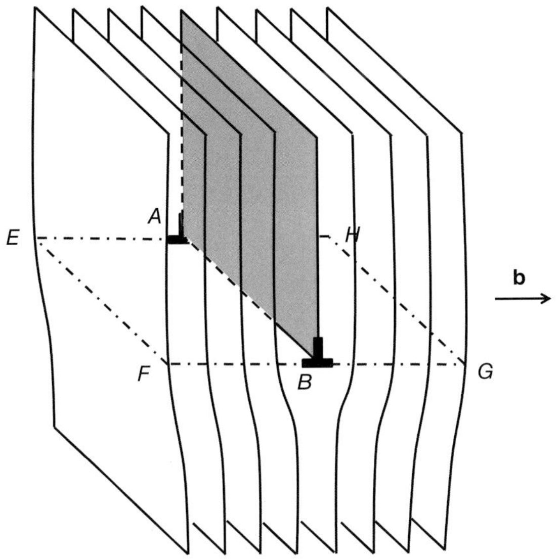

Figure 3.3 Projection diagram of an edge dislocation illustrating the line of the dislocation as *A*–*B* with slip on the *EFGH* plane. Away from the core of the dislocation, the crystal packing is not disturbed but along and near the core contact distances and orientations will be modified.

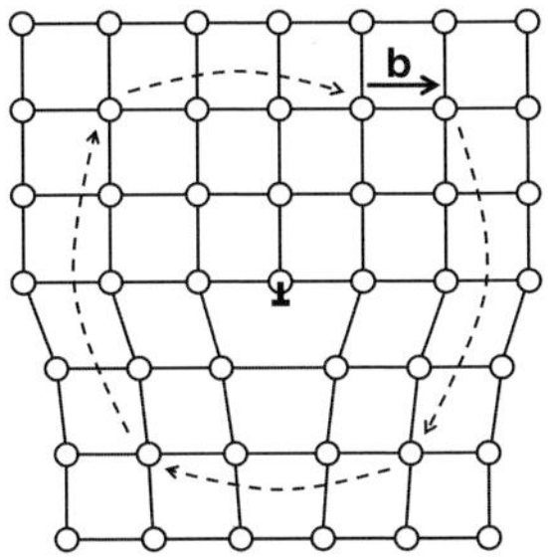

Figure 3.4 Illustration of the determination of the magnitude and direction of the Burgers vector associated with an edge dislocation defined by the failure to complete a closed circuit around the dislocation. A similar approach is used to define the magnitude and direction of a screw dislocation, see Figure 3.6.

2) *Line Imperfections*: Line imperfections, or dislocations, are linear regions of elastic distortion caused by atoms that are slightly displaced from their equilibrium lattice positions. The two basic types of dislocations are edge dislocation and screw dislocation [15]. An example of an edge dislocation is given in Figure 3.3. It is produced by the presence of an extra half plane of the atoms above the plane *EFGH*. The atoms near the lower edge of the half plane are elastically displaced to an arrangement indicated in the figure. The dislocation line is the line *AB*. Elsewhere, away from this line, the crystal retains its structural perfection.

A dislocation is characterized by a Burgers vector **b**, which represents the direction and magnitude of the elastic shear distortion. The Burgers vector is determined by viewing the dislocation from *A* to *B*. A Burgers circuit is then made by forming a closed circuit of an atom-to-atom path around the dislocation. Such a path is shown in Figure 3.4. When the structure is perfect, or contains only point imperfections, the Burgers circuit will close. When a dislocation is present, the Burgers circuit will not close, and the lattice vector between the start and end points is the Burgers vector for the dislocation. A dislocation is said to be perfect, or of unit strength, when the magnitude of the Burgers vector equals a unit lattice vector. A multiple strength dislocation, on the other hand, has a Burgers vector several times that of the lattice vector, whereas partial dislocations have a vector less than the unit lattice vector. The Burgers vector is perpendicular to the dislocation line *AB* in Figure 3.4, and this condition defines an edge dislocation. The Burgers vector of an edge dislocation is normal to the line of the dislocation.

The combination of the dislocation line and its perpendicular Burgers vector determines the slip plane on which the edge dislocation is constrained to move, or glide, under an applied shear stress. The movement, or slip, of the dislocation results in plastic deformation, and requires a lower energy than would have been predicted on the basis of the perfect structure. The result of the dislocation moving in its glide plane is indicated in Figure 3.5. This concept of slip enables the dislocation line to be defined as that line that separates the slipped from the unslipped portion of the slip plane.

Deformation studies to identify operative slip systems in organic crystals were reported in the 1960s and 1970s, and a revival occurred through the studies by Ramamurty, Desiraju, and others [28]. Even in isomorphous crystals, subtle changes in unit cell dimensions and angles as well as the nature of the intermolecular interactions across the slip plane (e.g., by calculation of attachment energies) can have noticeable effects. Such was the case for anthracene, anthraquinone, and *p*-terphenyl, each packing in the monoclinic space group $P2_1/a$ [29]. Finnie *et al.* reported a detailed study of the monoclinic form of paracetamol from microhardness indentation techniques [30], and nanoindentation has recently been applied to sucrose, aspirin, and other systems [31].

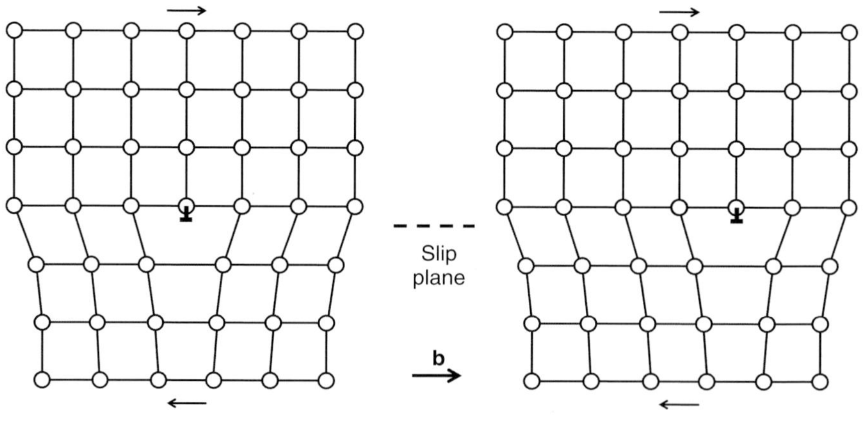

Figure 3.5 Illustration of the movement of a dislocation under an applied shearing stress.

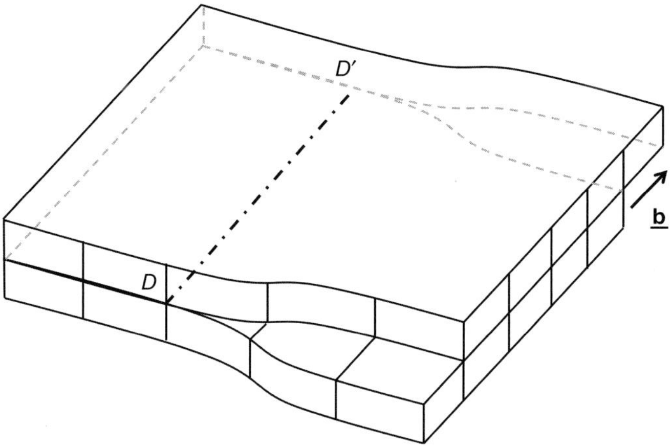

Figure 3.6 Screw dislocation model with the dislocation core D–D' running parallel to the Burgers vector.

The connection with desolvation mechanisms and indentation studies has been reviewed by Varughese *et al.* [28].

If the dislocation line AB in Figure 3.3 were imagined to bend in the slip plane to the point where the dislocation line runs parallel to the direction of the Burgers vector of the dislocation, this condition would describe a screw dislocation (Figure 3.6). The Burgers vector of a screw dislocation is parallel to the line of the dislocation. The structural error that arises in the screw dislocation is a spiral ramp of displaced atoms that advances one Burgers vector per circuit of the dislocation (Figure 3.7). Because the Burgers vector and dislocation line are parallel, they do not define a specific slip plane, as was the case for the edge dislocation. Consequently, screw dislocations are not constrained to move on a certain slip plane, but can move by cross-sliding on any facile slip plane. In general, a dislocation will be oriented perpendicular and parallel to the Burgers vector, meaning that it has both edge and screw characteristics. The elastic distortions around a dislocation produce a stress field that moves with the dislocation. For an edge dislocation, the stress field has both dilatational and shear components, whereas for a screw dislocation the stresses are purely shear and are parallel to the dislocation. The presence of these localized stress fields will mean that dislocations represent regions of stored elastic energy. This energy will vary markedly from region to region and with distance around the dislocation core [15].

3) *Planar Imperfections*: Planar imperfections are any surface across which the atom positions in a perfect crystal are not preserved. Two common planar faults found in crystals are twin planes and stacking faults. Another common defect is the grain boundary – the region between randomly oriented domains within the crystal. When the misorientation is small, these would be referred to as "low-angle grain boundaries" and consist of an array of dislocations. Twin planes separate the two parts of a twinned crystal, the parent

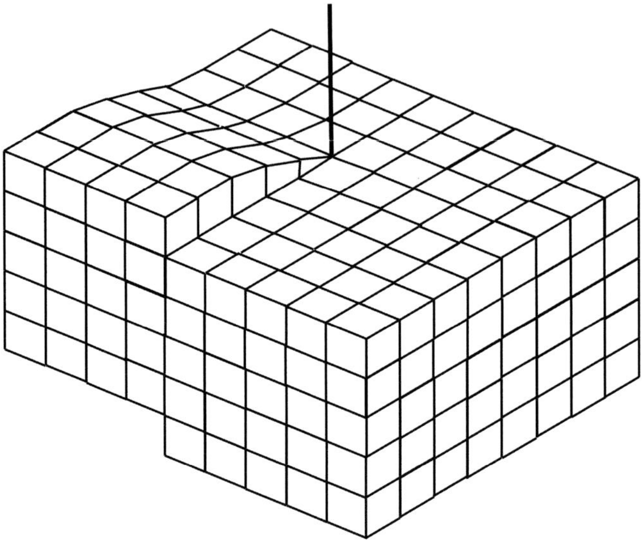

Figure 3.7 Screw dislocation emerging from the surface of a crystal, creating a continuous helical sheet. Such emergent dislocations are often cited as the location of growth spirals during crystal growth.

and its twin, which are related by symmetry. Twinning may be thought of as a homogeneous shear of the lattice with a rotation (unlike slip which simply moves one section of the crystal with respect to the one below) (Figure 3.8). The example of anthraquinone by Williams *et al.* resulted in a model of the possible alignment of the molecules at the (20-1) interface, clearly showing a markedly different intermolecular relationship at the interface [29].

Stacking faults are common in materials that have a strongly layered character. Graphite is a well-known material that contains such faults between domains of hexagonal and rhombohedral composition. Polytypism is another concept for such solids [32]. The dissociation of dislocations into partial dislocations can result in stacking faults with new intermolecular contacts – as will be discussed in the following in the case of 9-cyanoanthracene.

3.3 Role of Imperfections in Reactivity and Stability – Chemistry in the Perfect and Imperfect Lattice

The constraint imposed upon organic molecules as a result of being placed in the perfect and repeated environment of the crystalline phase raises many interesting questions for the organic solid-state chemist [33, 34]. There are likely to be basic differences in the reactivity of the crystal phase compared with the fluid and amorphous phases. This difference arises because the molecules in a crystal

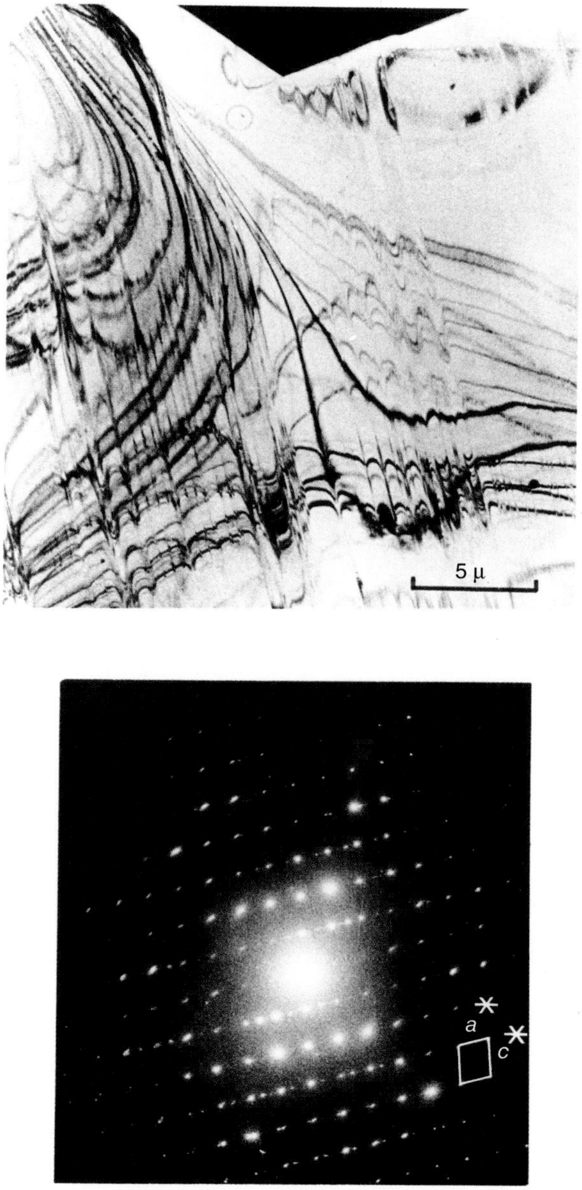

Figure 3.8 Electron micrograph and associated diffraction pattern for an organic crystal showing a high density of microtwins. At each twin interface the molecular packing will be different from that within the perfect bulk of the material.

occur in only a few possible arrangements and conformations and also because the geometry relating different molecules (space symmetry) is limited in the crystalline phase. It may also happen that a conformer is found in the crystal that does not persist, because of energetic reasons, in solution or perhaps in the amorphous phase.

The effect of fewer conformers existing in the solid state than in solution was demonstrated in early work concerning the double dehydrobromination of dimethyl meso β,β′-dibromoadipate [35]. When the reaction was carried out in solution, three reaction products were obtained. In contrast, when gaseous amines were passed over single crystals of this material, the only reaction product was the *trans,trans*-muconate, indicating that the conformation of the molecule was locked. For trans elimination to occur, the eliminating H and Br must be in an anti-periplanar arrangement (giving a centrosymmetric product). This was, in fact, in agreement with the conformation found in the crystal.

For reactions between two or more molecules, it is the relative arrangement of the molecules – the molecular packing – rather than molecular conformation that will dominate the reaction chemistry. For example, the solid-state chemistry of cinnamic acid and its derivatives, as well as numerous solid-state polymerization reactions, have been shown to be primarily controlled by packing in the perfect lattice.

In a series of papers, Schmidt, Cohen, and colleagues [36–38] established that the photochemical reactivity of crystals of cinnamic acid and various cinnamic acid derivatives was a function of the molecular registry, as revealed by single-crystal diffraction studies. The crystal structures of numerous substituted cinnamic acids can be categorized into three general types – α, β, and γ. Some derivatives can adopt multiple different crystal structures; for example, *o*-ethoxy-*trans*-cinnamic acid crystallizes in all three structural types. In each of the structure types, the molecules pack in one-dimensional stacks and form pairwise hydrogen-bonding interactions across centers of symmetry. Within the stacks, the molecules lie parallel with a distance on the order of 3.5 Å between the molecular planes. The three types differ, however, in the angle between this normal and the stack axis, in the repeat distance along this axis, and in the extent and type of overlap between adjacent molecules in the stack. The β-type resulted in the mirror (m) dimer; the α-type gave the centrosymmetric dimer (1-), and the γ-type was light-stable – see Figure 3.9.

As an extension of this lattice-controlled dimerization of cinnamic acids, the technique of solid-state polymerization was developed. The first description of an oriented polymer chain growth was published by Morawetz [39], who exposed single crystals of trioxane to formaldehyde vapor and showed that polyoxymethylene was formed on the surface of the trioxane in chains that were oriented parallel to the sixfold axis of the hexagonal trioxane crystals.

Hirshfeld and Schmidt [40] suggested that the necessary condition for matrix controlled polymerization is that no significant disruption of the molecular positions occurred on polymerization. They referred specifically to crystals of molecules with two potentially reactive centers which were so oriented that

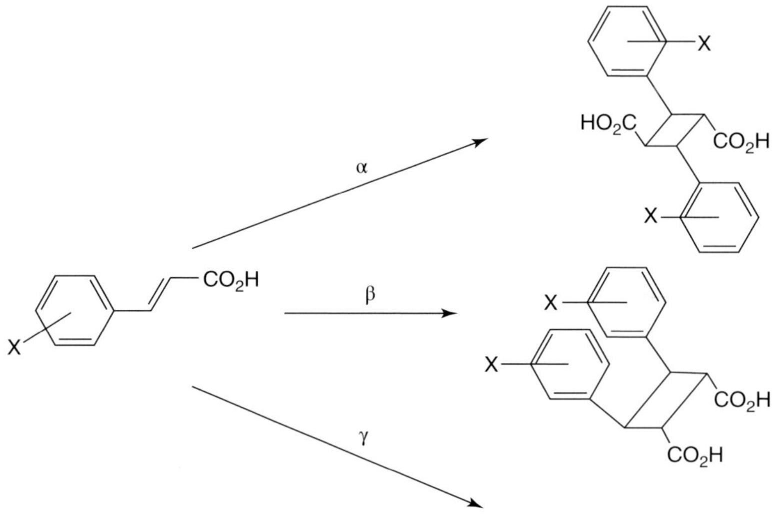

Figure 3.9 Schematic diagram of the possible outcome of UV irradiation in substituted cinnamic acids – centrosymmetric dimer, mirror-symmetric dimer, or photostability.

they could rotate in place to link up with their neighbors without any linear displacement of the molecular centers. Two groups of materials that explored this concept were the diacetylenes and divinyl monomers.

The work of Wegner [41] demonstrated that solid-state polymerization (thermal and photolytic) could be rationalized on the basis of crystal packing in the perfect lattice, the polymerization proceeding to high molecular weight products with little change in lattice parameters.

In the case of divinyl monomers, an important example was that of 2,5-distyrylpyrazine,which also readily polymerized in the solid state [42, 43] – with the final crystalline photoproduct having a molecular weight several times greater than the solution-grown polymer, which again could be rationalized on the basis of packing within the perfect lattice.

A recognized challenge, however, was found in the abnormal solid-state photochemical reactions of anthracene and 9-cyanoanthracene (among others) [44]. In these systems, analysis of the perfect lattice would suggest stability, as the molecules are inappropriately arranged to react (either by orientation or separation). Despite this, reaction did, in fact, take place.

One explanation was that those reactions that gave the non-topochemical product occurred preferentially at defect sites in the crystal. This was supported by the work of Thomas and his group [27], who separately etched and photodimerized, matched halves of cleaved melt-grown anthracene crystals (Figure 3.10). They found that the degree of correspondence between etch pits and dimerization centers was usually greater than 85%. For 9-cyananoanthracene, the observed product was rationalized in terms of stacking faults in the crystals generated by the dissociation of [1 0 0](2 2 1) dislocations (Figure 3.11) [45].

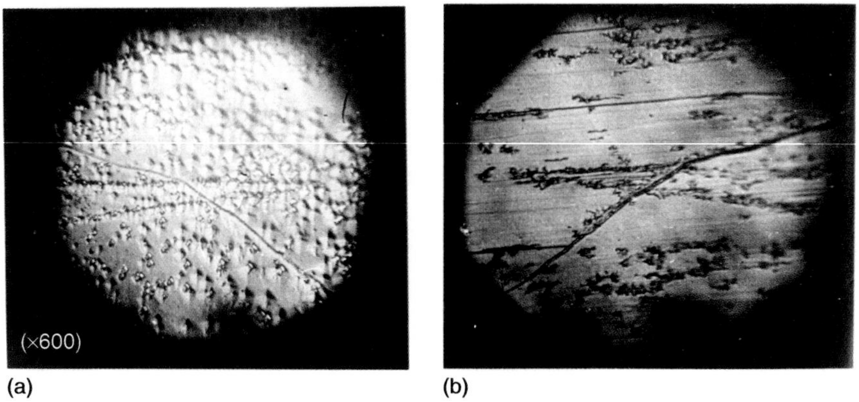

Figure 3.10 Optical microscope images of a cleaved crystal of anthracene showing correspondence between defects highlighted by chemical etching (a) and the corresponding photoinduced product distribution (b).

Recent micro-Raman studies of the photodimerization of 9-cyanoanthracene have demonstrated that there is a delay between the onset of a chemical reaction and the appearance of the product crystals – the initial reaction can occur within the crystal but no diffraction evidence appears until a reconstructive phase transition occurs [46].

In studies of dehydration, there seems to be less firm evidence for imperfections playing a direct role. Work by Sherwood [18] on the desolvation of paracetamol solvates suggested a lack of association of dislocations as the loci of desolvation in the bulk lattice. Such a desolvation process, however, will require a mechanism by which solvent molecules may migrate from the bulk to the surface. In such cases, it may be that the dislocations do not provide a pathway for diffusion of the solvent and that as a result the process initiates at the surface and then moves into the bulk.

3.4
Role in Physical Processes

It is well established from experimental observations of fast growth rates that, at levels of supersaturation significantly lower than expected from classical theories of crystal growth, emergent screw dislocations play an important role in crystal growth. Evidence from atomic force microscopy (AFM) measurements clearly demonstrates this fact [47].

In phase transformations, two alternative models can be cited. The first by Mnyukh [48, 49] is based on the presence of defects (such as micro-cavities) that exist between crystal domains. Migration of molecules across this interface allows the new phase to be formed [22].The model would not anticipate a structural relationship between the two phases.

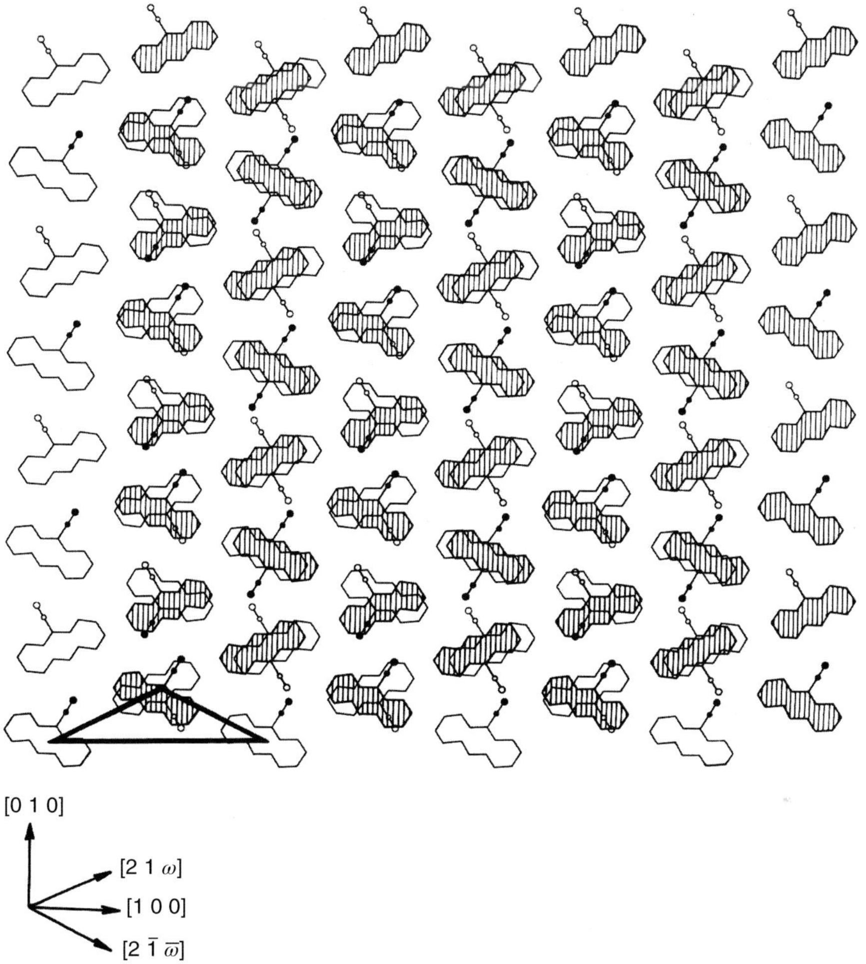

Figure 3.11 Schematic of the consequences of a stacking fault in 9-cyanoanthracene where layers are brought into contact such that the centrosymmetric dimer becomes a viable photoproduct.

In other instances, a direct comparison to the well-established mechanism in metals (a martensitic transformation) was suggested to occur [28, 50]. In crystals of 1,8-dichloro-10-methylanthracene, it was demonstrated that there was a well-defined crystallographic relationship between the so-called parent and daughter lattices (Figures 3.12 and 3.13). The transformation was believed to be stress-induced, with the transformation occurring across a specific and invariant plane (2.88,0,1). Transmission electron microscopy results suggested a possible role of dislocations running in the [0 0 1] direction. Noteworthy was the existence of the daughter phases within the parent with the interface common to both structures. In aspirin crystals, nanoindentation was used to explore

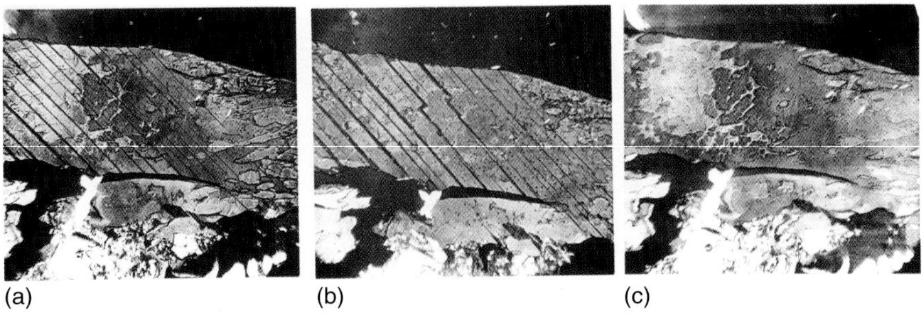

Figure 3.12 Development of lamellar faults in an organic crystal as the material is cooled from the melt. The faults disappear close to the melt (c) only to reappear at similar locations on recooling (a and b).

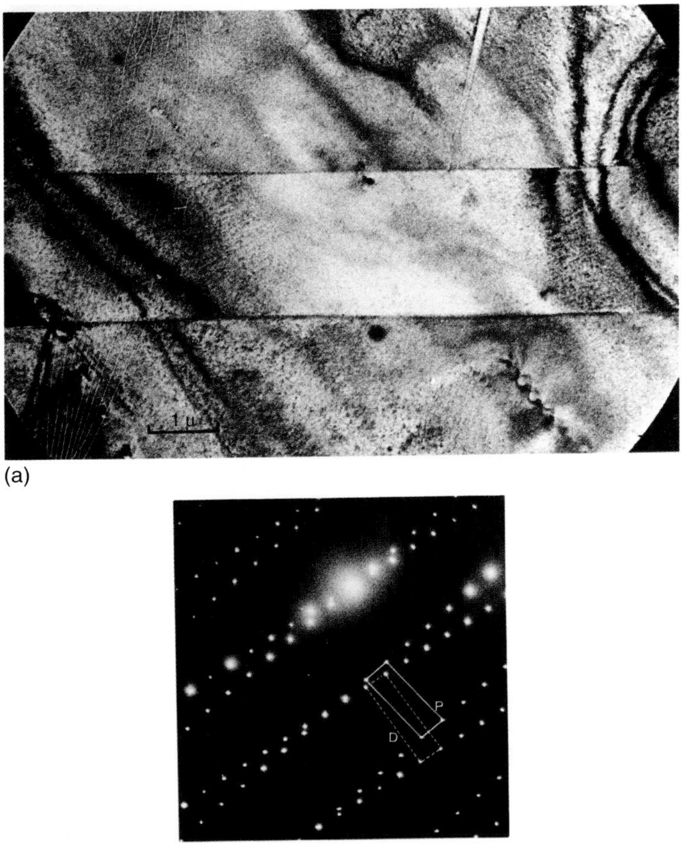

Figure 3.13 Image and diffraction pattern of a stress-induced phase in a crystal of 1,8-dichloro-10-methylanthracene (a). The interface is crystallographically defined, and the diffraction pattern (b) confirms a change in unit cell from one side of the interface to the other.

the stress-induced phase transformation between polymorphs I and II. The transformation was reported to occur via a slip mechanism [31].

Numerous studies have considered the role of point defects and line dislocations in explosive organic systems. Dislocations in particular were related to so-called *hot spots* in triggering explosive decomposition. Work by Kuklja using first-principles calculations indicated that in RDX (1,3,5-trinitroperhydro-1,3,5-triazine) and PETN (pentaerythritol tetranitrate) edge dislocations introduced electronic states within the optical bandgap of the solids [51, 52]. These states were mainly formed by molecular orbitals of critical bonds responsible for stability/instability. Molecular dynamics simulations have also been reported for HMX (octahydro-1,3,5,7-tetranitro-1,3,5,7-tetrazocine) in terms of vacancies and creation of structural voids [53]. In atomistic simulations of RDX, it was found that molecular flexibility played a critical role in the ductile behavior of the crystals [54]. In other atomistic simulations for RDX, good agreement with experiment confirmed the importance of the (1 0 0)[1 0 0] slip system.

In attempting to understand the inelastic deformation of molecular crystals, Lei and Koslowski have reported dislocation dynamics simulations for crystals of sucrose and paracetamol [55]. The simulations gave yield stress in good agreement with experimental observations as well as the anisotropy in mechanical response observed experimentally. Overall, the results indicated that dislocation dynamics was able to quantify and describe plastic deformation in molecular crystals.

3.5 Concluding Remarks

Early work on the importance of crystalline imperfections in organic molecular crystals was motivated by their possible role in photophysical phenomena – for example, energy (exciton) transfer. Molecules close to or at the defect had lower excitation energies. The importance of dislocations and slip systems with regard to mechanical properties was also studied and continues to be an area of importance. In pharmaceutical systems, it is likely that such imperfections may play an important role in the initial stages of reaction (thermal or otherwise), with subsequent stress fields generating additional sites of instability, but very little is known about the precise molecular dispositions at or close to the defects, either in terms of intermolecular contacts or possible changes in molecular conformation.

References

1. Newman, A. and Zografi, G. (2014) Critical considerations for the qualitative and quantitative determination of process-induced disorder in crystalline solids. *J. Pharm. Sci.*, **103**, 2595–2604.
2. Ward, G.H. and Schultz, R.K. (1995) Process-induced crystallinity changes in albuterol sulfate and its effect on powder physical stability. *Pharm. Res.*, **12**, 773–779.

3. Wildfong, P.L.D., Hancock, B.C., Moore, M.D., and Morris, K.R. (2006) Towards an understanding of the structurally based potential for mechanically activated disordering of small molecule organic crystals. *J. Pharm. Sci.*, **95**, 2645–2656.
4. Li, H., Stowell, J.G., He, X., Morris, K.R., and Byrn, S.R. (2007) Investigations on solid–solid phase transformation of 5-methyl-2-[(4-methyl-2-nitrophenyl)amino]-3-thiophenecarbonitrile. *J. Pharm. Sci.*, **96**, 1079–1089.
5. Chamarthy, S.P. and Pinal, R. (2008) The nature of crystal disorder in milled pharmaceutical materials. *Colloids Surf., A*, **331**, 68–75.
6. Li, H., Wen, H., Stowell, J.G., Morris, K.R., and Byrn, S.R. (2010) Crystal quality and physical reactivity in the case of flufenamic acid (FFA). *J. Pharm. Sci.*, **99**, 3839–3848.
7. Grant, D.J.W. and York, P. (1986) Entropy of processing: a new quantity for comparing the solid state disorder of pharmaceutical materials. *Int. J. Pharm.*, **30**, 161–180.
8. Descamps, M., Willart, J.F., Dudognon, E., and Caron, V. (2007) Transformation of pharmaceutical compounds upon milling and comilling: the role of T_g. *J. Pharm. Sci.*, **96**, 1398–1407.
9. Forcino, R., Brum, J., Galop, M., and Sun, Y. (2010) Using thermally stimulated current (TSC) to investigate disorder in micronized drug substance produced at different milling energies. *Pharm. Res.*, **27**, 2234–2241.
10. Clawson, J.S., Kennedy-Gabb, S., Lee, A.Y., and Copley, R.C.B. (2011) One-phase crystal disorder in pharmaceutical solids and its implication for solid-state stability. *J. Pharm. Sci.*, **100**, 4302–4316.
11. Otte, A. and Carvajal, M.T. (2011) Assessment of milling-induced disorder of two pharmaceutical compounds. *J. Pharm. Sci.*, **100**, 1793–1804.
12. Luisi, B.S., Medek, A., Liu, Z., Mudunuri, P., and Moulton, B. (2012) Milling-induced disorder of pharmaceuticals: one-phase or two-phase system? *J. Pharm. Sci.*, **101**, 1475–1485.
13. Tilley, R.D. (1998) *Principle and Applications of Chemical Defects*, Stanley Thorne Publishers, Cheltenham.
14. West, A.R. (2010) *Basic Solid State Chemistry*, 2nd edn, John Wiley & Sons, Ltd., Chichester.
15. Hull, D. and Bacon, D.J. (2011) *Introduction to Dislocations*, 5th edn, Elsevier, Amsterdam.
16. Shetty, M.N. (2013) *Dislocations and Mechanical Behaviour of Materials*, PHI Learning Private Limited, Delhi.
17. Cotterill, R. (2008) *The Material World*, 2nd edn, Cambridge University Press, Cambridge.
18. Sherwood, J.N. (2004) Fifty years as a crystal gazer: life as an imperfectionist. *Cryst. Growth Des.*, **4**, 863–877.
19. Somorjai, G.A. (1994) *Introduction to Surface Chemistry and Catalysis*, 2nd edn, John Wiley & Sons, Inc., New York.
20. Thomas, J.M. and Thomas, J.W. (1996) *Principles and Practice of Heterogeneous Catalysis*, Wiley-VCH Verlag GmbH, Weinheim.
21. Waldschmidt, A., Rietveld, I., Couvrat, N., Dupray, V., Sanselme, M., Berton, B., Nicolai, B., Mahe, N., Petit, S., Ceolin, R., and Coquerel, G. (2011) About aged heterogeneous liquid inclusions inside organic crystals in relation to crystal formation, structure, and morphology. *Cryst. Growth Des.*, **11**, 2580–2587.
22. Dunitz, J.D. (1995) Phase changes and chemical reactions in molecular crystals. *Acta Crystallogr., Sect. B: Struct. Sci.*, **B51**, 619–631.
23. Berendt, R.T. and Munson, E.J. (2011) Detection of chiral defects in crystalline organic solids using solid-state NMR spectroscopy. *J. Pharm. Sci.*, **100**, 1879–1891.
24. Thomas, J.M. (1974) Topography and topology in solid state chemistry. *Philos. Trans. R. Soc. London, Ser. A*, **277**, 251–287.
25. Welberry, T.R., Chan, E.J., Goossens, D.J., and Heerdegen, A.P. (2012) Diffuse scattering as an aid to the understanding of polymorphism in pharmaceuticals. *Metall. Mater. Trans. A*, **43**, 1434–1444.
26. Flack, H.D. (1970) Refinement and thermal expansion coefficients of the structure of anthrone (20.deg., -90.deg.)

and comparison with anthraquinone. *Philos. Trans. R. Soc. London, Ser. A*, **266**, 561–574.

27. Thomas, J.M. and Williams, J.O. (1972) *Surface and Defect Properties of Solids*, Royal Society of Chemistry, pp. 130–143.
28. Varughese, S., Kiran, M.S.R.N., Ramamurty, U., and Desiraju, G.R. (2013) Nanoindentation in crystal engineering: quantifying mechanical properties of molecular crystals. *Angew. Chem. Int. Ed.*, **52**, 2701–2712.
29. Williams, J.O., Adams, I., and Thomas, J.M. (1969) Relation between dislocations and molecular structure in aromatic crystals. *J. Mater. Sci.*, **4**, 1064–1068.
30. Finnie, S., Prasad, K.V.R., Sheen, D.B., and Sherwood, J.N. (2001) Microhardness and dislocation identification studies on paracetamol single crystals. *Pharm. Res.*, **18**, 674–681.
31. Varughese, S., Kiran, M.S.R.N., Solanko, K.A., Bond, A.D., Ramamurty, U., and Desiraju, G.R. (2011) Interaction anisotropy and shear instability of aspirin polymorphs established by nanoindentation. *Chem. Sci.*, **2**, 2236–2242.
32. Day, G.M., Trask, A.V., Motherwell, W.D.S., and Jones, W. (2006) Investigating the latent polymorphism of maleic acid. *Chem. Commun.*, 54–56.
33. Ramamurthy, V. and Venkatesan, K. (1987) Photochemical reactions of organic crystals. *Chem. Rev.*, **87**, 433–481.
34. Singh, N.B., Singh, R.J., and Singh, N.P. (1994) Organic solid state reactivity. *Tetrahedron*, **50**, 6441–6493.
35. Green, B.S., Lahav, M., and Rabinovich, D. (1979) Asymmetric synthesis via reactions in chiral crystals. *Acc. Chem. Res.*, **12**, 191–197.
36. Cohen, M.D. and Schmidt, G.M.J. (1964) Topochemistry. I. A survey. *J. Chem. Soc.*, 1996–2000.
37. Cohen, M.D., Schmidt, G.M.J., and Sonntag, F.I. (1964) Topochemistry. II. The photochemistry of trans-cinnamic acids. *J. Chem. Soc.*, 2000–2013.
38. Schmidt, G.M.J. (1971) Photodimerization in the solid state. *Pure Appl. Chem.*, **27**, 647–678.
39. Morawetz, H. (1977) in *Reactivity of Solids* (eds J. Woods, O. Lindqvist, C. Helgesson, and N.-G. Vannerberg), Plenum Press, New York, pp. 457–474.
40. Hirshfeld, F.L. and Schmidt, G.M.J. (1964) Topochemical control of solid-state polymerization. *J. Polym. Sci.*, **2**, 2181–2190.
41. Wegner, G. (1977) Solid-state polymerization mechanisms. *Pure Appl. Chem.*, **49**, 443–454.
42. Nakanishi, H., Hasegawa, M., and Sasada, Y. (1972) Four-center type photopolymerization in the crystalline state. V. X-ray crystallographic study of the polymerization of 2,5-distyrylpyrazine. *J. Polym. Sci., Part A-2*, **10**, 1537–1553.
43. Nakanishi, H., Jones, W., Thomas, J.M., Hasegawa, M., and Rees, W.L. (1980) Topochemically controlled solid-state polymerization. *Proc. R. Soc. London, Ser. A*, **369**, 307–325, 306 plates.
44. Craig, D.P. and Sarti-Fantoni, P. (1966) Photochemical dimerization in crystalline anthracenes. *Chem. Commun.*, 742–743.
45. Thomas, J.M., Evans, E.L., and Williams, J.O. (1972) Microscopic studies of enhanced reactivity at structural faults in solids. *Proc. R. Soc. London, Ser. A: Math. Phys. Eng. Sci.*, **331**, 417–427.
46. Salzillo, T., Zaccheroni, S., Della Valle, R.G., Venuti, E., and Brillante, A. (2014) Micro Raman investigation of the photodimerization reaction of 9-cyanoanthracene in the solid state. *J. Phys. Chem. C*, **118**, 9628–9635.
47. Chow, E.H.H., Bučar, D.-K., and Jones, W. (2012) New opportunities in crystal engineering – the role of atomic force microscopy in studies of molecular crystals. *Chem. Commun.*, **48**, 9210–9226.
48. Mnyukh, Y.V. (1979) Molecular mechanism of polymorphic transitions. *Mol. Cryst. Liq. Cryst.*, **52**, 467–503.
49. Mnyukh, Y.V. (1979) Polymorphic transitions in crystals: kinetics. *Mol. Cryst. Liq. Cryst.*, **52**, 505–521.
50. Jones, W., Thomas, J.M., and Williams, J.O. (1975) Electron and optical microscopic studies of a stress-induced

phase transition in 1,8-dichloro-10-methylanthracene. *Philos. Mag.*, **32**, 1–11.

51. Kuklja, M.M. (2001) Electronic structure modifications induced by nanosize lattice imperfections in molecular crystals. *Comput. Nanosci.*, **2**, 65–68.
52. Kuklja, M.M. and Kunz, A.B. (2001) Electronic structure of molecular crystals containing edge dislocations. *J. Appl. Phys.*, **89**, 4962–4970.
53. Duan, X.H., Li, W.P., Pei, C.H., and Zhou, X.Q. (2013) Molecular dynamics simulations of void defects in the energetic material HMX. *J. Mol. Model.*, **19**, 3893–3899.
54. Munday, L.B., Mitchell, R.L., Knap, J., and Chung, P.W. (2013) Role of molecule flexibility on the nucleation of dislocations in molecular crystals. *Appl. Phys. Lett.*, **103**, 151911-1–151911-4.
55. Lei, L. and Koslowski, M. (2011) Mesoscale modeling of dislocations in molecular crystals. *Philos. Mag.*, **91**, 865–878.

4
Observation and Characterization of Crystal Defects in Pharmaceutical Solids

Mark D. Eddleston and William Jones

4.1
Introduction

The importance of understanding the solid-state behavior of pharmaceutical compounds and formulations is universally recognized. Different crystal forms display different properties, and the process of identifying all possible forms of an active pharmaceutical ingredient (API), and selecting one that displays good chemical and physical stability, is an important step toward ensuring that patients are supplied with a drug product that is consistently both safe and efficacious. To this end, there has been a large volume of research into property modification through salt and co-crystal formation and into methods of predicting and screening for new polymorphic forms of compounds. The unexpected crystallization of polymorphic forms of marketed compounds does still occasionally occur [1, 2], causing turmoil for both patients and the scientists involved in drug development. In general, however, the polymorphic behavior of pharmaceuticals is increasingly better understood, rigorously investigated during drug development, and strictly regulated by licensing authorities such as the Food and Drug Administration (FDA).

In stark contrast, the nature and abundance of crystal defects in pharmaceutical type materials has not been extensively researched and is not well understood. Defects are not routinely investigated during the drug development process, and there is no requirement from the FDA to do so. This is perhaps surprising, given that it is widely recognized that crystal defects influence the solid-state properties of organic solids in general and of pharmaceuticals and excipients in particular [3–13]. As described in detail in the previous chapter, defects can affect a whole range of pharmaceutically relevant material properties such as crystal growth, dissolution rate, chemical stability (and therefore shelf-life), physical stability, bioavailability, hygroscopicity, dehydration, and compaction behavior. Various types of defects are likely to be introduced both during crystallization and during widely employed processing steps such as milling, drying, granulation, and

Disordered Pharmaceutical Materials, First Edition. Edited by Marc Descamps.

tableting [3–14]. In fact, any modifications to manufacturing processes, even if small, could change the type and amount of defects present in the resulting crystals and, therefore, affect the behavior of a drug substance or drug product. Indeed, it has been suggested that the undesirable batch-to-batch variability regularly observed during the production of pharmaceuticals could be due to differences in the defect contents of batches [9].

This leads to a question of why crystal defects are not routinely investigated during pharmaceutical product development. Rather than being a simple oversight, this situation primarily results from the fact that the analytical tools that are widely available to the pharmaceutical scientist, while ideal for characterizing crystal form, are not designed for the observation and identification of defects. As a result, there is rarely any mention of crystal defects in the many reviews of pharmaceutical analytical methods [15–17], and in literature sources where the important influence of crystal defects on the properties of pharmaceutical materials is discussed, there is seldom any indication of how these defects might be described or indeed characterized [5, 6].

There are, however, many areas of industry and research, such as metallurgy and geology, where the analysis of defects is performed routinely and has proven to be critical in understanding a wide range of material properties including metal fatigue, phase-change propensity, magnetism, and superconductivity. It is the aim of this chapter to show that the analysis of crystal defects is also widely applicable in a pharmaceutical context. We will introduce the main techniques that are used for studying defects, summarize the range of information that these techniques can be expected to yield, outline possible limitations of these methods when applied to organic compounds, and give examples of their use for the characterization of defects in pharmaceutical samples. The chapter will be split into three sections: First is the section describing techniques that can be used to observe defects within crystals, which will focus on transmission electron microscopy (TEM) and X-ray topography. Second is a section on techniques for visualizing defects that are emergent at crystal surfaces, which will focus on atomic force microscopy (AFM) and chemical etching. Third is a section on techniques that do not allow the visualization of defects directly, but can give an indication of the amount of defects that are present in a material. The techniques described in this chapter are equally applicable to other sectors where the solid-state behavior of organic compounds is important, such as paints and pigments, agrochemicals, and organic semiconductors.

4.2
Techniques for Characterizing Defects within Crystals

This section will focus on methods of observing and characterizing defects that are present in the bulk of crystals. The two key techniques that will be introduced are TEM and X-ray diffraction topography, both of which utilize diffraction contrast as a means of imaging defects.

Figure 4.1 Images showing (a) part of a TEM sample grid supporting crystals of theophylline, (b) a TEM sample holder, and (c) a TEM instrument.

4.2.1 Transmission Electron Microscopy

4.2.1.1 Introduction to Transmission Electron Microscopy

TEM [18, 19] has been developed as an analytical tool for over 50 years and is widely used for the characterization of inorganic samples such as metals and minerals [19–24]. In a typical TEM instrument, a coherent beam of electrons is generated, passed through a sample, and then focussed on a screen at the bottom of the instrument, as shown in Figure 4.1. Samples for TEM are supported on a circular mesh grid, ~3 mm in diameter, which is usually made of a metal such as copper, and often coated with a film of amorphous carbon. The grid is then placed into a sample holder and inserted into the instrument. Only a small region of the sample grid is viewed at any one time, but the grid can be moved within the instrument so that the whole area of the grid can be analyzed. Various types of holders are available, such as double tilt or tilt-and-rotate, to enable orientation of the specimen with respect to the electron beam. Also, variable temperature holders can allow cooling of the specimen, typically to 100 K (although some modified microscopes have been used to study specimens at ~10 K).

Some electrons will pass straight through the sample, while others will be scattered. The scattering gives rise to contrast in the image, showing differences in sample thickness, density, and composition. TEM has a high spatial resolution, better than 0.1 nm for modern aberration-corrected instruments, allowing imaging of the arrangement of atoms or molecules in the crystal lattice (high-resolution electron microscopy, HREM) [25]. This ability is unlikely to be widely applicable to pharmaceutical samples however, as they would be damaged by the electron beam under the high flux conditions that are used for HREM. Transmission electron microscopes also have a diffraction mode, and electron diffraction is analogous to single-crystal X-ray diffraction; but as the wavelength of electrons (0.0197 Å at 300 kV) is 2 orders of magnitudes shorter than that of X-rays, the Bragg scattering angles are correspondingly smaller. Diffraction patterns are usually obtained by aligning the electron beam with a crystallographic axis, and because samples are not perfectly flat and Bragg angles are small, the resulting diffraction pattern will contain reflections from each of the crystal planes that are perpendicular to the crystallographic axis. Electron diffraction patterns can, as a result, be obtained from very small amounts of sample (<1 fg) [26], and are characteristic of the crystallographic axis and crystal structure from which they were recorded. The patterns can therefore be used for crystal form identification and for measuring the orientation of crystals with respect to the electron beam. It is also possible to solve crystal structures using electron diffraction, but the process is not as straightforward as with single-crystal X-ray diffraction due to difficulties in obtaining accurate reflection intensities because the scattering is no longer kinematic [27]. Switching between TEM imaging and diffraction modes is easily achieved by varying the current through the electron lenses, allowing corresponding image and diffraction data to be obtained, and is a particular strength of the technique. A typical TEM image, showing crystals from a sample of a 1 : 1 caffeine/theophylline co-crystal [28], is given in Figure 4.2 along with the corresponding electron

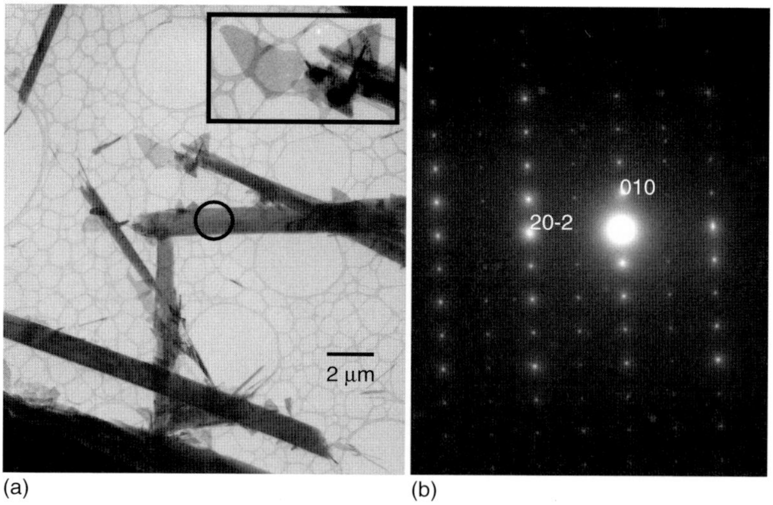

Figure 4.2 (a) TEM image showing lath-shaped crystals of form I of a caffeine/theophylline co-crystal [28] supported on a lacy film of amorphous carbon. The dark region at the bottom left of the image is one of the copper bars of the sample support grid. (b) Corresponding electron diffraction pattern (from the region circled in the TEM image). Adapted from Eddleston *et al.* [28]. Reproduced with permission of Royal Society of Chemistry.

diffraction pattern recorded from the area circled in the image. The co-crystal has a lath-shaped crystal habit and particle sizes ranging from submicrometer to tens of micrometers. The dark lines on these crystals are bend (or extinction) contours and mark regions where the orientation of the crystals with respect to the electron beam fulfills the conditions for Bragg diffraction to occur. Smaller crystallites with a distinctive triangular plate-like habit decorate the laths (see the inset). This triangular habit is characteristic of form II of theophylline, indicating that the sample contains a mixture of crystal forms [29, 30]. The diffraction pattern was indexed to the <101> zone axis of form I of the caffeine/theophylline co-crystal (details of a typical indexing process are given elsewhere) [24]. The combination of crystal habit and diffraction information enabled the determination that the long edge of the crystal lies in the (010) plane. A second diffraction pattern (not shown), recorded from the <001> zone axis of this crystal, demonstrated that the dominant crystal face is (001) and that the crystals are elongated in the [100] direction.

4.2.1.2 **Application of Transmission Electron Microscopy to the Analysis of Pharmaceutical Samples**

Historically, TEM has only rarely been applied to the analysis of pharmaceutical samples, usually for the imaging of an API encapsulated in liposomal, micellar, and polymeric formulations [32–34]; the use of TEM to characterize the solid form of an API itself has been very limited. This probably stems from two potential difficulties with the analysis of organic, molecular materials: sample preparation, and beam damage [35]. Because of the strong interaction of the electron beam with the sample [19], it is required that specimens be very thin (<500 nm even for light organic compounds) [35]. It can be difficult to prepare crystals of this size, and while thin films of metals and minerals can be prepared from larger crystals by focused-ion-beam milling or mechanical polishing, the fragile nature of pharmaceutical molecules often makes this impractical. Additionally, in an electron beam, crystals of organic compounds can undergo bending, amorphization, sublimation, and even melting.

Importantly, strategies have been developed to deal with both sample preparation and beam damage. For example, crystal growth on a water surface has been found to be a good way of preparing plate-like crystals that are sufficiently thin for TEM analysis [36], as have crystallization from the melt [24] and vapor deposition [19, 29]. For compounds that naturally crystallize with a plate-like habit, conventional solution growth can be used, as shown with the crystal of (*RS*)-ibuprofen sodium dihydrate in Figure 4.3a. In addition, small particles can be readily prepared by milling, as shown for a 1 : 1 theophylline/L-malic acid co-crystal in Figure 4.3b [24].

Where it is desirable to analyze crystals in their native form, sample preparation can simply comprise a very gentle crushing of crystals between two glass slides, which avoids the risk of inducing a change in crystal form. Thicker crystals can also be analyzed by TEM, but they tend to appear dark as they are too thick for many electrons to penetrate, meaning that only particle size and habit are likely to be

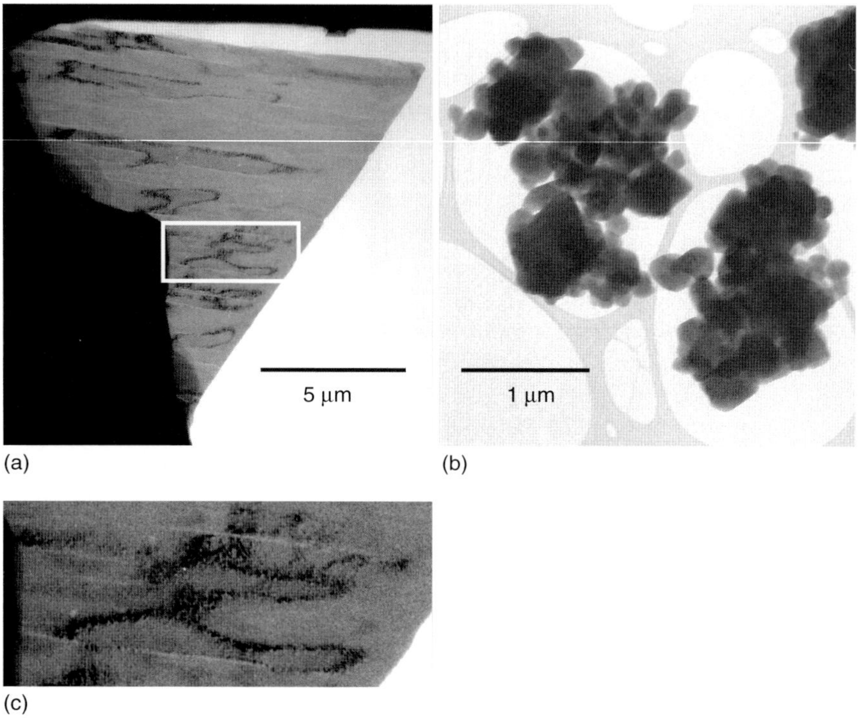

Figure 4.3 TEM images of (a) a crystal of (*RS*)-ibuprofen sodium dihydrate grown by evaporative crystallization from ethanol and (b) submicrometer-sized crystallites of a 1 : 1 cocrystal of theophylline/L-malic acid prepared by liquid-assisted grinding with nitromethane. The white lines running horizontally in the crystal of (*RS*)-ibuprofen sodium dihydrate are cracks, indicating that the crystal has lost water while being held under vacuum in the TEM instrument. The black lines are bend contours, and have a rough appearance due to the presence of crystal defects. (c) Expanded view of the region highlighted in (a).

derivable. It is, however, often possible to obtain electron diffraction patterns (and hence perform crystal phase identification) from such crystals by directing the electron beam to the edges and corners where they are thinner, as demonstrated with a paracetamol crystal in Figure 4.4.

It has been demonstrated that electron beam damage can be reduced by decreasing the flux of electrons through the sample, by increasing the accelerating voltage, or by analyzing samples at low temperatures [23, 30, 35, 37]. Most pharmaceutical samples are sufficiently stable under such conditions to allow useful analysis to be performed before significant deterioration is observed. Other strategies for avoiding beam damage include working with a defocused beam [38, 39], and rastering a focused beam across a sample when generating images so that each region of the sample is exposed only transiently to the beam (scanning transmission electron microscopy (STEM) analysis) [40]. It should be noted, however, that the high-vacuum conditions in TEM instruments are rarely

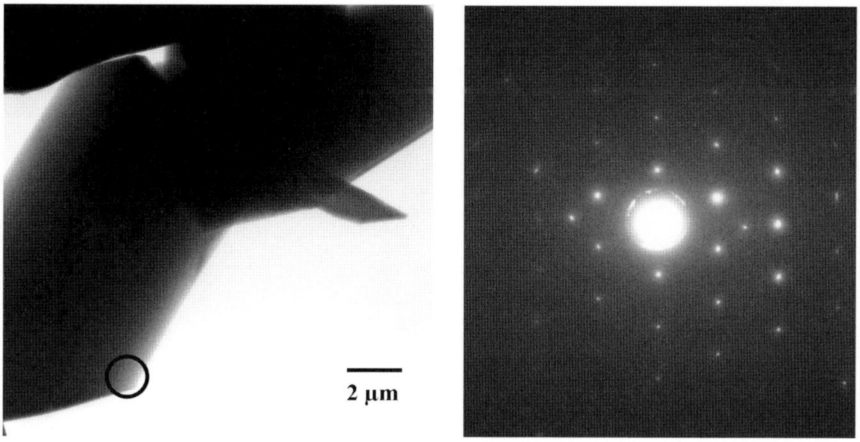

(a) (b)

Figure 4.4 (a) TEM image of crystals of paracetamol with a thickness ≫ 1 μm. It is evident that the crystals have a hexagonal habit and lengths of ~20 μm, but no further detail can be discerned. (b) Corresponding selected area electron diffraction pattern, from the corner of the left-hand crystal (circled), which could be unambiguously indexed to the <21−1> zone axis of form I of paracetamol, thus identifying the polymorphic form of this crystal.

suitable for the study of hydrates and solvates (Figure 4.3a) unless the samples can be cooled to low temperature before exposure to the vacuum.

Through the use of such strategies for reducing electron beam damage, several examples of pharmaceutical characterization using TEM have recently been reported, and the full potential of the technique is beginning to be realized. For example, TEM has been used for imaging crystals of compounds such as dipyridamole [33] and paclitaxel (a compound for which interactions with human intravenous immunoglobulin have also been studied) [41, 42], and liquid crystals of fenoprofen [43]. Imaging with TEM has advantages such as a resolution approximately an order of magnitude greater than with scanning electron microscopy (SEM). Furthermore, as TEM data is taken from electrons that have passed through particles, information from the entire cross section is obtained, making it complementary to other high-magnification imaging techniques such as SEM and AFM, where only surfaces are visualized. In addition, image (real space) and diffraction (reciprocal space) information from pharmaceutical crystals can be combined, making it possible to map crystal structure to crystal habit (see, e.g., Figure 4.2) [30]. This mapping is extremely useful when it is necessary to modify the crystal habit of an API away from problematic needles or plates, and avoids the difficulty of growing large single crystals, which would be a requirement for a similar analysis using X-ray diffraction.

Electron diffraction patterns have been used as a "fingerprint tool" to identify the polymorphic form of pharmaceutical compounds such as paracetamol [30, 44]. With TEM, crystal phase identification requires far less material than is needed for a powder X-ray diffraction (PXRD) measurement, and analysis can be performed on individual crystals. The ability to study crystals one by one during TEM analysis means that all of the phases present in samples can, in theory, be identified, even if they are present in low proportions [45, 46]. This identification is aided by the imaging mode, as crystals of the various forms that are present in a sample can often be distinguished by their different morphologies (as shown in Figure 4.2) [30, 45]. An even more sophisticated approach to phase identification, involving the use of precession electron diffraction [27, 47] and tomographic reconstruction of the reciprocal lattice (automated electron diffraction tomography) [48], has recently been used in a study on the growth of caffeine crystals. Needle-like crystals of form I of caffeine have been observed to initially precipitate as hexagonal plates before growing rapidly in a direction perpendicular to this face, sometimes forming hollow, tubular structures [49]. The study used TEM to determine that the hexagonal plates forming at the initial stages of crystal growth are also form I of caffeine [48].

The combination of precision electron diffraction and tomographic reconstruction can also be applied to the determination of crystal structures, removing the need to grow large crystals as would be required for X-ray methods, and has been used to obtain structures of organic compounds such as 9,9-bianthracene-10-carbonitrile and Pigment Red 53 [40, 50]. In a separate study, a combination of electron diffraction and PXRD was used to determine the unit cell parameters of two polymorphs of roxifiban [51, 52]. Recently, a novel approach to crystal structure determination combining electron diffraction and crystal structure prediction was used to identify a new polymorphic form of theophylline. This analysis was performed on a pictogram-sized crystallite and despite the new polymorph existing as a minor component in a mixture with form II of theophylline [44, 46].

Finally, TEM has been found to be an ideal technique for the characterization of submicrometer- and nanometer-sized crystallites of pharmaceuticals, yielding a range of useful information such as particle size, particle morphology, and crystal phase [30, 45]. Interest in submicrometer-sized particles of pharmaceuticals is increasing as a result of their enhanced properties such as solubility [53–56], and TEM can complement the methods that are currently used for their characterization such as SEM, infrared (IR), and dynamic light scattering [24, 57, 58]. The ability to obtain diffraction data will be particularly useful when small particle size leads to very broad lines in the powder X-ray pattern of a sample.

4.2.1.3 **Characterization of Defects Using Transmission Electron Microscopy**

Crystal defects can be observed using a transmission electron microscope as they give rise to diffraction contrast [24, 59]. As described in the previous chapter, defects cause crystal planes in the vicinity to become distorted and oriented differently than in the bulk of the crystal. The difference in orientation means that, with careful alignment of the electron beam, Bragg diffraction can be induced locally at defect sites, and not in the bulk of the crystal. By using an aperture to selectively cut out diffracted beams, defects will appear as dark regions on the resulting image. In addition, thin organic crystals are often bowed, rather than completely

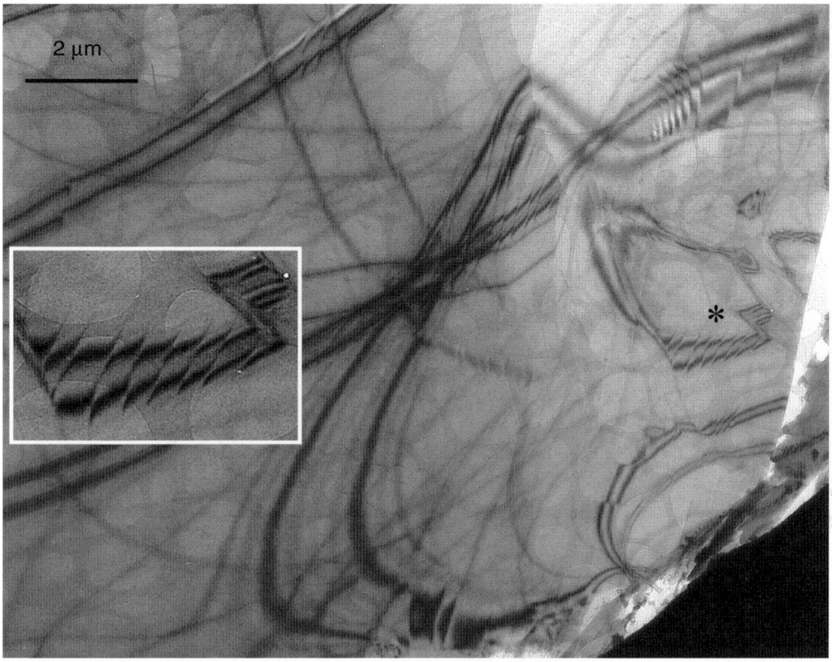

Figure 4.5 TEM image of a crystal of theophylline grown by crystallization from nitromethane. The dark lines on the crystal are bend contours, and are disturbed wherever they cross regions of crystal where defects are located, such as that marked with an asterisk (this area is expanded in the inset).

flat, and the presence of crystal defects can be identified through localized distortions to bend contours, which appear on TEM images in areas of a crystal where the Bragg diffraction conditions are met [23]. At every point along a bend contour, the orientation between the crystal and electron beam is constant, and in the vicinity of defects, crystal planes have a different orientation to that in the bulk crystal, which causes a bend contour to have a discontinuous appearance in this region. The effect of the presence of crystal defects on the appearance of bend contours can be seen in the image of a theophylline crystal in Figure 4.5, in the region marked with an asterisk (this area is magnified in the inset), where the bend contour appears to be cut in several places. When crystals contain a high defect density, bend contours can be heavily disrupted (see Figure 4.3a).

Using TEM it is possible to distinguish different types of defects within crystals [19]. The defects in the theophylline crystal in Figure 4.5, for example, are characteristic of edge or screw dislocations and run for several micrometers parallel to the plate face of the crystal (in the [010] crystallographic direction). By varying the orientation of a crystal with respect to the electron beam and observing how this changes the amount of contrast that a dislocation generates, it is possible to determine the Burger's vector of the dislocation [59], giving a full description of the nature of the dislocation. Stacking faults can often be identified from the characteristic fringes of intensity that they generate in TEM images [24, 60]; or they can even be observed directly in high-resolution images of beam-stable specimens [29]. Grain boundaries and twinning in crystals can be detected from the way bend contours will abruptly end at the boundary between two differently oriented crystalline domains. Further information from twinned crystals, such as the relative orientations of the domains, can be extracted from electron diffraction patterns [60]. It should be noted that the distortions caused by point defects in organic crystals are usually too small to be observed by TEM.

There are several reported examples of the use of TEM to study the nature and quantity of defects in crystals of organic compounds including anthracene [61], pyrene [35], and *p*-terphenyl [36]. Furthermore, it has been identified that polymorphic conversions in 1,8-dichloro-10-methylanthracene, *p*-dichlorobenzene, and pyrene initiate at crystal defects and are martensitic in character [62–64].

4.2.1.4 **Examples of the Study of Defects in Pharmaceutical Crystals Using Transmission Electron Microscopy**

Two crystals of theophylline are shown in Figure 4.6 to demonstrate that different crystallization conditions can influence the amount of defects that are present in

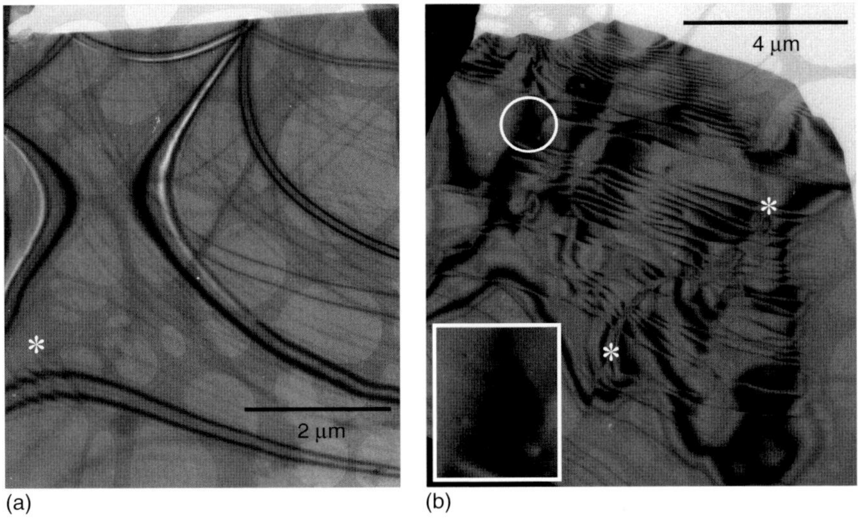

Figure 4.6 TEM images of plate-like crystals of form II of theophylline crystallized at different rates. (a) A crystal prepared by cooling a 2.5 mg ml^{-1} solution of theophylline in nitromethane (the white lines running next to bend contours signify that some parts of the crystal are out of focus). There is only one small region where a bend contours are disrupted by the presence of defects (highlighted with an asterisk). (b) A crystal grown more rapidly by cooling a 5.0 mg ml^{-1} solution of theophylline in nitromethane. The crystal is highly defective, as evidenced by the heavy disruption to bend contours, with many linear (edge or screw) dislocations running approximately horizontally across the crystal in the [010] direction. A further defect is highlighted between the two asterisks, and runs roughly perpendicular to the others. The small dots in the circled region (magnified in the inset) are points where linear dislocations emerge from the crystal surface.

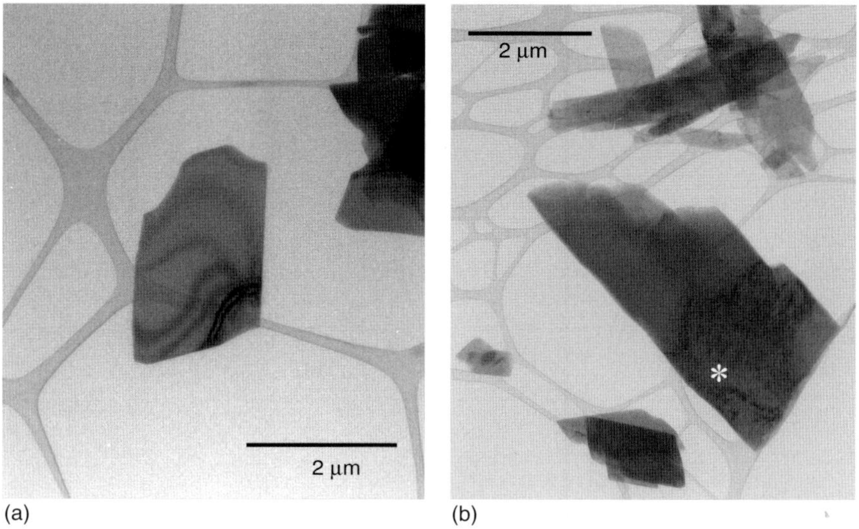

Figure 4.7 TEM images of plate-like crystals of acetazolamide crystallized from acetone. (a) Bend contours on the crystal in this image are continuous, suggesting few defects are present in the crystal. (b) The largest crystal in this image shows significant disruption to bend contours in the region marked with an asterisk, indicating a high defect density.

crystals of a pharmaceutical compound. Both crystals are form II of theophylline, but the rate of crystal growth was higher for that in Figure 4.6b than it was for that in Figure 4.6a, and this has led to a dramatic increase in the number of defects that are evident by TEM.

A similar observation was made for crystals of acetazolamide grown from acetone (Figure 4.7). Although these crystals are all from the same batch, it is evident that the crystal shown in Figure 4.7a has much less defects than those in Figure 4.7b.

The defects shown in Figures 4.6 and 4.7 are primarily linear screw and edge dislocations, but 2D planar defects such as stacking faults and grain boundaries can also be identified in pharmaceutical crystals by TEM. For example, the electron diffraction pattern in Figure 4.8a, recorded from a crystal of a 5-fluorouracil/phenazine co-crystal grown by fast evaporative crystallization from methanol [45], shows extensive streaking, indicative of the presence of stacking faults. In Figure 4.8b, crystals of form I of paracetamol containing discrete crystalline domains that meet at grain boundaries are shown [30].

Importantly, the analysis of defects by TEM can aid in the interpretation of the solid-state behavior of pharmaceutical materials. Crystals of form II of theophylline grow from several solvents with a characteristic triangular habit (Figure 4.9a). If these crystals are left in the solvent, over time they fracture in a direction parallel to the base of the triangles, resulting in trapezoidal fragments (Figure 4.9b). Interestingly, this fracturing does not occur along an

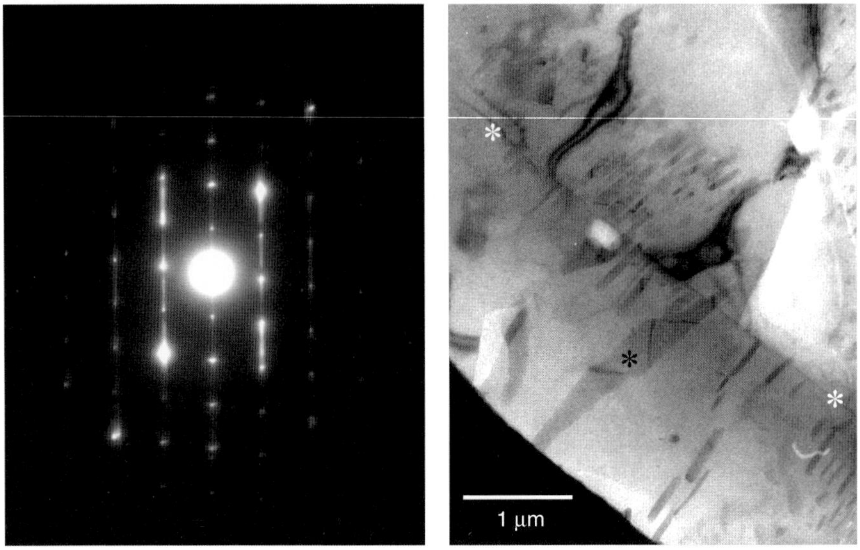

(a) (b)

Figure 4.8 TEM evidence for planar defects in pharmaceutical crystals. (a) Electron diffraction pattern from a 5-fluorouracil/phenazine co-crystal where reflections are streaked in one direction, suggesting the presence of stacking faults in the crystal. (b) TEM image showing a plate-like particle of paracetamol crystallized from acetic acid. There is a line marked with two white asterisks, running from the upper left to the lower right on the image, which bend contours do not cross, suggesting it is a grain boundary. There are also several small, discrete, aligned crystalline domains within the larger paracetamol particle, such as that marked with a black asterisk (it is also possible that these domains are small crystallites growing epitaxially on the surface of the larger crystal).

(a) (b)

Figure 4.9 Polarized light microscopy images of triangular crystals of theophylline (a) immediately after growth and (b) after storage in solvent for several weeks. The crystals fracture over time along the direction of dislocations to give trapezoidal shaped crystallites.

obvious cleavage plane. TEM analysis of theophylline crystals revealed dislocations that commonly run in the [010] crystallographic direction, parallel to the direction of fracture, indicating that they likely have a role in the fracture process [30].

While TEM is arguably the most powerful technique available for identifying and characterizing crystal defects, it should be noted that TEM defect analysis is limited to crystals with a thickness of less than 1 μm, for which diffraction contrast can be observed, meaning that it is unlikely to become routinely applicable to the study of defects in pharmaceutical samples.

4.2.2 X-ray Diffraction Topography

4.2.2.1 Introduction to X-ray Topography

Another technique that exploits diffraction contrast for the imaging of crystals is X-ray diffraction topography (usually shortened to X-ray topography) [65–67], where a small coherent beam of approximately parallel X-rays is directed onto a crystal oriented in such a way that the Bragg diffraction conditions are met for a chosen set of crystal planes. The diffracted beam is directed onto photographic film (or a charge-coupled device (CCD) camera), and an image of the crystal is generated by rastering both the crystal and the detector with respect to the X-ray beam. Dark regions on the image show regions of the crystal where lattice type and/or orientation differs from that in the bulk of the crystal. A schematic of a typical X-ray topography arrangement is shown in Figure 4.10.

The first topographic images of crystals using X-rays were generated in the 1930s [69], and the term "X-ray diffraction topography" was coined in 1944 [70]. The experimental arrangements that are now typically used for laboratory X-ray topography experiments were developed during the 1940s and 1950s [71–74]. It should be noted that synchrotron X-ray sources have also been exploited for topographical analyses, reducing experiment times from hours to minutes, and offering the possibility of using multiwavelength "white" radiation so that diffraction from multiple sets of crystal planes can be probed simultaneously [67, 75]. A comprehensive review of the development of X-ray topography has been given elsewhere [65].

As X-rays cannot be focused, in an X-ray topography instrument it is not possible to magnify images. The resolution of the images is therefore dependent on the spread of the X-ray beam, both before reaching the sample and after diffraction, and on the distance between the source and the detector. Typically, the best resolutions achievable are of the order of 1 μm, which is coincidentally about the best resolution that can be achieved on an X-ray photographic film (pixel sizes of X-ray CCD cameras are approximately an order of magnitude greater) [65, 76, 77]. As a result, it is necessary to use relatively large crystals for generating an image in order that areas of different contrasts will be apparent, and closely spaced features will be indistinguishable. The low resolution of X-ray topography is a weakness of the technique [78].

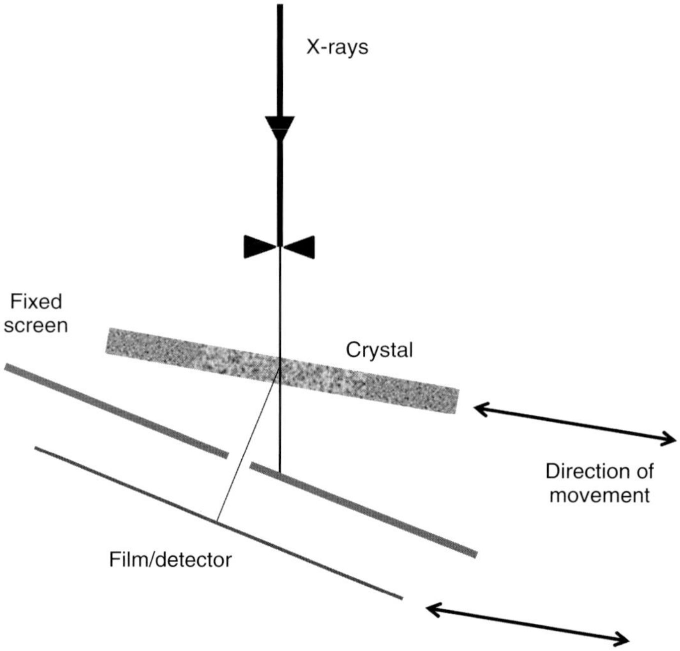

Figure 4.10 Schematic of a typical X-ray topography instrument (based on a diagram by Lang [68]).

4.2.2.2 **Characterization of Defects Using X-ray Topography**

Defects are visible in X-ray topographs as a result of diffraction contrast, as is the case for TEM images. As described above, in an X-ray topography experiment, a crystal is oriented such that the Bragg diffraction conditions are met for one set of lattice planes, and the diffracted beam is used to generate an image. In the presence of defects, there will be localized distortion of the lattice planes, taking them out of the Bragg condition, and giving rise to a dark area on the resulting image.

There are many studies where X-ray topography has been applied to the investigation of crystal defects. For example, various properties of ice crystals – including defect densities at the surface and in the bulk of crystals, the generation of defects at impurities, and the movement of dislocations within crystals – have been determined using this technique [79]. Defects in the room-temperature polymorph of *p*-terphenyl have been characterized, and the polymorphic transformation on cooling was investigated, revealing the existence of two types of domains in the low-temperature form [80, 81]. A phase transition has also been studied in a crystal of copper formate tetrahydrate (Figure 4.11) [78]. X-ray topographs taken during the transformation show the movement of the boundary, which is associated with high strain and is evident as a dark band running across the crystal.

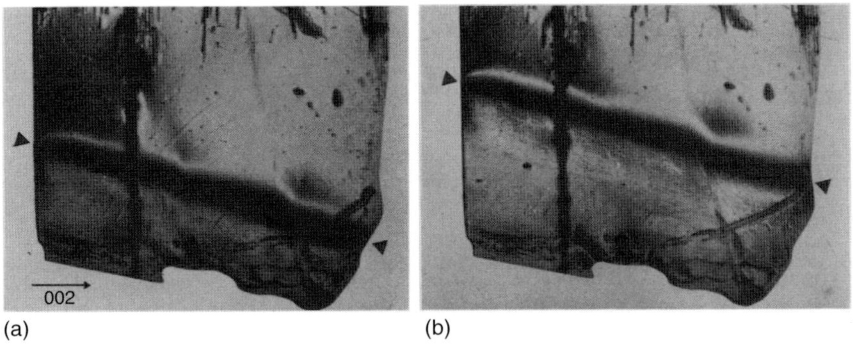

Figure 4.11 X-ray topographs of a phase transition in a crystal of copper formate tetrahydrate at −38 °C. The dark region marked with arrows is the phase boundary, and appears dark because of the high strain caused by two different crystal phases being in contact. The crystal is 9.5 mm across. Adapted from Klapper *et al.* [78]. Reproduced with permission of Elsevier.

4.2.2.3 Application of X-ray Topography to the Analysis of Pharmaceutical Samples

X-ray topography has yet to find widespread use for the analysis of pharmaceutical crystals or, indeed, for the analysis of organic crystals in general. This may in part be due to concerns arising during the early development of the technique that organic samples would be too fragile to withstand the intense X-ray beam [82]. Subsequent studies have, however, shown that organic crystals are usually sufficiently stable for typical X-ray topographic studies, even with the use of synchrotron radiation [82, 83].

The few studies conducted to date on pharmaceutical samples, primarily protein crystals, have employed X-ray topography to gain an increased understanding of the introduction of defects during crystal growth [84, 85] and of the influence of defects on crystal dissolution rates [86]. For example, tetragonal lysozyme crystals have been found to contain relatively few defects if grown under uniform conditions [87]. In contrast, if the crystals are grown in the presence of impurities, or if changes are made to crystallization conditions such as temperature, pH, or salt concentration during crystal growth, there is a dramatic increase in the number of defects observed by X-ray topography [87, 88]. In addition, X-ray topographic analysis of the dehydration of lysozyme crystals has revealed that the loss of water occurs at a front that propagates inward from the external surfaces, introducing cracks and defects into the crystal as it proceeds [89]. Crystals of a second protein, canavalin, have been found to contain high defect concentrations even if grown under uniform conditions [90].

The dissolution of both columnar and plate-like crystals of paracetamol was investigated by measuring rates of erosion at various faces of crystals [91]. With columnar crystals, the dissolution rate at the {001} faces was found to be greater than that at the {110} faces. In contrast, the rate of dissolution of the {110} faces of the plate-like crystals was greater than that of the {001} faces. An explanation for this unexpected observation was obtained from X-ray topography studies, which showed that the {110} growth sectors of the plate-like crystals had a much

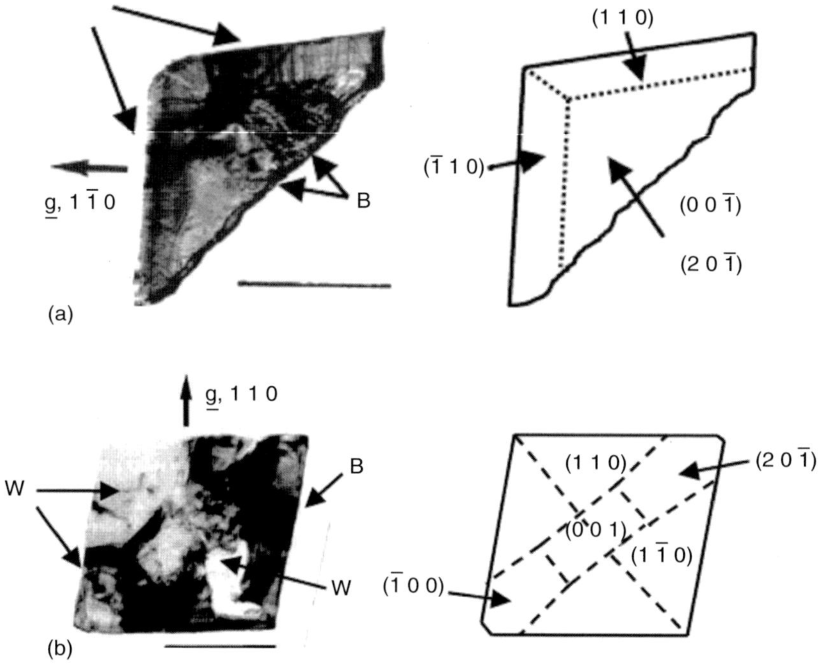

Figure 4.12 X-ray topographs of (a) a fragment of a columnar (rod-like) crystal and (b) a plate-like crystal of form I of paracetamol shown alongside schematics, which indicate where different growth sectors of the crystals are located. Areas of extended strain/high defect content appear black on the images (marked B), or white if the strain is particularly high (marked W). Dislocations are marked with a D, and the scale bars represent 2.5 mm. The {001} growth sector of the columnar crystal is significantly more strained than the corresponding region of the plate-like crystal. In contrast, there are more defects in the {110} sector of the plate-like crystal than in that of the columnar crystal. The density of defects in the respective crystal growth sectors was correlated with dissolution rates. Adapted from Prasad *et al.* [91]. Reproduced with permission of Elsevier.

greater density of defects than the corresponding sector in columnar crystals (Figure 4.12).

4.2.3 Other Significant Methods for Characterizing Defects within Crystals

Optical microscopy, an analytical tool that is widely available to pharmaceutical scientists, has limited applicability to the identification of crystal defects. While point defects, dislocations, and stacking faults would not be expected to be apparent in optical images, the technique can be used to identify gross flaws such as voids and inclusions, twinned crystals, and crystals that are comprised of multiple domains. A similar range of capabilities is associated with laser scattering tomography [92, 93], a technique that enhances contrast from inclusions and voids, making them easier to detect [92], thus having advantages over X-ray topography for the observation of inclusions in protein crystals, which are often highly imperfect with large mosaic spreads, and where diffraction contrasts can be small, making inclusions difficult to identify [94, 95]. Furthermore, dislocations have been identified indirectly in lysozyme crystals through the observation of precipitates decorating these defects [94]. Similarly, strained regions of crystals resulting from the presence of crystal defects have been identified using such techniques as IR polarized light microscopy [96], optical birefringence imaging, and laser confocal differential interference contrast microscopy [97], but the resolution is severely limited because of the wavelength of the applied radiation.

Although crystal defects cannot be directly observed using optical microscopy, the presence of defects can be inferred from their influence on the behavior of crystals, as has been demonstrated during studies on the desolvation of pharmaceutical hydrates [5]. For example, loss of water from needle-shaped caffeine hydrate crystals is accompanied by a change in appearance from clear to frosty. It was observed that dehydration initially occurs at the ends of crystals, with a water loss front then progressing to the center of the crystal in the direction parallel to water channels in the crystal structure. Importantly, it was found that cutting the ends of crystals, an action that was taken to deliberately introduce crystal defects, caused a significant increase in the rate at which the dehydration and corresponding change in appearance from clear to frosty occurred [98]. In addition, the reversible, temperature-dependent $\alpha-\beta$ polymorphic conversion in large single crystals of *p*-dichlorobenzene has been found to be influenced by defects. This transformation was observed to occur by growth of one or more daughter single crystals within the parent crystal. Interestingly, when multiple form change cycles were performed, it was noted that the point in the crystals from which the daughter phase grew remained constant, indicating that nucleation was occurring at specific defect sites within the crystals [31].

4.3 Techniques for Characterizing Defects Emergent at Crystal Surfaces

4.3.1 Atomic Force Microscopy

4.3.1.1 Introduction to Atomic Force Microscopy

AFM is a relatively recent development, first reported in 1986 [99–101]. AFM instruments feature a sharp tip, which is scanned across the surface of the sample under analysis in a manner analogous to a record player needle reading a record (though typically the sample is rastered in linear paths with respect to the tip rather than being rotated). The AFM tip is located at the end of a cantilever onto which a beam of laser light is directed. Whenever the tip encounters a change in the height of the sample surface, the cantilever is bent and the laser beam is displaced. By recording this displacement, a 3D topographical map of the sample surface can be constructed. AFM is so sensitive to changes in height that atomic

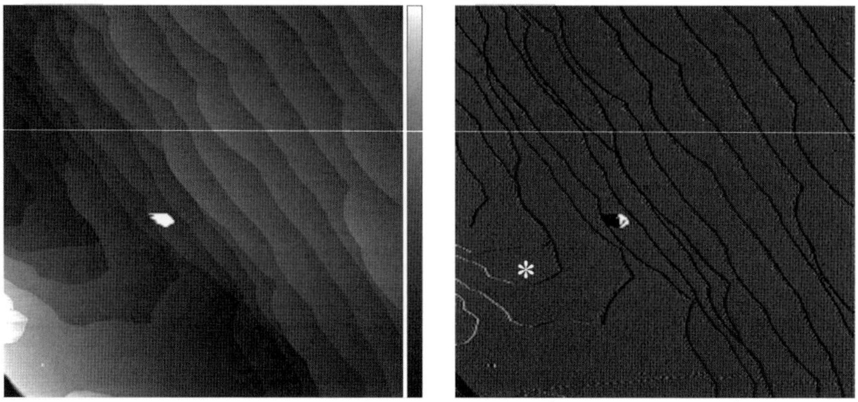

(a) (b)

Figure 4.13 AFM (a) height and (b) amplitude images of a 5 μm^2 region of the surface of a crystal of fenofibrate recorded in tapping mode. The lines on the images mark the edges of layers of fenofibrate molecules. There is an impurity particle on the surface, which appears as a bright white region on both images. The height scale for the left-hand image is 12 nm.

resolution has been achieved for materials such as graphite [102]. Though atomic resolution is unlikely to be possible with pharmaceutical crystals, it is routinely possible to record AFM images showing steps corresponding to the edges of individual layers of molecules (Figure 4.13a). The force of interaction between the tip and sample can also be measured through the generation and analysis of force versus distance plots (force curves).

There are three principal methods of operation with AFM imaging: contact mode, intermittent contact mode (tapping mode), and noncontact mode. In contact mode, the tip and sample are always in contact while the surface is probed, whereas in intermittent contact mode the tip is oscillating up and down as it crosses the sample, potentially causing less damage to crystal surfaces. Noncontact mode requires high-vacuum conditions, which can destabilize sample surfaces, especially those of organic compounds that are prone to sublimation. In the intermittent contact mode, when the tip moves from a lower region of sample to a higher one, the oscillation amplitude decreases, and when moving from a higher to lower region the amplitude increases (there is a feedback loop to ensure that when a change in amplitude is detected an adjustment is made to restore the original amplitude). This information is recorded as an amplitude signal, and can be used to show whether the tip has moved upward or downward on encountering each of the features on the crystal surface. One way of visualizing AFM amplitude images is to imagine that light is shining from either the left or right side of the image, and features will either be *lit up*, or *in shadow*, depending on their orientation. This effect is evident in Figure 4.13b, which is an AFM amplitude image showing lines that correspond to the edges of individual layers of fenofibrate molecules. In the region marked with an asterisk, there is an edge that curves around, meaning that as the AFM tip images this part of the sample, tracing in a horizontal

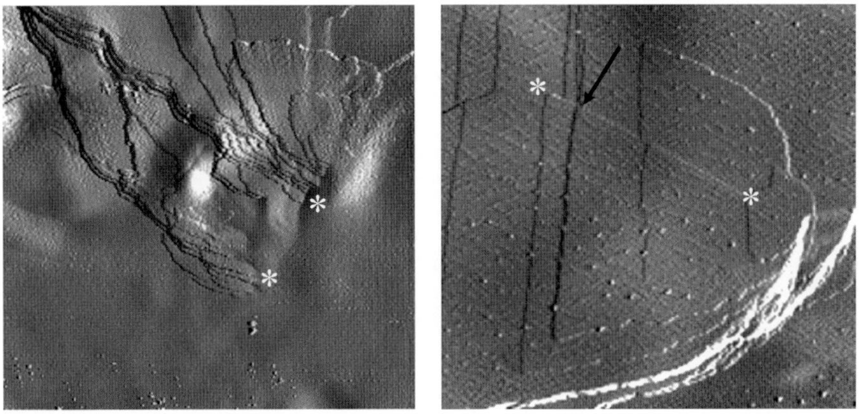

Figure 4.14 AFM amplitude images of (100) surfaces of crystals of form II of theophylline. Both images show a 5.5 μm^2 area, and the lines on the images are the edges of individual layers of theophylline molecules. (a) The asterisks highlight screw dislocations that are located at points where the edges of layers disappear into the crystal surface. (b) The line marked with asterisks is the edge of a layer of theophylline molecules that, between the arrow and the right-hand asterisk, is below the crystal surface, meaning that it must be an edge dislocation in this region.

direction, it will encounter this edge twice, first dropping from a higher crystal layer to the one below, and then rising back up to the higher layer. As a result, this edge appears light to one side of the asterisk and dark to the other.

Reviews of the pharmaceutical uses of AFM are given elsewhere [101, 103–105].

4.3.1.2 Characterization of Defects Using Atomic Force Microscopy

As individual layers of molecules can be observed at crystal surfaces by AFM, it is an ideal technique for observing emergent defects. Screw dislocations, for example, can be observed either directly [106], appearing as a region of crystal at which the edge of a layer of molecules abruptly disappears into the surface (Figure 4.14a), or through the characteristic "hillocks" that result from the growth of new layers of molecules at the dislocation. Subsurface edge dislocations can also be detected [107], as shown in Figure 4.14b. Macroscopic inclusions at crystal surfaces and planar defects such as stacking faults are readily identified [11, 108, 109], but point defects are not typically observable unless the individual molecules are large enough that their positions can be detected by AFM [110, 111], as shown for the protein thaumatin in Figure 4.15a [11].

4.3.1.3 Examples of the Study of Defects in Pharmaceutical Crystals Using Atomic Force Microscopy

AFM is probably the method that has been most widely applied to the study of defects in pharmaceutical crystals, in a majority of cases to aid understanding of the growth of protein crystals.

(a) (b)

Figure 4.15 (a) AFM image showing individual molecules on the surface of a thaumatin crystal. Point defects (vacancies) appear as dark regions. (b) A double screw dislocation on the surface of a crystal of canavalin. Malkin *et al.* [11]. Reproduced with permission of Elsevier.

Defect densities at the surfaces of protein crystals have been calculated for various proteins using AFM, and found to vary enormously. The density of screw dislocations on canavalin crystals, for example, was estimated to be $10^5 - 10^6$ cm^{-2}, a value several orders of magnitude greater than for typical small-molecule crystals [109, 112]. A double screw dislocation on a canavalin crystal surface is shown in Figure 4.15b. In stark contrast, crystals of other proteins such as thaumatin and catalase were found to be completely free of screw dislocations [112].

Interestingly, these differences in dislocation densities have a profound influence on the growth mechanism of protein crystals. The addition of new layers of molecules has been found to occur by a 2D nucleation mechanism for most proteins, including thaumatin and catalase, whereas for canavalin the growth of new layers of molecules initiates at screw dislocations [112, 113]. Similarly, AFM of insulin crystals revealed the presence of screw dislocations, and, again, this was found to be where the growth of new layers of molecules occurs [114, 115].

Uric acid crystals are often one of the components of kidney stones, and a study has been conducted with the aim of understanding the mechanism by which these crystals grow [116]. AFM was used to follow the development of crystals in real time, allowing growth rates to be calculated and revealing that crystals typically have small numbers of screw dislocations on the dominant (100) face at which layers of molecules initiate. In a separate study, L-cysteine crystals, another component of kidney stones, were studied by AFM and also found to have a screw dislocation growth mechanism [117]. Furthermore, it was found that on addition of analogs of L-cysteine (L-cysteine methylester or L-cysteine dimethylester) to the crystallization medium, the crystal growth rate at the screw dislocations, as measured by AFM, was dramatically reduced, indicating a potential pathway for kidney stone prevention.

4.3.2 Surface Etching

4.3.2.1 Introduction to Chemical Etching

Chemical etching is simply a process where a liquid is applied to a particle surface to induce partial dissolution. The liquid, or etchant, could be water, an organic solvent, a solution containing a dissolved solid, an acid/base, or even a chiral liquid [118, 119].

Etching was initially used to study crystal symmetry [120], but it was subsequently realized that etch pits are likely to form preferentially at the sites of emergent dislocations [120], which are high-energy sites on a crystal surface. The link between etch pits and sites of emergent dislocations was elegantly proven by cleaving a crystal of calcite into two halves, thus bisecting dislocations present in the crystal and showing that etch pits on the two halves of the crystal corresponded. Furthermore, if stress was temporarily applied to a calcite crystal during etching, which would be expected to cause a dislocation to move across the crystal surface, some etch pits stopped becoming deeper while new etch pits started to grow [120].

4.3.2.2 Characterization of Defects Using Chemical Etching

A key advantage of etching for the study of crystal defects is that it does not require any specialist instrumentation for its use. While optical microscopy is not an appropriate technique for observing defects directly, it can be used to view etch pits on crystal surfaces, and hence to observe defects indirectly. Other methods that have been used for visualizing etch pits are interference microscopy, SEM, and AFM [119, 121, 122]. The shape of an etch pit can yield information such as the type and direction of the corresponding dislocation [119, 123], and is dependent on the nature of the etchant used in its generation [120].

Etching has been extensively applied to the study of crystal defects, and rather than attempt to summarize this work here, we will describe three studies where etching has been used to investigate the influence of crystal defects on a pharmaceutically relevant particle property. First, defects have been linked to the dissolution rate of potassium perchlorate crystals [124]. The crystals in question were prepared in a column of silica gel, and it was noted that the growth rate of crystals at the top of the column was greater than that of crystals at the bottom. As a result, the crystals from the top of the column were found to be more defective, as evidenced by a higher density of etch pits when subjected to a chemical etchant. Furthermore, the dissolution rates the potassium perchlorate crystals from different parts of the column were measured, using standard United States Pharmacopeia (USP) conditions and equivalent sample weights and surface areas, and it was found that those from the top of the column dissolved fastest [124]. Second, a large crystal of a hydrated alum was cleaved in order to investigate the dehydration behavior [125]. One half of the crystal was etched in order to locate dislocations, while the other half of the crystal was held under vacuum. It was observed that the sites of nucleation of the anhydrous alum on one half of crystal almost exactly matched the pattern of etch pits on the other half of the crystal [125].

Third, a similar approach has been used to investigate the chemical reactivity of anthracene [123]. A large, melt-grown crystal of anthracene was cleaved, with one half being etched to locate emergent dislocations, and the other half irradiated with UV light. Crystallites of anthracene dimer were observed on the irradiated half of the crystal in largely the same positions as etch pits on the other half.

4.3.2.3 Examples of the Study of Defects Emergent at the Surfaces of Pharmaceutical Crystals Using Chemical Etching

Despite being a relatively straightforward method for studying crystal surfaces, there are very few examples of the application of etching to pharmaceutical samples. On a few occasions, the etch patterns produced by partial dissolution of crystal surfaces have been studied to gain a better understanding of dissolution processes. For example, etch patterns on the (100) surface of aspirin and the (010) surface of glycine resulting from partial dissolution in several solvents were measured using AFM and compared with computational models of surface dissolution [121]. The shape of etch pits on the dominant surface of paracetamol crystals was found to be strongly dependent on the solvent of dissolution, in contrast to phenacetin where etch pits were elongated in the [001] direction for all solvents investigated [126]. A comparison of the etch pits that form on the surfaces of two polymorphs of paracetamol, forms I and II, has also been made [127]. A further study into paracetamol dissolution investigated the etch patterns that formed on several crystal surfaces of form I. Although differences in etch pit densities were noted, there was also an isotropic distribution of pits across the crystal surfaces, suggesting that they were formed by 2D nucleation rather than specifically at the sites of emergent dislocations (Figure 4.16) [91].

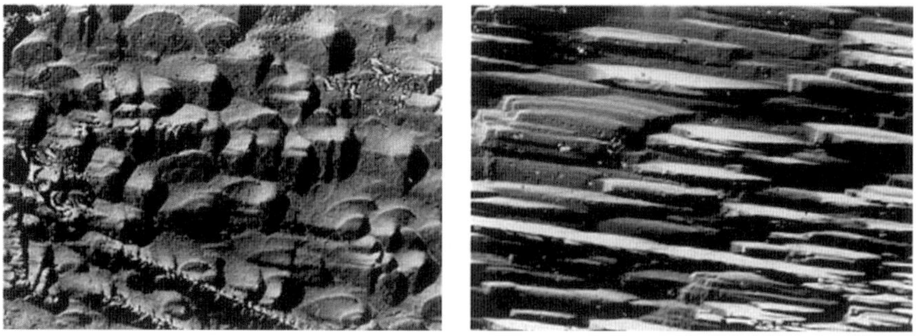

(a) (b)

Figure 4.16 (a and b) Etch pits on the (201) and (001) surfaces of paracetamol after the application of a slightly undersaturated paracetamol solution. The etch pit densities were calculated to be 4×10^5 and 1×10^5 cm^{-2}, respectively. Prasad *et al.* [91]. Reproduced with permission of Elsevier.

4.3.3
Other Relevant Methods for Characterizing Defects Emergent at Crystal Surfaces

It has been reported that the emergence of defects at crystal surfaces can be detected by laser scattering tomography [128]. In addition, it may also be possible to observe defects using SEM, an analytical method that is widely available in the pharmaceutical sector, appearing as a small region of contrast at the point at which they meet a crystal surface [129]. The growth hillocks associated with screw dislocations are also sometimes evident with SEM [130].

4.4
Techniques for Quantifying Defect Densities within Crystals

A number of methods have been reported that, while not enabling defects to be observed and characterized directly, can be used to measure the relative numbers of defects in crystals. These methods typically probe deviations in crystal lattice orientations or increases in the total energies of crystals that result from the presence of defects. For example, reflections in PXRD patterns are broadened as a result of crystal defects, especially those at high angles [4, 13, 131]. The introduction of crystal defects into raffinose pentahydrate crystals during dehydration has been studied using this effect, and it was even possible to distinguish defect content from amorphous content in the resulting material [13]. A variation on this theme, grazing incidence (or glancing angle) diffraction [132, 133], where PXRD traces are recorded at different depths through a sample, has been used to investigate how uniformly defects are introduced into crystals during tableting, revealing that crystals at the surfaces of tablets tend to be more defective than those located toward the center [4]. In a separate study, the generation of crystal defects during the dehydration of α-lactose monohydrate was quantified by measuring the corresponding changes in X-ray intensity [14]. The introduction of defects into crystals has also been studied using thermal methods, for example, differential scanning calorimetry or differential thermal analysis, to measure changes in the entropy of samples after processing relative to a crystalline reference sample (entropy of processing) [9, 10]. The entropy of processing was measured for two polymorphs of chloramphenicol palmitate, forms A and B, before and after grinding [10]. Form A, the stable polymorph, had the lower initial entropy as might be expected, but on grinding the introduction of defects and disorder was much more pronounced in this form than in metastable form B, as evidenced by a greater increase in entropy of disorder.

Though point defects are rarely directly observable in organic crystals, the quantification of such defects in crystals has been widely achieved using positron annihilation [134–136]. Additionally, for the specific case of point defects, where molecules adopt a different conformation to that seen in the bulk crystal, the analytical tools of PXRD, differential scanning calorimetry, and solid-state

NMR have been used to investigate batch-to-batch variability in defect contents, exploiting the influence of these defects on lattice parameters, enthalpy of fusion, and melting point [3].

It was observed, during a study that investigated the generation of crystal defects during compaction and cryo-milling, that solid-state NMR relaxation times become shorter as crystal defects and amorphous content in crystals increase [7], though it has yet to be determined whether this effect could be used to quantify defect densities in crystals.

4.5
The Complementarity of Techniques for Characterizing Defects

It should be noted that, when investigating the influence of crystal defects on the kinetics of processes such as chemical degradation or polymorphic form changes, it is not necessarily the total number of defects within a crystal that is the important parameter. More critical could be the presence, or absence, of particular defects at which molecules are locally perfectly aligned to react, or are arranged as they would be in a different polymorphic form of the compound, and so provide nucleation sites at which processes can initiate [123]. In general, techniques that can be used to observe and characterize defects are not ideal for quantifying defect densities as, in the case of TEM and AFM for example, only small amounts of crystal are sampled, and with X-ray topography and chemical etching individual defects are not resolved. Clearly, there are advantages to the use of a combination of techniques that can be used to determine both the type of defects that are present in a material and the total number of defects.

There is an obvious complementarity between TEM and X-ray topography for the analysis of defects within crystals in that the former can be applied only to small crystals whereas the latter requires large crystals. In addition, the high resolutions possible with TEM allow characterization of individual defects in small regions of sample, while X-ray topography can often give a better overview of the distribution of defects in crystals as a whole. Similarly, at surfaces, AFM is ideal for identifying individual defects, whereas etching is better for studying defect distributions over whole crystals. X-ray topography also offers benefits over TEM for the analysis of solvates (which would desolvate in the vacuum of the TEM instrument) and samples for which beam damage is a particular issue.

Combining bulk and surface techniques can also be useful, as shown in Figure 4.17. The TEM image of a crystal of form I of nifedipine in Figure 4.17a, taken with the beam aligned with the [31] direction, shows a bend contour disrupted by defects that run approximately parallel to the dominant (100) plate face of the crystal. On studying the (100) face by AFM, features characteristic of emergent screw dislocations were observed (Figure 4.17b). These screw dislocations point in the [31] direction, perpendicular to the direction of the dislocations seen by TEM, and so are unlikely to be the same type, meaning that

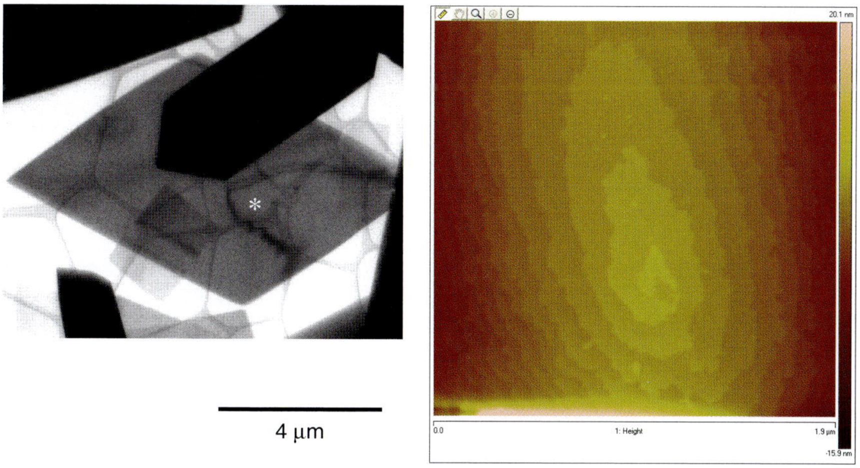

Figure 4.17 Analysis of defects in crystals of form I of nifedipine prepared by evaporation from ethanol. (a) TEM image showing a rectangular plate crystal with disruption to a bend contour in the region marked with an asterisk, indicating the presence of defects that run parallel to the crystal surface. (b) AFM height image of the (100) surface showing a screw dislocation hillock at the surface of a nifedipine crystal. Each step was measured to be approximately 12 Å in height, corresponding to one layer of molecules. The screw dislocation is emergent from the crystal surface at the center of the hillock. The defects in (a) and (b) are different, showing the importance of applying multiple techniques to the identification of defects in crystals.

a better understanding of defects in nifedipine crystals was gained by using TEM and AFM in conjunction.

4.6 Summary and Outlook

This chapter has presented the principal techniques for the characterization of crystal defects, namely TEM, X-ray topography, AFM, and chemical etching, and described their application for the characterization of pharmaceutical samples.

In practical terms, it is not yet clear how widely applicable each of these methodologies will prove to be for pharmaceutical samples, and they should be used in combination as it is unlikely that any one in isolation will give a comprehensive picture of the type and amount of defects that are present in a given batch of crystals.

Increased use of these techniques within the pharmaceutical sector would certainly bring benefits. From a fundamental point of view, there is the potential to transform our understanding of both the nature of defects in the crystals of pharmaceutical type molecules and their influence on the solid-state behavior of pharmaceutical samples. Additionally, if it becomes routinely possible to measure variability in the defect contents of different batches of drug substance or drug product and correlate this variability with differences in the solid-state properties of the batches, it could enable scientists to achieve a new level of control over drug product behavior.

Acknowledgment

The authors are grateful to the EPSRC and the EU INTERREG IVA 2 Mers-Seas-Zeeën Cross-border Cooperation Programme for funding this work.

References

1. Goldbeck, G., Pidcock, E., and Groom, C. (2011) Polymorphism in Pharmaceutical Solids, 2nd Edn (ed H.G. Brittain), Informa Healthcare, Inc., USA, New York, USA.
2. Bauer, J., Spanton, S., Henry, R., Quick, J., Dziki, W., Porter, W., and Morris, J. (2001) Ritonavir: an extraordinary example of conformational polymorphism. *Pharm. Res.*, **18**, 859–866.
3. Byard, S.J., Jackson, S.L., Smail, A., Bauer, M., and Apperley, D.C. (2005) Studies on the crystallinity of a pharmaceutical development drug substance. *J. Pharm. Sci.*, **94**, 1321–1335.
4. Koivisto, M., Heinaenen, P., Tanninen, V.P., and Lehto, V.-P. (2006) Depth profiling of compression-induced disorders and polymorphic transition on tablet surfaces with grazing incidence X-ray diffraction. *Pharm. Res.*, **23**, 813–820.
5. Byrn, S.R. (1982) *Solid-State Chemistry of Drugs*, Academic Press, New York.
6. Qiu, Y., Chen, Y., and Zhang, G.G.Z. (2009) *Developing Solid Oral Dosage Forms: Pharmaceutical Theory and Practise*, Academic Press, Burlington, MA.
7. Lubach, J.W., Xu, D., Segmuller, B.E., and Munson, E.J. (2007) Investigation of the effects of pharmaceutical processing upon solid-state NMR relaxation times and implications to solid-state formulation stability. *J. Pharm. Sci.*, **96**, 777–787.
8. Huettenrauch, R. and Fricke, S. (1981) The influence of lattice defects on the course and rate of drying of particulate solids. *Int. J. Pharm. Technol. Prod. Manuf.*, **2**, 35–37.
9. Grant, D.J.W. and York, P. (1986) Entropy of processing: a new quantity for comparing the solid state disorder of pharmaceutical materials. *Int. J. Pharm.*, **30**, 161–180.
10. Duddu, S.P. and Grant, D.J.W. (1995) The use of thermal analysis in the assessment of crystal disruption. *Thermochim. Acta*, **248**, 131–145.
11. Malkin, A.J., Kuznetsov, Y.G., and McPherson, A. (1996) Defect structure of macromolecular crystals. *J. Struct. Biol.*, **117**, 124–137.
12. Bates, S., Zografi, G., Engers, D., Morris, K., Crowley, K., and Newman, A. (2006) Analysis of amorphous and nanocrystalline solids from their X-ray diffraction patterns. *Pharm. Res.*, **23**, 2333–2349.
13. Bates, S., Kelly, R.C., Ivanisevic, I., Schields, P., Zografi, G., and Newman, A.W. (2007) Assessment of defects and amorphous structure produced in raffinose pentahydrate upon dehydration. *J. Pharm. Sci.*, **96**, 1418–1433.
14. Huttenrauch, R. and Keiner, I. (1979) Producing lattice defects by drying processes. *Int. J. Pharm.*, **2**, 59–60.
15. Brittain, H.G., Bogdanowich, S.J., Bugay, D.E., DeVincentis, J., Lewen, G., and Newman, A.W. (1991) Physical characterization of pharmaceutical solids. *Pharm. Res.*, **8**, 963–973.
16. Storey, R.A. and Ymen, I. (2011) *Solid State Characterization of*

Pharmaceuticals, Blackwell Publishing, West Sussex.
17. Ohannesian, L. and Streeter, A.J. (2002) *Handbook of Pharmaceutical Analysis*, Marcel Dekker, New York.
18. Brydson, R. (2011) *Aberration-Corrected Analytical Electron Microscopy*, 1st edn, John Wiley & Sons, Ltd., Chichester.
19. Reimer, L. and Kohl, H. (2008) *Transmission Electron Microscopy: Physics of Image Formation*, 5th edn, Springer, New York.
20. Hirsch, P., Cockayne, D., Spence, J., and Whelan, M. (2006) 50 years of TEM of dislocations. Past, present and future. *Philos. Mag.*, **86**, 4519–4528.
21. Wilkes, P. (1973) *Solid State Chemistry in Metallurgy*, Cambridge University Press, Cambridge.
22. McLaren, A.C. (1991) *Transmission Electron Microscopy of Minerals and Rocks*, 2nd edn, Cambridge University Press, Cambridge.
23. Champness, P.E. (1977) Transmission electron microscopy in earth science. *Annu. Rev. Earth Planet. Sci.*, **5**, 203–226.
24. Czichos, H., Saito, T., and Smith, L. (eds) (2006) *Springer Handbook of Materials Measurement Methods*, Springer.
25. Urban, K.W. (2009) Is science prepared for atomic-resolution electron microscopy? *Nat. Mater.*, **8**, 260–262.
26. Thomas, J.M. and Midgley, P.A. (2004) High-resolution transmission electron microscopy: the ultimate nanoanalytical technique. *Chem. Commun.*, 1253–1267.
27. Vincent, R. and Midgley, P.A. (1994) Double conical beam-rocking system for measurement of integrated electron diffraction intensities. *Ultramicroscopy*, **53**, 271–282.
28. Eddleston, M.D., Lloyd, G.O., and Jones, W. (2012) Cocrystal dissociation and molecular demixing in the solid state. *Chem. Commun.*, **48**, 8075–8077.
29. Roy, C., Vega-Gonzalez, A., and Subra-Paternault, P. (2007) Theophylline formulation by supercritical antisolvents. *Int. J. Pharm.*, **343**, 79–89.
30. Eddleston, M.D., Bithell, E.G., and Jones, W. (2010) Transmission electron microscopy of pharmaceutical materials. *J. Pharm. Sci.*, **99**, 4072–4083.
31. Kitaigorodskii, A.I., Mnyukh, Y.V., and Asadov, Y.G. (1965) Relations for single-crystal growth during polymorphic trans- formation. *Phys. Chem. Solids*, **26**, 463–472.
32. Almgren, M., Edwards, K., and Karlsson, G. (2000) Cryo transmission electron microscopy of liposomes and related structures. *Colloids Surf., A*, **174**, 3–21.
33. Tang, Y., Liu, S.Y., Armes, S.P., and Billingham, N.C. (2003) Solubilization and controlled release of a hydrophobic drug using novel micelle-forming ABC triblock copolymers. *Biomacromolecules*, **4**, 1636–1645.
34. Kim, I.-S. and Kim, S.-H. (2003) Development of polymeric nanoparticulate drug delivery systems: evaluation of nanoparticles based on biotinylated poly(ethylene glycol) with sugar moiety. *Int. J. Pharm.*, **257**, 195–203.
35. Jones, W. and Thomas, J.M. (1979) Applications of electron microscopy to organic solid-state chemistry. *Prog. Solid State Chem.*, **12**, 101–124.
36. Jones, W., Thomas, J.M., Williams, J.O., and Hobbs, L.W. (1975) Electron microscopic studies of extended defects in organic molecular crystals. I. p-Terphenyl. *J. Chem. Soc., Faraday Trans. 2*, **71**, 138–145.
37. Ohno, T., Sengoku, M., and Arii, T. (2002) Measurements of electron beam damage for organic crystals in a high voltage electron microscope with image plates. *Micron*, **33**, 403–406.
38. Morniroli, J.P. (2006) CBED and LACBED characterization of crystal defects. *J. Microsc.*, **223**, 240–245.
39. Morniroli, J.P., Houdellier, F., Roucau, C., Puiggali, J., Gesti, S., and Redjaimia, A. (2008) LACDIF, a new electron diffraction technique obtained with the LACBED configuration and a Cs corrector: comparison with electron precession. *Ultramicroscopy*, **108**, 100–115.
40. Kolb, U., Gorelik, T.E., Mugnaioli, E., and Stewart, A. (2010) Structural

characterization of organics using manual and automated electron diffraction. *Polym. Rev.*, **50**, 385–409.
41. Liu, Y., Yang, Z., Du, J., Yao, X., Zheng, X., Lei, R., Liu, J., Hu, H., and Li, H. (2008) Interaction of Taxol with intravenous immunoglobulin: an inhibition of Taxol from crystallizing in aqueous solution. *Int. Immunopharmacol.*, **8**, 390–400.
42. Szebeni, J., Alving, C.R., Savay, S., Barenholz, Y., Priev, A., Danino, D., and Talmon, Y. (2001) Formation of complement-activating particles in aqueous solutions of Taxol: possible role in hypersensitivity reactions. *Int. Immunopharmacol.*, **1**, 721–735.
43. Rades, T. and Mueller-Goymann, C.C. (1997) Electron and light microscopical investigation of defect structures in mesophases of pharmaceutical substances. *Colloid. Polym. Sci.*, **275**, 1169–1178.
44. Eddleston, M.D., Hejczyk, K.E., Bithell, E.G., Day, G.M., and Jones, W. (2013) Polymorph identification and crystal structure determination by a combined crystal structure prediction and transmission electron microscopy approach. *Chem.–Eur. J.*, **19**, 7874–7882.
45. Delori, A., Eddleston, M.D., and Jones, W. (2013) Cocrystals of 5-fluorouracil. *CrystEngComm*, **15**, 73–77.
46. Eddleston, M.D., Hejczyk, K.E., Bithell, E.G., Day, G.M., and Jones, W. (2013) Determination of the crystal structure of a new polymorph of theophylline. *Chem.–Eur. J.*, **19**, 7883–7888.
47. Eggeman, A.S. and Midgley, P.A. (2012) in *Advances in Imaging and Electron Physics* (ed P.W. Hawkes), Elsevier Ltd, pp. 1–63.
48. Gorelik, T.E., Sarfraz, A., Kolb, U., Emmerling, F., and Rademann, K. (2012) Detecting crystalline nonequilibrium phases on the nanometer scale. *Cryst. Growth Des.*, **12**, 3239–3242.
49. Eddleston, M.D. and Jones, W. (2010) Formation of tubular crystals of pharmaceutical compounds. *Cryst. Growth Des.*, **10**, 365–370.
50. Gorelik, T., Schmidt, M.U., Bruning, J., Beko, S., and Kolb, U. (2009) Using electron diffraction to solve the crystal structure of a laked azo pigment. *Cryst. Growth Des.*, **9**, 3898–3903.
51. Li, Z.G., Harlow, R.L., Foris, C.M., Li, H., Ma, P., Vickery, R.D., Maurin, M.B., and Toby, B.H. (2002) New applications of electron diffraction in the pharmaceutical industry: polymorph determination by using a combination of electron diffraction and synchrotron X-ray powder diffraction techniques. *Microsc. Microanal.*, **8**, 134–138.
52. Li, Z.G., Harlow, R.L., Foris, C.M., Li, H., Ma, P., Vickery, R.D., Maurin, M.B., and Toby, B.H. (1999) Polymorph determination for the GP IIb/IIIa antagonist, roxifiban, using a combination of electron diffraction and synchrotron X-ray powder diffraction techniques. *J. Pharm. Sci.*, **88**, 297–301.
53. Katteboinaa, S., Chandrasekhar, V.S.R., and Balaji, S. (2009) Drug nanocrystals: a novel formulation approach for poorly soluble drugs. *Int. J. PharmTech. Res.*, **1**, 682–694.
54. Colombo, I., Grassi, G., and Grassi, M. (2009) Drug mechanochemical activation. *J. Pharm. Sci.*, **98**, 3961–3986.
55. Zhang, H., Wang, D., Butler, R., Campbell, N.L., Long, J., Tan, B., Duncalf, D.J., Foster, A.J., Hopkinson, A., Taylor, D., Angus, D., Cooper, A.I., and Rannard, S.P. (2008) Formation and enhanced biocidal activity of water-dispersable organic nanoparticles. *Nat. Nanotechnol.*, **3**, 506–511.
56. Bucar, D.-K. and Macgillivray, L.R. (2007) Preparation and reactivity of nanocrystalline cocrystals formed via sonocrystallization. *J. Am. Chem. Soc.*, **129**, 32–33.
57. Signorell, R., Kunzmann, M.K., and Suhm, M.A. (2000) FTIR investigation of non-volatile molecular nanoparticles. *Chem. Phys. Lett.*, **329**, 52–60.
58. Chu, B. and Liu, T. (2000) Characterization of nanoparticles by scattering techniques. *J. Nanopart. Res.*, **2**, 29–41.
59. Putnis, A. (1992) *Introduction to Mineral Sciences*, Cambridge University Press, Cambridge.

60. Cowley, J.M. (1993) *Electron Diffraction Techniques*, vol. **2**, Oxford University Press, Oxford.
61. Parkinson, G.M., Thomas, J.M., Goringe, M.J., and Smith, D.A. (1979) Planar faults in crystalline anthracene. *Chem. Phys. Lett.*, **63**, 436–437.
62. Jones, W., Thomas, J.M., and Williams, J.O. (1975) Electron and optical microscopic studies of a stress-induced phase transition in 1,8-dichloro-10-methylanthracene. *Philos. Mag.*, **32**, 1–11.
63. Reynolds, P.A. (1977) Martensitic phase transitions in molecular crystals – p-dichlorobenzene. *Acta Crystallogr. A*, **A33**, 185–191.
64. Jones, W. and Cohen, M.D. (1977) Transmission electron microscopic study of the low temperature phase transformation in pyrene. *Mol. Cryst. Liq. Cryst.*, **41**, 103–107.
65. Raghothamachar, B., Dhanaraj, G., Bai, J., and Dudley, M. (2006) Defect analysis in crystals using X-ray topography. *Microsc. Res. Tech.*, **69**, 343–358.
66. Tanner, B.K. (1977) in *Specialist Periodical Reports, Surface and Defect Properties of Solids*, vol. **6** (eds M.W. Roberts and J.M. Thomas), Royal Society of Chemistry, London, UK, pp. 280–307.
67. Bonse, U. (2008) in *Developments in X-Ray Tomography VI* (ed S.R. Stock), SPIE, p. 596.
68. Lang, A.R. (1959) The projection topograph: a new method in x-ray diffraction microradiography. *Acta Crystallogr.*, **12**, 249–250.
69. Berg, V.W. (1931) Uber eine rontgenographische methode zur untersuchung von gitterstorungen an kristallen. *Naturwissenschaften*, **19**, 391–396.
70. Ramachandran, G.N. (1944) X-ray topographs of diamond. *Proc. – Indian Acad. Sci., Sect. A*, **19A**, 280–292.
71. Lang, A.R. (1957) A method for the examination of crystal sections using penetrating characteristic X-radiation. *Acta Metall.*, **5**, 358–364.
72. Lang, A.R. (1958) Direct observation of individual dislocations by X-ray diffraction. *J. Appl. Phys.*, **29**, 597–598.
73. Bond, W.L. and Andrus, J. (1952) Structural imperfections in quartz crystals. *Am. Mineral.*, **37**, 622–632.
74. Barrett, C.S. (1945) A new microscopy and its potentialities. *Trans. Am. Inst. Min. Metall. Pet. Eng.*, **1865**, 15–65.
75. Tuomi, T., Naukkarinen, K., and Rabe, P. (1974) Use of synchrotron radiation in X-ray diffraction topography. *Phys. Status Solidi A*, **25**, 93–106.
76. Lovelace, J.J., Soares, A.S., Bellamy, H.D., Sweet, R.M., Snell, E.H., and Borgstahl, G.E.O. (2004) First results of digital topography applied to macromolecular crystals. *J. Appl. Crystallogr.*, **37**, 481–485.
77. Wako, K., Kimura, K., Yamamoto, Y., Sawaura, T., Shen, M., Tachibana, M., and Kojima, K. (2012) Digital topography with an X-ray CCD camera for characterizing perfection in protein crystals. *J. Appl. Crystallogr.*, **45**, 1009–1014.
78. Klapper, H., Roberts, K.J., Goetz, D., and Herres, N. (1983) X-ray topographic investigations of phase transitions in crystals. *J. Cryst. Growth*, **65**, 621–636.
79. Jones, S.J. and Gilra, N.K. (1973) X-ray topographical study of dislocations in pure and hydrogen fluoride-doped ice. *Philos. Mag.*, **27**, 457–472.
80. Dudley, M., Disalvo, R., Hou, S.Y., Foxman, B.M., and Jones, W. (1992) Characterization of defects in p-terphenyl single crystals. *Mol. Cryst. Liq. Cryst.*, **211**, 35–42.
81. Dudley, M., Disalvo, R., Wu, J., Gordon-Smith, D., and Jones, W. (1992) Synchrotron topography observations of a low temperature phase transition in an organic crystal. *Mol. Cryst. Liq. Cryst. Sci. Technol., Sect. A*, **211**, 43–49.
82. Roberts, K.J., Sherwood, J.N., Bowen, D.K., and Davies, S.T. (1983) A synchrotron radiation X-ray topographic study of the beam stability and perfection of some organic single crystals. *Mater. Lett.*, **2**, 104–110.

83. Vetter, W.M., Gallagher, D.T., and Dudley, M. (2002) Synchrotron white-beam X-ray topography of ribonuclease S crystals. *Acta Crystallogr., Sect. D: Biol. Crystallogr.*, **D58**, 579–584.
84. Stojanoff, V. and Siddons, D.P. (1996) X-ray topography of a lysozyme crystal. *Acta Crystallogr., Sect. A: Found. Crystallogr.*, **A52**, 498–499.
85. Koishi, M., Ohya, N., Mukobayashi, Y., Koizumi, H., Kojima, K., and Tachibana, M. (2007) Observation of clear images of dislocations in protein crystals by synchrotron monochromatic-beam X-ray topography. *Cryst. Growth Des.*, **7**, 2182–2186.
86. Raghavan, S.L., Ristic, R.I., Sheen, D.B., and Sherwood, J.N. (2002) Dissolution kinetics of single crystals of α-lactose monohydrate. *J. Pharm. Sci.*, **91**, 2166–2174.
87. Dobrianov, I., Finkelstein, K.D., Lemay, S.G., and Thorne, R.E. (1998) X-ray topographic studies of protein crystal perfection and growth. *Acta Crystallogr., Sect. D: Biol. Crystallogr.*, **D54**, 922–937.
88. Caylor, C.L., Dobrianov, I., Lemay, S.G., Kimmer, C., Kriminski, S., Finkelstein, K.D., Zipfel, W., Webb, W.W., Thomas, B.R., Chernov, A.A., and Thorne, R.E. (1999) Macromolecular impurities and disorder in protein crystals. *Proteins Struct. Funct. Genet.*, **36**, 270–281.
89. Dobrianov, I., Kriminski, S., Caylor, C.L., Lemay, S.G., Kimmer, C., Kisselev, A., Finkelstein, K.D., and Thorne, R.E. (2001) Dynamic response of tetragonal lysozyme crystals to changes in relative humidity: implications for post-growth crystal treatments. *Acta Crystallogr., Sect. D: Biol. Crystallogr.*, **D57**, 61–68.
90. Dobrianov, I., Caylor, C., Lemay, S.G., Finkelstein, K.D., and Thorne, R.E. (1999) X-ray diffraction studies of protein crystal disorder. *J. Cryst. Growth*, **196**, 511–523.
91. Prasad, K.V.R., Ristic, R.I., Sheen, D.B., and Sherwood, J.N. (2002) Dissolution kinetics of paracetamol single crystals. *Int. J. Pharm.*, **238**, 29–41.
92. Donecker, J. and Naumann, M. (2002) Laser scattering tomography for crystal characterization: quantitative approaches. *Cryst. Res. Technol.*, **37**, 147–157.
93. Alhalaweh, A. and Velaga, S.P. (2010) Formation of cocrystals from stoichiometric solutions of incongruently saturating systems by spray drying. *Cryst. Growth Des.*, **10**, 3302–3305.
94. Sato, K., Fukuba, Y., Mitsuda, T., Hirai, K., and Moriya, K. (1992) Observation of lattice defects in orthorhombic hen-egg white lysozyme crystals with laser scattering tomography. *J. Cryst. Growth*, **122**, 87–94.
95. Helliwell, J.R. (1988) Protein crystal perfection and the nature of radiation damage. *J. Cryst. Growth*, **90**, 259–272.
96. Bond, W.L. and Andrus, J. (1956) Photographs of the stress field around edge dislocations. *Phys. Rev.*, **101**, 1211.
97. Sazaki, G., Tsukamoto, K., Yai, S., Okada, M., and Nakajima, K. (2005) In situ observation of dislocations in protein crystals during growth by advanced optical microscopy. *Cryst. Growth Des.*, **5**, 1729–1735.
98. Byrn, S.R. and Lin, C.-T. (1976) The effect of crystal packing and defects on desolvation of hydrate crystals of caffeine and L-(-)-1,4-cyclohexadiene-1-alanine. *J. Am. Chem. Soc.*, **98**, 4004–4005.
99. Binnig, G., Quate, C.F., and Gerber, C. (1986) Atomic force microscope. *Phys. Rev. Lett.*, **56**, 930–933.
100. Ward, M.D. (2001) Bulk crystals to surfaces. Combining X-ray diffraction and atomic force microscopy to probe the structure and formation of crystal interfaces. *Chem. Rev.*, **101**, 1697–1725.
101. Chow, E.H.H., Bucar, D.-K., and Jones, W. (2012) New opportunities in crystal engineering – the role of atomic force microscopy in studies of molecular crystals. *Chem. Commun.*, **48**, 9210–9226.
102. Binnig, G., Gerber, C., Stoll, E., Albrecht, T.R., and Quate, C.F. (1987) Atomic resolution with atomic force microscope. *Europhys. Lett.*, **3**, 1281–1286.
103. Roberts, C.J. (2009) Applications of atomic force microscopy in pharmaceutical research. *Microsc. Anal.*, **23**, 11–14.

104. Bowen, W.R. and Hilal, N. (2009) *Atomic Force Microscopy in Process Engineering*, Butterworth-Heinemann, Oxford.

105. Turner, Y.T.A., Roberts, C.J., and Davies, M.C. (2007) Scanning probe microscopy in the field of drug delivery. *Adv. Drug Delivery Rev.*, **59**, 1453–1473.

106. Teng, H.H., Fenter, P., Cheng, L., and Sturchio, N.C. (2001) Resolving orthoclase dissolution processes with atomic force microscopy and X-ray reflectivity. *Geochim. Cosmochim. Acta*, **65**, 3459–3474.

107. Yamanaka, K., Ogiso, H., and Kolosov, O. (1994) Ultrasonic force microscopy for nanometer resolution subsurface imaging. *Appl. Phys. Lett.*, **64**, 178–180.

108. Ko, T.P., Kuznetsov, Y.G., Malkin, A.J., Day, J., and McPherson, A. (2001) X-ray diffraction and atomic force microscopy analysis of twinned crystals: rhombohedral canavalin. *Acta Crystallogr., Sect. D: Biol. Crystallogr.*, **D57**, 829–839.

109. McPherson, A., Malkin, A.J., Kuznetsov, Y.G., and Koszelak, S. (1996) Incorporation of impurities into macromolecular crystals. *J. Cryst. Growth*, **168**, 74–92.

110. Malkin, A.J., Kuznetsov, Y.G., Lucas, R.W., and McPherson, A. (1999) Surface processes in the crystallization of turnip yellow mosaic virus visualized by atomic force microscopy. *J. Struct. Biol.*, **127**, 35–43.

111. McPherson, A. (2001) Atomic force microscopy applications in macromolecular crystallography. *Acta Crystallogr., Sect. D: Biol. Crystallogr.*, **D57**, 1053–1060.

112. Malkin, A.J., Kuznetsov Yu, G., Land, T.A., DeYoreo, J.J., and McPherson, A. (1995) Mechanisms of growth for protein and virus crystals. *Nat. Struct. Biol.*, **2**, 956–959.

113. McPherson, A., Malkin, A.J., and Kuznetsov, Y.G. (1995) The science of macromolecular crystallization. *Structure*, **3**, 759–768.

114. Yip, C.M. and Ward, M.D. (1996) Atomic force microscopy of insulin single crystals: direct visualization of molecules and crystal growth. *Biophys. J.*, **71**, 1071–1078.

115. Gliko, O., Reviakine, I., and Vekilov, P.G. (2003) Stable equidistant step trains during crystallization of insulin. *Phys. Rev. Lett.*, **90**, 1–4.

116. Sours, R.E., Zellelow, A.Z., and Swift, J.A. (2005) An in situ atomic force microscopy study of uric acid crystal growth. *J. Phys. Chem. B*, **109**, 9989–9995.

117. Rimer, J.D., An, Z., Zhu, Z., Lee, M.H., Goldfarb, D.S., Wesson, J.A., and Ward, M.D. (2010) Crystal growth inhibitors for the prevention of L-cystine kidney stones through molecular design. *Science*, **330**, 337–341.

118. Sangwal, K. (1987) *Etching of Crystals. Theory, Experiment, and Application*, North-Holland Physics Publishing, Amsterdam.

119. Thomas, J.M., Renshaw, G.D., and Roscoe, C. (1964) Production of dislocation etch pits on calcite using optically active etchants. *Nature* (London, United Kingdom), **203**, 72–73.

120. Keith, R.E. and Gilman, J.J. (1960) Dislocation etch pits and plastic deformation in calcite. *Acta Metall.*, **8**, 1–10.

121. Wen, H., Li, T., Morris, K.R., and Park, K. (2004) Dissolution study on aspirin and α-glycine crystals. *J. Phys. Chem. B*, **108**, 11219–11227.

122. Kuroiwa, D. (1969) Surface topography of etched ice crystals, observed by a scanning electron microscope. *J. Glaciol.*, **8**, 475 483.

123. Thomas, J.M. and Williams, J.O. (1971) Dislocations and the reactivity of organic solids. *Prog. Solid State Chem.*, **6**, 119–154.

124. Burt, H.M. and Mitchell, A.G. (1981) Crystal defects and dissolution. *Int. J. Pharm.*, **9**, 137–152.

125. Galwey, A.K., Spinicci, R., and Guarini, G.G.T. (1981) Nucleation and growth processes occurring during the dehydration of certain alums: the generation, the development and the function of the reaction interface. *Proc. R. Soc. London, Ser. A*, **378**, 477–505, 476 plates.

126. Vasilchenko, M.A., Shakhtshneider, T.P., Naumov, D.Y., and Boldyrev, V.V. (1999) The morphology of etch pits during thermal treatment of drugs and its dependence on the features of their crystallochemical structure. *J. Therm. Anal. Calorim.*, **57**, 157–164.
127. Mikhailenko, M.A., Drebushchak, T.N., Shakhtsneider, T.P., and Boldyrev, V.V. (2004) Etching patterns on the monoclinic and orthorhombic paracetamol. *Arkivoc*, **12**, 156–169.
128. Gonda, T., Kakiuchi, H., and Moriya, K. (1990) In situ observation of internal structure in growing ice crystals by laser scattering tomography. *J. Cryst. Growth*, **102**, 167–174.
129. Milne, R.H., Hembree, G.G., Drucker, J.S., Harland, C.J., and Venables, J.A. (1993) Surface studies in UHV SEM and STEM. *J. Microsc.*, **170**, 193–199.
130. Chen, G.C., Li, B., Yan, Z.Q., and Lu, F.X. (2012) Single crystalline diamond grown by repositioning substrate in DC arcjet plasma enhanced chemical vapor deposition. *Diamond Relat. Mater.*, **21**, 83–87.
131. Snyder, R.L., Fiala, J., and Bunge, H.J. (1999) *Defect and Microstructure Analysis by Diffraction*, Oxford University Press.
132. Debnath, S., Predecki, P., and Suryanarayanan, R. (2004) Use of glancing angle X-ray powder diffractometry to depth-profile phase transformations during dissolution of indomethacin and theophylline tablets. *Pharm. Res.*, **21**, 149–159.
133. van Brussel, B.A. and De Hosson, J.T.M. (1994) Glancing angle X-ray diffraction: a different approach. *Appl. Phys. Lett.*, **64**, 1585–1587.
134. Connors, D.C. and West, R.N. (1969) Positron annihilation and defects in metals. *Phys. Lett. A*, **30**, 24–25.
135. Seeger, A. (1973) Investigation of point defects in equilibrium concentrations with particular reference to positron annihilation techniques. *J. Phys. F*, **3**, 248–294.
136. Eldrup, M., Lightbody, D., and Sherwood, J.N. (1981) The temperature dependence of positron lifetimes in solid pivalic acid. *Chem. Phys.*, **63**, 51–58.

5 "Enantiomeric Disorder" Pharmaceutically Oriented

Gerard Coquerel and Rui Tamura

5.1 Introduction

In this chapter, we present the specific features of a chiral molecule when it is disordered. We show the impact of chirality on various levels of disorder. Of course, when a pure chiral molecule is subjected to disorder in the solid state, all the concepts, that is, glass transition, α relaxation, β relaxation, plastic crystals, dynamic disorders such as rotatory phases, mesophases (i.e., liquid crystals), static disorders, misalignment, modulated crystal structures, lack of crystallinity, and so on, are the same as those valid for any organic molecule. Here, the special relationship between one enantiomer and its antipode will be detailed in the context of disorder in the solid state. After a brief clarification of the jargon used in this domain, this chapter details three levels of disorder involving chiral molecules: (i) amorphous mixtures, (ii) substitution of one enantiomer by its antipode in crystallized phases, and (iii) epitaxy and multiepitaxy between crystals of antipodes.

Even though the asymmetric character of the molecule stems from axial chirality (aR–aS), or Δ–Λ complexes, or several stereogenic centers (e.g., RRS–SRR), the enantiomers will be simply labeled *R* and *S*.

5.2 Introduction and Lexicon of Specific Terms Used among Chiral Molecules and Chiral Molecular Associations

The chiral molecules considered here are non-superimposable mirror images of molecules. For example, couples of enantiomers or antipodes are couples of 3D objects that differ from their image through inversion symmetry (Figure 5.1).

Note: Couples of enantiomers have then numerous identical features such as volume, molecular weight, temperatures, as well as enthalpies of fusion, sublimation, and ebullition, C_p versus T function, glass transition, solubility in achiral solvents, pK_a, and so on. In other words, all non-vectorial

Disordered Pharmaceutical Materials, First Edition. Edited by Marc Descamps.

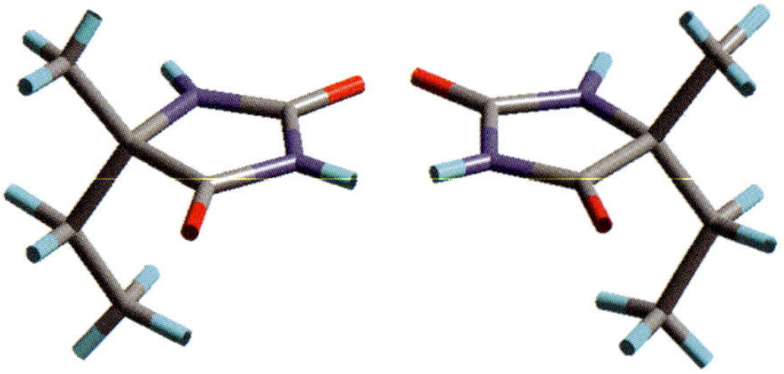

Figure 5.1 Antipodes of 5-ethyl-5methylhydantoin.

intensive properties are simply identical, and all vectorial properties differ only by their signs; that is, their absolute values are identical.

Racemization: Under a specific environment, the two mirror image molecular objects can interconvert through a dynamic equilibrium: this process is named *racemization*. Some molecules spontaneously racemize at room temperature and/or at specific pH, while some others do under the action of a catalyst, and others are hardly racemizable.

Enantiomeric excess: Symbolized as ee, it is defined as $ee = (R - S)/(R + S)$. This quantifies the amount of pure enantiomer that is in excess from the 50 : 50 mixture of enantiomers.

Racemic composition: Composition whose ee = 0; it corresponds to the 50 : 50 mixture of enantiomers.

Racemic compound: Intermediate compound between the (−) and (+) enantiomers whose composition is <1 − 1>. When this phase is not a salt, it can be considered as a co-crystal [1]. Most of the time, it is a stoichiometric compound. Symmetrical solid solutions around the 1 − 1 composition are rare, see Section 3.2.

Twin: In the context of this chapter, it is a geometrical association of crystals of one enantiomer. The extrapolation of the point lattice of one body matches well or fairly well with the point lattice of the other body. Twins can form by penetration (rare for enantiomers) or by contact (i.e., the twin boundary is a plane). The occurrence of a twin can be poor, but a genuine twin is not a random association of crystals, and therefore it has to be differentiated from "accidents during crystal growth." A twin can be repeated in the same particle (i.e., polysynthetic twin).

Epitaxy: In the context of this chapter, it is the same as a twin except that the two bodies correspond to the two antipodes.

5.3 Restrictions in Symmetry Operations Inside Crystal Lattices with an Enantiomeric Excess Different from Zero

Any inversion symmetry operator switches one enantiomer into its antipode. Therefore, the following improper axes are forbidden in a crystal structure that deviates from the 50 : 50 composition: (−1), that is, center of symmetry; (−2), that is, mirror; all glide mirrors and (−3), (−4), and (−6). Eleven point groups containing direct symmetry elements are therefore compatible with a nonracemic composition. Similarly, 65 space groups only – the Schoencke space groups – exist for any crystal whose composition deviates from ee = 0 [2].

For “small” organic molecules and their metal-organic counterparts, several statistical studies reveal that $P2_12_12_1$ (about 58%), $P2_1$ (about 30%), $P1$, $C2$, and $P2_12_12$ account for 95% of the occurrence [3]. This shows the strong deviation from an even distribution. It noteworthy that those space groups are compatible with the low symmetry usually observed for crystals of “small” organic material. For proteins, the molecules are associated with a large number of water molecules and sometimes with themselves to give *n*-mers (e.g., dimers, tetramers, etc.), which build an oblong asymmetric unit more compatible with crystal lattices having higher symmetry operations such as tetragonal, trigonal, and hexagonal.

By contrast, the racemic compound is, most of the time, a stoichiometric compound that crystallizes in one or several (if polymorphism exists) centrosymmetric space groups. *P21/c*, *C2/c*, *P*-1, and *Pbca* are the most popular space groups observed for racemic compounds. This behavior is compatible with the Kitaigorodsky statement [4] on preferred symmetry elements among organic molecules, that is, in decreasing order, translations, center of symmetry, glide mirrors, and direct axes [5].

5.4
Impact of Chirality on Phase Diagrams and the Gibbs–Scott Phase Rule

Prior to analyzing properly the heterogeneous equilibria involving couples of enantiomers (with or without an additional component such as a nonchiral solvent or co-crystal former), it is necessary to know whether the system undergoes spontaneous racemization. Because of the kinetics associated with the interconversion equilibrium, the following two extreme situations can be defined:

1) *Fast racemizable enantiomers:* The system is deemed to evolve toward a single chirality in the solid state if there is no stable partial solid solution between the enantiomers [6] and if it is a conglomerate-forming system. The spontaneous break in symmetry is accelerated by gentle attrition [7] or by ultrasound [8]. By contrast, the liquid phase remains racemic (ee = 0). If it is a racemic-compound-forming system, both the liquid and the solid state will not show any evolution versus time.
2) *Non racemizable enantiomers:* The couple of independent components leads to a symmetrical binary system with reference to the racemic composition [9]. The perfect symmetry between the two mirror-imaged components introduces a modification of the Gibbs phase rule to the Gibbs–Scott phase rule [10].

$$v = \frac{n_2}{2} + n_1 + 2 - \frac{\varphi_2}{2} - \varphi_1 \tag{5.1}$$

where

n_1 and φ_1 are the number of independent components and the number of phases that are not symmetrical

n_2 and φ_2 are the number of independent components and the number of phases that are symmetrical. This relation is applicable only if $\varphi_2 \geq 2$ [9].

5.4.1
Level 1: Amorphous Mixtures: Complete Spatial Disorder

As in any organic compound, the amorphous state of chiral molecules is most of the time accessible by one or several of the following methods: melt-quenching, freeze-drying, high-energy milling below T_g, fast desolvation of solvate, precipitation by anti-solvent addition, and so on.

We will examine here the Gordon–Taylor law (excess entropy neglected) in any R–S binary system: the symmetry between the two components induces a degenerate situation, which is detailed as follows:

$$\text{Gordon – Taylor law}: T_g(X) = \frac{xT_{gS} + (1-x)KT_{gR}}{x + (1-x)K} \quad (5.2)$$

In that particular case, the constant K equals 1 as

$$K = \frac{\Delta C_{pR}}{\Delta C_{pS}} = 1 \quad (5.3)$$

for enantiomers whatever be T.

Then,

$$T_g(X) = T_{gS} = T_{gR} \quad (5.4)$$

that is, T_g should be independent of x!

Experimental confirmation has been obtained by Gallis *et al.* [11] with their systematic study on limonene (Figure 5.2).

As clearly illustrated in Figure 5.3a and b, whatever the enantiomeric excess of a mixture of enantiomers, the glass transition temperature remains constant.

Chirality has also an important impact on the ability of molecules to crystallize. Many series of organic compounds (e.g., alkylamide derived from *trans*-1,2-*bis*(amino)cyclohexane) show that a racemic mixture has a much higher propensity to crystallize than its enantiomeric counterpart. This fact illustrates,

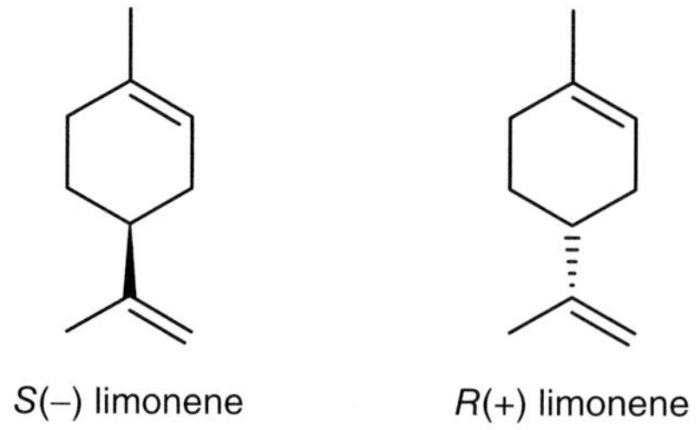

Figure 5.2 Developed formulae of the enantiomers of limonene.

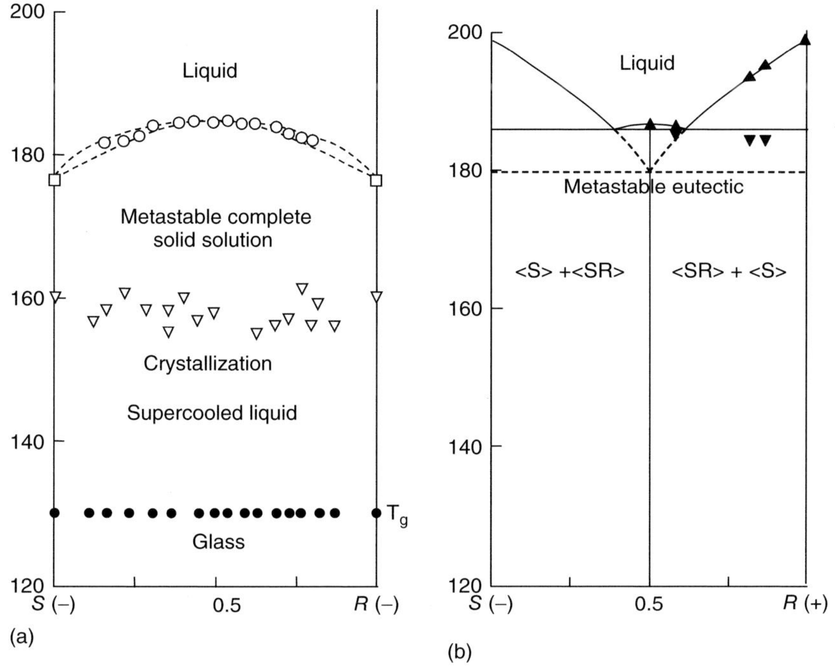

Figure 5.3 (a) Temperature of the glass transition (T_g) versus composition (i.e., composition) and the metastable equilibria leading to a single solid solution between the chiral components. (b) Stable equilibria between enantiomers of limonene. (Reproduced from Gallis *et al.* [11], with permission of Royal Society of Chemistry.)

again, the preference for a racemic compound to form, rather than solid solutions, conglomerates or any mixed situations. Consequently, on one hand, when a chiral molecule is found to be very difficult to crystallize, it is worth trying to crystallize its racemic compound. In case of success, the crystals of the racemic compound can be used as a template to trigger the crystallization of the pure enantiomer. On the other hand, gel formation, even in organic solvents, has also been shown to depend on the enantiomeric excess [12–17]. For instance, it was also established that only nonracemic samples of *para*-tolyl glycerol ether are able to produce supramolecular gels [18].

Partially amorphous materials, once again chiral organic materials (including active pharmaceutical ingredients (APIs)), can show like any other molecular compound variable degrees of crystallinity depending on the crystallization conditions and conditions of storage.

5.4.2
Level 2: In Crystallized Phases (Solid Solutions)

As already stressed, every enantiomer can undergo several types of static or dynamic disorders like any organic molecule. We will differentiate between these disorders as those affecting the chirality and those with a limited impact, if any, on the asymmetric character of the components.

5.4.2.1 Disorders Not Affecting the Stereogenic Centers (Nontreated)

As examples, here we list some reference codes ("refcodes" extracted from the Cambridge Structural Database) illustrating these common disorders among chiral organic molecules and/or their crystallization nonchiral partners (here solvent molecules):

- ADOSUH and ADOTIW, solvent molecule of the solvates are disordered.
- *Rac* EFIWIZ (anticoagulant) isopropanol molecule of the solvated structure is disordered.
- JAPHIR (pure enantiomer C24 is disordered on two sites with 0.8 and 0.2 occupancy factor)
- Activity on dopamine and serotonin receptors.
- *Rac*-LEQGAP chlorine atom and methoxy group are disordered over two positions.
- *R*(+) Cryspine A (Picrate) LEYPOF show disorder on oxygen atom of the NO_2 groups.
- The piperidine ring of SIWCAC pure enantiomer is disordered.

5.4.2.2 Disorders Affecting the Handedness of the Molecules (Examples)

Schematic binary systems between nonracemizable enantiomers are displayed in Figure 5.4a–f. They are illustrative of various degrees of disorder among mixtures of crystallized enantiomers.

There is a continuum between a complete solid solution (Figure 5.4a) and the conglomerate (i.e., eutectic between pure components (Figure 5.4d)). In order to highlight the difference between the cases depicted in Figure 5.4c,d, Figure 5.5a,b shows the evolution of the Gibbs function versus composition at three temperatures. In Figure 5.5b, the Gibbs functions associated with the solid phases (the pure enantiomers) are so steep that they are considered vertical.

Because the molecular volumes of the two antipodes are identical, the solid solution is of substitution type. In Figure 5.5a, a single crystallized phase whose composition spans from pure *S* to pure *R* can be written $<S_xR_{1-x}> 0 \leq x \leq 1$, which means that whatever the overall composition of the mixture, there is a statistical occupancy of the crystallographic sites. For the racemic composition, the solid solution can exhibit a higher global symmetry than for any other composition. This has been discussed in detailed in a previous review [9]. For a given couple of chiral amines B_R and B_S, an ordering of set of possible salts with different acidic components *Ai* can be considered by the volume of *Ai*; the probability of having solid solutions such as those depicted in Figure 5.4a,b sharply increases above a threshold of the ratio (volume *Ai*/volume of B_R).

For several practical reasons, the possibility of full deracemization [19], resolution by preferential crystallization [20], or quantitative recovery of the enantiomeric excess, the case shown in Figure 5.4d is of particular interest. This

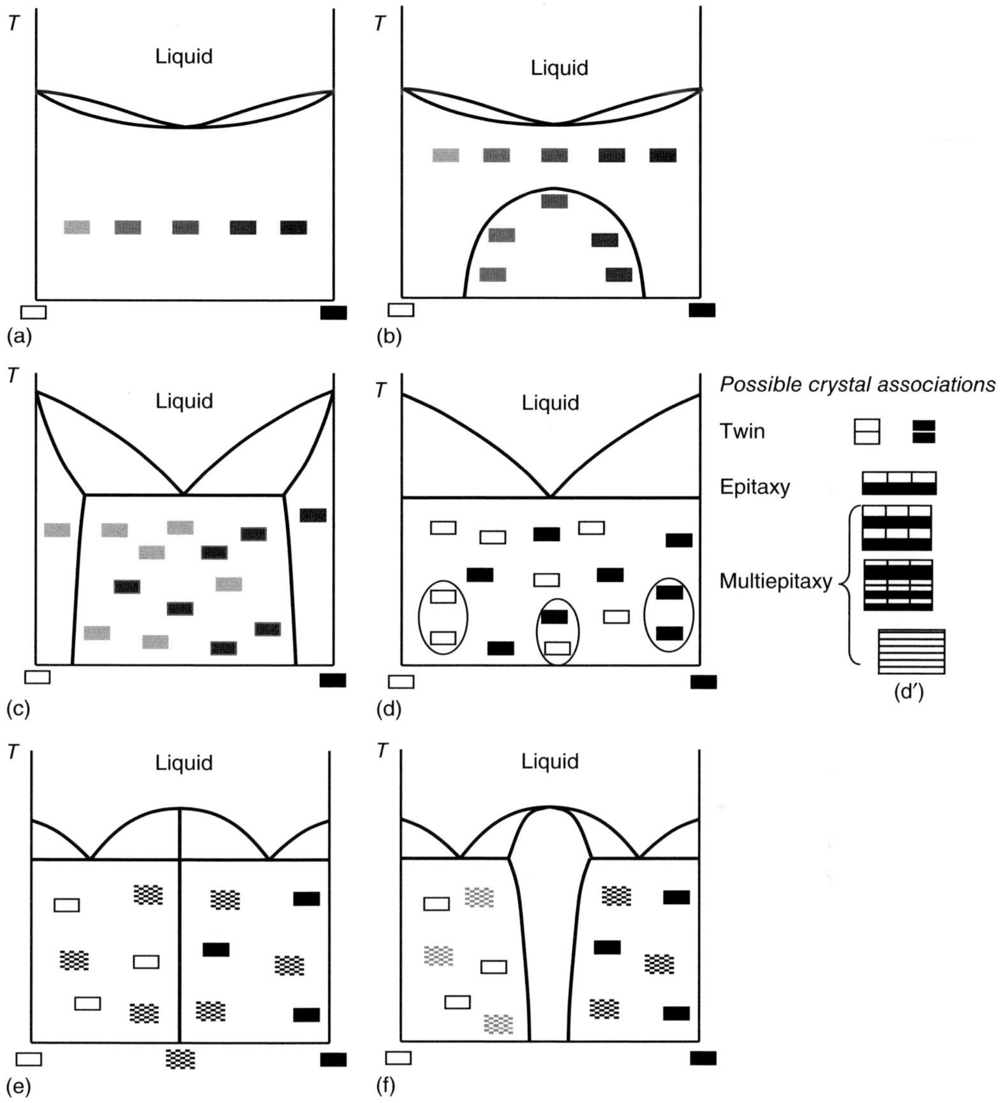

Figure 5.4 Schematic representations of binary systems between nonracemizable enantiomers. Note the perfect symmetry with reference to the racemic composition. The level of gray is indicative of the amount of *R* in the solid phase (i.e., black box means pure *R*). (a) Complete solid solution with a minimum at $x = 0.5$, that is, ee $= 0$. Complete solid solution with maximum for $x = 0.5$ exits as well. (b) Same as (a) at high temperature but a miscibility gap in the solid state existing at low temperature. (c) Eutectic (i.e., conglomerate) with symmetrical partial solid solutions. (d) Eutectic (i.e., conglomerate) without any detectable partial solid solution. (d′) Same as (d) except that every pure particle of one enantiomer can be twinned and/or associated with a foreign particle through epitaxies between particles of opposite chirality. (e) Binary system with a stoichiometry racemic compound. (f) Same as (e) except that the intermediate compound is a nonstoichiometric racemic compound.

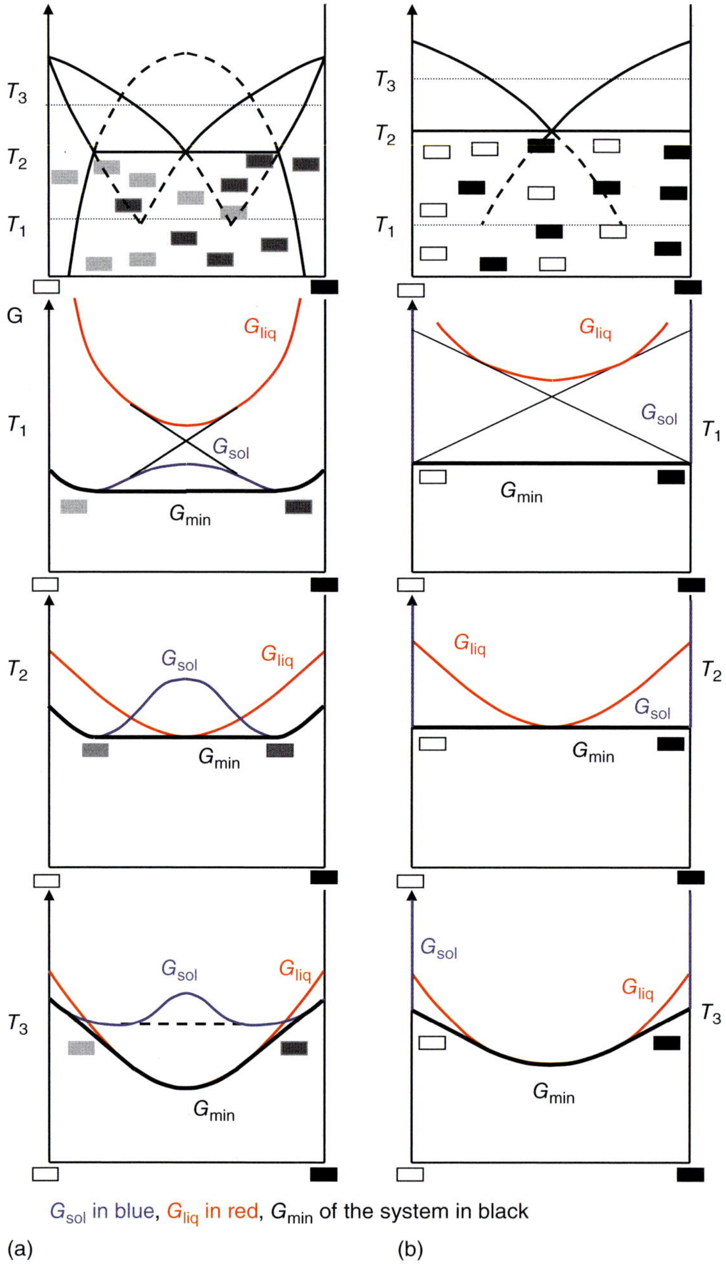

Figure 5.5 Free enthalpy functions at various temperatures (T_1, T_2, and T_3) for (a) eutectic with partial solid solution (Figure 5.4c) and (b) eutectic without detectable solid solution (Figure 5.4d).

behavior corresponds to a full chiral discrimination in the solid state. Usually it represents between 1% and 10% of the derivatives that can be formed with the crystallizing partners: ionic counterparts (salts), solvent molecules (solvates or heterosolvates), and co-crystal formers (co-crystals).

By contrast, Figure 5.4e depicts the overwhelming occurrence in binary systems of enantiomers. The <1−1> compound is by far favored in about 90–95% of the cases. Many scientists have tried to justify this prevalence [21] on thermodynamic and crystallographic grounds. From a crystallographic perspective, this fact must be associated with the high preference for centrosymmetric associations, which is accessible only to racemic compounds that crystallize in popular space groups such as $P2_1/c$ (identical to: $P2_1/a$ and $P2_1/n$), P-1, $C2/c$, $Pbca$, and so on.

This property is used to pre-screen conglomerates within a large series of derivatives. Indeed, any material crystallizing in one of the 92 centrosymmetric space groups does not show any second harmonic generation effect (SHG) [22]. Interestingly, the behavior illustrated in Figure 5.4f is of very poor occurrence [23]. This points out the poor frequency of disorder in composition associated with a racemic compound.

Other cases have been observed or predicted, but they constitute a minor proportion even though they are of fundamental importance for the understanding of crystal packing versus the enantiomeric excess.

(+) and (−) *cis-π*-Camphanic acid constitutes one unusual example (Figure 5.6). On increasing the temperature, one can observe in sequence (i) a complete discrimination in the solid state, that is, a stable conglomerate without any detectable partial solid solution; (ii) a complete solid solution; (iii) plastic crystals; and (iv) an isotropic liquid.

(+) and (−) Diprophylline, a drug prescribed for the treatment of apnea and bronchospasm, deserves attention too. A thorough analysis of the heterogeneous equilibria (Figure 5.7) reveals that – due to polymorphism – this couple of

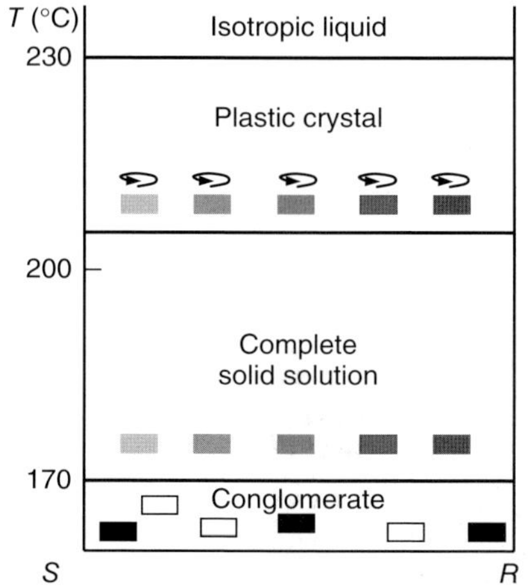

Figure 5.6 Phase diagram for *cis-π*-camphanic acid. Adapted from [3]. The biphasic domains between the labeled phases are not represented. Boxes keep the same symbolism as in Figure 5.4.

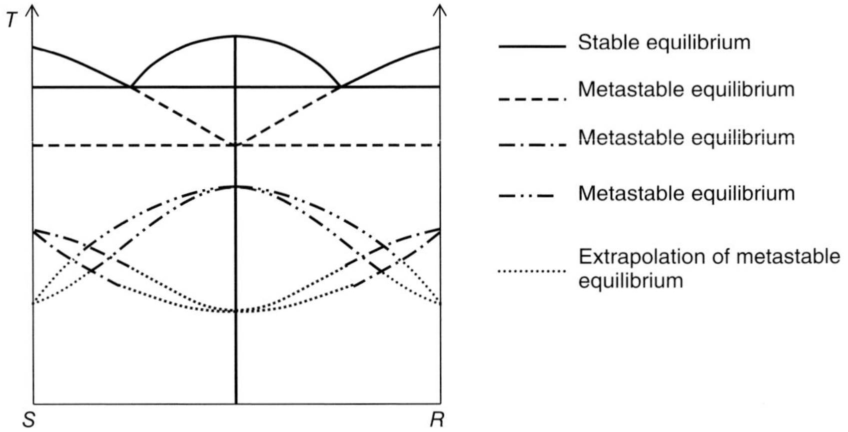

Figure 5.7 Schematic (not to scale) representation of the binary system between (+) and (−) antipodes of diprophylline.

enantiomers leads to two complete solid solutions: one with a maximum and the other with a minimum (see Figure 5.4a), a racemic compound (as schematically depicted in Figure 5.4e) and a conglomerate (Figure 5.4d) [24]. The latter was proven by achievement of the chiral resolution by using a seeded isothermal preferential crystallization process in a particular solvent. The energetic barrier between the different conformations appearing in these different solid phases might justify, at least partially, this unusual variety of behaviors.

The system of L- and D-carvone [25] (Figure 5.8) exemplifies Figure 5.4b. Thus, there is complete miscibility in the solid state from pure L- to pure D- as long as the temperature is above T_c (about 215 K). For temperatures below 215 K, there are symmetrical domains of partial solid solutions: the lower the temperature, the greater the chiral discrimination. Nevertheless, diffusion in the solid state slows down rapidly as T decreases, and the practical attainment of a 5–95% biphasic domain will require very long annealing.

In 2006, there appeared in the *Journal of the American Chemical Society* (JACS) a study [26] on Tazofelone, a chiral API used as an anti-inflammatory drug in bowel diseases. This couple of enantiomers offers a nice mixed situation. Unfortunately, the interpretation of the experimental data shows very little respect to W. Gibbs (Figure 5.9). Figure 5.10 is a reinterpretation of the data extracted from this publication. As the two stoichiometric polymorphs of the racemic compound have their melting points above the maximum melting temperature of the solid solution, the stable solid solution cannot cross the whole domain in composition. Thus, there is a need to have on both sides of the binary system two symmetrical, stable partial solid solutions. It is likely that these partial solid solutions are limited at high temperature by two symmetrical peritectic transitions (at T_p, see Figure 5.10). The polymorphic transition at about 138 °C between RACI and RACII corresponds to a usual enantiotropy.

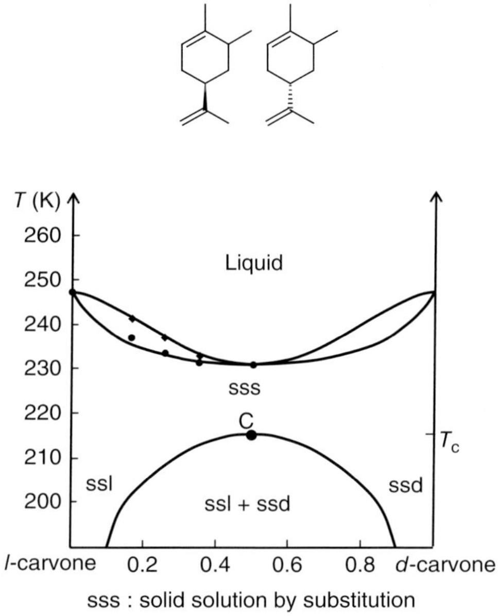

Figure 5.8 Binary system between *l* and *d* carvone, showing a complete miscibility in the solid state above the critical point *C*. Below T_c, there is a miscibility gap in the solid state; that is, the chiral discrimination increases as *T* decreases.

Figures 5.9 and 5.10 show, respectively, the main features of the binary phase diagram between (*S*) and (*R*) Tazofelone as it appeared in JACS [26] and a tentative reinterpretation of the data.

As mentioned previously, any crystallization partner might induce a redistribution of the enantiomers in new crystal lattices. Figure 5.11 illustrates the role of the solvent via the formation of mirror-imaged solvates in preventing the formation of a solid solution between enantiomers of a precursor of *rac*-modafinil, a drug for the prevention of sleepiness, and its active enantiomer armodafinil.

There is a sharp contrast between the complete solid solution in the binary system of the two enantiomers and the stable conglomerate of monohydrates (−) and (+) at room temperature. In the former situation, the two antipodes are distributed at random in a single crystal. In the latter, a single water molecule per solute molecule is enough to bring about complete chiral discrimination in the solid state (i.e., formation of a conglomerate [27]). The water molecule is involved in an electrostatic relay between the anion and the cation (Figure 5.12).

It would be too naive to think that a H-bond is a guarantee to avoid solid solution formation! (+) and (−) Ethanolammonium mandelate [28] and (+) and (−) ethanolammonium 3-chloromandelate [29] are examples of partial solid solutions (e.g., disorder among crystallographic site occupancy) in which the stereogenic centers are involved in H-bonds. Actually, by exchanging the hydrogen and the hydroxyl moiety bonded to the asymmetric carbon, it has been possible to show a kind of geometrical frustration. For a given crystallographic site, it is possible

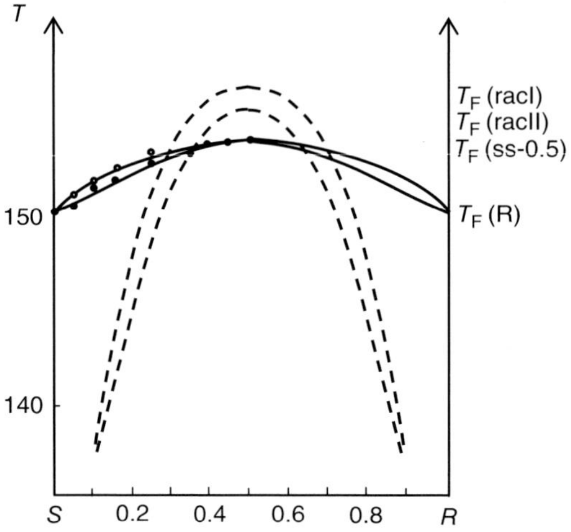

Figure 5.9 Schematic representation of (-); (+) Tazofelone binary system from reference [26].

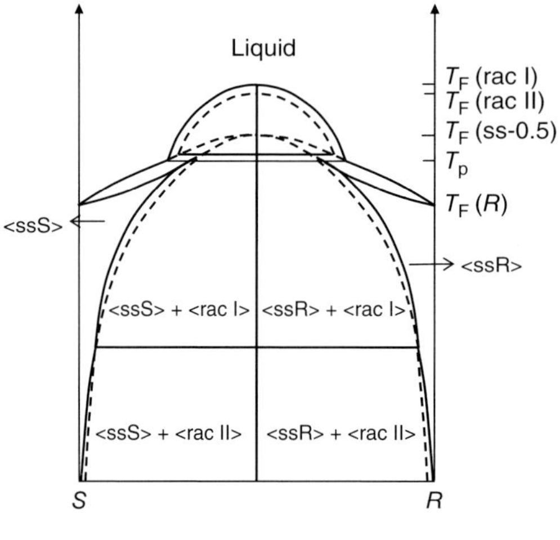

Figure 5.10 Tentative re-interpretation of (-); (+) Tazofelone binary system.

to locate the (+) and the (−) enantiomers with different H-bonds involving the hydroxyl groups.

Another extensive study on the solid state of an API used in the treatment of angina and hypertension, namely carvedilol phosphate hemihydrate, by Vogt *et al.* [30] shows a complete solid solution of the hemihydrate with a particular symmetry for the racemic composition.

Figure 5.13 shows a schematic ternary isotherm of (*S*) clopidogrel hydrogeno-sulfate (Plavix) and (*R*) clopidogrel hydrogeno-sulfate in methanol at 20 °C (G. Coquerel, unpublished results). The stable equilibria involve a large biphasic domain with the stoichiometric racemic compound and its saturated solution.

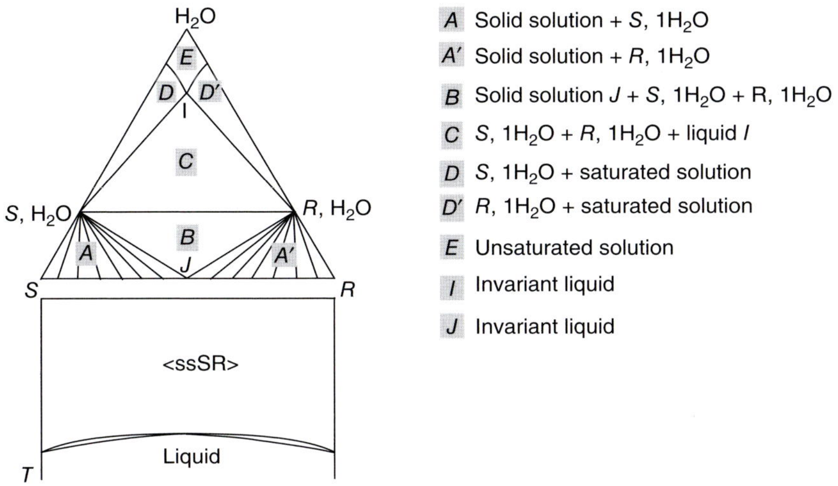

Figure 5.11 Schematic water and (+) and (−) ethanolammonium modafinate ternary system. Reproduced from Wermester *et al.* [27], with permission of Royal Society of Chemistry.

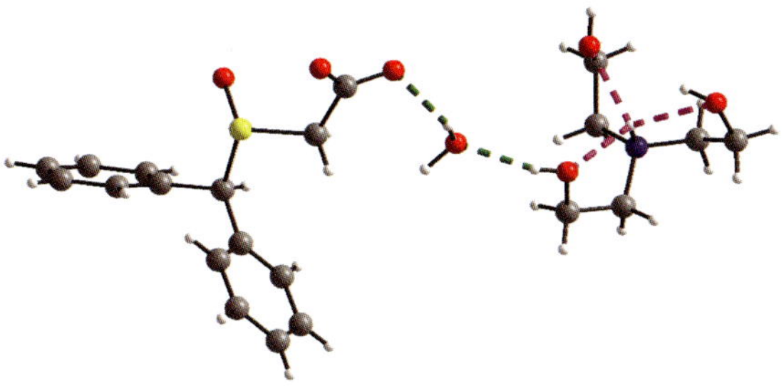

Figure 5.12 Schematic representation of the electrostatic relay of the water molecule between the carboxylate and the triethanolammonium cation in the crystal structure.

The two other symmetrical biphasic domains have very limited areas and involve form I of the pure enantiomers and their respective saturated solutions.

In contrast to stable equilibria, metastable equilibria involve a nonstoichiometric compound whose composition can exceed 20% ee. This phase can be obtained when the kinetics is the key factor of the crystallization (e.g., precipitation). The crystallographic data collected in Table 5.1 are also consistent with the possible enlargement in composition of the intermediate compound or with a complete solid solution between the two enantiomeric form I. The parameters are similar, and the space group of the pure enantiomers (with two independent molecules in the asymmetric unit: $Z' = 2$) is a sub-space group of the racemic compound. A statistical occupation of the crystallographic sites is therefore conceivable for the metastable phase.

Starting from an overall composition represented by point E (about 50% ee), the mixture is homogenized at high temperature and cooled down to 20 °C rather

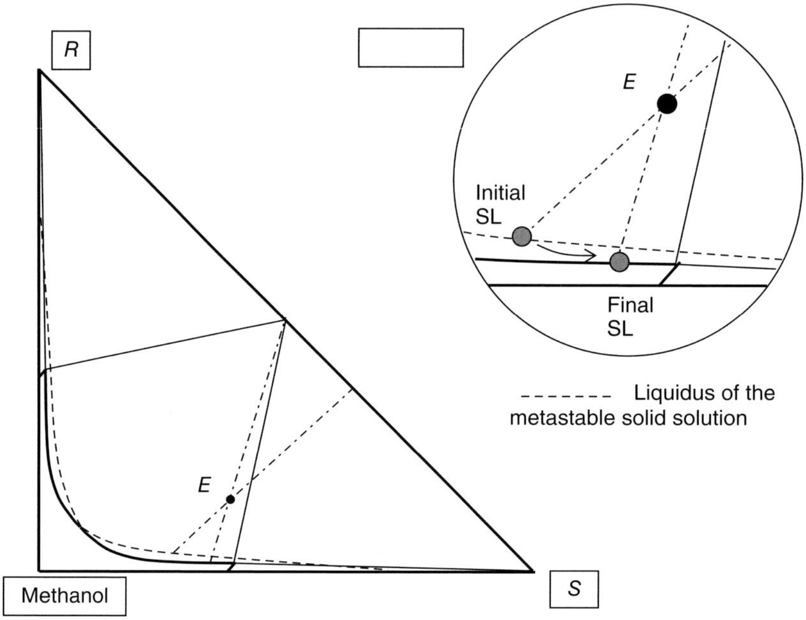

Figure 5.13 Stable and metastable equilibria between *S* and *R* hydrogeno-sulfate of clopidogrel in methanol at 20 °C (out of scale isotherm). The inset represents the detail of the irreversible evolution of the saturated liquid from the metastable equilibrium to the stable equilibrium.

Table 5.1 Crystallographic parameters and space groups of *S* and *R* hydrogeno-sulfate of clopidogrel form I and the racemic compound stable at 20 °C.

	Racemic compound	Enantiomer form I
a (Å)	10.25	10.442
b (Å)	15.43	15.295
c (Å)	12.56	12.620
β (°)	110.0	113.20
SG	$P2_1/c'$	$P2_1$

rapidly without any seeding. The phase that is likely to crystallize first will be the solid solution. In other words, the liquid phase will show a limited enrichment in the enantiomer in excess. If the system is left under agitation, an irreversible evolution will take place, as illustrated in the inset of Figure 5.13. The stoichiometric compound will substitute the solid solution, and simultaneously the saturated solution will significantly increase its enantiomeric excess.

Unfortunately, the components cannot undergo fusion because of a concomitant chemical degradation. Several binary systems can be proposed to

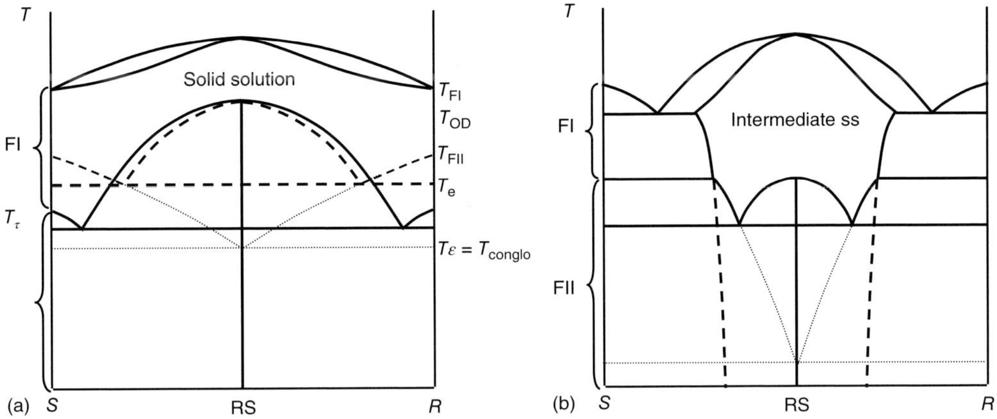

Figure 5.14 (a,b) Hypothetical binary systems between *S* and *R* hydrogeno-sulfate of clopidogrel. For both systems, there is an enantiotropy relationship between form I and form II of the enantiomers. Panel (a) illustrates a complete solid solution between the two form I enantiomers and no solid solution between their forms II. Panel (b) illustrates a limited domain of solid solution of the racemic compound at high temperature. In both hypotheses, the stoichiometric racemic compound undergoes an order–disorder transition when it transforms into a solid solution at high temperature.

rationalize the experimental facts; two among the simpler ones are represented in Figure 5.14.

Starting from a racemic-compound-forming system, the formation of co-crystal can also lead to the formation of solid solutions. This has been exemplified by a study on the binary system *R* and *S* (ibuprofen)$_2$(4,4-dipyridyl) by Zhang and coworkers [31].

5.5 Competitions between Solid Solutions (Impact of Polymorphism on Solid Solutions) Application: Preferential Enrichment

Phase transition between two polymorphs of certain racemic or nearly racemic solid solutions induced a unique spontaneous enantioresolution phenomenon, which was referred to as *preferential enrichment*.

5.5.1 Discovery and Feature of Preferential Enrichment

Preferential enrichment was discovered during the development of a certain antiallergic drug, suplatast tosylate [(±)-ST], which has been used as a potent inhibitor of IgE synthesis without suppressing IgM and IgG [32, 33]. For example, whenever the racemic sample of ST was synthesized, the enantiomeric excess

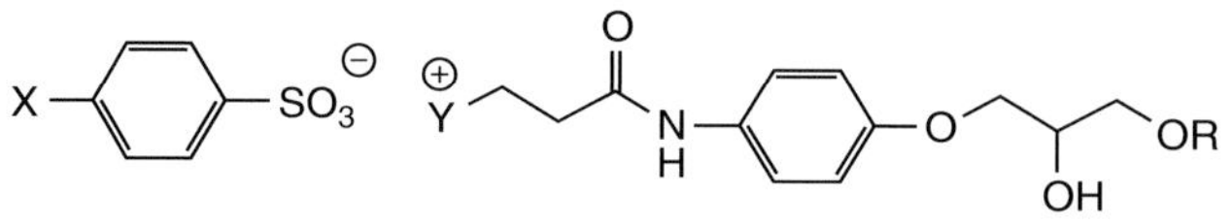

ST (X = Me, Y = SMe_2, R = Et)

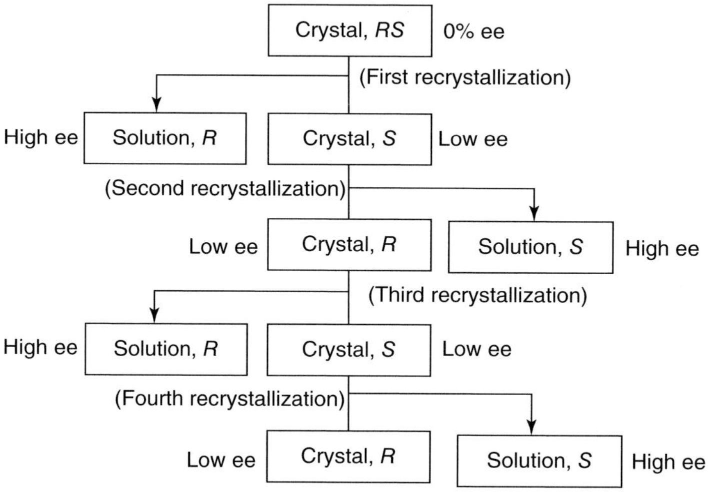

Figure 5.15 Principle and feature of preferential enrichment in the case of substantial enrichment of *R* enantiomer in solution after the first recrystallization of (±)-ST.

values of the solid samples obtained after recrystallization from ethanol differed; they were distributed over a narrow range of less than 10%. By systematic studies on the relationship between the ee value and chirality in the supernatant solution and those in the deposited crystals, a unique spontaneous enantioresolution phenomenon was demonstrated to occur (Figure 5.15) [34]. In fact, recrystallization of (±)-ST from ethanol under 25-fold supersaturated conditions led to substantial enrichment of either the *R* or *S* enantiomer in the mother liquor. At the same time, slight enrichment of the opposite enantiomer occurred in the deposited crystals. More interestingly, whenever the slightly *R* (or *S*)-enriched sample of less than 10% ee was recrystallized from ethanol, substantial enrichment of the same *R* (or *S*) enantiomer in the mother liquor and slight enrichment of the opposite *S* (or *R*) enantiomer in the deposited crystals concurrently occurred with full reproducibility.

5.5.2
Crystalline Nature of the Compounds Showing Preferential Enrichment

The crystalline nature of ST and its analogous compounds that showed preferential enrichment was identified as a solid solution of the racemic compound type

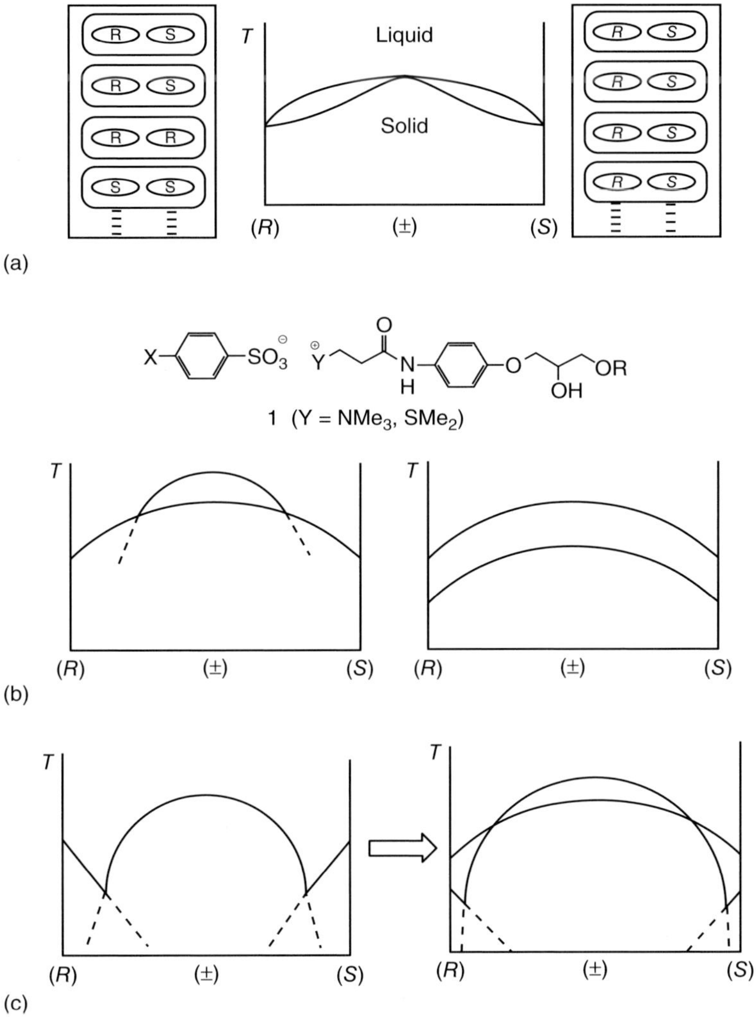

Figure 5.16 Melting point diagrams. (a) A fairly or highly ordered solid solution (racemic compound type). (b) Two typical diagrams of a solid solution with an appropriate polymorphism capable of showing preferential enrichment with respect to compounds 1. (c) Transformation of a racemic compound into another racemic compound with a high eutectic ee value and an appropriate polymorphism by crystal engineering so as to induce preferential enrichment. For clarity, only the temperatures of end of fusion are shown in (b) and (c).

by X-ray crystallography, comparison of X-ray diffraction data, and differential scanning calorimetry (DSC) analysis (Figure 5.16a) [35–40]. Furthermore, there always existed two polymorphs whose free energy difference was small enough to allow a polymorphic transition to proceed at a moderate rate during crystallization [34–36] (Figure 5.16b).

5.5.3
Mechanism and Requirements

The proposed overall mechanism of preferential enrichment, which is consistent with five requirements listed below, is illustrated in Figure 5.17 [37–40]. This is the case where a considerable enrichment of the *R* enantiomer occurs in the mother liquor by crystallization from the slightly *R*-rich (around 5% ee) supersaturated solution. In solution, homochiral 1D *R* and *S* chains are formed preferentially, which aggregate to form a γ-form supramolecular cluster and undergoes a phase transition to give the metastable γ-form crystals with fair disorder. Since the *R* enantiomer is slightly in excess in the solution, the resulting metastable γ-form crystals become slightly *R*-rich, too. The next step is the solid-to-solid polymorphic transition initiated by the rearrangement of the hydrogen bonds in the crystal lattice. After the polymorphic transition of the γ-form into a more stable form which consists of dimer chains, partial crystal disintegration occurs at the incomplete dimer sites in the transformed crystals, followed by redissolution of the excess *R* enantiomer into solution until the deposited crystals are slightly enriched with the opposite *S* enantiomer.

Requirements:

1) Sufficient solubility difference: much higher solubility of the enantiopure sample than that of the corresponding racemic sample, allowing the preferential formation of homochiral 1D *R* and *S* chains even in the racemic solution;
2) Occurrence of an appropriate solid-to-solid polymorphic transition during crystallization;
3) Unique crystal structures in the metastable and transformed crystals;
4) Partial crystal disintegration in the transformed crystals after polymorphic transition;
5) Deposition of nonracemic solid solutions capable of memorizing the chiral symmetry breaking.

Based on the mechanism of this polymorphic transition, induction of preferential enrichment could be accomplished by controlling the mode of the polymorphic transition by minor molecular modification or with appropriate seed crystals [41, 42].

5.5.4
Extension to Racemic Compound Crystals

Very recently, preferential enrichment was found to be applicable to a racemic compound that has a high eutectic ee value and can satisfy the above five requirements (Figure 5.16c). For example, preferential enrichment was observed for alanine (eutectic ee: 56%) and leucine (86%) by carefully choosing the kinetic crystallization conditions so as to induce an appropriate polymorphic transition, although the efficiency of resolution was very low [43]. Furthermore, we

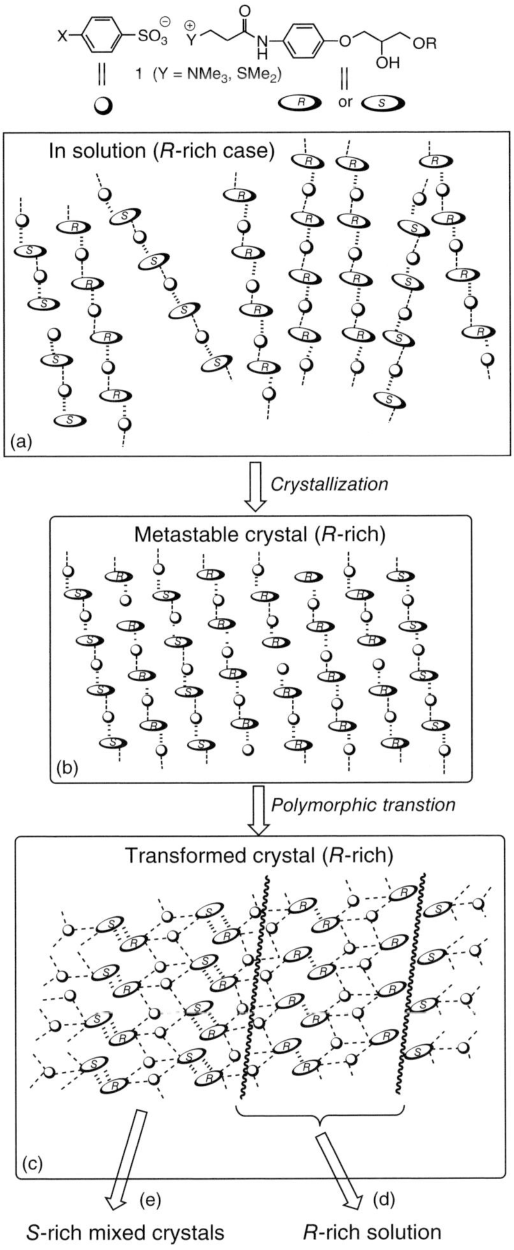

Figure 5.17 Mechanism of preferential enrichment in the case of crystallization from the supersaturated solution of slightly *R*-rich compounds 1. Homochiral 1D molecular association (a) in solution and (b) in the metastable γ-form crystal, polymorphic transition to give (c) stable crystals composed of dimer chains, followed by (d) crystal disintegration and selective redissolution of the excess *R* enantiomer into the solution, and (e) deposition of nonracemic solid solutions enriched with the opposite *S* enantiomer. Dotted and dashed lines denote hydrogen bonds.

have found that among the three typical nonsteroidal anti-inflammatory drugs (NSAIDs) ibuprofen, ketoprofen, and naproxen, only ketoprofen (eutectic ee: 82%) satisfies all the five requirements and, indeed, shows a good preferential enrichment phenomenon [44]. Notably, the 1 : 1 co-crystal (eutectic ee: 99% or >95%) of (DL)-phenylalanine or (DL)-arginine and fumaric acid, respectively, which has been considered to be a racemic compound [45], behaved like a solid solution and satisfied the five requirements to show an efficient preferential enrichment phenomenon under kinetic crystallization conditions, whereas free (DL)-phenylalanine or (DL)-arginine, a racemic compound, failed to exhibit preferential enrichment [46, 47].

Thus, it is concluded that (i) it is of great advantage to use a two-component co-crystal to induce preferential enrichment, because a solid-to-solid polymorphic transition involving a molecular rearrangement or rearrangement of hydrogen bonds in the crystal lattice is feasible with high probability, and (ii) the strategy to transform a racemic compound into another racemic compound with a high eutectic ee value and an appropriate polymorphism by crystal engineering is useful for the induction of preferential enrichment (Figure 5.17c).

5.6
Disorder at Level 3 Multiepitaxy between Enantiomers

Level 3 of disorder: epitaxy between enantiomers crystallizing as conglomerates.

In a sense, epitaxy between enantiomers is not a disorder affecting conglomerate-forming systems *per se.* Nevertheless, it could be also considered that this phenomenon creates particles close to a racemic composition built from a disordered stacking of homochiral fragments (Figure 5.4d′). Similar to twins (for homochiral fragments), epitaxy between enantiomers is not a fortuitous association of crystals but a stacking of particles according to a geometrical law with some possible degrees of misalignment.

So far, less than 20 examples have been reported in the literature [48–59], but we think that the occurrence of this phenomenon has been underestimated and a significant number of preferential crystallizations associated with poor results might simply be due to the geometrical assembly of antipodal fragments. Indeed, the heterochiral association at crystal interface is favored because of the following:

1) The enantiomers have exactly the same crystallographic parameters.
2) The enantiomers have also the same space groups, for example, $P2_12_12_1$ (*S*) and $P2_12_12_1$ (*R*) or $P2_1$ (*S*) and $P2_1$ (*R*), or enantiomorphous space groups $P3_1$ (*S*-modafinic acid) $P3_2$ (*R*-modafinic acid) or $P4_12_12$ (Maribavir) $P4_32_12$ for its non-active antipode [60].
3) At the epitaxial interface, the 2D racemic structure is not limited to symmetries belonging to the 65 space groups, but the whole set of symmetries compatible with a crystal lattice are possible including inversion

symmetries, that is, center of symmetry (−1) and glide mirrors (translation associated to −2) most probably but also inversion axes −3, −4, and −6 are possible. This is why the term *twinning by inversion* (50−50 inversion twin) is sometimes used [61].

These can explain why sometime epitaxy can be revealed "by surprise," when considering the Flack parameter [62], which departs significantly from zero when calculating also carefully the error on that value. The crystals grown in a stagnant racemic solution exhibit the shape of perfect single crystals [56−63], whereas they are actually multiple heterochiral associations of homochiral domains. It can also be seen as a "moyen détourné" by Nature to go back to heterochiral assemblage and a return to the prevalence of inversion symmetries. In the extreme, a racemic compound can be seen as a repetition of the epitaxial law in every plane; for example, two enantiomeric structures $P2_12_12_1$ (*S*) and $P2_12_12_1$ (*R*) can be merged into a $P2_1/b2_1/c2_1/a$ racemic compound.

The "good" reasons given above are valid for every couple of enantiomers crystallizing as a conglomerate; the open question is therefore: why do we observe that phenomenon with a limited frequency? Most of the time, molecular modeling reveals a small to a very small difference in energy between the "artificial" racemic compound and the corresponding homochiral packing [57]. In order to make those differences significant, progress in computational energy calculations is needed.

Epitaxy between antipode is a drawback for chemists interested in enantiomeric purification and resolution by preferential crystallization (i.e., entrainment). One way to circumvent (most of the time partially only!) this problem is to apply a very energetic stirring of the suspension. This mechanical treatment avoids the supersaturation of the counter enantiomer, in the vicinity of the growing surface, to exceed the threshold above which there is a 2D nucleation of the antipode. Repetition of this phenomenon (multiepitaxy) leads to near-racemic single particles despite the stability of the conglomerate.

Bis-Tetralone [63] is another such example associated with atropo-enantiomerism in which the *S* and the *R* molecules form a stable conglomerate. The two mirror-related crystal lattices ($P4_12_12$ and $P4_32_12_1$) form multiepitaxy. Interestingly, the "single-crystal-like" particles have the tendency to show a morphology consistent with the (222) point group rather than the expected (422) point group. In addition, measurements of the intensity of the re-emitted beam in SHG versus composition confirm that the homochiral particles make some repeated associations different from a physical random mixture. For the first time, a particular variant of preferential crystallization (ASPreCISE) [64] has been able to circumvent the difficulties of a preparative resolution by entrainment.

Surprisingly, to the best of our knowledge, no report has been issued on mastering this phenomenon: that is to say, in controlling the thickness – that is, the periodicity – of the stacking homochiral domains. Some physical properties can be expected from that control such as an enhancement of the SHG activity for non-phase-matching compounds.

5.7
Conclusion and Perspectives

In addition to all kinds of disorders affecting the organic solid state in general and the APIs in particular, the presence of chirality induces some specific phenomena/properties that deserve attention. This chapter has highlighted three levels of disorder:

Level 1: Amorphous chiral material. Because the perfect symmetry between the two components; it is shown that the glass transition temperature should remain constant whatever the enantiomeric excess of the mixture.

Level 2: Solid solution among enantiomers. The exact match between the molecular volumes of the *R* and *S* components makes favorable the substitution of one enantiomer by its antipode. In addition, the two components share the same moieties; just the spatial distribution of the substituents around the stereogenic center(s) differs.

Various situations have been detailed: from complete solid solutions to partial solid solutions. Thermodynamic and kinetic competitions between the stoichiometric racemic compound and solid solutions have been highlighted. In case of solid solutions, the racemic composition can exhibit extra symmetries, simply because the inversion symmetry operations are possible through an average global symmetry.

Then the full discrimination in the solid state results from the competition between solid solution, the crystallization of the stoichiometry racemic compound, and the 50 : 50 mixture of pure enantiomers labeled conglomerate. The winner of this competition – that is, the most stable phase – can be temperature dependent. In other words, the *G* function of those three possibilities might cross at different temperatures.

A quasi-full chiral discrimination in the liquid state can also be obtained by a subtle interplay between stable and metastable solid solutions around the racemic composition and if other conditions for preferential enrichment application are fulfilled.

Level 3: Epitaxy and conglomerates. The identity of the crystallographic parameters between the two antipodes favors the epitaxy and multi-epitaxy between particles of opposite chirality. Up to now, less than 20 examples have been reported, but it is likely that this phenomenon has been overlooked and some failures or poor productivities of preferential crystallization might be due to those phenomena.

One can see the interface of the two particles of opposite handedness as a bidimensional racemic compound. The open fundamental question is, why does this 2D racemic compound not expand in the third dimension?

In this chapter, the relationships between disorders in the organic solid state and chirality have been treated in a qualitative way. This is because the molecules were considered to be chiral or achiral. A stereogenic center such as R_1R_2CHD will be most probably associated with a number of disorders in comparison to other more conventional cases where the four substituents of the stereogenic center are clearly different in a chemical sense. Intuitively, one can perceive the need for quantification of chirality. This quantitative descriptor is necessary for a full understanding of the connection between disorder and the asymmetric character of the molecules. Mathematicians (e.g., Hausdorff), physicists, and chemists have proposed different approaches of this difficult problem; almost an "attempt to the impossible" [65]. In our opinion, the quantification of chirality must be associated with the context in which it expresses itself [66].

Acknowledgments

We thank Dr Marie-Noelle Petit for a substantial help in the illustrations of this chapter.

References

1. Coquerel, G. (2012) in *Pharmaceutical Salts and Co-Crystals*, vol. **13** (eds J. Wouters and L. Quéré), RCS, p. 300.
2. Levkin, P.A., Torbeev, V.Y., Lenev, D.A., and Kostyanovsky, R.G. (2006) in *Topics in Stereochemistry*, vol. **25** (eds S.E. Denmark and J.S. Siegel), John Wiley & Sons, Inc.
3. Jacques, J., Collet, A., and Wilen, S.H. (1994) *Enantiomers, Racemates and Resolutions*, Krieger Publishing Company, Malabar, FL.
4. Kitaigorodskii, A.I. (1961) *Organic Chemical Crystallography*, Springer, New York.
5. Coquerel, G. and Amabilino, D.B. (2009) in *Chirality at the Nanoscale* (ed. D.B. Amabilino), Wiley-VCH Verlag GmbH, Weinheim, p. 305.
6. Gonella, S., Levilain, G., and Coquerel, G. (2011) *J. Therm. Anal. Calorim.*, **103**, 125.
7. Viedma, C. (2004) *J. Cryst. Growth*, **261**, 118.
8. Rougeot, C., Guillem, F., Plaquevent, J.-C., and Coquerel, G. (2015) *Cryst. Growth Des.*, **15**, 2151.
9. Coquerel, G. (2000) *Enantiomer*, **5**, 481.
10. Scott, R.L. (1977) *J. Chem. Soc., Faraday Trans. 2*, **3**, 356.
11. Gallis, H.E., van der Miltenburg, J.C., and Oonk, H.A.J. (2000) *Phys. Chem. Chem. Phys.*, **2**, 5619.
12. Hanabusa, K., Kato, T., Katsuna, T., and Ukon, M. (1998) *Adv. Mater.*, **10**, 606.
13. Snijder, C.S., de Jong, C., Meetsma, A., van Bolhuis, F., and Feringa, B.L. (1995) *Chem. Eur. J.*, **1**, 594.
14. Hanabusa, K., Yamada, M., Kimura, M., and Shirai, H. (1996) *Angew. Chem. Int. Ed. Engl.*, **35**, 1949.
15. Lu, R., Bao, C., Jin, M., Song, Z., Yang, X., Song, D., Xu, T., Liu, G., and Zhao, Y. (2007) *Tetrahedron*, **63**, 7443.
16. Fuhrhop, J.H., Schneider, P., Rosenberg, J., and Boekema, E. (1987) *J. Am. Chem. Soc.*, **109**, 3387.
17. Chandrasekhar, S. (1992) *Liquid Crystals*, Cambridge University Press.
18. Bredikhin, A.A., Bredikhina, Z.A., Zakharychev, D.V., Gubaidullin, A.T., and Fayzullin, R.R. (2012) *CrystEngComm*, **14**, 648.
19. Noorduin, W.L., Vlieg, E., Kellog, R.M., and Kaptein, B. (2009) *Angew. Chem. Int. Ed.*, **48**, 9600.
20. Coquerel, G. (2007) in *Novel Optical Resolution Technologies*, vol. **269** (eds K. Sakai, N. Hirayama, and R. Tamura), Springer, Berlin, p. 1.
21. Pratt Brock, C., Schweizer, W.B., and Dunitz, J.D. (1991) *J. Am. Chem. Soc.*, **113**, 9811.

22. Galland, A., Dupray, V., Berton, B., Morin, S., Sanselme, M., Atmani, H., and Coquerel, G. (2009) *Cryst. Growth Des.*, **9**, 1851.
23. Kaemmerer, H., Lorenz, H., Black, S.N., and Seidel-Morgenstern, A. (2009) *Cryst. Growth Des.*, **9**, 1851.
24. Brandel, C., Amharar, Y., Rollinger, J., Griesser, U., Cartigny, Y., Petit, S., and Coquerel, G. (2013) *Mol. Pharmaceutics*, **10** (10), 3850–3861.
25. Gallis, H.E., Bougrioua, F., Oonk, H.A.J., van Ekeren, P.J., and van Miltenburg, J.C. (1996) *Thermochim. Acta*, **274**, 231.
26. Huang, J., Chen, S., Guzei, I.A., and Yu, L. (2006) *J. Am. Chem. Soc.*, **128**, 11985.
27. Wermester, N., Lambert, O., and Coquerel, G. (2008) *CrystEngComm*, **10**, 724.
28. Wermester, N., Aubin, E., Pauchet, M., Coste, S., and Coquerel, G. (2007) *Tetrahedron: Asymmetry*, **18**, 821.
29. Taratin, N.V., Lorenz, H., Kotelnikova, E.N., Glikin, A.E., Galland, A., Dupray, V., Coquerel, G., and Seidel-Morgenstern, A. (2012) *Cryst. Growth Des.*, **12**, 5882.
30. Vogt, F.G., Copley, R.C.B., Mueller, R.L., Spoors, G.P., Cacchio, T.N., Carlton, R.A., Katrincic, L.M., Kennady, J.M., Parsons, S., and Chetina, O.V. (2010) *Cryst. Growth Des.*, **10**, 2713.
31. Chen, S., Xi, H., Henry, R.F., Marsden, I., and Zhang, G.G.Z. (2010) *CrystEngComm*, **12**, 1485.
32. Cheng, X.-M. (1996) in *Annual Reports in Medicinal Chemistry*, vol. **31** (ed. J.A. Bristol), Academic Press, San Diego, CA.
33. Koda, A., Yanagihara, K., and Matsuura, N. (1991) *Agents Actions*, **34**, 369.
34. Ushio, T., Tamura, R., Takahashi, H., and Yamamoto, K. (1996) *Angew. Chem. Int. Ed.*, **35**, 2372.
35. Takahashi, H., Tamura, R., Fujimoto, D., Lepp, Z., Kobayashi, K., and Ushio, T. (2002) *Chirality*, **14**, 541.
36. Tamura, R., Fujimoto, D., Lepp, Z., Misaki, K., Miura, H., Takahashi, H., Ushio, T., Nakai, T., and Hirotsu, K. (2002) *J. Am. Chem. Soc.*, **124**, 13139.
37. Tamura, R., Iwama, S., and Gonnade, R.G. (2011) *CrystEngComm*, **13**, 5269.
38. Tamura, R., Iwama, S., and Takahashi, H. (2010) *Symmetry*, **2**, 112.
39. Tamura, R., Takahashi, H., Fujimoto, D., and Ushio, T. (2007) *Top. Curr. Chem.*, **269**.
40. Tamura, R. and Ushio, T. (2004) in *Enantiomer Separation: Fundamentals and Practical Methods* (ed. F. Toda), Kluwer Academic Publishers, Dordrecht.
41. Horiguchi, M., Okuhara, S., Shimano, E., Fujimoto, D., Takahashi, H., Tsue, H., and Tamura, R. (2008) *Cryst. Growth Des.*, **8**, 540.
42. Tamura, R., Mizuta, M., Yabunaka, S., Fujimoto, D., Ariga, T., Okuhara, S., Ikuma, N., Takahashi, H., and Tsue, H. (2006) *Chem. Eur. J.*, **12**, 3515.
43. Iwama, S., Horiguchi, M., Sato, H., Uchida, Y., Takahashi, H., Tsue, H., and Tamura, R. (2010) *Cryst. Growth Des.*, **10**, 2668.
44. Gonnade, R.G., Iwama, S., Sugiwake, R., Manoj, K., Takahashi, H., Tsue, H., and Tamura, R. (2012) *Chem. Commun.*, **48**, 2791.
45. Klussman, M., Izumi, T., White, A.J.P., Armstrong, A., and Blackmond, D.G. (2007) *J. Am. Chem. Soc.*, **129**, 7657.
46. Gonnade, R.G., Iwama, S., Mori, Y., Takahashi, H., Tsue, H., and Tamura, R. (2011) *Cryst. Growth Des.*, **11**, 607.
47. Iwama, S., Kuyama, K., Mori, Y., Manoj, K., Gonnade, R.G., Suzuki, K., Hughes, C.E., Williams, P.A., Harris, K.D.M., Veesler, S., Takahashi, H., Tsue, H., and Tamura, R. (2014) *Chem. Eur. J.*, **20**, 10343.
48. Furberg, S. and Hassel, O. (1950) *Acta Chem. Scand.*, **4**, 1020.
49. Martin, R.H. and Marchant, M.J. (1974) *Tetrahedron*, **30**, 343.
50. Green, B.S. and Knossow, M. (1981) *Science*, **214**, 795.
51. Davey, R.J., Black, S.N., Williams, L.J., McEwan, D., and Sadler, D.E. (1990) *J. Cryst. Growth*, **102**, 97.
52. Potter, G.A., Garcia, C., McCague, R., Adger, B., and Collet, A. (1996) *Angew. Chem. Int. Ed. Engl.*, **35**, 1666.
53. Berfeld, M., Zbaida, D., Leiserowitz, L., and Lahav, M. (1999) *Adv. Mater.*, **56**, 328.

54. Zbaida, D., Lahav, M., Drauz, K., Knaup, G., and Kottenhahn, M. (2000) *Tetrahedron*, **56**, 6645.
55. Addadi, L., Weinstein, S., Gati, E., Weissbuch, I., and Lahav, M. (1982) *J. Am. Chem. Soc.*, **104**, 4610.
56. Beilles, S., Cardinaël, P., Ndzié, E., Petit, S., and Coquerel, G. (2001) *Chem. Eng. Sci.*, **56**, 2281.
57. Gervais, C., Beilles, S., Cardinaël, P., Petit, S., and Coquerel, G. (2002) *J. Phys. Chem. B*, **106**, 646.
58. Torbeev, V.Y., Lyssenko, K.A., Kharybin, O.N., Antipin, M.Y., and Kostyanovsky, R.G. (2003) *J. Phys. Chem. B*, **107**, 13523.
59. van Eupen, J., Westheim, R., Deij, M., Meekes, H., Bennema, P., and Vlieg, E. (2009) *Int. J. Pharm.*, **368**, 146.
60. Coquerel, G., Levilain, G., Petit, M.-N., and Coste, S. (2011) Patent US20120178709.
61. Flack, H.D. and Bernardinelli, G. (2008) *Chirality*, **20**, 681.
62. Parsons, S., Pattison, P., and Flack, H.D. (2012) *Acta Crystallogr., Sect. A: Found. Crystallogr.*, A68, 6, 736.
63. Mahieux, J., Sanselme, M., Harthong, S., Melan, C., Aronica, C., Guy, L., and Coquerel, G. (2013) *Cryst. Growth Des.*, **13** (8).
64. Levilain, G. and Coquerel, G. (2010) Patent WO2011/07330.
65. Fowler, P.W. (2005) *Symmetry Cult. Sci.*, **16**, 321.
66. Coquerel, G. (2004) *J. Phys. IV France*, **113**, 11.

6
Conformational Disorder and Atropisomerism in Pharmaceutical Compounds

Attilio Cesàro, Barbara Bellich, Giovanna Giannini, and Alessandro Maiocchi

6.1
Premise: Conformational Energy Barriers in Flexible Molecules

There are many simple molecules of pharmaceutical relevance that show complex conformational disorder in the liquid and solid states. Conformational disorder is a major impediment to crystallization and generally favors the formation of amorphous materials. The obvious outcome is a substantial change in the material's physicochemical properties and in particular in the equilibrium between solution saturation and the solid state. The industrially relevant open issue deals with the full understanding of crystallization conditions and the prediction of the polymorphic forms for new (and old) drugs.

An attempt to analyze supersaturation and crystal growth during batch crystallization from solutions containing multiple conformers has been mechanistically presented in the literature on the basis of the so-called approach of the single right conformer [1]. Literature also reports on the strategies for controlling crystallization of conformationally flexible molecules, including conformational mimicry, solvent-mediated self-assembly, and template growth. In addition, the relation between conformational disorder and crystal polymorphism is another relevant issue and is nowadays the specific goal of extensive modeling with the use of hybrid *ab initio* computer simulations [2, 3].

A particular category of conformational isomerism has been named "atropisomerism," a term introduced by Kuhn to describe molecules with a chiral axis maintained by hindered rotation about single bonds [4]. Thus, atropisomerism can be attributed to a type of conformational or rotational isomerism in which the atropisomers can be experimentally separated if the barrier to rotation is large enough. The conventional energy barrier for atropisomers to be isolated at room temperature is approximately 25 kcal mol^{-1}. The most important issue is that, when the barrier to rotation between the interconverting atropisomers is very high, the interconversion is hindered to the point where they can be considered as thermodynamically distinct molecules. Under conditions of intermediate values of the energy barrier, the apparent solubility of a crystalline form containing one

Disordered Pharmaceutical Materials, First Edition. Edited by Marc Descamps.

atropisomer form only can sum up over all the several conformationally distinct molecules.

However, the simple and common example of the implication of conformational disorder in the crystallization process comes from the polymer field. Historically, the explicit connection between polymorphism and molecular conformation, referred to as *conformational polymorphism* [5], was apparently first made by Corradini [6]. Therefore, some effects of conformational flexibility on crystallization, namely conformational polymorphism and reduction of crystallization tendency, can be analyzed using examples from the literature, starting from principles of crystallization of polymer chains. The preferred molecular conformations observed in several polymorphic systems are usually correlated with the nature of the forces present in the crystals [7], keeping in mind that the molecular size and flexibility may affect the extent of crystallization and the dimension of crystals.

Thus, the problem of conformational flexibility in pharmaceutical compounds involves not only the thermodynamic principle of stability and the multiplicity of several distinct families of conformational states [8] but also the timescale for their interconversion. Besides the general principles and rules, some typical case studies provide a better working hypothesis on the research approaches and achievements in controlling strategies for these phenomena.

6.2
Conformational Topology and Crystallization of Chain Molecules

Most linear polymers are highly flexible molecules, which means they can exist in a very large number of conformations. The number of distinct conformations grows so rapidly with chain length that beyond degrees of polymerization of about 10–20, the geometry of polymer molecules can only be described statistically. The reason why chain molecules are so flexible is that the potential energy of a molecule depends on the relative positions of its atoms.

For a section of a linear chain molecule such as polyethylene (see the schematic representation of butane in Figure 6.1), the geometry can be specified in terms of bond lengths, bond angles (θ), and dihedral (torsion) angles (ϕ). Changing the geometry of a molecule away from its minimum energy causes its potential energy to rise. For most organic molecules, such as that of Figure 6.1, the change in the potential energy as a function of a bond angle (θ) or bond length is very rapid, while for a dihedral angle (ϕ) the energy change is often on the order of some kilocalorie

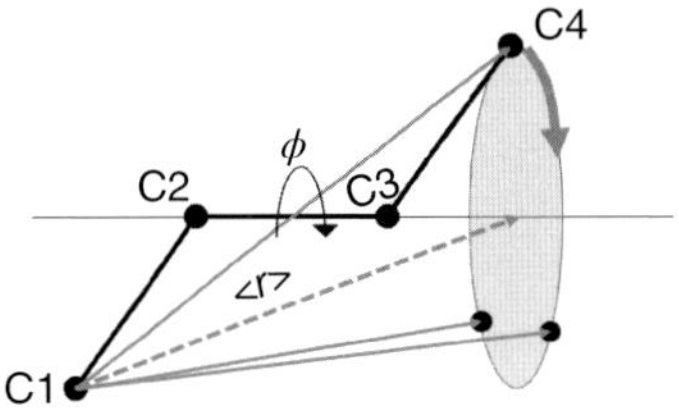

Figure 6.1 Schematic geometry of a portion of a polyethylene chain showing the parameters and the topological changes upon rotation about the central bond.

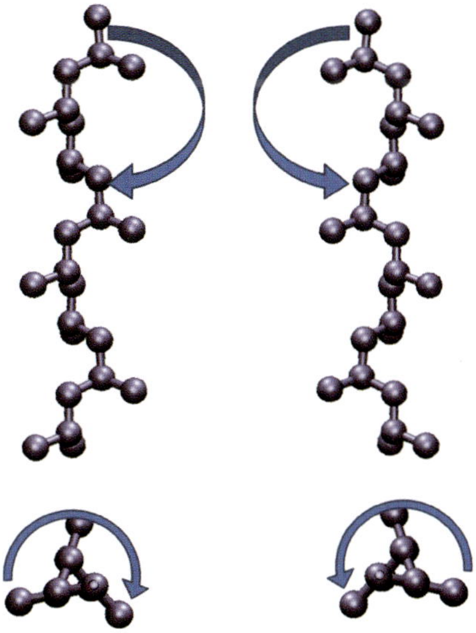

Figure 6.2 The two ternary helical conformations of isotactic polypropylene. The two forms are isoenergetic (enantiomers) and co-crystallize.

per mole. Thus, large changes in the chain geometry and the end-to-end distance r are obtained with even small changes in energy.

In addition to the extended conformation, as that shown in Figure 6.1 for C1–C2–C3–C4, simple vinyl polymers such as isotactic polypropylene may adopt two "enantiomeric" helical conformations as shown in Figure 6.2. The reason for the quotation marks is that the energetic barrier for changing the conformation is low and, upon melting, the two conformational states equally distribute along each polymer chain. Therefore, this is a typical example of helical chirality, which, however, is stable only in the solid state and completely disappears or equilibrates in solution.

In general, it is possible to state that the flexibility of polymers is almost exclusively associated with the torsional degrees of freedom. This, in turn, becomes the major obstacle for the crystallization of chain molecules, since for a polymeric chain the conformational equivalence along the sequence of monomers is a prerequisite for crystallization in lamellae crystals. There are two other conditions in polymers that concern the quite obvious regularity in the monomer's chemical and stereochemical constitution. A typical and seminal example of these requirements is provided by the polypropylene chain, as schematically represented in Figure 6.2. The regularity of the chemical constitution means that the chain is made up only by the propylene monomer polymerized in a strictly head-to-tail orientation, while the stereochemical requirement is characteristic, for example, of an isotactic chain. Constitutional and configurational order is thus a necessary, but still not sufficient, condition for crystallization, since the isotactic polypropylene chains can still be in a statistically disordered state. Indeed, what is additionally required is a driving force for the change of the rotational angles toward a regular conformation that enters the elementary cell.

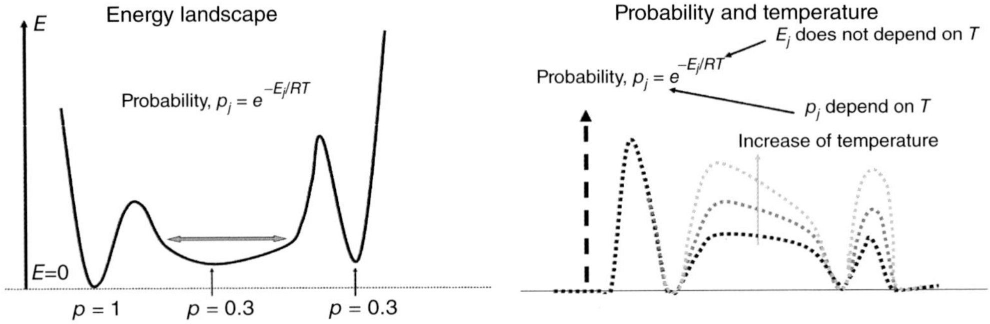

Figure 6.3 Schematic representation of the energy landscape for conformational conversion, showing the relation between energy and probability, including the effect of temperature.

Note that the energy itself is, by definition, temperature-independent. Therefore, it is the probability of a conformational state (of a family of conformational states) that depends on the temperature. This concept is related to the Boltzmann distribution and is illustrated in Figure 6.3, where the schematic energy landscape with three minima and the associated probabilities of the minima are represented. The figure also illustrates how, upon increasing the temperature, the minima with lower probability become more populated. It is immediately apparent that, at room temperature, bond angles and bond distances are very stiff, that is, they are confined to a narrow interval around their equilibrium positions. On the other hand, dihedral angles can change from the primary to the secondary minima rather easily, since the energy difference is often quite small, and the barrier is not too high, either.

In a statistically disordered state, all accessible conformational states are populated according to the Boltzmann distribution, which is intrinsically determined in a single isolated chain by the polymer's chemical structure and the temperature. By cooling the liquid phase (either pure liquid or solution), the conformational states with the lowest energy become more favored and the chain assumes a more ordered shape. The energetic and entropic contributions to obtain a sufficiently long, ordered segment is the crucial factor in the kinetics of polymer crystallization [9]. The implicit assumption made in the above description is that conformational states with high energy can convert into those with lower energy by overcoming the energy barriers between these states.

The above examples show how the interplay between conformational states and, therefore, of conformational energies determines the possibility to reach the crystalline conformation, but it cannot ensure that this state will be totally reached. Chain molecules are, indeed, the most evident examples of the coexistence of ordered crystalline domains interdispersed in a disordered amorphous material (Figure 6.4). Without entering into a kinetic-thermodynamic model that justifies this coexistence, let us simply state here that the higher the crystallization temperature, the larger the size of the crystallites and the higher the fraction of the crystalline material. As a practical consequence, the melting temperature of a

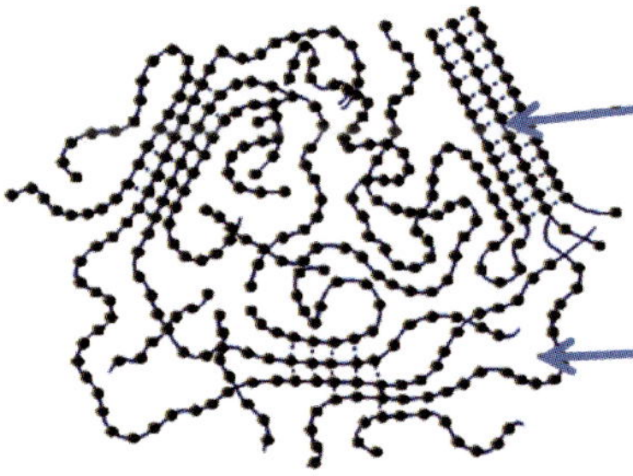

Figure 6.4 Schematic representation of a solid semicrystalline polymer: crystalline and amorphous forms are shown.

(semi)crystalline polymer depends on the temperature of its crystallization process, that is, the thermal history [10].

There are many relevant examples of polymorphic and mesomorphic phases in the crystalline state of polymers [11]. Conformational polymorphism is present either with difference in the chain conformational state or with different topological arrangement of the chains having the same conformational symmetry. In principle, all polymers that can adopt a helical conformation will give an arrangement in the solid state in the crystalline form. However, only those with a stereoregular sequence of monomers will eventually crystallize with a high degree of crystallinity, such as cellulose, amylose, and stereoregular poly(L-lactide). Still, whenever there is a multiplicity of conformational states separated by even modest energy barriers, then for both thermodynamic and kinetic reasons it can be difficult to reach a long-range crystalline state. An obvious parallel can be drawn between the concepts above summarized for polymers and the crystallization of conformationally disordered small molecules.

6.3 Conformational Polymorphism and Crystallization of Flexible Molecules

6.3.1 Conformational Polymorphism

Since the seminal work by Bernstein [12], the industrial interest in polymorphism has rapidly grown because it impacts pharmaceutical form discovery, patenting, and drug formulation [13–15]. However, although most of these concepts have received long-standing attention by polymer scientists, it has been often stressed in the literature that the subject of conformational polymorphism has not been always properly addressed, especially in terms of definitions. It has been recently suggested [5] that even the definitions of terms like *conformation* and conformer must be properly addressed, as they have been often interpreted and applied in a number of different ways. Referring to the IUPAC definitions (see also Figures 6.1 and 6.2), a conformation is "the spatial arrangement of the atoms affording distinction between stereoisomers which can be interconverted by rotations about formally single bonds," whereas a conformer "is one of a set of stereoisomers, each of which is characterized by a conformation corresponding to a distinct potential energy minimum" [16]. The definition justifies the historical use of the "rotational isomeric state model" [17] in the computational description by allowing every bond to adopt only one of a defined set of rotational states (i.e., dihedral angles). These states usually correspond to minima of potential energy, for instance the trans, gauche(+), and gauche(−) conformations.

Indeed, the common accepted meaning of polymorphism is that of different crystalline modifications of the same chemical substance. It has been often considered as an understatement because this definition of crystalline polymorphs implies only that they are characterized by different space groups, since the molecular packing mode is changed. A particular case is termed *conformational polymorphism* when different conformers of the same molecule occur in different crystal forms (Figure 6.5), or different conformers exist in the same unit cell.

The phenomenon of conformational polymorphism and its energetic background have been deeply investigated in the last decades by using extensive computational approaches and crystal structure prediction investigations. Indeed, not only the conformational space of the molecule must be accurately explored in physically realistic conditions but also the search for multiple minima and not the global minimum alone in the lattice energy must be carried out. Notably, conformer and lattice energy have been found compensatory in 16 polymorphic systems out of 23 structures analyzed in a recent overall survey by Nangia [18].

Two different views have been provided on the chance that conformationally flexible molecules might exist as polymorphs. On one hand, it has been argued that when different conformers lead to different crystal structures, the effective concentration of the conformer that leads to the observed form(s) is lowered in the Boltzmann population, which reduces the degree of supersaturation [8]. For example, crystallization of carbohydrates and alditols is affected by the relative energies of anomers in solution. Following this hypothesis, conformationally flexible molecules in the Cambridge Structural Database have been analyzed, and the conclusion is that the likelihood of similar or different conformers in polymorphic crystals is comparable; that is, conformational flexibility neither hinders nor favors polymorph frequency [19]. On the other hand, an analysis of the same issue from an energetic viewpoint led Nangia to conclude that conformational flexibility in organic molecules increases the likelihood of polymorphism. The conclusion of the physical organic approach to understanding conformational polymorphism was based on two points [18]. First, when several conformers are available in the crystallization milieu (solution or melt phase), they can form different hydrogen-bond synthons and close-packing motifs. Second, the intra and intermolecular energy compensation reduces total crystal energy differences, which increases the likelihood of polymorphism.

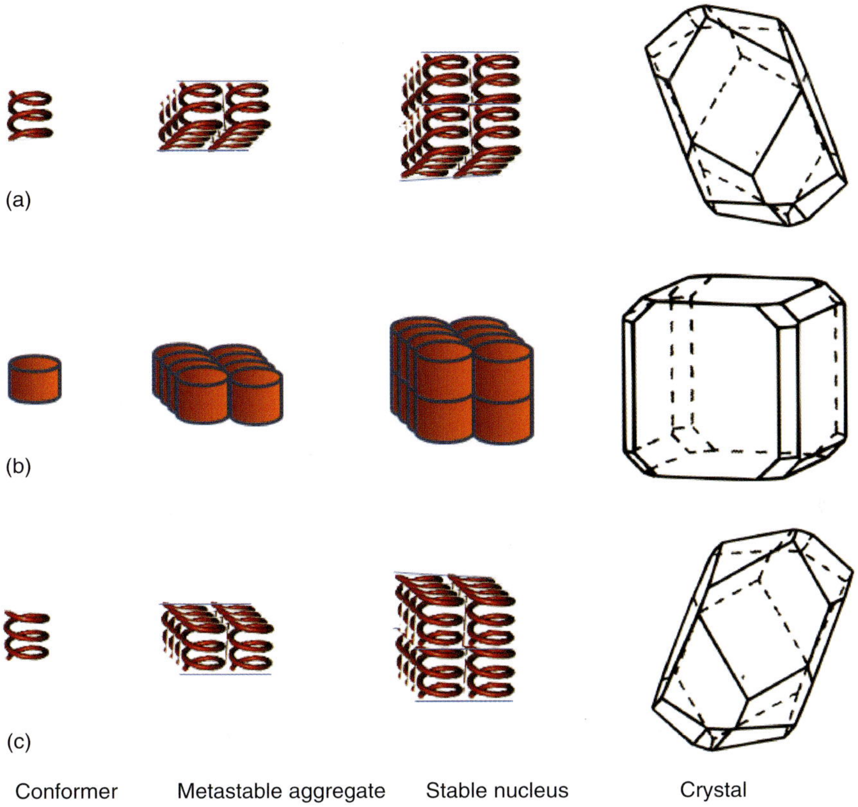

Figure 6.5 Example of solution crystallization of a conformationally flexible molecule. In the schematic representation, the molecules are assumed to exhibit a right (a) and a left (c) chiral conformation (the two enantiomeric helices) and a trans (b) conformation (the cylinder), giving different final crystal forms.

6.3.2
Flexibility and Crystallization

Whenever conformational flexibility exists, two effects can occur on crystallization, namely conformational polymorphism and reduction of crystallization tendency. On one hand, in the presence of conformational polymorphism, the preferred molecular conformations are correlated with the nature of the forces present in the crystals. On the other hand, for conformationally flexible molecules a reduction of crystallization tendency arises, in general, from the presence of multiple conformers in the crystallizing media. The question is whether molecules that populate distinct conformations having different energy and geometry may co-crystallize, or they interconvert into a single, well-defined conformational state. From the thermodynamic viewpoint, unless strong interactions are established between the organic solute molecules and the solvent, the most common trend is that an increase in solubility with the temperature is displayed. In some specific cases, however, it has been reported that for systems having low energy barriers between conformers, crystal forces can modify the conformation of the molecule into the one that is suitable for surface integration. Thus, both the energy difference between the conformers (thermodynamic control) and the energy barrier (kinetic control) must be taken into account [20].

The literature reports that certain molecules may not be able to overcome the energy barriers at the crystallization temperature, and, in some instances, they also crystallize in high-energy conformations. In the past, these peculiarities have been treated mainly from the phenomenological viewpoint. Nowadays, special techniques, including conformational mimicry, solvent-mediated self-assembly, and templated growth, have been devised to introduce molecular-level control into the crystallization process [8]. In addition, the potential of controlling the crystallization of conformationally flexible molecules has attracted growing attention of theoretical approaches, although the success depends on how close the molecular conformation in the solid state is to a gas-phase conformation, and, second, how sensitive the lattice energy search is to the packing-induced differences in the solid state. Although the use of highly accurate models for both the inter and the intramolecular forces would be an enormous improvement over the traditional force fields, the extension of this scheme to large organic molecules is currently still demanding from the computational viewpoint [21].

Schematically, in many cases the energy barrier between conformers is less than $10\,\text{kcal mol}^{-1}$, and it is expected, therefore, that more or less rapid interchange of conformational states facilitates conformational conversion from solution conformations to the crystal conformation. Under these circumstances, the impact of multiple conformations in solution on the crystallization kinetics is usually considered negligible, although crystalline conformational polymorphs can be generated by changing the crystallization parameters [22]. It follows that the transition between conformers is frequently considered as a nonlimiting step and is thus hidden in the determination of crystallization kinetics. The correct assessment of the solution conformational population is, therefore, the first necessary step for a clear picture of the crystallization mechanism.

The effects of conformational flexibility on crystallization can be appreciated from the inspection of Figure 6.1. Since multiple conformers (1, 2, etc.) exist in solution, the "right" conformer (1) that gives a desired crystal, say Polymorph 1, is "diluted," which in turn reduces the degree of supersaturation and the tendency of crystallization [23]. If the "right" conformer has high energy in solution, crystallization will slow down even more in proportion to the unfavorable Boltzmann population factor. As the crystallization begins, the solution must be "restocked" with the crystallizing conformer, at a rate that is a function of the barrier of conformer conversion and the temperature. It is easy to imagine that an unfavorable combination of these factors will lead to significantly decreased crystallization tendency of conformationally disordered molecules.

The effect just described can be also viewed in the reverse mode as the result of conformationally flexible molecules on the stability of crystals. The presence of other conformers in solution (or melt) has the same effect as impurities in that they cause a depression of the melting or dissolution temperature. Thermodynamically, the presence of other conformers stabilizes the solution (or melt), thus shifting the dissolution or melting equilibrium in favor of the solution (or melt).

6.3.3
Prediction of Conformational Polymorphs

The variability of molecular conformation in solution and the relative crystal lattice energies often suggest the following rules for conformational polymorphism: Facile isomeric conversion in solution and very similar crystal energies mean that more than one molecular conformation may crystallize out simultaneously to give concomitant conformational polymorphs. Depending on the geometry of a particular conformation, different specific interactions (e.g., H-bond or other weak interactions) and nonpolar packing motifs may lead to a metastable crystal structure. Therefore, any change in the molecular conformation will alter the strength of the weak direct and van der Waals interactions and, in turn, the preferred crystal-packing motif. The energy penalty in the molecular conformation is thus compensated by lattice energy gain from intermolecular interactions and close packing. It is clear that the main difficulties in predicting the structures of conformationally flexible molecules are (i) the most stable solution conformation may not result in the most stable crystal structure; (ii) the number of low-energy metastable conformations can be extremely high for effective simulations; and (iii) in principle, both conformation and lattice energy contributions to crystal structure stabilization must be "simultaneously" taken into account.

An emblematic literature example is that of aspirin [24] because of its therapeutic and commercial importance. Aspirin is a simple, flexible molecule with three main torsion angles, as shown in Figure 6.6.

Therefore, it is a good candidate for crystal structure prediction studies, and it has been heavily studied and frequently crystallized under a variety of conditions. The fairly successful crystal structure prediction was possible, however, only because the energy gap between the few lowest energy structures and the other hypothetical structures was larger than the changes in relative lattice energies U_{latt} produced by replacing the *ab initio* optimized conformation with the observed molecular structure. The considerable variations in some physical properties, such as morphology and dissolution rates, observed for aspirin have been often taken as possible evidence for a new polymorph; however, despite various efforts, until now only one crystal structure has been established by diffraction, and it is unlikely that there are other stable polymorphs.

In another family of molecules, that of highly flexible alcohols and sugars, the impressive series of crystal structure prediction studies led to the conclusion that the force field accuracy was the major issue. Indeed, the development of a highly accurate *ab initio* based model potential for calculating the intermolecular lattice energy was attempted, which generally gave superior results in crystal structure prediction studies compared to standard empirical force fields [24]. Still, a number

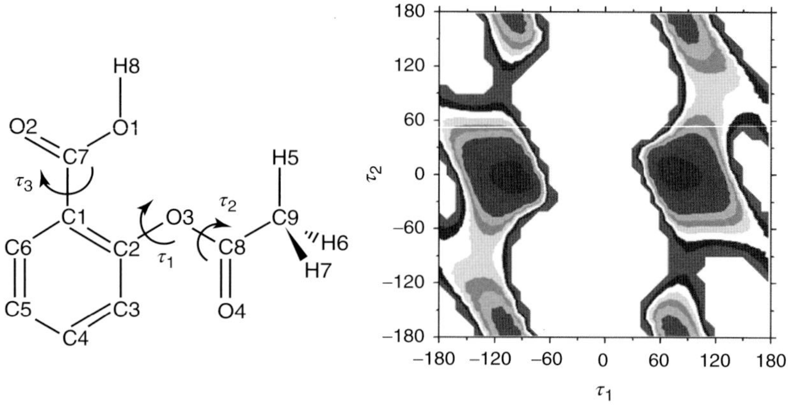

Figure 6.6 Molecular structure and conformational energy landscape of aspirin as a function of angles τ_1 and τ_2 (all other parameters fixed as in the conformer closest to that of the crystal structure); Adapted from Ouvrard. and Price [24]. Reproduced with permission of American Chemical Society.

of semiempirical rules, not surprisingly, govern this area. Indeed, the conformational flexibility of alditol carbon chains is evident from the variety of straight and bent geometries observed in the crystals. The correlation between conformation and configuration is based on the repulsion of parallel second-neighbor C−OH interactions [25]. This rule implies that certain alditols (e.g., mannitol) preferentially assume extended chain conformation as that in aqueous solution or melt conformation, while others (e.g., sorbitol) take on bent-chain conformations. This rule summarizes a body of crystallographic data and is substantiated by a series of solution NMR studies [26], but has a singular exception involving sorbitol. The crystallization of cyclic carbohydrates, either monomers or dimers (ribose, glucose, maltose, lactose, etc.), has similar features to that of alditols and is presented in the following section. In principle, in the same way the conformational equilibrium affects crystallization tendency, so should the configurational equilibrium. A particular feature of carbohydrates is that conformational freedom of disaccharides is commonly shown in the Ramachandran maps by minima that usually, but not always, include the crystalline conformation [27].

6.4 Conformational Flexibility of Ring Molecules: Carbohydrates

Carbohydrate monomers belong to a particular family of molecules characterized by a large number of conformational isomers in solution. Indeed, they not only exhibit several stable conformational states in the cyclic form but also exist in solution as mixtures of anomers (configurational isomers). The occurrence of this phenomenon is due to the opening of the ring structure in the aqueous

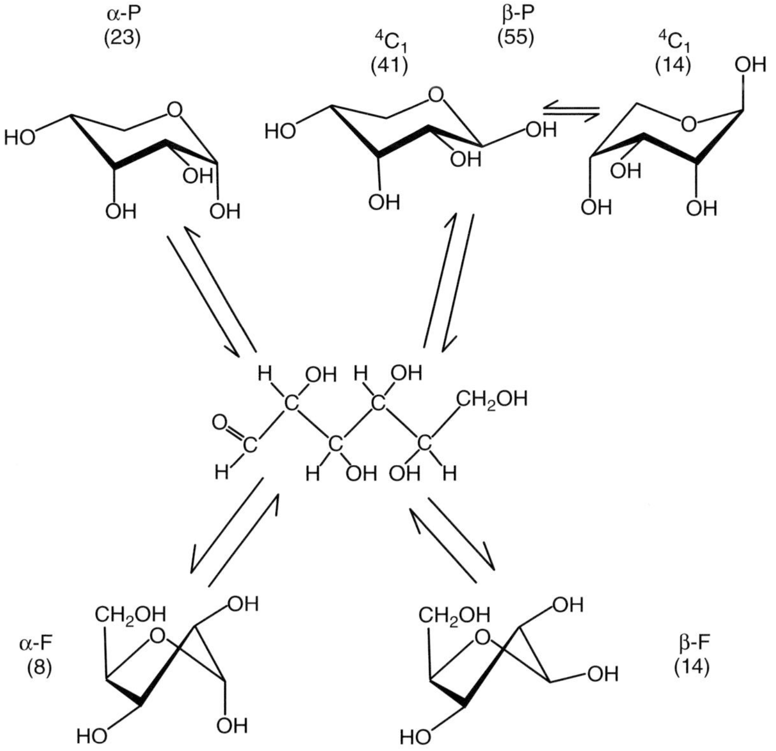

Figure 6.7 Schematic representation of the complex equilibria of ribose in water. The percentage compositions are reported for the relevant conformations (the drawings do not necessarily correspond to the minimum conformational energy). Upper and lower numbers indicate the positions of C1 and C4 relative to the ring plane.

solution and is particularly complex in the example of D-ribose [28]. The linear open chain can fold into five- or six-atom rings (called furanose (F) and pyranose (P) forms, respectively), and each of them may be locked in the α- or β-anomer. Figure 6.7 shows the selected low-energy isomers of D-ribose, with their percentage composition in brackets. For the monosaccharide D-ribose, in addition to the ring-opening interconversion, the additional equilibrium between $^4C_1$ and $^1C_4$ is shown for the β-anomer.

Partially simpler, a glucose solution at equilibrium contains almost exclusively a mixture of the β-anomer (64%) and the α-anomer (36%), both of which can produce crystals (α-form as a monohydrate and an anhydrate and β-form as an anhydrate). Despite the greater solution stability of the β-anomer [26], α-D-glucose hydrate is the stable crystal form below 50 °C, which is less soluble. Above 50 °C, the α-anhydrate form is obtained, and at still higher temperatures β-D-glucose is formed [29].

Similar to glucose, for the disaccharide lactose the equilibrium anomeric composition is 63% β-form and 37% α-form in aqueous solutions. An α-monohydrate normally crystallizes from water, and a β-anhydrate precipitates above 93.5 °C [29]. By spray-drying a solution prepared with a commercial anhydrous lactose (20% α), an amorphous solid containing 24% α is obtained, which crystallized in humid atmosphere into a mixture of 29% α-monohydrate and 71% β-anhydrate [30]. In this example, the change in anomer composition is small (5%) during crystallization, perhaps because of the slow rate of anomer conversion in the amorphous solid. Consequently, the crystallization of one anomer is independent of the other, and both crystallize. This feature is in contrast with the outcome of lactose crystallization from solutions. Because the anomer conversion is rapid in solutions, the nucleation of one anomer can convert all molecules into crystals of that anomer.

6.5
Hindered Conformational Isomerism: Atropisomerism

The butane molecular unit depicted in Figure 6.1 can be described in terms of three distinct conformers, two of them characterized as chiral images. However, their rapid interconversion ensures thermodynamic equilibration and absence of any detectable presence of enantiomeric forms. However, this axiom is valid inasmuch as the energy barrier separating the two enantiomers is sufficiently small. The fact that a large number of molecules possess enantiomeric forms separated by barriers with increasing values of energy makes necessary a formal, though arbitrary, definition of this limit. The word "atropisomerism" was introduced by Kuhn to identify a class of monomeric substances characterized by the presence of hindered conformational rotation [4]. In this case, the important step required for converting all molecules into the same conformational state is that the rate of conversion, kinetically speaking, be high enough with respect of the cooling rate and the rate of formation of nuclei; otherwise, the crystalline state may result from a co-crystallization of the several stereoisomers. The phenomenon of atropisomerism has been recently reproposed to the scientific community in a couple of reviews and articles [31]. It implies that conformational interconversion is not rapid, with times that may reach minutes or years. The aim of this section is to illustrate this particular phenomenon by focusing on the implication of the physicochemical properties of pharmaceutical materials, and to provide a few example cases that highlight the importance and the need for new categories also in terms of regulatory statements.

By using the concept of hindered conformational rotation, it is easy to understand why atropisomerism is often defined as another type of enantiomerism. However, it is a chirality in molecules that are devoid of stereogenic centers and that have a long interconversion time. The steric hindrance, most commonly about a single bond, is generated by bulky groups or substituents, generally on aromatic rings. Notably, crystal structures of atropisomers are quite rare (a survey in 2004 of over 300 000 entries gave only four hits). In some cases, a pair of atropisomers crystallize concomitantly, yielding two distinct needle-like crystals that can be distinguished in orthoscopy (quoted in [32]). However, atropisomerism does not always arise because of steric hindrance; it can also be caused by noncovalent intramolecular bonds, such as a hydroxyl group that connects via a hydrogen bond to an oxygen atom of either an amide or a sulfonamide. Finally, two tautomers can be crystallized separately, separating the two entities of the same compound that coexist in solution (although in this case, strictly speaking, the two primary structures are different).

Perhaps, the most striking and easily visualized example of conformational enantiomers, as quoted in many textbooks, is that of hexahelicene, made up of six benzene rings condensed in a nonlinear fashion (Figure 6.8). The unique helical, nonplanar π-electron system and the very high rotational values drew considerable attention to helicene and derivatives since its first synthesis in 1955. Crystalline forms of hexahelicene were initially believed to be characterized either by co-crystallization of the two enantiomeric species or by disordered co-crystallization. More recently, alternating stacked layers of crystals of D- and L-forms were recognized [33].

However, atropisomers can, in general, slowly crystallize if more conformational equilibria exist, and only one enantiomer or the two chiral enantiomers can actually be in the unit cell.

Another relevant example is that of a solid-state transition involving atropisomers presented for a molecular triad of the type A–B–A [34], in which two phenyl-substituted groups (A) are linked to a central aromatic or heteroaromatic unit (B). Such a molecule can exist as *syn* or anti conformers or atropisomers, having a C or S shape, respectively (this geometric feature has also been termed in

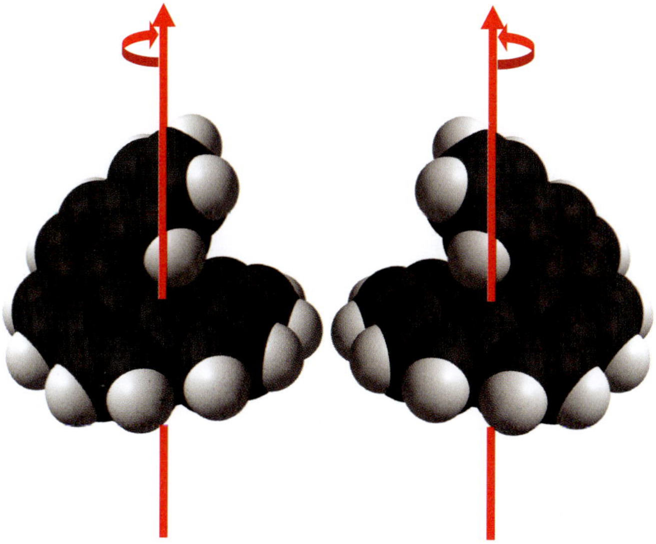

Figure 6.8 Molecular conformation of the two enantiomeric forms of hexahelicene, devoid of carbon atom stereocenters.

the past as *cis* and *trans*). Given the energy barrier, the transition from one form (anti) to the other (syn) is quite hampered at room temperature, either in solution (atropisomerism) or in the solid state, but not at sufficiently high temperature. Indeed, the crystalline polymorph syn can be crystallized at room temperature, but it converts into the anti form upon increasing the temperature. In addition, the specific compound quoted (anti) was crystallized as a pseudo-racemate (solid solution of enantiomers), a rather uncommon behavior for chiral compounds in the solid state.

Some ambiguity is intrinsic in the term atropisomerism, insofar as the quantification of the interconversion time is concerned, because the original boundary timescale between atropisomers and conformers (and therefore between configuration and conformation) is given by an arbitrary definition [4] that atropisomers are conformers that interconvert with a half-life of more than 1000 s at a given temperature. A schematic picture showing the correspondence between molecular structures, energy barrier, and half-life timescale is provided in Figure 6.9 with reference to the average instrumental timescale.

No reference is therefore made to a specific instrumental timescale that could either distinguish or not between different atropisomeric forms. Indeed, most spectroscopic methods are able to identify different forms having interconversion half-life of even less than 100 s, which is usually taken as the thermodynamic detection limit. The statement above is essential for separating the solution interconversion from the occurrence of solid-state transitions from one form to another.

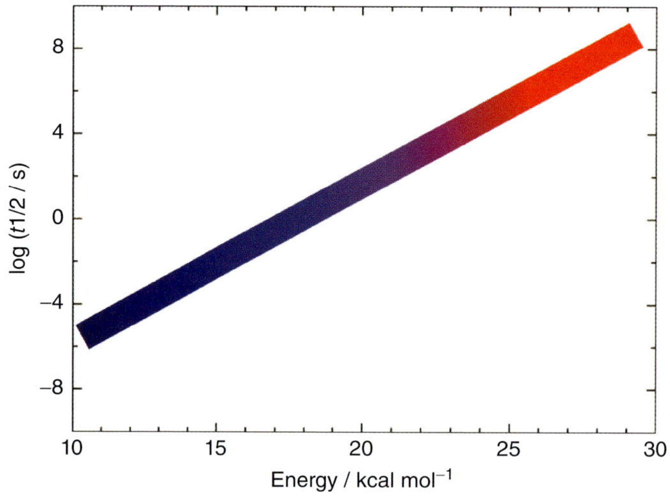

Figure 6.9 Naive correlation between timescale and energy of molecular processes such as that of molecular interconversion of atropisomers, with increasing energy and interconversion time from blue to red.

6.5.1 Atropisomerism in Iodinated Contrast Media (ICM)

This last section is dedicated to a class of molecules that, although not true drugs, are highly prevalent in radiology as contrast media for angiography. The global contrast media market boasted of a total value of $6.2 billion in 2012 and is expected to register a growth rate of about 7% per year. Indeed, contrast media signify a vital requirement of medical imaging, as these agents provide better images of body structures and organs when visualized during medical procedures such as X-ray, computed tomography (CT), or magnetic resonance imaging (MRI). In the current use, X-ray contrast media solutions have high concentrations (up to 1 M), owing to the need to increase as much as possible the contrast of soft tissues in the images for diagnostic purposes.

The most common molecular motif for an X-ray contrast agent is the one containing three iodine atoms covalently bound to an aromatic ring buried by other polar substituents that increase water solubility. The three iodine atoms covalently linked in positions 2, 4, and 6 (Figure 6.10) make the rotation of substituents around the single bond aril–N and aril–CO hindered, giving rise to atropisomerism. Indeed, due to the steric hindrance of iodine atoms, the rotation of the three hydrophilic side chains, depending on the groups present, is hampered, and therefore (metastable) conformational isomers that are not easily convertible (atropisomers) can be simultaneously detected in solution as different forms [35].

The nomenclature of the several atropisomers is illustrated for the specific case of two structurally similar molecules, namely iopamidol and iomeprol, which are structural isomers. Figure 6.11 reports the several possible atropisomers of ICM (iodinated contrast media) that can be present in solution as a consequence of the hampered rotation around single bonds linked to the aromatic ring bearing three iodine atoms, responsible of the steric hindrance.

A peculiar example is represented by the two ICM iomeprol and iopamidol (Figure 6.12). They are structural isomers that differ in the position of a methyl

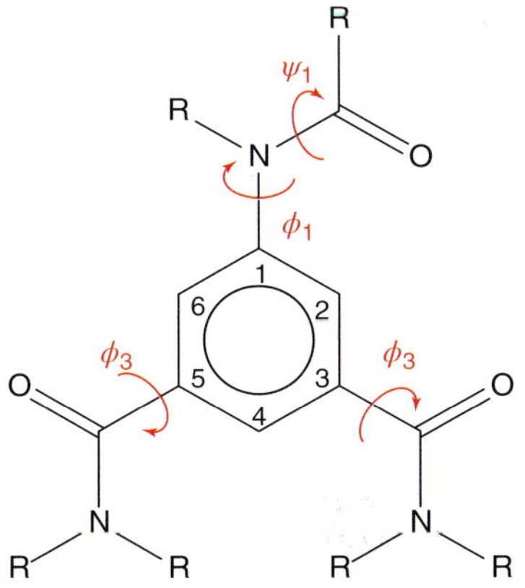

Figure 6.10 General scheme of the three hindered rotations in ICM molecules (the three arrows identify the rotations about the single bonds).

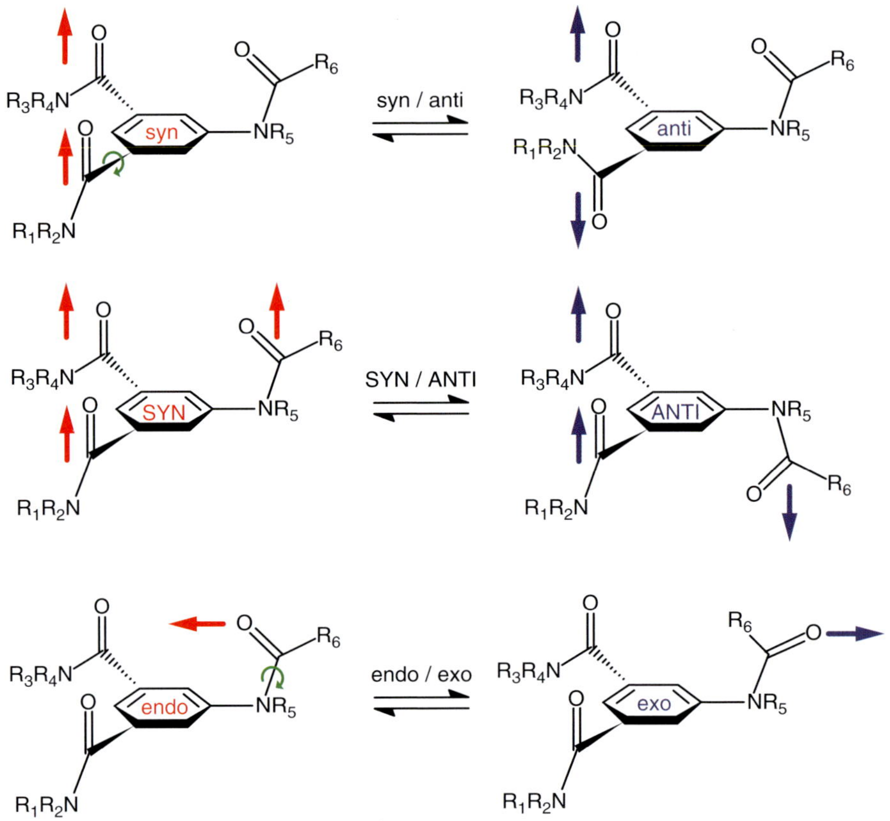

Figure 6.11 Representation of the isomeric conformations of ICM:
• syn/anti: generated by the two possible relative orientations of the two carbonyl in positions 3 and 5 with respect to the aromatic ring plane;
• SYN/ANTI: generated by the two possible orientations of the anilido carbonyl with respect to the other carbonyls when the molecule is in syn form (this atropisomerism does not exist in the anti conformer);
• endo/exo: generated by the *cis–trans* equilibrium in the anilido moiety. (endo in case the carbonyl group is pointing inwards and *exo* in case it is pointing outwards).
In addition, *Z*/*E* equilibria at the level of the amidic bonds are present in the branches in positions 3 and 5. Adapted with permission from [39]. Copyright 2015, American Chemical Society.

group in the aril–N side chain and in the C chain length in the two aril–CO side chains (not directly relevant for the phenomenon of atropisomerism). In case of iomeprol, the methyl group is bonded to the N atom in the anilido carbonyl group, while in the case of iopamidol this group is linked to the C-terminal of the same side chain. The small difference in the position of the methyl group has important implications in the possible atropisomeric conformations in solution.

Indeed, it is worth mentioning that for iopamidol the rotation of the side chain around the aril–N bond is not hampered because of the presence of a H atom

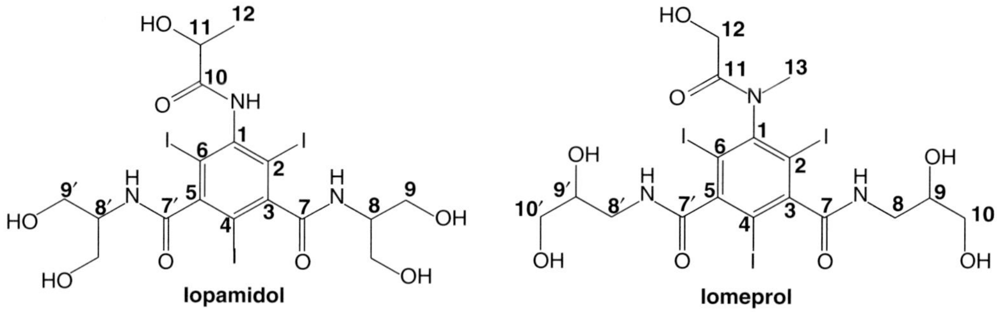

Figure 6.12 Molecular structures of iopamidol and iomeprol with carbon atom numbering; "primed" notation refers to symmetrical side arms.

instead of the $-CH_3$. This means that for iopamidol the atropisomeric distinction SYN/ANTI is not made. Thus the number of possible atropisomeric conformations is lower, since the only two possible forms are syn and anti. On the contrary, iomeprol is characterized by all the three mentioned hindered rotations of the aril–CO and aril–N single bonds. Therefore a higher number of possible atropisomeric forms are expected.

The atropisomerism phenomenon, studied by NMR [36] and by high-performance liquid chromatography (HPLC) [37], revealed the presence of multiple conformers interconverting with energetic barriers ranging from 17 to 20 $kcal\,mol^{-1}$ [38]. A more detailed study on iopamidol and iomeprol and on their precursors by NMR provided the relative population of each forms [39], as given in Table 6.1.

It is worth mentioning that atropisomeric forms that do not exist in solution, due to the rapid interconversion time, are, on the contrary, significant in the solid state. Such an example, iopamidol, does not show atropisomerism in the aril N side chain, but in the solid state this chain can be SYN or ANTI when the other two symmetric side chains are of course in the syn conformation. This issue is governed by the thermodynamic stability of the two forms. Furthermore, the solid state is driven by the intermolecular association and hydration phenomena occurring in solution of ICM, characterized by a balance of polar and nonpolar interactions. The nature of the molecular interactions, as well as some apparent metastability of these systems, is still a matter of debate. One important aspect is that molecules in

Table 6.1 Relative populations of atropisomers in solution of iopamidol and iomeprol.

Atropisomer forms	Iopamidol	Iomeprol
syn:anti	29:71	33:67
SYN:ANTI	Not present	50:50
endo:exo	Not present	23:77
E:Z	Not present	Not present

solution at high concentration form "flickering clusters" with a peculiar dynamic aggregation at a borderline state between supersaturation and primary nucleation. Therefore, at this stage, small perturbations of the system can cause ordered aggregation of the molecules in solution, giving rise to a primary nucleation, with dramatic consequences on the stability of the solution.

Given the kinetic hampering of the atropisomeric forms in solution, one would expect that iopamidol crystallization behavior could be fairly complicated and that the right conformer "anti" could be the most probable form in the solid crystalline state. To date, three crystalline polymorphs, two hydrate and one anhydrate forms, have been isolated. Among these forms, only one (a pentahydrate) shows the anti form in the crystalline packing, while both the anhydrate and the monohydrate have the syn form.

It should also be noted that in the solid form, the additional isomeric ANTI/SYN is also possible, although freely rotating in solution. Starting with each of the three known polymorphs, it is worth mentioning that several other polymorphs are obtained by thermal treatment, thereby giving more than 10 polymorphs that are not yet fully classified.

Several polymorphs are also known for iomeprol; however, a more complex disorder needs to be illustrated. Iomeprol structure has two enantiomeric centers, and the commercial product is a mixture of three enantiomers, rr, ss, and rs (in the ratio of 1 : 1 : 2). Each of them is a chemically distinct species giving all combination of anti/syn, ANTI/SYN, and exo/endo atropisomeric forms. Thus, a total of 72 enantio-atropisomer species could be present in solution, some of them at very low percentage according to NMR studies [39]. As a matter of fact, iomeprol crystallization from aqueous solution is not known, and crystallization in the form of anhydrate occurs only from hydroalcoholic solution. This crystalline form shows thermal interconversion to other polymorphs.

In conclusion, both iopamidol and iomeprol show a rich variety of atropisomeric forms in solution, and therefore concentrated solutions are nominally oversaturated in the presence of slow interconverting atropisomers. This is one of the factors hampering their crystallization from aqueous solutions.

6.6
Conclusion

The phenomenon illustrated in Figure 6.5 concerns a system with pathways originating from different conformers and leading to different mature crystals. Thus, the tendency of crystallization may be significantly reduced by conformational flexibility. Since flexible molecules exist in solutions or melts as mixtures of energetically similar conformers, the process of crystallization must select the "right" conformers from among the "wrong" ones, a difficulty not encountered by rigid molecules and analogous to that faced by the crystallization of enantiomers from a racemate [40]. Thus, the crystallization mechanism can be described as a sequential multistep process. In solution, molecules first associate into molecular clusters

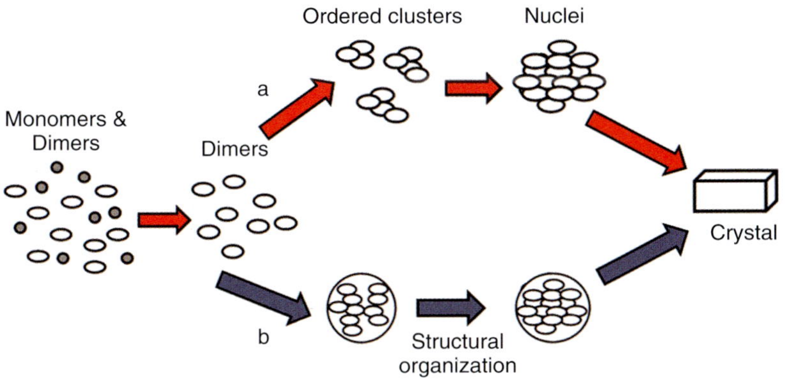

Figure 6.13 Schematic mechanism of crystal formation from solution aggregates (a) via assembly of ordered clusters and nuclei or (b) via random aggregation and structural organization into nuclei.

(pre-nucleation aggregates), whose structure often resembles that of the mature crystal. These pre-nucleation aggregates then assemble into crystal nuclei, and crystal nuclei finally grow into mature crystals. Conformational flexibility introduces potential complications to the crystallization process. For crystallization from solution, the ordering mechanism of conformationally disordered molecules may be schematically illustrated by a process in which either the final conformation entering in the crystal is selected at the stage of nucleation or the aggregation of statistically disordered conformations is the first step promoting subsequent ordering. This is schematically illustrated in Figure 6.13 with two different (competing) pathways that may present further differences. The final crystalline form may differ not only in the mode of packing but also in molecular conformation (conformational polymorphism). In all cases, there is a reduced tendency of crystallization, since the process of crystallization must proceed by incorporating a selected conformer among the mixtures of energetically similar conformers. Some atropisomeric molecules are a clear example of the competition between interconversion and crystallization.

In summary, conformational flexibility of organic molecules apparently introduces more structural options in the crystallization but also potential complications, such as conformational polymorphism and reduced crystallization tendency. Although the variety of conformers and the selected polymorph could be anticipated to some extent from the nature of crystal forces, it has been shown that it is extremely difficult to predict straightforwardly the crystal conformation and polymorphism for conformationally flexible molecules. Given these problems of multiple conformers, also the reduced crystallization tendency of conformationally flexible molecules should be approached with greater caution and with attention to each stage of the crystallization process.

As a final comment, it should be recalled that crystallization of polymorphs depends on the solvent conditions [41]. Indeed, it is often neglected that the nature of the solvent can have a substantial effect in controlling the conformational populations. Therefore, the solute–solvent interaction has to be explored in terms of changes of conformational population, its temperature dependence, and solubility curve of the selected polymorph, in order to be able to control polymorph crystallization.

Acknowledgments

Beneficial discussion with Dr F. Uggeri is warmly acknowledged. The chapter has been prepared under a research collaboration with Bracco Imaging spa, Italy.

References

1. Derdour, L. and Skliar, D. (2014) *Chem. Eng. Sci.*, **106**, 275–292.
2. Byrn, S.R., Pfeiffer, R.R., Stephenson, G., Grant, D.J.W., and Gleason, W.B. (1994) *Chem. Mater.*, **6** (8), 1148–1158.
3. Delaney, S.P., Pan, D., Galella, M., Yin, S.X., and Korter, T.M. (2012) *Cryst. Growth Des.*, **12** (10), 5017–5024.
4. Oki, M. (1983) *Top. Stereochem.*, **14** (1), 1–81.
5. Cruz-Cabeza, A.J. and Bernstein, J. (2013) *Chem. Rev.*, **114** (4), 2170–2191.
6. Corradini, P. (1973) *Chem. Ind. (Milan)*, **55**, 122.
7. Price, S.L. (2008) *Acc. Chem. Res.*, **42**, 117.
8. Yu, L., Reutzel-Edens, S.M., and Mitchell, C.A. (2000) *Org. Process Res. Dev.*, **4** (5), 396–402.
9. Mandelkern, L. (1956) *Chem. Rev.*, **56** (5), 903–958.
10. Wunderlich, B. (1989) *Pure Appl. Chem.*, **61** (8), 1347–1351.
11. Allegra, G. and Meille, S.V. (2004) *Macromolecules*, **37** (9), 3487–3496.
12. Bernstein, J. (1987) in *Organic Solid State Chemistry* (ed G.R. Desiraju), Elsevier, Amsterdam, pp. 471–518.
13. Hilfiker, R. (ed) (2006) *Polymorphism in the Pharmaceutical Industry*, Weinheim, Wiley-VCH Verlag GmbH.
14. Brittain, H.G. (2007) *J. Pharm. Sci.*, **96**, 705–728.
15. Moulton, B. and Zaworotko, M.J. (2001) *Chem. Rev.*, **101**, 1629–1658.
16. Moss, G.P. (1996) *Pure Appl. Chem.*, **68**, 2193.
17. Flory, P.J. (1969) *Statistical Mechanics of Chain Molecules*, John Wiley & Sons, Inc., New York.
18. Nangia, A. (2008) *Acc. Chem. Res.*, **41** (5), 595–604.
19. Bernal, I. (2008). Is Polymorphism Caused by Molecular Conformational Changes?. In *Models, Mysteries and Magic of Molecules* (pp. 137-165). Springer Dordrecht, Netherlands.
20. Bernstein, J. and Hagler, A.T. (1978) *J. Am. Chem. Soc.*, **100** (3), 673–681.
21. Price, S.L. (2014) *Chem. Soc. Rev.*, **43** (7), 2098–2111.
22. Mangin, D., Puel, F., and Veesler, S. (2009) *Org. Process Res. Dev.*, **13** (6), 1241–1253.
23. Derdour, L. and Skliar, D. (2012) *Cryst. Growth Des.*, **12** (11), 5180–5187.
24. Ouvrard, C. and Price, S.L. (2004) *Cryst. Growth Des.*, **4** (6), 1119–1127.
25. Jeffrey, G.A. and Kim, H.S. (1970) *Carbohydr. Res.*, **14** (2), 207–216.
26. Angyal, S.J. and Le Fur, R. (1980) *Carbohydr. Res.*, **84** (2), 201–209.
27. French, A.D., Kelterer, A.M., Johnson, G.P., Dowd, M.K., and Cramer, C.J. (2000) *J. Mol. Graphics Modell.*, **18** (2), 95–107.
28. Šišak, D., McCusker, L.B., Zandomeneghi, G., Meier, B.H., Bläser, D., Boese, R. *et al.* (2010) *Angew. Chem. Int. Ed.*, **49** (26), 4503–4505.
29. Roos, Y. (1993) *Carbohydr. Res.*, **238**, 39–48.
30. Schmitt, E.A., Law, D., and Zhang, G.G. (1999) *J. Pharm. Sci.*, **88** (3), 291–296.

31. LaPlante, S.R., Fader, L.D., Fandrick, K.R., Fandrick, D.R., Hucke, O., Kemper, R. *et al* (2011) *J. Med. Chem.*, **54** (20), 7005–7022.

32. Ochsenbein, P. and Schenk, K.J. (2006) Crystallography for polymorphs, Chapter 6, in *Polymorphism: in the Pharmaceutical Industry* (ed R. Hilfiker), Wiley-VCH Verlag GmbH.

33. Green, B.S. and Knossow, M. (1981) *Science*, **214** (4522), 795–797.

34. Einhorn, C., Durif, A., Averbuch, M.T., and Einhorn, J. (2001) *Angew. Chem. Int. Ed.*, **40** (10), 1926–1929.

35. Pitré, D. and Felder, E. (1980) *Invest. Radiol.*, **15**, S301–S309.

36. Bradamante, S. and Vittadini, G. (1987) *Magn. Reson. Chem.*, **25** (4), 283–292.

37. Gallotti, A., Uggeri, F., Favilla, A., Cabrini, M., and de Haën, C. (1994) *Eur. J. Radiol.*, **18**, S1–S12.

38. Krause, W., & Schneider, P. W. (2002). In *Contrast Agents II* (pp. 107–150). Springer Berlin, Heidelberg.

39. Fontanive, L., D'Amelio, N., Cesàro, A. *et al.* (2015) *Mol. Pharm.*, **12**, 1939–1950.

40. Bernstein, J. (1987) *Org. Solid State Chem.*, **32**, 471–517.

41. Threlfall, T. (2000) *Org. Process Res. Dev.*, **4** (5), 384–390.

7
Tautomerism in Drug Delivery

Zaneta Wojnarowska and Marian Paluch

According to common knowledge, tautomerism is observed when one chemical compound is represented by two or more molecular structures that are related by an intramolecular movement of hydrogen between different polar atoms and the rearrangement of double bonds [1]. The most commonly observed examples of this phenomenon are reversible transformations between ketone and enol, amide and imidic acid, lactam and lactim, or enamine and imine forms. On the other hand, the tautomeric reactions in which the heterocyclic ring is opened and closed is usually called ring–chain tautomerism or mutarotation in the case of carbohydrates chemistry.

Tautomerization attracts the attention of scientists from many disciplines, including physics, organic and biochemistry, and pharmaceutical science. It is of great importance especially in drug industry because there are a number of active pharmaceutical ingredients (APIs), biopharmaceuticals, and chemical excipients (e.g., saccharides) that readily convert into other isomers when their crystalline structure is lost. According to the literature data, 26% of commercially available APIs reveal the ability to exist in more than one chemical form (Figure 7.1) [2].

In this group one can find

- medicines from sulfonamide class (a structural class of compounds broadly prevalent in the pharmaceutical field), for example, sulfadoxine (antibiotic) [3], indapamide [4] (thiazide diuretics), glipizide, glibenclamide (GCM), glimepiride, gliclazide (commonly prescribed hypoglycemic agents) [5], sildenafil [6] (a pharmaceutical used to treat erectile dysfunction), etoricoxib [7] (COX-2 inhibitor) or piroxicam [8] (non-steroidal anti-inflammatory drug);
- pyrimidine derivatives being components of barbitals [9] (e.g., phenobarbital) or biopharmaceuticals [10] (containing nucleobases);
- porphyrins that undergo NH tautomerism [11];
- antibiotics from the tetracycline group that can be presented by even 10 tautomers [12], as well as macrolides such as erythromycin [13] and many others.

The isomeric structures of several APIs are depicted in Figures 7.2 and 7.3.

The proton transfer reaction is of a great importance, especially in view of the growing interest in amorphous drugs, characterized by better solubility and

Disordered Pharmaceutical Materials, First Edition. Edited by Marc Descamps.

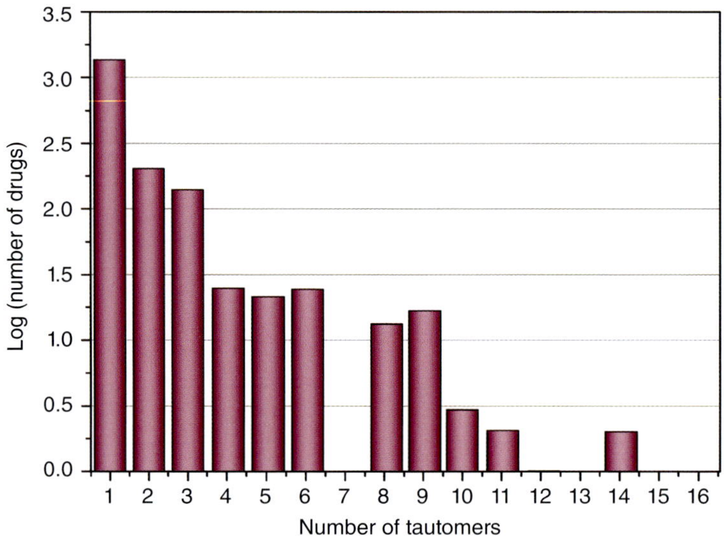

Figure 7.1 Frequency distribution of tautomers of a marketed drug. Adapted from Martin [2]. Reproduced with permission of Springer.

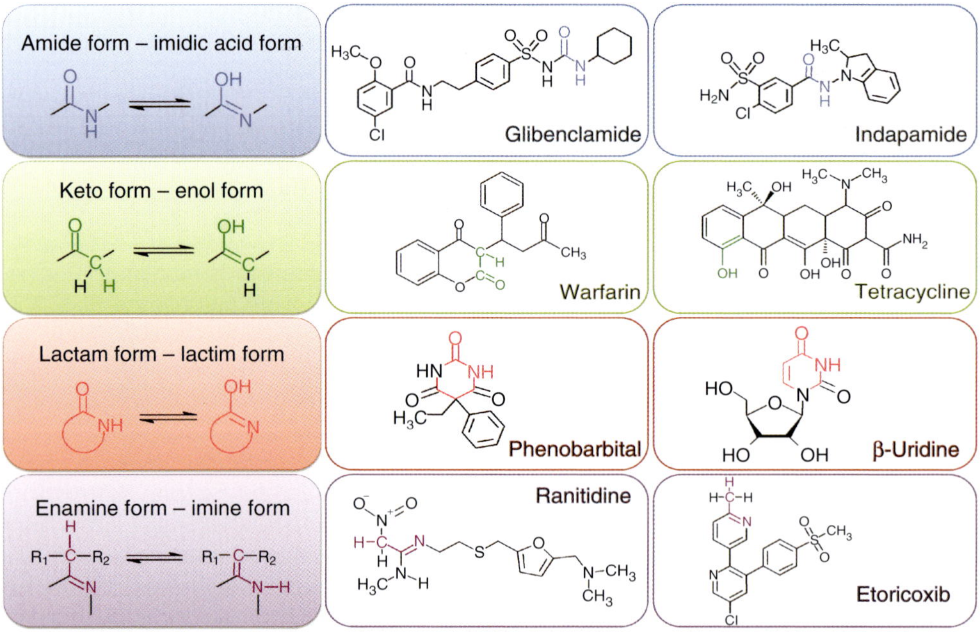

Figure 7.2 Most commonly observed tautomeric pairs. Taken from [4, 5, 7, 9, 12, 14–16].

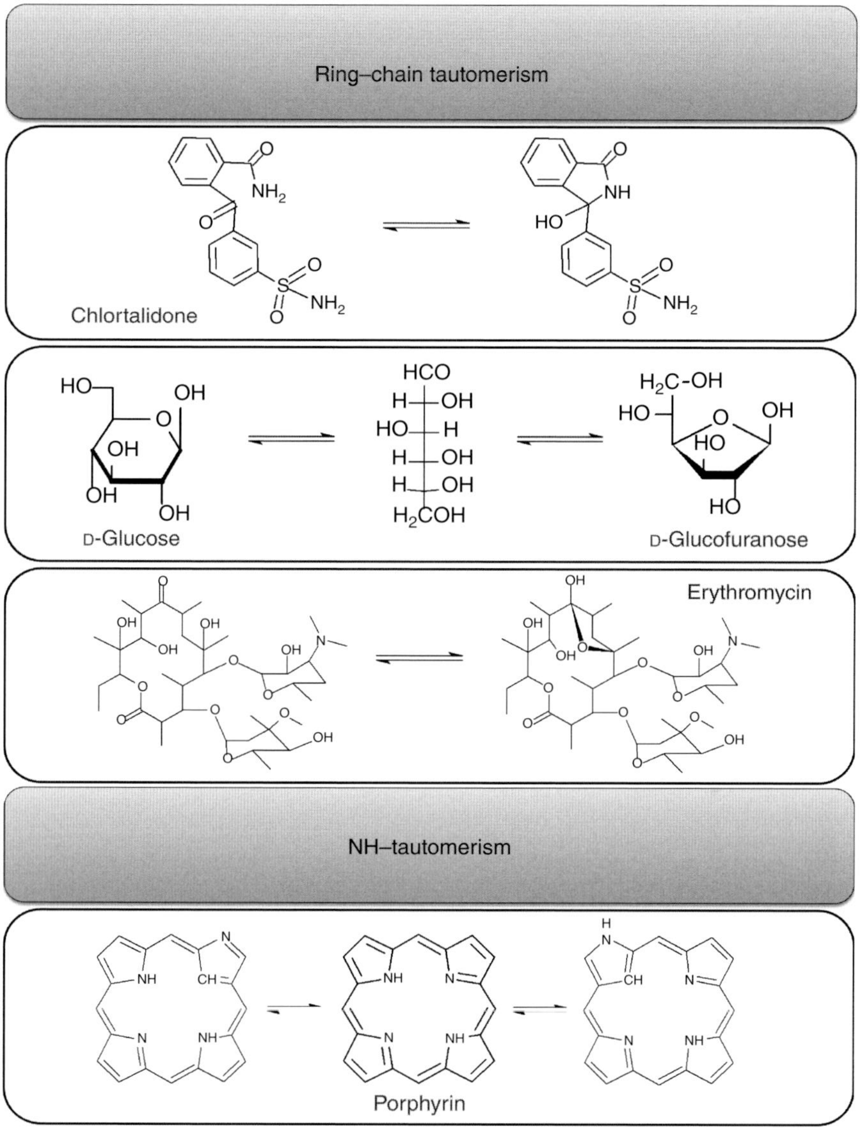

Figure 7.3 Examples of ring–chain and NH tautomerism [13, 17, 18].

bioavailability in comparison to their crystalline counterparts [19]. In the last years, it has been shown that preparation of amorphous pharmaceuticals, in which proton migration takes place, leads to a state that is chemically different from its crystalline counterpart. This phenomenon can have positive as well as negative impacts on the quality of the amorphous drug. On one hand, it has been reported that the less stable isomers may improve the drug solubility and, consequently, enhance the bioavailability of the active substance. This is

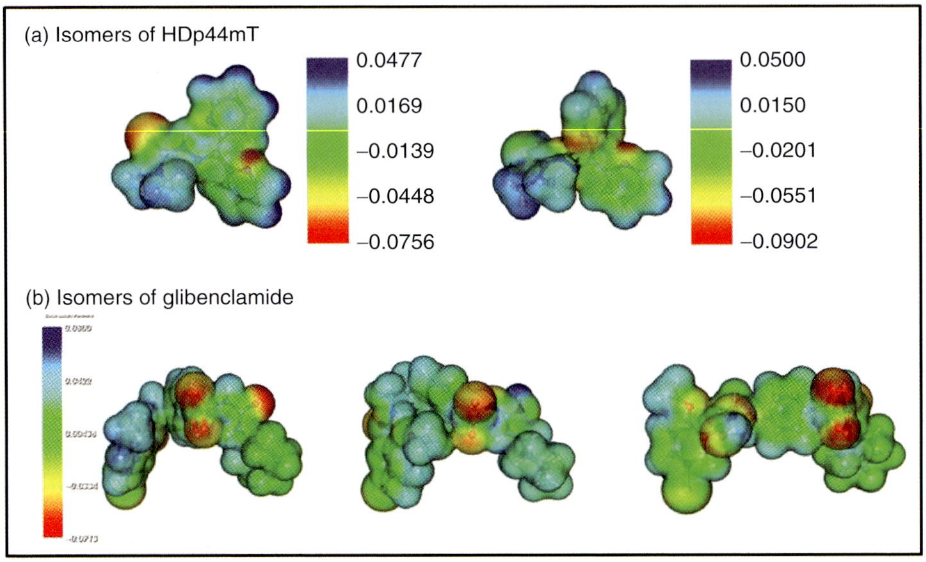

Figure 7.4 Electrostatic potentials for one amide and two imidic acid forms of glibemclamide and two tautomers of HDp44mT compound. Adapted from Wojnarowska *et al.* [21]. Reproduced with permission of American Institute of Physics.

possible because various tautomers of a pharmaceutical compound usually have different molecular fingerprints, hydrophobicities, as well as the ability to gain or lose a proton as reflected in their different pK_a values [20]. Additionally, in many cases the occurrence of two or more tautomers in the sample strongly reduces the tendency for the drug to recrystallize [5]. On the other hand, it has been conclusively demonstrated that isomers of chemical compounds are characterized by different 3D shapes and electrostatic properties as presented in Figure 7.4 for HDp44mT and glibenclamide molecules. [21] (see Figure 7.4).

As a result, they are often characterized by different abilities to form hydrogen bonds and various affinities to the active protein cell. Consequently, one cannot exclude that the biological activity of a given isomer may be completely different from that expected. Additionally, it has been reported many times that the tautomer ratio is strongly dependent on the environmental conditions. Even a small increase or decrease in temperature, compression of the sample, or change of the solvent or pH value can dramatically modify the tautomer concentration [22]. Such changes complicate the assignment of physical properties to a specific chemical structure and the identification of the bioactive species from the tautomeric mixture. In this context, one can mention erythromycin, one of the most frequently prescribed macrolide antibiotics. In hydrogen-bonding solvents or in the amorphous state, erythromycin exists in three tautomeric forms (a ketone form and two cyclic hemiketals), which are characterized by different pharmaceutical activities. It was proven that the ketone is the bioactive form, which binds to bacterial ribosomes and inhibits protein synthesis, while the cyclic hemiketals are completely inactive [13]. What is more, in the gastrointestinal tract the amount of inactive tautomers can reach even 20%, and therefore erythromycin must be taken in a large dosage to achieve optimal therapeutic effect. Moreover, it cannot be excluded that inactive tautomers can cause a number of drug side effects. The cited examples clearly demonstrate that tautomerization can cause significant changes in the physicochemical properties of drugs and consequently affect their biological activity. That is why many experimental and theoretical studies in the last decade were dedicated to understanding the proton transfer reaction in pharmaceutical materials. However, the existing literature data reveal that it is very hard to find appropriate experimental techniques that would allow rapid and reliable determination of tautomerization kinetics and, consequently, predict the equilibration time of pharmaceutical substances at various thermal and chemical conditions.

Recently, it was demonstrated that these goals can be achieved by means of broadband dielectric spectroscopy (BDS). A great advantage of this technique is the possibility to study the various molecular aspects of the tautomerization process. Using this method, one can investigate the effect of proton transfer reaction on the molecular dynamics of amorphous pharmaceuticals over a wide temperature range. It is also possible to examine the influence of tautomer concentration on the physical stability of glassy systems. Moreover, dielectric data enable us to determine the activation energy of proton transfer reaction as well as recognize the real timescale of tautomerism of drugs at various temperatures. This knowledge, which is impossible to achieve using the other experimental methods, could be crucial in the pharmaceutical industry because it is relevant for determining the expiration date as well as the best storage conditions of pharmaceutical materials. Another advantage of the BDS technique is that it can be easily extended to studies at elevated pressures and, consequently, to determine the changes in tautomer concentration during the tableting process.

7.1
Broadband Dielectric Spectroscopy as a Powerful Tool for Investigating the Tautomerization Process in Condensed Materials

Since proton migration is possible only under certain conditions, for example, in solution, the gas phase, or in the amorphous liquid state, generally to observe the isomer transformation by means of BDS technique, first the crystalline material has to be dissolved or converted into the amorphous form. However, it should be noted that in solution the tautomerization reaction may be faster than the time required for temperature stabilization [23]. That is why it is more convenient to study the isomerization reaction in the supercooled liquid state where the proton transfer usually takes much more time than in the normal liquid state. Additionally, this procedure enables us to focus only on the temperature influence on the tautomer equilibrium.

From the present knowledge, it is understood that proton transfer reaction begins at the melting point of a given material [5]. This means that, when the crystalline structure is lost, the less stable isomers appear in the sample. When a

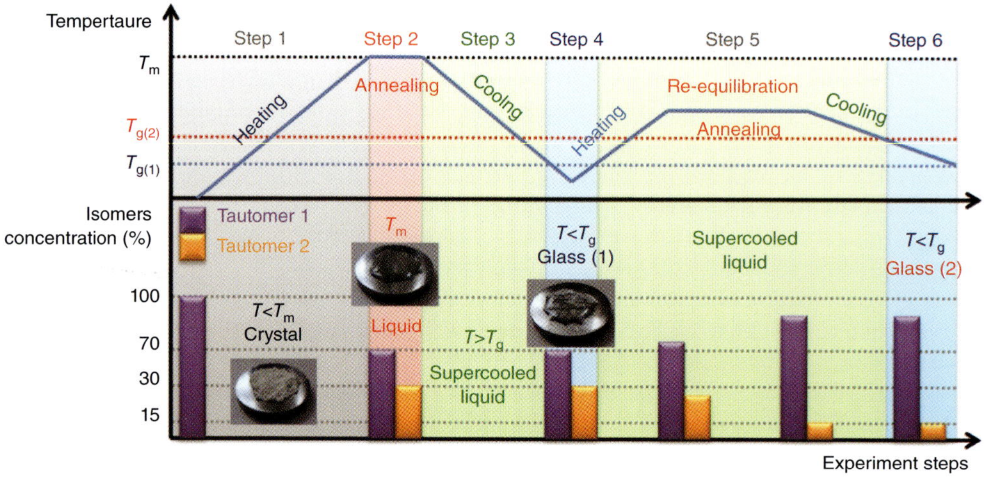

Figure 7.5 Possible equilibrium states of an organic compound (which exists in two tautomeric forms 1 and 2) at several different temperatures. After annealing at $T > T_g$, samples differ from each other in their tautomer concentration.

liquid is next cooled rapidly below its melting point and the viscosity of the system increases, the isomer concentration achieved at T_m becomes "frozen" below T_g. However, when the temperature is changed, the system's re-equilibration begins; that is, the tautomer ratio is changing dynamically. Depending on the value of the actual temperature, this process may take even a few days despite the fact that the sample is thermally equilibrated. The variations in tautomer concentration during melting–quenching–annealing steps are schematically presented in Figure 7.5.

In the literature, one can find many experimental reports indicating that different tautomers of given compound are very often characterized with different chemical and physical properties such as the dipole moment or the molecular relaxation time [24]. Consequently, amorphous materials with different isomer concentrations may be characterized by various T_g values, dc conductivity, or average dipole moment. Since all these properties can be determined from dielectric measurements, the BDS technique can be effectively used for detecting the tautomerization process and monitoring its kinetics.

The experiments performed so far for several pharmaceutically important materials reveal that tautomerization reaction can be easily detected by means of dielectric experiments if the proton hopping requires a conformational change in molecular structure or when it involves a ring–chain equilibrium. Such conditions are satisfied for some drugs (e.g., indapamide [4], GCM [21], and glimepiride), monosaccharides (e.g., D-fructose [25], D-ribose [26], D-glucose, L-, and D-fucose [27, 28]), nucleosides (e.g., β-uridine [29]), and nucleobases. According to the literature data, the preceding conformational interconversion is necessary for proton migration to take place in the case of hypoglycemic agents with two tautomerization centers (e.g., GCM or glimepiride), but it is not needed for amide–imidic acid transformation in the heterocyclic ring of DNA and RNA nucleobases. At the same time, it is quite simple to detect and describe the proton transfer reaction of the mentioned antidiabetic drugs, while the tautomerization of DNA and RNA compounds can be easily overlooked. This is because, in the first case, the proton migration is accompanied by the time evolution of τ_α, dc conductivity, and ε_s, while in the second case it refers only to the small changes of ε_s. Interestingly, tautomeric reactions in which a heterocyclic ring is opened and closed usually involve a significant change in the dipole moment as well as T_g (~10 K) of glass-forming liquids. Consequently, the ring–chain tautomerism of carbohydrates can be easily monitored by means of dielectric spectroscopy in contrast to the anomeric transformation of monosaccharides, which does not change at all the dielectric response of the sample (e.g., in the case of D-glucose). Furthermore, experiments performed so far indicate that, generally, amorphous materials with a higher concentration of the less stable tautomers exhibit a lower T_g value. As presented in Figure 7.6, the freshly quenched, imide-rich GCM and pyranose-rich D-fructose are characterized by a T_g that is 11 K lower than that determined for equilibrated materials (with higher concentration of amide and furanose forms, respectively). Interestingly, the milled drug GCM, which is characterized by a small amount of less stable imidic-acid isomers (about ~5%), has also the highest T_g values (see the green points in Figure 7.6). The experimental data presented in Figure 7.6 indicate that the tautomer equilibrium can be also reflected in changes of the dynamic fragility m_P. Consequently, the

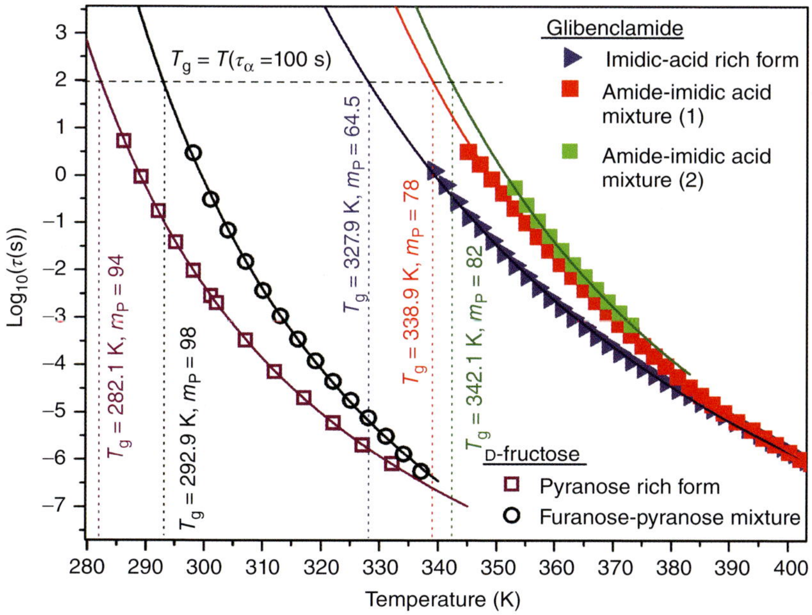

Figure 7.6 Temperature dependences of the structure relaxation times determined for D-fructose and the drug glibenclamide. Solid lines are fits of the VFT equation to the experimental data. The experimental points were taken from [21, 30].

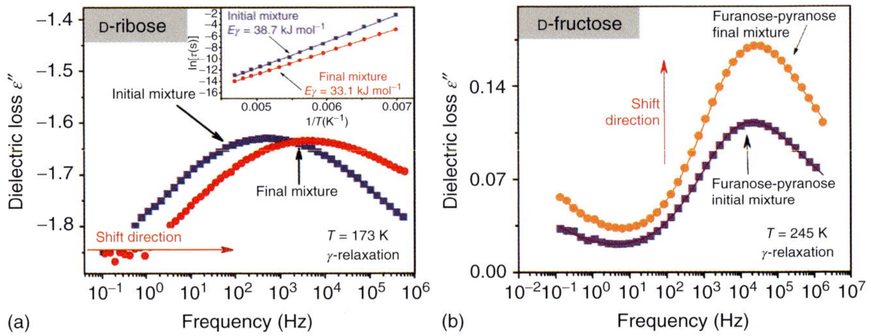

Figure 7.7 Comparison of γ-relaxations observed in the freshly quenched and equilibrated samples of D-ribose and D-fructose. In the inset, the activation energies of both freshly quenched and equilibrated samples of D-fructose are presented. (a,b) Adapted from Wlodarczyk *et al.* [26, 30]. Reproduced with permission of American Institute of Physics; American Chemical Society.

analysis of m_P values at various isobaric conditions may be employed to designate the direction of proton transfer reaction under high-pressure conditions. Such experiments performed for a GCM sample have been presented in [31].

However, α-relaxation is not the only one parameter that can be sensitive to the isomerization process. In some cases, freshly quenched and equilibrated samples may differ from each other in secondary relaxations. Such a phenomenon can be observed, for example, in the case of monosaccharides (D-ribose and D-fructose). Because the furanose ring is more flexible and has more internal degrees of freedom than the pyranose ring, the change of the γ-relaxation character is expected during mutarotation. To verify this assumption, two dielectric runs are required, namely before and after sample equilibration, both at the desired temperature below the liquid-to-glass transition. The first measurement has to be performed immediately after quenching and the second one about 2 days of equilibrating sample at a temperature higher than T_g. In Figure 7.7, one can see the results of the described procedure. It is easily seen that the amplitude of the γ-process of D-fructose has increased by approximately 50% during mutarotation, whereas the relaxation time did not change at all (see Figure 7.7b) [30]. A very different effect was found in the case of D-ribose: the dielectric strength is almost the same while the maximum of the γ-mode is shifted by about one decade toward higher frequencies (see Figure 7.7a). Moreover, the activation energy has decreased from 38 to 33 kJ mol^{-1} [26].

7.2
Tautomerization Kinetics of Supercooled Pharmaceuticals

To monitor the kinetics of proton transfer reaction in pharmaceutically important materials using the BDS technique, several isothermal, time-dependent dielectric measurements of fast-quenched material have to be performed. In this context, one should note that tautomerization is a thermally controlled reaction. Consequently, with decreasing temperature, the equilibration time gets longer, and close to the liquid–glass transition, it may take even a couple of days. However, it does not create any technical problems because the dielectric experiments are completely automated and the measurements can be easily controlled even for a long time.

If the various tautomers of a given compound differ from each other in their T_g values, during the sample equilibration carried out at constant temperature the fluctuation of viscosity takes place. As a consequence, the structural relaxation observed in the dielectric spectrum shifts toward lower frequencies, which means that the α-relaxation time gets longer. On the other hand, the changes of the average dipole moment value of the sample observed during tautomerization are reflected in the variation of static permittivity (ε_s).

The representative dielectric loss spectra of GCM (a popular antidiabetic drug) and D-ribose (being a part of biologically important molecules, especially ribonucleic acid RNA, and energy-storing molecule, adenosine triphosphat ATP) evolving with time at isothermal conditions are depicted in Figure 7.8. As can be easily seen, in both cases the structural relaxation peak shifts toward lower frequencies during annealing. Consequently, τ_α recorded in the final equilibrium state

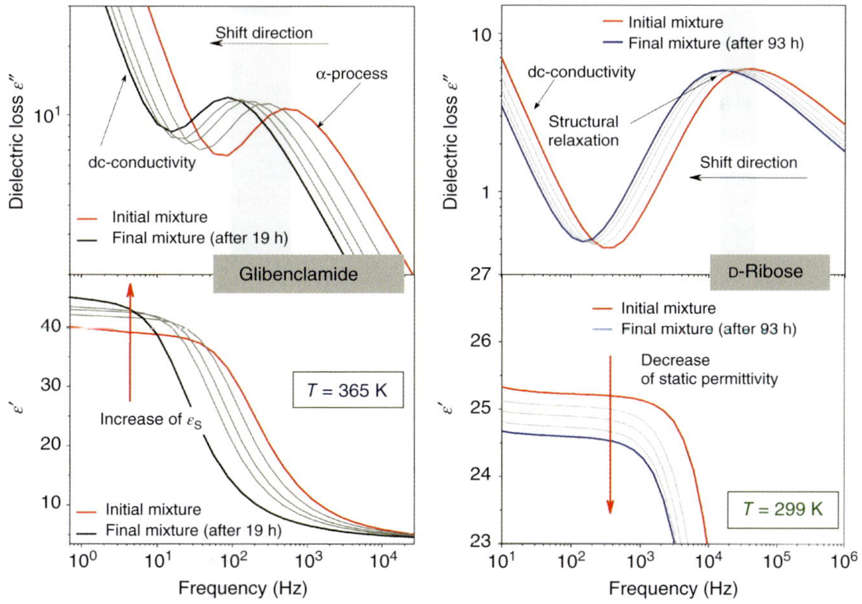

Figure 7.8 Time evolution of kinetic parameters relaxation times and static permittivity observed for glibenclamide (a) and D-ribose (b). Initial and final states are plotted as bold solid lines, while the selected intermediate states are depicted by thin solid lines. Partially reproduced from [21, 26].

gets longer, in comparison with that determined from the peak position of freshly quenched samples, and at the same time the dc conductivity decreases. Simultaneously, both samples are characterized with different static permittivity behavior – while in the case of the drug GCM, ε_s slightly increases, the opposite trend is observed for D-ribose.

Here, it should be noted that if one knows the behavior of the static permittivity of the sample and the dipole moment of every tautomer, it is possible to predict the tautomerization direction in supercooled liquid state. This is possible because the low-frequency limiting value of $\varepsilon'(f)$, that is, the static permittivity ε_s, is governed by the average dipole moment (μ^2) and the number of dipoles (N):

$$\Delta\varepsilon = \varepsilon_s - \varepsilon_\infty \approx N \cdot \mu^2 \tag{7.1}$$

Thus, the greater the average μ of the molecules, the higher the value of ε_s. This relation was recently applied to predict the mutarotation direction for two monosaccharides [25, 26]. In the crystalline state, D-ribose and D-fructose occur only in the β-pyranose form. However, during the melting process, when the mutarotation process is induced, β-pyranoses begin to transform to the less stable furanose forms. To determine the direction of the mutarotation process in the supercooled liquid state, first the dipole moments of various tautomers were calculated by means of density functional theory (DFT), which is one of the most popular quantum mechanical methods. The results of DFT calculations were next compared with the BDS experimental data. It was found that both the increase of static permittivity in time observed for D-fructose and the decrease of ε_s for D-ribose are related to the furanose-to-pyranose transformations. The same procedure was also used to describe the direction of proton transfer reaction in β-uridine, the nucleoside being a component of RNA [29]. Based on the analysis of the dielectric spectra collected at isothermal conditions, supplemented by quantum mechanical calculations, it was found that high temperature shifts the equilibrium toward the less stable lactim tautomeric species that are responsible for point mutations in the RNA chain and consequently accumulation of abnormally constructed proteins (see Figure 7.9).

Next valuable piece of information on the isomerization kinetics of pharmaceutically important compounds, coming from dielectric experiments, concerns the energy barrier of the proton transfer reaction. Intuitively, the higher the activation energy necessary for proton hopping, the lower the probability of the tautomerization process. Consequently, based on the value of E_a it is possible to pre-identify the isomerization abilities of pharmaceutical materials. This knowledge is especially important in light of the recent research showing that the existence of a chemical compound in the keton, amide, or lactam form does not guarantee that the tautomeric transformation will proceed.

To determine the value of E_a of given tautomeric conversion, first the kinetic curves have to be constructed and the rate constants k in various temperature conditions have to be determined.

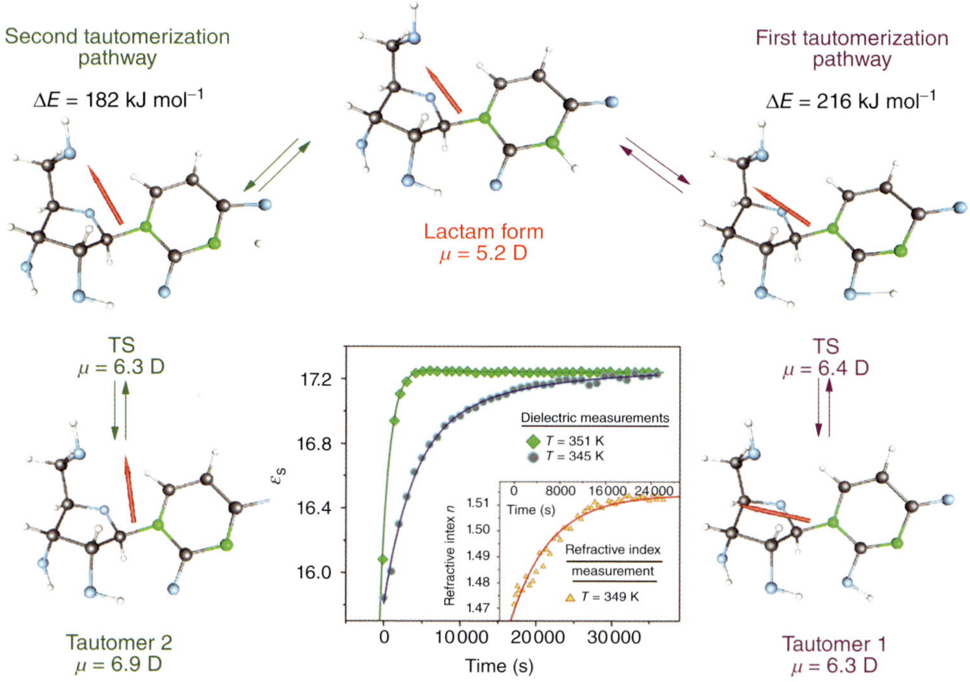

Figure 7.9 Dipole moment of possible β-uridine tautomers. The inset panel presents the time evolution of static permittivity of β-uridine during annealing at 345 and 351 K. Adapted from Wojnarowska *et al.* [29]. Reproduced with permission of American Chemical Society.

To describe the tautomerization kinetics of the GCM hypoglycemic agent, the changes of relaxation times and dc conductivity with time were monitored. On the other hand, in the case of D-ribose the behavior of the relaxation times and static permittivity was analyzed. In all these cases, to compare the kinetic curves ($\tau_\alpha(t)$, $\varepsilon_s(t)$, and $\sigma_{dc}(t)$) measured at different temperature conditions on one graph, the degree of reaction (α) was calculated as follows:

$$\alpha = \frac{p - p_0}{p_f - p_0} \tag{7.2}$$

For every temperature, p_0 was the first measured value of τ_α, σ_{dc}, or ε_s after temperature stabilization while p_f means the value measured in the final equilibrium state. The representative kinetic curves constructed for the drug GCM and D-ribose are depicted in Figure 7.10. As can be seen, the $\tau_\alpha(t)$ and $\varepsilon_s(t)$ dependences obtained for D-ribose have exponential character. It means that the product concentration is raised to the first power. That is why these experimental data can be satisfactorily parameterized by the first-order kinetic equation

$$\alpha = 1 - A\cdot\exp(-kt) + C \tag{7.3}$$

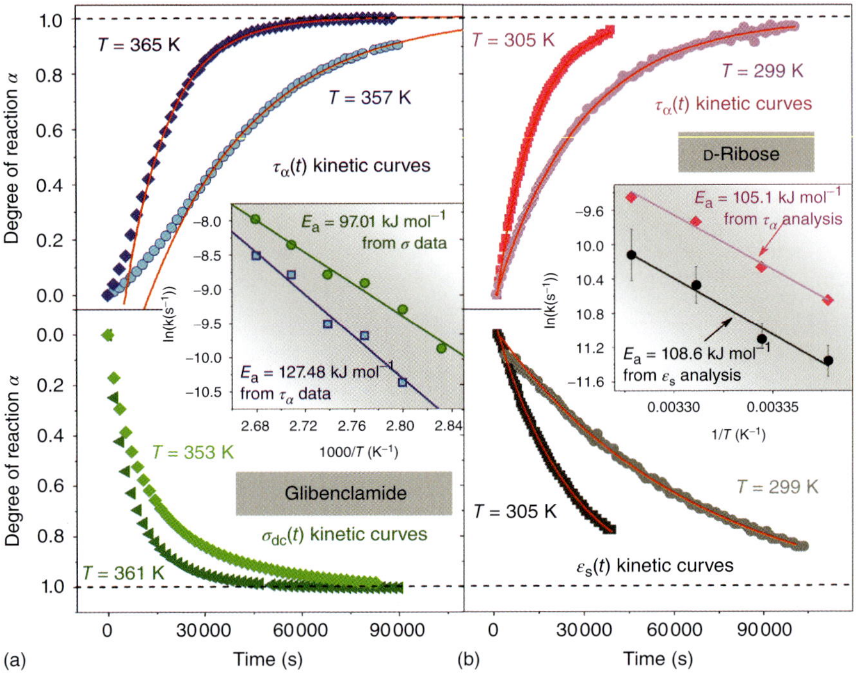

Figure 7.10 (a) Selected kinetic curves for the drug glibenclamide representing $\tau_\alpha(t)$ and $\sigma_{dc}(t)$ dependences. (b) Selected kinetic curves for D-ribose representing $\tau_\alpha(t)$ and $\varepsilon_s(t)$ dependences. As can be seen, the degree of reaction can vary from 0 to 1, where 0 denotes the beginning and 1 denotes the end of the reaction. Inset panels show the temperature dependence of the rate constants. Solid lines denote the fits of the Arrhenius equation to the experimental data. Partially taken from [21, 26].

where α is the degree of the reaction (concentration equivalent), t is time, A is a pre-exponential factor, C is an additional constant, and k is the rate of reaction – one of the most important parameters describing the kinetics of chemical reactions. However, one should note that sometimes a single-exponential time dependence cannot describe the experimental data satisfactorily. Such situation has been observed for $\varepsilon_s(t)$ kinetic curves of the nucleoside β-uridine as well as for τ_α kinetic curves measured in the supercooled liquid state of the drug GCM. In the case of RNA component, a very good fit was achieved using a superposition of two first-order kinetic equations:

$$\varepsilon_s = A_1 \cdot \exp(-k_1 \cdot t) + A_2 \cdot \exp(-k_2 \cdot t) + C \tag{7.4}$$

Consequently, the values of two rate constants, k_1 and k_2, were determined for each temperature. This analysis reveals that, during isothermal equilibration, two different tautomeric conversions of the uridine take place simultaneously.

On the other hand, $\tau_\alpha(t)$ dependences of GCM exhibit the characteristic "incubation" time, especially close to the liquid–glass transition when the tautomerization reaction is relatively slow. As a consequence, the single exponential function is not able to parameterize the early stage of the tautomerization process and the fitting procedure has to be applied in the limited range of time. Interestingly, a similar pattern of behavior of the kinetic curves has been also observed during the dielectric studies of the mutarotation process in D-fructose as well as D- and L-fucose [27]. Additionally, the investigations by means of a rheology technique performed for GCM and D-fructose have shown that the characteristic sigmoidal shape of $\tau_\alpha(t)$ dependence is a part of the double-exponential behavior of the kinetic curves (Figure 7.11) [32]. That is why the time-dependent BDS experimental data for GCM and D-fructose can be also described using Eq. 7.4. This result indicates that the first step of the proton transfer in GCM and the mutarotation process in some monohydrates is too fast to be detected by monitoring of time evolution of τ_α even when the experiments are performed close to the liquid–glass transition. But at the same time, this technique can be successfully applied to describe the second slower stages of isomeric transformation of the mentioned materials.

Using the values of k determined from the fitting procedure at various temperature conditions, we are able to calculate the activation energy of the tautomeric conversion. As can be easily seen in the insets to Figure 7.10, the logarithm of the rate constant is linearly related to $1/T$. Thus, from the fit of Arrhenius equation (Eq. 7.5) to the experimental data, the activation energy barrier of tautomerization and mutarotation processes can be easily determined.

$$\ln k = \ln A - \frac{E_\mathrm{a}}{RT} \tag{7.5}$$

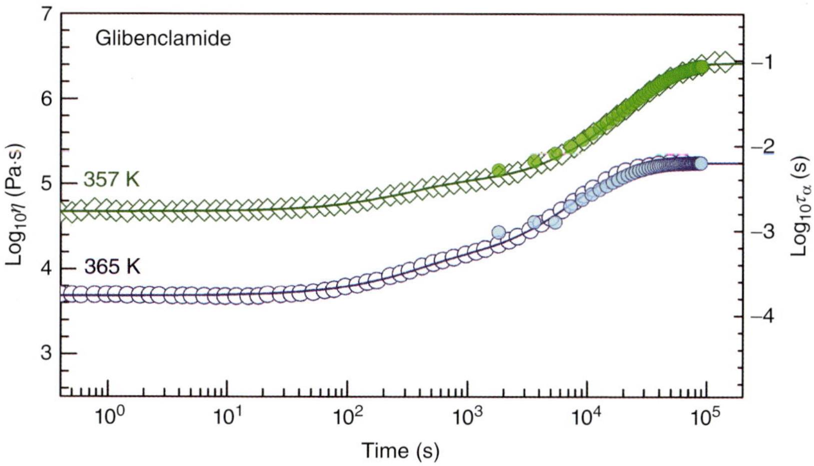

Figure 7.11 Open symbols: $\eta(t)$ kinetic curves recorded during isothermal rheological measurements. Solid symbols: $\tau_\alpha(t)$ kinetic curves from dielectric measurements. Solid lines: fits of rheological data using Eq. (7.4).

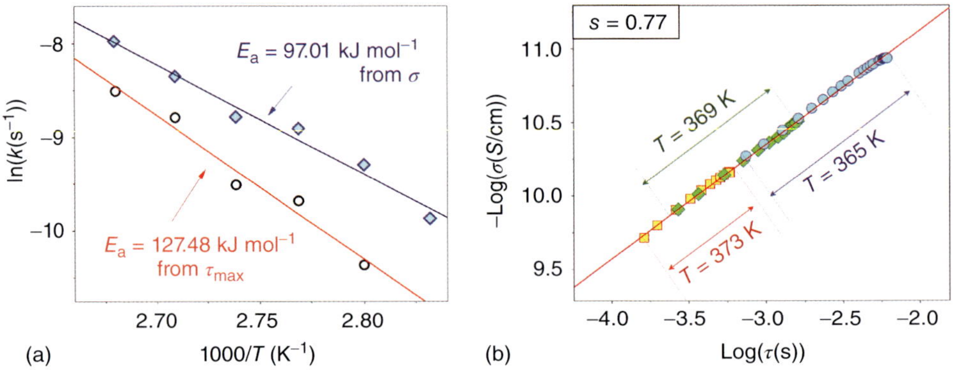

Figure 7.12 (a) Temperature dependence of the rate constant. Solid lines, red, and blue, respectively, denote the fits of the Arrhenius equation (7.5) to the experimental points determined on the basis of α-relaxation time and dc conductivity analysis. (b) dc Conductivity versus relaxation time in double logarithmic scale for three representative time-dependent isothermal measurements.

As presented in Figure 7.10, the values of E_a of furanose–pyranose transformation in D-ribose obtained from relaxation time and static permittivity analysis are practically the same. Thus, in this case, to establish the rate of the ring–chain tautomerism at different temperature conditions, analysis of the evolution in time of only one of mentioned parameters is sufficient.

On the other hand, the activation energies of imidic acid to amide conversion of the GCM molecule, determined on the basis of $\tau_\alpha(t)$ and $\sigma_{dc}(t)$ dependences, slightly differ from each other (see Figure 7.12).

Thus, the question arises: which value $E_a^{\tau\alpha}$ or $E_a^{\sigma dc}$ is correct? To solve this problem, an equation, analogous to the fractional Debye–Stokes–Einstein law, which relates the rate constant determined on the basis of $\tau_\alpha(t)$ and $\sigma_{dc}(t)$ dependences, was proposed [21]: $k_\sigma k_\tau^s \cong$ const (where s is the slope of log $\tau_\alpha = f(\log \sigma_{dc})$ curve) and the relation between both values of E_a was derived: $E_a^{\sigma dc} = sE_a^{\tau\alpha}$. Since the value of the fractional exponent s for the drug GCM = 0.77, the decoupling phenomenon between the values of E_a determined from the analysis of relaxation times and dc conductivity can be easily explained.

Here, it is worth noting that the experiments of tautomerization kinetics performed so far for various chemical compounds clearly indicate that, generally, the values of the activation energy of given isomeric conversion determined from dielectric measurements are in good agreement with theoretical predictions as well as with pressure–temperature–volume (PTV), rheological, infrared experimental data, and refractive index measurements (see Figure 7.13) [28, 32].

It is well known that the knowledge of the kinetic parameters, that is, the rate constant k, the half-time of tautomerization process $t_{1/2} = \ln(2)/k$, and its activation energy barrier E_a, is one of the keys to describe the mechanisms of proton transfer reaction in solid phases. In the case of amorphous drugs that show a tendency to exist in the less stable isomeric forms, measurements of these

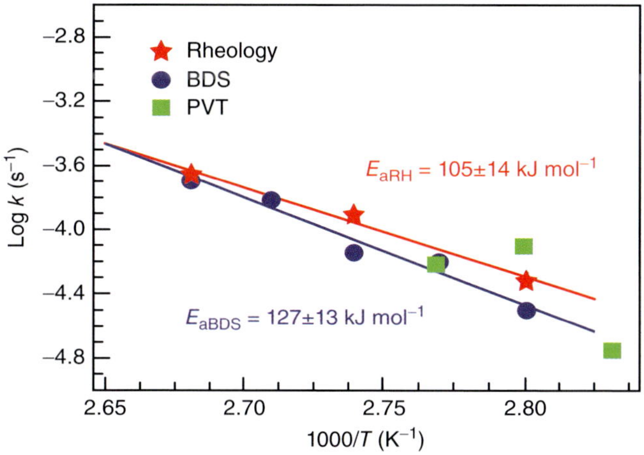

Figure 7.13 Temperature dependence of the rate constants determined from dielectric (solid circles), rheological (solid stars), and PVT measurements (solid squares). Solid lines denote the Arrhenius fits of the experimental points determined on the basis of viscosity and structural relaxation times analysis, respectively.

parameters are needed for the determination of their accurate treatment conditions. This follows from the fact that a change in temperature or humidity usually leads to a different tautomer equilibrium and consequently may affect the solubility and bioavailability of the amorphous pharmaceutical. Moreover, if one knows the slope of ln $k = f(1/T)$ dependence, it is possible to estimate the half-life time of a tautomerization reaction under any thermal conditions and therefore predict the physical stability of amorphous drug, for example, at room temperature.

In the literature, one can find many experimental reports where a higher stability of binary mixtures in the amorphous state was postulated. That is why one can expect that glassy materials in which the less stable isomers are dispersed in the stable matrix should not recrystallize during a typical shelf-life. The experiments performed on one of the most commonly prescribed diuretic agent indapamide [4] as well as the hypoglycemic drugs GCM [5] and glimepiride [33] strongly support this assumption. In the above-mentioned pharmaceuticals, there were no signs of crystallization even after 2 year of storage at room temperature. Moreover, from the time-dependent dielectric measurements performed for the vitrified drug GCM, it was found that the half-life time of conversion of imidic acid to amide at room conditions may be even 3 years. Consequently, one can expect that the amorphous GCM sample may be stored for 5 years without any sign of crystallization.

Acknowledgment

This work was funded by a research project within the program OPUS 3 supported by the Polish National Science Center (No.DEC-2012/05/B/ST3/02837).

References

1. Smith, M.B. and March, J. (2001) *Advanced Organic Chemistry*, 5th edn, Wiley-Interscience, New York, pp. 1218–1223.
2. Martin, Y.C. (2009) *J. Comput. Aided Mol. Des.*, **23**, 693–704.
3. Braschi, I., Paul, G., Gatti, G., Cossi, M., and Marchese, L. (2013) *RSC Adv.*, **3**, 7427–7437.
4. Wojnarowska, Z., Grzybowska, K., Hawelek, L., Dulski, M., Wrzalik, R., Gruszka, I., Paluch, M., Pienkowska, K., Sawicki, W., Bujak, P., Paluch, K.J., Tajber, L., and Markowski, J. (2013) *Mol. Pharm.*, **10** (10), 3612–3627.
5. Wojnarowska, Z., Grzybowska, K., Adrjanowicz, K., Kaminski, K., Paluch, M., Hawelek, L., Wrzalik, R., Dulski, M., Sawicki, W., Mazgalski, J., Tukalska, A., and Bieg, T. (2010) *Mol. Pharm.*, **7** (5), 1692–1707.
6. Kolodziejczyk, K., Paluch, M., Grzybowska, K., Grzybowski, A., Wojnarowska, Z., Hawelek, L., and Ziolo, J.D. (2013) *Mol. Pharm.*, **10** (6), 2270–2282.
7. Kaushal, A.M., Chakraborti, A.K., and Bansal, A.K. (2008) *Mol. Pharm.*, **5** (6), 937–945.
8. Sheth, A.R., Lubach, J.W., Munson, E.J., Muller, F.X., and Grant, D.J.W. (2005) *J. Am. Chem. Soc.*, **127** (18), 6641–6651.
9. Claramunt, R.M., Herranz, F., Dolores Santa María, M., Pinilla, E., Torres, M.R., and Elguero, J. (2005) *Tetrahedron*, **61** (21), 5089–5100.
10. Rozwarski, D.A., Grant, G.A., Barton, D.H.R., Jacobs, W.R.J. Jr., and Sacchettini, J.C. (1998) *Science*, **279**, 98–102.
11. Abraham, R.J., Hawkes, G.E., and Smith, K.M. (1974) *Tetrahedron Lett.*, **15** (16), 1483–1486.
12. Duarte, H.A., Carvalho, S., Paniago, E.B., and Simas, A.M. (1999) *J. Pharm. Sci.*, **88**, 111–120.
13. Iuliucci, R.J., Clawson, J., Hu, J.Z., Solum, M.S., Barich, D., Grant, D.M., and Taylor, C.M.V. (2003) *Solid State Nucl. Magn. Reson.*, **24**, 23.
14. Porter, W.R. Tautomers of Warfarin. Personal communication. August 2009.
15. Mirmehrabi, M., Rohani, S., Murthy, K.S.K., and Radatus, B. (2004) *J. Cryst. Growth*, **260** (3–4), 517–526.
16. Paneth, P. (2010) *Kinetics and Dynamics: From Nano- to Bio-Scale*, Springer, New York.
17. Temperini, C., Cecchi, A., Scozzafava, A., and Supuran, C.T. (2009) *Bioorg. Med. Chem.*, **17**, 1214–1221.
18. Salas, M., Viñuela, E., and Sols, A. (1965) *J. Biol. Chem.*, **240**, 561–568.
19. Lepek, P., Sawicki, W., Wlodarski, K., Wojnarowska, Z., Paluch, M., and Guzik, L. (2013) *Eur. J. Pharm. Biopharm.*, **83** (1), 114–121.
20. Katritzky, A.R., Hall, C.D., El-Gendy, B.E.M., and Draghici, B. (2010) *J. Comput.-Aided Mol. Des.*, **24** (6-7), 475–484.
21. Wojnarowska, Z., Wlodarczyk, P., Kaminski, K., Grzybowska, K., Hawelek, L., and Paluch, M. (2010) *J. Chem. Phys.*, **133**, 094507.
22. Jones, R.C., Herasymchuk, K., Mahdi, T., Petrov, A., Resanović, S., Vaughan, D.G., Lough, A.J., Quail, J.W., Koivisto, B.D., Wylie, R.S., and Gossage, R.A. (2013) *Org. Biomol. Chem.*, **11** (21), 3484–3493.
23. Elguero, J., Marzin, C., Katritzky, A., and Linda, P. (1976) *Advances in Heterocyclic Chemistry. Supplement 1: The Tautomerism of Heterocycles*, Academic Press, New York.
24. Lumbroso, H., Curé, J., and Butler, R.N. (1980) *J. Heterocycl. Chem.*, **17** (7), 1373–1376.
25. Wlodarczyk, P., Kaminski, K., Dulski, M., Haracz, S., Paluch, M., and Ziolo, J. (2010) *J. Non-Cryst. Solids*, **356** (11–17), 738–742.
26. Wlodarczyk, P., Kaminski, K., Haracz, S., Dulski, M., Paluch, M., Ziolo, J., and Wygledowska-Kania, M. (2010) *J. Chem. Phys.*, **132**, 195104.
27. Wlodarczyk, P., Cecotka, A., Adrjanowicz, K., Kaminski, K., and Paluch, M. (2013) *J. Phys.: Condens. Matter*, **25**, 375101.
28. Kossack, W., Kipnusu, W.K., Dulski, M., Adrjanowicz, K., Madejczyk, O., Kaminska, E., Mapesa, E.U., Tress, M., Kaminski, K., and Kremer, F.

(2014) *J. Chem. Phys.*, **140**, 215101. doi: 10.1063/1.4880718

29. Wojnarowska, Z., Paluch, M., Wlodarczyk, P., Dulski, M., Wrzalik, R., and Roland, C.M. (2012) *J. Phys. Chem. Lett.*, **3**, 2288.
30. Wlodarczyk, P., Kaminski, K., Paluch, M., and Ziolo, J. (2009) *J. Phys. Chem. B*, **113** (13), 4379–4383.
31. Wojnarowska, Z., Adrjanowicz, K., Kaminski, K., Hawelek, L., and Paluch, M. (2010) *J. Phys. Chem. B*, **114** (46), 14815–14820.
32. Wojnarowska, Z., Paluch, M., and Pionteck, J. (2011) *J. Chem. Phys.*, **135**, 214506.
33. Wlodarczyk, P., Paluch, M., Wojnarowska, Z., Hawelek, L., Kaminski, K., and Pilch, J. (2011) *J. Phys.: Condens. Matter*, **23** (42), 425901.

8
Disorders in Pharmaceutical Polymers

Emeline Dudognon and Sheng Qi

The size of macromolecules and the variety of their possible architectures make polymers a separate class of materials. This is mainly due to the degree of disorder they bring, which gives polymers very specific chemical and physical properties. The use of polymers in the pharmaceutical industry is usually discussed in terms of their origin, architecture, processability, hydrophobicity, gelation ability, biocompatibility, and biodegradability. In this chapter we discuss the performance and applications of pharmaceutical polymers in relation to their structure and degree of disorder.

For this purpose, we will first give an overview of the main characteristics of polymers in terms of their architecture, physical states, and mobility. The aim here is not to describe all the physical properties of polymers, as many excellent books exist that can be consulted for a complete description [1–3]. Here we will highlight the different origins and types of disorders that can be found in polymers, namely structural disorders (architecture, configurations, conformations), which directly influence the physical state, and dynamic disorders, that is, the different types of mobility that characterize the amorphous state (supercooled liquid and glass).

Pharmaceutical polymers are widely used in solid dosage forms such as drug/excipient amorphous solid dispersions, which are mixtures of polymers and low molecular weight (MW) drugs. In this chapter, we will pay particular attention to the influence of drug incorporation on the polymer mobility and the issues of the limits of polymer–drug miscibility and solubility. Finally, in the last part, we will discuss the relationship between the disorder and the physical properties and behavior of pharmaceutical polymers when used in solid dosage forms through a few illustrative examples involving the most widely used polymers.

Disordered Pharmaceutical Materials, First Edition. Edited by Marc Descamps.

8.1 Polymers Architectures – Structural Disorders

8.1.1 Architectures

Many different types of architectures can be found in polymers. For a complete description, readers may refer to IUPAC recommendations for macromolecular nomenclatures, but one can roughly distinguish the following:

- Linear polymers, which consist of a linear succession of repetitive monomer units. However, perfectly linear polymers can be obtained only by following very specific conditions, and branching often occurs during polymerization. These branches appear as irregularities of the structure.
- Branched polymers, which refer to highly branched polymers, such as comb polymers or star polymers (singular branch points from which linear chains emanate).
- Networks are polymers with a three-dimensional structure resulting from covalent cross-links of chains.
- Copolymers are polymers that can be derived from more than one species of the monomer [4]. The distribution of monomer species can be random (classified as unspecified, statistical, or random copolymers) or alternating [5]. If there is a linear arrangement of the blocks, with each block comprising units derived from a characteristic species of monomers, one speaks of block copolymers. If the chain backbone is constituted by one polymer with covalently bonded side chains of the other polymer, one speaks of graft copolymers [6] (cf. Figure 8.1).
- Polysaccharides are linear or branched polymers constituted of monosaccharide units. For example, cellulose and starch are constituted of glucose units linked at the 1, 4 positions, but it is a β linkage for cellulose and an α linkage for starch.
- Polypeptides are constituted of a defined series of amino acids –(CO–NH–CHR)–. Depending on the nature of the residue R, 20 common types of amino acids exist, which means that polypeptides are a kind of very complex copolymers. As proteins are a very specific subject, it will not be developed further here.

8.1.2 Local Structure of Linear Polymers

8.1.2.1 Stereoregularity and Configurations

When polymer chains are constituted of asymmetric monomeric units, these constitutive units are usually linked in a head-to-tail manner: $-CH_2-CHR-CH_2-CHR-$. It is most frequently the case for energetic reasons, but some faults can happen with head-to-head placements ($-CH_2-CHR-CHR-CH_2-$), leading to positional isomerism. When these faults are dominant, the polymers are irregular. Isomerism can also be due to the existence of a chiral carbon in the constitutive unit, as it can be found under two enantiomorphic forms. When two consecutive

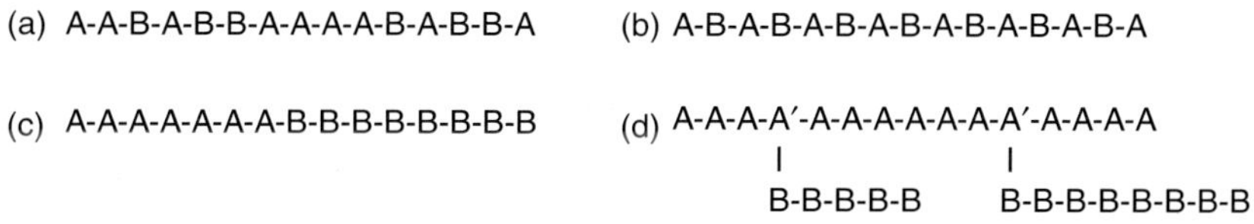

Figure 8.1 Different types of copolymers AB. (a) Random copolymer. (b) Alternating copolymer. (c) Block copolymer. (d) Graft copolymer.

monomeric units are considered, they constitute either a mesomeric (m) dyad if both chiral carbons are under the same configuration, or a racemic (r) dyad if both chiral carbons are under opposite configurations. When three consecutive monomeric units are considered, three types of triads can be formed: mm, mr (or rm), or rr. When the whole polymeric chain is considered, if it is constituted of mm triads, the configuration of the chain is named isotactic; if it is constituted of rr triads, the configuration of the chain is syndiotactic, and if the distribution of triads is random, then the configuration of the chain is atactic (Figure 8.2).

In the case of isotactic or syndiotactic configurations, polymers are perfectly stereoregular but, on the contrary, atactic configurations lead to irregular polymers.

8.1.2.2 **Conformations of Chains**

In order to describe the conformational structure of polymer chains in space, the chain can be considered as a mechanical system constituted of N rigid-rod-like segments of length l and $N-1$ joints (N represents the number of carbon–carbon links).

In the statistical freely jointed model, the assumption is made that these segments have no volume and can cross or occupy the same position in space. The end-to-end distance of the chain, h, can then be estimated. The mean square of this end-to-end distance is given, for high N values, by $\langle h^2 \rangle = Nl^2$. This distance is proportional to the number of monomeric units of the chain and, therefore, to the molecular weight of the polymer. Moreover, it is N times smaller than the

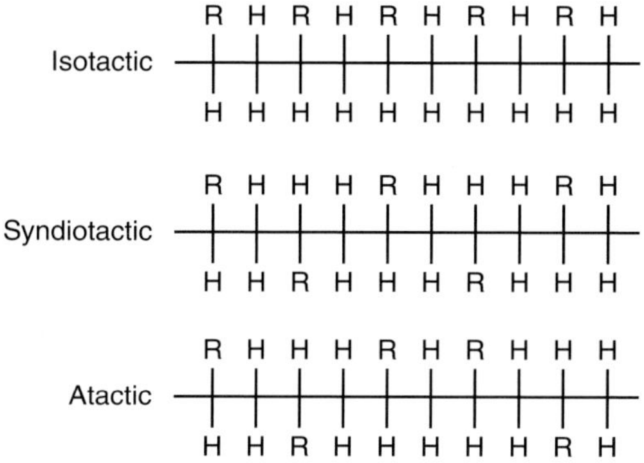

Figure 8.2 Schematic Fisher representation of the different configurations of a vinyl polymer.

square of the chain length in complete extension (N^2l^2), which indicates that the chain is strongly folded, forming a random coil.

The radius of gyration R_g is defined as the radius of the spherical surface with a uniform superficial mass density that would have the same moment of inertia as the considered chain. For high N values, the mean square of R_g is one-sixth of the mean square of the end-to-end distance of the chain.

In the model of hindered rotating chain, the rigid segments have a volume and their orientations are not random: the carbon–carbon link is represented by a segment of 1.54 Å in length, and the valence angle θ between two consecutive segments is 109.5°. Because of the tetrahedral angle ϕ, some mobility persists. However, preferred positions are imposed as a result of possible interactions between electronic orbitals and between atoms that are not directly linked.

These preferred conformations, corresponding to energy wells, are trans t (or anti) ($\phi = 180°$), gauche plus g_+ ($\phi = 60°$), and gauche minus g_- ($\phi = -60°$). A change of conformation implies a jump of the energetic barrier ΔE between theses orientations, which are not energetically equivalent if the side groups are too bulky or carry charges. The flexibility (modification of conformations) of the chain that is constituted of a series of t, g_+, and g_- conformations thus depends on the temperature and the energy $k_B T$ it brings to the system. It implies, first, that if ΔE is high, a high temperature is necessary to make the chain flexible, leading to a high value of the glass transition temperature. Second, for low ΔE values, transitions between conformations can easily occur: a large number of conformations are possible, and one can only speak of a statistical conformation of the chain. On the contrary, for high ΔE, chains are constituted of regular series of conformations, and one can define a global chain conformation. It should be noted that, in this case, transitions from one conformation of the chain to another one can still be observed for some polymers.

Regular Conformations of Chains

Regular conformations of chains can be found for polyolefins of high symmetry, such as polyethylene $-(CH_2-CH_2)_n-$ or polytetrafluoroethylene $-(CF_2-CF_2)_n-$ (PTFE), which adopt transplanar (zig-zag) conformations: the chain is in complete extension, constituted of series of trans orientations of the C–C links (even though in the case of PTFE the bulkiness of fluorine atoms brings about a small warping, which leads to a slight helicity of the chain) [7]. In the case of monosubstituted polyolefins, regular conformations of chains are found for stereoregular polymers. Because of steric hindrances, for syndiotactic polymers the series of rr triads will lead to chains in complete extension (transplanar conformations), whereas for isotactic polymers the series of mm triads will lead to chains in helicoïdal conformations with a number of monomeric units per turn depending of the steric hindrance of the side group [8].

In summary, in the case of stereoregular chains, the adopted regular conformations will lead to some crystallization of the chains. But if local conformations are of comparable energies or if the polymer is atactic, the random series of local conformations will lead to a disordered global conformation of chains, which will form random coils: the polymer is in a disordered or amorphous state. One notable exception is the atactic poly(vinyl alcohol) (PVA), which can crystallize because the replacement of H atoms by OH groups (which nearly have the same size) does not prevent a close regular packing [9].

8.2 Structural States and Phases Transitions

Because of their length, polymeric chains are highly entangled and can exist only in liquid or solid states. In the liquid state, movements of chain segments are possible but imply the movement of neighboring segments. Moreover, molecular interactions, which can be of intra or intermolecular nature, develop between adjacent segments. The self-diffusion rate is thus smaller than for small molecules, and the viscosity is higher. Therefore, under a fast cooling rate, this disordered liquid state can easily be frozen and the polymer can become vitreous. As said earlier, in the case of stereoregular chains, some crystallization can occur under cooling, but faults are numerous, all the more the molecular weight, and thus the length of chains, is high. Complete crystallization can therefore never be reached, and polymers are partly crystalline and partly amorphous, under a semicrystalline state, with the crystalline and amorphous parts being linked to each other.

8.2.1 Amorphous Phase

In the liquid state, thermal energy allows the movements of the chain segments. As the temperature decreases and the material condenses, movements become more and more difficult, and the system needs longer time to explore all possible conformations till the timescale of the observation prevents this complete exploration. The system then falls out of equilibrium and vitrifies: the disordered liquid state is frozen in a glassy solid state. This thermodynamic transition from a metastable liquid state to the nonequilibrium glassy state is the glass transition. In a differential scanning calorimetry (DSC) experiment, upon reheating, the recovery of all degrees of freedom occurs with the release of the molecular mobility, whose timescale corresponds to that of the experiment (fixed by the heating rate). This recovery induces an endothermic C_p jump (Figure 8.3).

Depending on the desired properties, the glass transition temperature T_g is defined as the onset, the mid-point, or the end of this C_p jump. The glass transition is obviously a kinetic process: the higher the cooling rate, the higher the glass transition temperature. Moreover, an isotherm below T_g gives the out-of-equilibrium system time to relax toward the equilibrium state and its enthalpy decreases toward the enthalpy of this equilibrium state (which would be reached at infinite time) and the material densifies: it undergoes a structural relaxation phenomenon, also called physical aging. For a given annealing time, this process is all the more pronounced at a temperature close to T_g, typically $20° - 15°$ below it. Upon

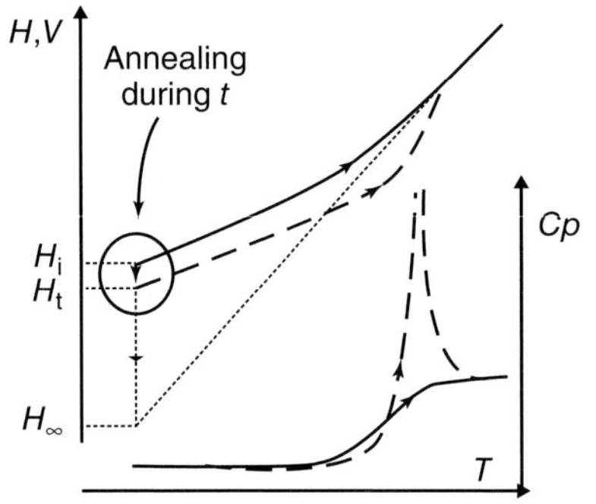

Figure 8.3 Schematic representation of the evolution of the enthalpy H or the volume V of an amorphous polymer upon heating, and corresponding evolution of the heat capacity C_p, after quenching from the molten state (solid line), and after annealing below T_g (dashed line).

reheating, it shows as an overshoot superimposed on the C_p jump: as the increasing temperature allows the system to evolve, the enthalpy of the corresponding equilibrium state is caught up, leading to an additional peak on $C_p = dH/dT$ (slope of the enthalpy evolution) (Figures 8.3 and 8.9, scan a). The more important is the ageing process (the lower is the reached enthalpy), the stronger the overshoot.

Another point that should be underlined is that, for linear polymers, the glass transition temperature also depends on the polymer molecular weight M_n (see, e.g., the case of poly(vinyl pyrrolidone) [10]). The end of the chain has a higher mobility than the rest of the chain and, therefore, as the molecular weight increases (the length of the chains increases), this increase in mobility becomes more and more insignificant. The overall mobility of the chains decreases, and the glass transition increases till it reaches a plateau (Figure 8.4). Models have been developed to describe this evolution, the simplest one being the Fox and Flory law [11]:

$$T_g(M_n) = T_{g\infty} - K/M_n \quad (8.1)$$

where $T_{g\infty}$ is the value at the plateau, and K is a constant. In a more general way, all parameters that tend to hinder molecular mobility on a large scale tend to increase T_g, such as lower flexibility, higher interactions between chains, bulkier side groups, and higher cross-linking.

8.2.2 Crystalline Structures

For semicrystalline polymers, three scales of crystalline organization have to be considered: the crystalline structure at the scale of a few angstroms (crystalline cell), the texture due to the formation of crystalline lamellae at a wider scale, and the arrangement of these lamellae in more complex elements (morphology).

In the crystalline cell, chains are packed together parallel to each other. The length of the cell parallel to the chain axis is defined by the repeating monomeric unit. Usually, the conformation of the isolated chain is retained (helicoïdal, trans-planar, etc.) but it can be modified when the compactness is enhanced by the establishment of intermolecular interactions. It is the case, for example, of polar polymers, such as poly(ethylene terephthalate) (PET), or polymers that are able to develop hydrogen bonds, such as polyamides [12] and polypeptides.

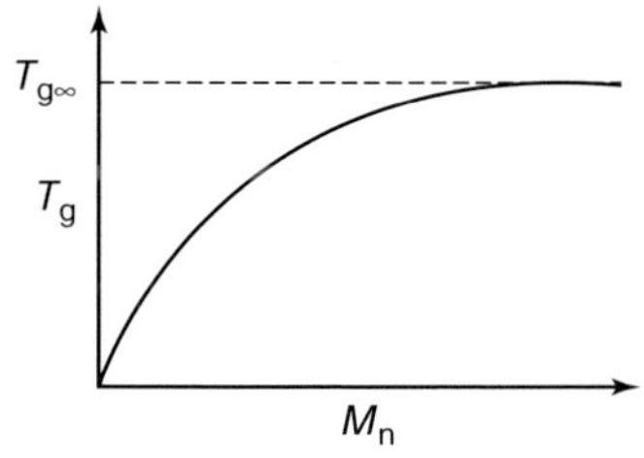

Figure 8.4 Schematic representation of the evolution of the glass transition temperature versus the molecular weight of a polymer.

The crystallites formed by these crystalline cells are small, with a mean thickness between 10 and 100 Å. Surprisingly, the chain axis is approximately perpendicular to the larger sides of the crystallites, which implies that chains are folded and form crystalline lamellae. Several models have been proposed to describe such arrangements, such as the fringed micelle model [13], the folded chain model, and the switchboard model [14]. The general picture is represented in Figure 8.5: crystallites are formed by folded polymeric chains that can pass in and out of several crystallites, and the folds are more or less regular depending on crystallization conditions.

The observation, by optical microscopy under cross polarizers, of semicrystalline polymers obtained by the usual cooling from the molten state shows crystalline structures in the form of spheres called spherulites (Figure 8.6). They are characterized by extinction crosses with the main axis parallel to the polarizer directions and are constituted of fibrils that radially grow from a common center. These fibrils are, in fact, crystalline lamellae: whatever their radius, their molecular axis is nearly parallel to the tangential plane and the radial direction corresponds to the direction of maximum growth (Figure 8.6). It should be noted that the lamellae are not necessary planar and sometimes they look more like ribbons, as twists occur upon growth, which give rise to additional extinction rings [16, 17]. Inside the spherulites, between lamellae, the amorphous zone remains, but some polymer chains can pass in and out several lamellae, and some thin crystalline structures in the form of threads have been observed, called "intercrystalline links" [18], that bond lamellae to each other and are constituted of polymer chain backbones parallel to the length. Upon crystallization, radial growth of crystalline lamellae first occurs. At a fixed temperature, the growth rate is constant, and growth stops when the available space is filled. The boundary between adjacent spherulites is either planar when nuclei appear at the same time, or hyperbolic when nuclei progressively appear. The spherulites'size can vary between a tenth of a micrometer and several millimeters, depending of the crystallization conditions and the

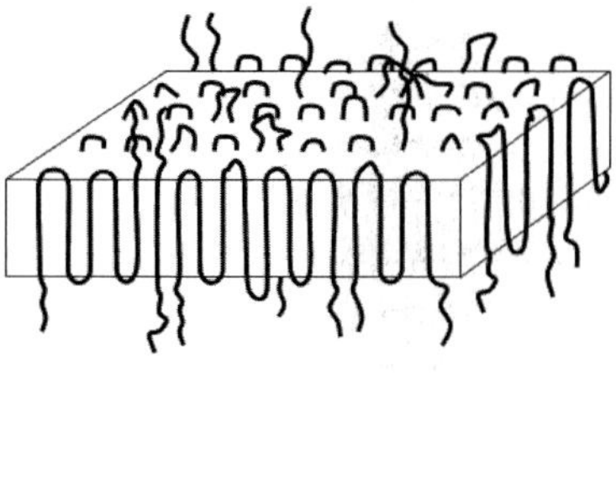

Figure 8.5 Schematic representation of a crystalline lamella.

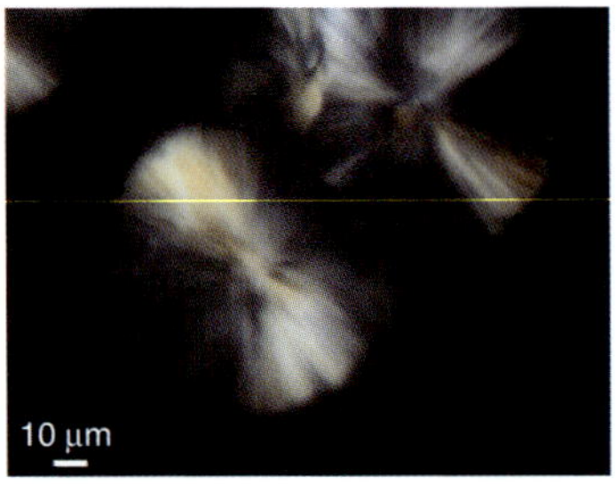

(a)

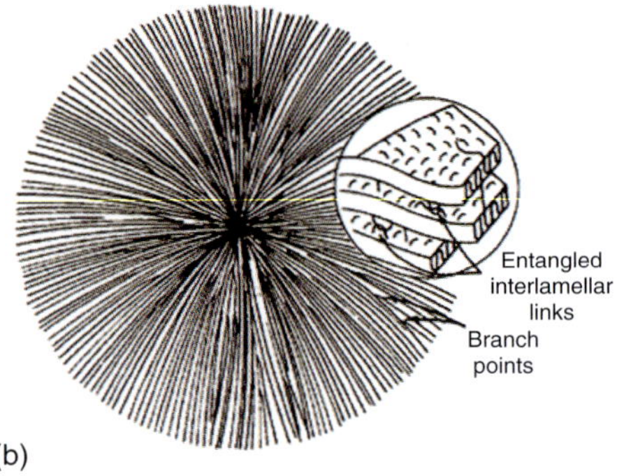

(b)

Figure 8.6 (a) polarized light micrograph of spherulites of poly(lactic acid) recorded at 125 °C upon crystallization from the melt. (b) Schematic representation of a spherulite. Adapted from Hoffman *et al.* [15]. Reproduced with permission of Springer.

material. For higher molecular weight or lower stereoregularity, the growth rate is smaller. Once the growth of spherulites stops, then the molten polymer between the fibrils crystallizes.

When the molten polymer is subjected to a mechanical stress – as is the case for numerous manufacturing processes such as injection molding, extrusion, film blowing, or spinning – the polymer chains are oriented and the following crystallization can be influenced. The obtained crystalline morphologies vary from deformed spherulites, microfibrils, or fibers with the chain axis parallel to the flow direction, to a "shish-kebab" organization. This peculiar crystalline arrangement is constituted of extended polymer chains forming a microfibril-like structure (shish) on which folded chain lamellae (kebabs) grow perpendicularly [19] (Figure 8.7). The crystalline morphology depends on the shearing rate. The same thing occurs when a semicrystalline polymer is submitted to stretching processes.

Fibers can also be obtained from natural polymers, such as cellulose derivatives or proteins, but their morphology is much more complex. For example, in the case of cellulose, the crystalline parts are strongly oriented as a result of the establishment of intra and intermolecular hydrogen bonds forming microfibrils composed of approximately 30–40 cellulose chains. These microfibrils aggregate into fibrils, which organize themselves at larger scale into rigid fibers. For natural fibers of animal origin composed, for example, of keratin [21] or collagen [22], helixes combine themselves into superhelixes, leading, on a wider scale, to highly crystallized micro and macro-fibrils.

According to classical nucleation theory, the nucleation rate per volume unit, N, can be express as

$$N = N_0 \exp\left(-\frac{\Delta G^*}{RT}\right) \exp\left(-\frac{\Delta G_\mathrm{d}}{RT}\right) \tag{8.2}$$

where N_0 is a constant, ΔG^* is the thermodynamic barrier that results from the competition between the crystal formation and the creation of a liquid/crystal interface, and ΔG_d is the diffusion barrier. ΔG^* increases as the undercooling (temperature gap compared to melting temperature) increases, whereas ΔG_d

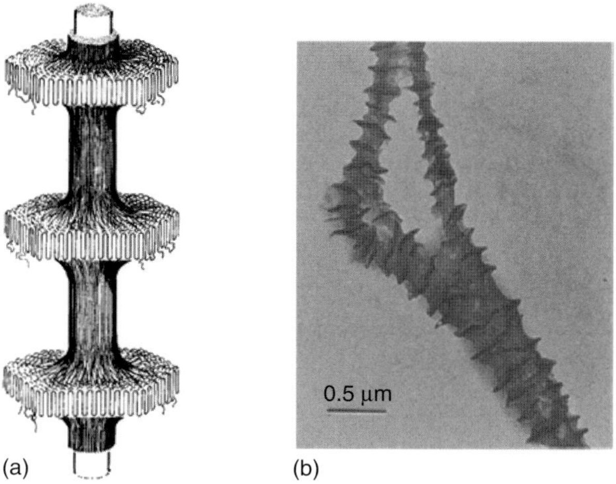

(a) (b)

Figure 8.7 "Shish-kebab" structure. (a) Molecular model Adapted from Pennings *et al.* [20]. Reproduced with permission of Springer. (b) Transmission electron micrograph of "shish-kebab" structure of polyethylene. Adapted from Pennings *et al.* [19]. Reproduced with permission of Springer.

decreases as the temperature approaches the glass transition and the molecular mobility decreases. The nucleation rate thus goes through a maximum, and typically the most favorable undercooling for nucleation varies from 10° to 50°. The growth rate also passes through a maximum, limited at low temperature by the low mobility.

The global kinetic of crystallization can be analyzed with an Avrami law, but several years ago, Lauritzen and Hoffman developed a theory to describe the kinetics of lamellae crystallization [23, 24]. The idea is that the formation of a lamella proceeds by secondary nucleation: a chain recovers by successive foldings of an already formed crystal (substrate). This crystallization proceeds by the most rapid mechanism rather than one that would lead to the equilibrium structure (extended-chain crystal). The growth rate depends on the rate of surface nucleation (secondary nucleation), i, and the rate of recovery of the substrate, g (Figure 8.8). Hoffman describes three different regimes of crystallization depending on these relative rates (which vary with temperature). The change of regimes can induce a change of morphology [25, 26].

Following the Lauritzen–Hoffman theory, the variation of Gibbs free energy associated with the formation of a lamella of thickness l, length and width x [15] is given by

$$\Delta G = 4xl\sigma + 2x^2\sigma_e - x^2 l \Delta G_{L-C} \tag{8.3}$$

where σ and σ_e are the interfacial Gibbs free energy per unit surface of, respectively, the lateral faces and fold faces. $\Delta G_{L-C} = G_L$-G_C is the difference in Gibbs free energy between the liquid and the crystal per unit volume, and is expressed as

$$\Delta G_{L-C} = \Delta H_m - T\Delta S_m$$

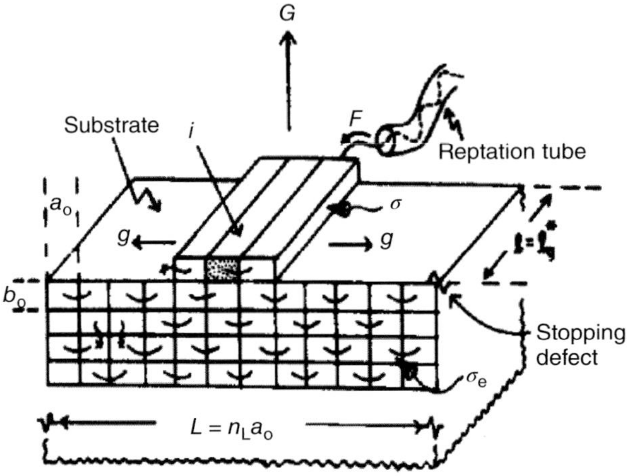

Figure 8.8 Model for surface nucleation and growth: schematic representation of a substrate (length *L*) on which a chain deposits by successive foldings of length *l*. *G* is the growth rate of the crystal, *i* is the nucleation rate, and *g* is the substrate completion rate. Adapted from Hoffman and Miller [24]. Reproduced with permission of Elsevier.

where ΔH_m and ΔS_m are, respectively, the melting enthalpy and entropy per unit volume.

At the melting temperature $T_m{}^\circ$, assuming that the C_p of the liquid and of the solid are equal

$$\Delta H_m = T_m{}^\circ \Delta S_m \quad \text{and} \quad \Delta G_{L-C} = \Delta H_m \left(\frac{T_m{}^\circ - T}{T_m{}^\circ}\right) \tag{8.4}$$

When the lamella melts, at a temperature T_m, $\Delta G = 0$. Assuming that l is small compared to x (general case), it leads to $2\sigma_e = l\Delta G_{L-C}$. Replacing ΔG_{L-C}, the equation of Gibbs–Thomson is obtained:

$$\mathrm{T_m} = \mathrm{T_m}^\circ \left[1 - \frac{2\sigma_e}{l\Delta H_m}\right] \tag{8.5}$$

It underlines the fact that the melting temperature of the lamella depends on its thickness l. As the thickness tends toward infinity (perfect crystal), T_m tends toward $T_m{}^\circ$. For polymers, one thus defines $T_m{}^\circ$, which is the theoretical melting temperature of the polymer that would be 100% crystalline, and T_m, which is the real melting temperature of the semicrystalline polymer.

Considering that this lamella of thickness l has been formed at the crystallization temperature T_c

$$l \approx \gamma \frac{2\sigma_e T_m{}^\circ}{(T_m{}^\circ - T_c)\Delta H_m} \tag{8.6}$$

where γ is a constant >1, experimentally found to be close to 2. The thickness of the lamella thus depends on its crystallization temperature. This leads to the relation between T_m and T_c, the Hoffman–Weeks equation:

$$T_m = T_m{}^\circ \left(\frac{\gamma - 1}{\gamma}\right) + \frac{T_c}{\gamma} \tag{8.7}$$

When a semicrystalline polymer is heated in a DSC experiment from a low temperature, the C_p jump characteristic of the glass transition of the glassy parts is observed. Above the jump, the amorphous parts are liquid but still maintained by the crystalline regions. At higher temperatures, the endothermic peak corresponding to the melting occurs (Figure 8.9). In contrast to that in small molecules, this peak is quite large due to the lack of perfection of the crystallites. As we have seen, the melting temperature depends on the crystallization conditions: the lower the crystallization temperature, the smaller the thickness of lamellae, and the lower the melting point (Figure 8.9, inset). Multiple melting peaks can even be observed, which do not result from polymorphism but from the heterogeneity of lamellae thickness resulting from crystallization that has occurred at different temperatures (Figure 8.9, scan c). Upon heating, the crystalline structure can also reorganize with melting of small crystallites, recrystallization at higher temperature, followed by corresponding melting. The obtained DSC scans obviously depend on the heating and cooling rates. It should be noted that, when a polymer is quenched from the liquid state at a rate that prevents crystallization, this crystallization can occur on the following heating as soon as the mobility allows it, that is, just above the glass transition (cold crystallization).

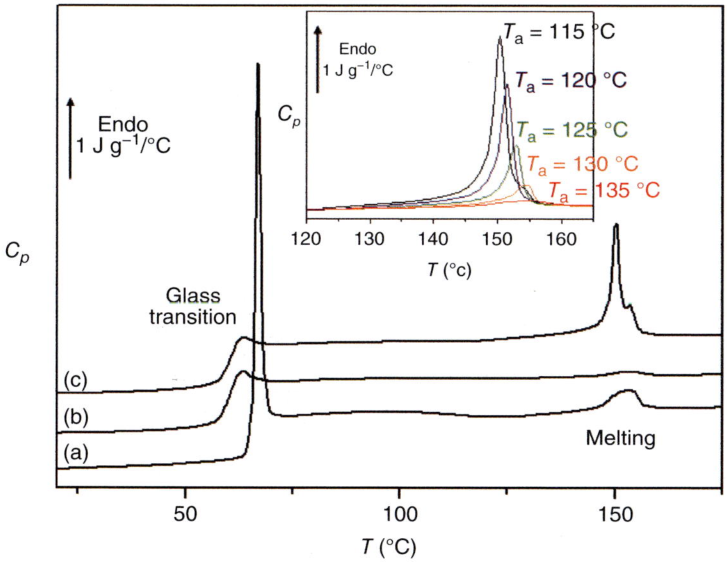

Figure 8.9 DSC scans recorded at 10 °C min^{-1} of poly(lactic acid) (a) as received, (b) after quenching from the melt at 20 °C min^{-1}, and (c) after successive annealings of 30 min upon cooling at 135, 125, and 115 °C. The insert shows the melting endotherms obtained after annealing of 60 min upon cooling at the indicated temperatures.

8.2.3
The Case of Copolymers: Organized Amorphous Structure

Because of their varied architectures (statistical, graft, block), copolymers are found in numerous morphologies. Here we cannot review all of them, but we will focus on block copolymers for which interesting morphologies can be observed as blocks of different chemical species are linked together.

Usually, a homogeneous phase can be obtained at low temperatures, but at higher temperatures segregation can occur. However, since the species are linked, this segregation develops only on a microscopic scale, on the order of the block size. Since these segregated domains are of the same size, they can arrange in an ordered manner. This leads to a phase transition from a disordered (homogeneous) state to an ordered one. The shape and the arrangement of these micro-phase-separated structures depend on the size of the domain, that is, the fraction of each species (Figure 8.10) [27]. In the case of an amorphous A-*block*-B copolymer, for a low A fraction, amorphous spherical A domains develop in an amorphous B matrix and these spherical domains are arranged in a body-centered cubic lattice. At higher A fraction, A blocks regroup in amorphous cylindrical domains into the amorphous B matrix. These cylinders form a hexagonal lattice. For equal A and B fractions, the structure is lamellar, forming a layered lattice. For higher A fractions, the same types of structures are observed but with inverted phases (inclusions of B domains in A matrix). In some cases, intermediate structures have been observed, where both phases interpenetrate as a bicontinuous gyroid structure. When the number of chemical species increases (as for triblock copolymers), the developed structures are more complex.

It should be added that if one block can crystallize, the situation is even more complex. Two cases can be encountered: First, the domains containing the crystallizable block can completely crystallize, forming, for example, in the case of layered structures, an alternation of amorphous one-block type and crystalline other-block type layers. The crystallinity can thus be adjusted by varying the fraction of the crystallizable species, which changes the physical properties (thermal, mechanical properties, solubility) (e.g., the case of poly(ethylene vinylacetate) (EVA) [28, 29]). These types of structures allow, for example, obtaining materials with mechanical strength (brought by the crystalline layers) but also some rubber

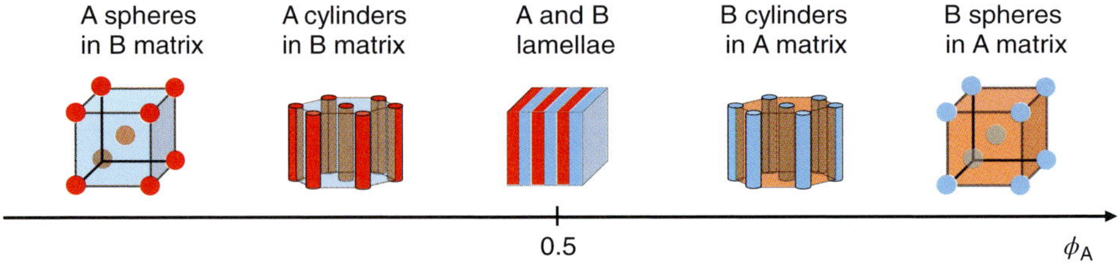

Figure 8.10 Schematic representation of the micro-phase-separated structures of A–B diblock copolymers versus the A volume fraction ϕ_A.

elasticity (see the following part of this chapter) if the working temperature is above the glass transition temperature of the amorphous layers. Second, the domains containing the crystallizable block can only partially crystallize, leading to even more complex structures.

8.3
Dynamic Disorders

The mechanical behavior of polymers is intrinsically bound to their structural state. Amorphous polymers, which are brittle at low temperatures in the glassy state, soften when they are heated above their glass transition upon the glass–rubber transition. This macroscopic behavior results from different types of mobilities that are released as the temperature increases.

8.3.1
Glass–Rubber Transition Behavior

In the usual mechanical testing, one measures either the strain resulting from an applied stress (creep experiment), or the stress necessary to obtain a given strain (stress relaxation experiment). The mechanical parameter that describes the stiffness of a material is the modulus, which is the ratio between the stress and the strain in a stress relaxation experiment (its reciprocal, the compliance J, is obtained from a creep experiment). It should be noted that it can be an elastic modulus (E), a shear modulus (G), or a compression modulus (K) depending of the type of deformation mode [3]. For an amorphous polymer, the value of the modulus evolves with temperature and time, as schematically represented in Figure 8.11.

Four different regions can be distinguished:

- At low temperatures, below T_g, in the glassy state, the polymer is stiff and brittle. At a given temperature, it follows an elastic behavior: G is constant and its value is approximately 10^9 Pa. Polymer chains are entangled, and the mobility is so low that only very localized motions can occur on the experiment timescale. These are localized or secondary relaxations.
- At higher temperatures, the G value abruptly drops (3 orders of magnitude over a few tens of degrees) in the viscoelastic region. When passing the glass transition, molecular mobility is released, allowing cooperative, large-amplitude motions. As this relaxation process is related to the glass transition, it is named the "main" or "α relaxation." On the stress–strain curves, a yield point can be observed (marking the occurrence of irreversible deformation) just before the break: portions of polymer chains align, releasing the stress.
- The modulus then reaches a plateau value: it is the rubbery-elastic plateau region. Under a small stress, the polymer can be stretched as an elastomer and comes back to its initial length when the stress is removed. Large-amplitude motions occur between the fixed points of the physical temporary network formed by the entanglements of chains.

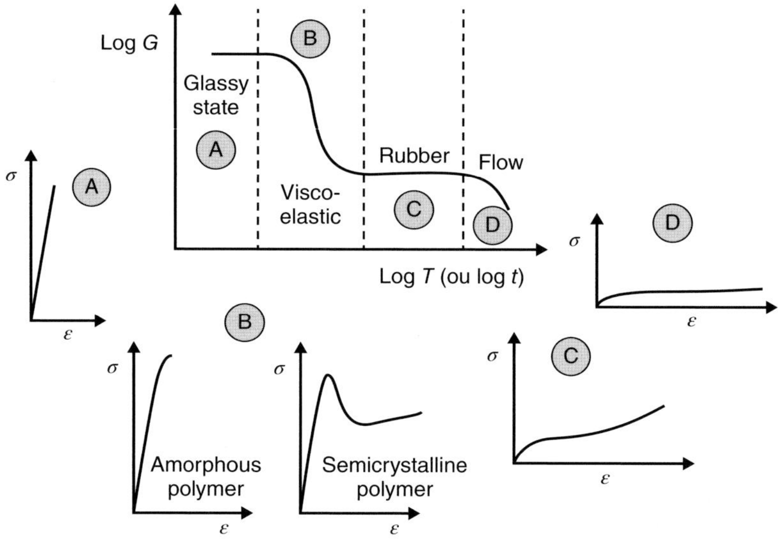

Figure 8.11 Schematic evolution of the shear modulus versus temperature (or time) revealing the different regions: (A) the glassy state, (B) the viscoelastic state, (C) the rubbery state, and (D) the flow state. Typical stress σ versus shear ε curves obtained in these zones are also shown.

- Finally, at higher temperatures, the chains can move by wiggling (reptation) and disentanglement, leading to viscous flow of the polymer and the loss of stiffness: it is the terminal flow region. For the hot-melt extrusion process, this temperature domain has to be reached. For cross-linked polymers (for which disentanglement is not possible), this region does not exist. For linear amorphous polymers, the temperature at which the disentanglement of chains occurs depends on the molecular weight: the higher the molecular weight, the longer the rubbery-elastic plateau (Figure 8.12). This plateau exists only above a minimum molecular weight M_c, known as the critical molecular weight at the entanglement limit.

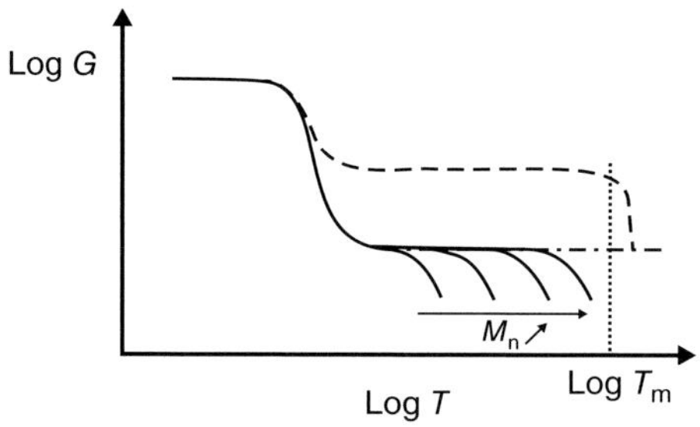

Figure 8.12 Schematic evolution of the shear modulus versus temperature for different types of polymers: amorphous with different molecular weights (solid lines), network (dashed dotted line), and semicrystalline (dashed line).

In the case of semicrystalline polymers, above T_g the mobility is released in the amorphous parts but the crystalline parts act as fixed points, reducing the overall mobility and maintaining the mechanical strength. Therefore, the higher the crystalline rate, the lower the drop of G (Figure 8.12). In mechanical tests, a ductile behavior is observed: after the yield point, large deformations take place without any increase of stress (plastic flow), which could be due to a slipping of crystalline lamellae in the direction of the stress, followed by an uncoiling of the chains [30]. In this temperature range, cold drawing of the polymer can be realized. It should be noted that the ductile behavior is all the more enhanced (better shock resistance) when the number of spherulites is high (and their size is small). At the melting point, all the mobility is released and the chains flow.

8.3.2 Mobilities

The macroscopic mechanical behavior results from molecular motions at the microscopic scale. The response of the material to the mechanical perturbations depends on the timescale characteristic of molecular movements. At a given temperature, the more delocalized the motion, the longer the time needed for this motion to occur. The characteristic time τ also depends on the temperature: the higher the temperature, the faster the motion. This characteristic time can be determined from a dynamic mechanical experiment. Rather than applying a static mechanical perturbation (stress or strain) to the polymer, it is submitted to a small dynamic stress (or strain) of a chosen frequency. The recorded response of the material to this perturbation (relaxation) depends on the frequency of the oscillating stress (or strain): only movements faster than the periodic variations will occur. At a given temperature, by varying the pulsation ω (or the frequency $F=\omega/2\pi$) of the mechanical field, the timescale of the movements can be deduced from the evolution of the storage (G') or the loss (G'') part of the complex modulus $G^*(\omega) = G'(\omega) + iG''(\omega)$ (or from the storage J' or the loss J'' part of the complex compliance $J^*(\omega) = J'(\omega) - iJ''(\omega) = G^{*-1}$) [3]. Relaxation processes appear as a step on G' (or J') and a peak on G'' (or J''). The characteristic relaxation time τ of these processes can be deduced from the frequency of appearance ($\tau = 1/\omega$). The same type of experiments can be realized using a dielectric probe instead of a mechanical one (see Chapter 11). The sample is submitted to a dynamic electrical field, and the response of the polymer to this perturbation is recorded. The evolution of the complex dielectric permittivity ε^* can also be divided into a storage component ε' and a loss component ε'' according to $\varepsilon^*(\omega) = \varepsilon'(\omega) - i\varepsilon''(\omega)$. A relaxation process appears as a step on $\varepsilon'(\omega)$ curve and a peak on $\varepsilon''(\omega)$ curve (Figure 8.13a). By varying the temperature, the evolution of the relaxation time can be obtained, and the relaxation map characteristic of the polymer can be constructed (Figure 8.13b).

8.3.2.1 Localized Processes

Localized motions concern motions over a characteristic length smaller than 1 nm. They are of intramolecular origin and can involve fluctuations of localized

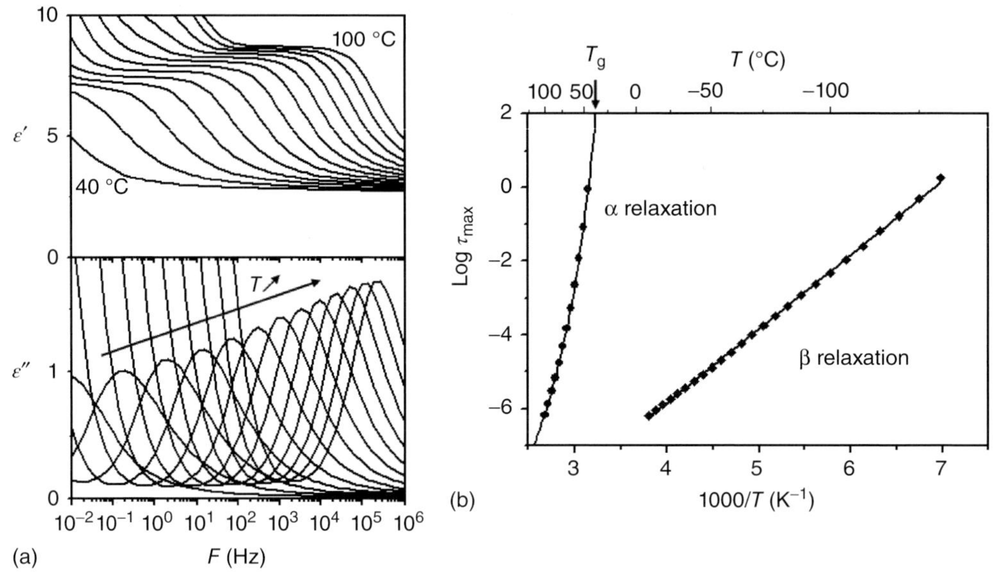

Figure 8.13 (a) Evolution versus the frequency of the applied electrical field of the real (ε') part and the imaginary (ε'') part of the dielectric permittivity of poly(vinyl acetate), showing the main α process. Spectra are recorded every 5 °C from 40 to 100 °C. (b) Deduced relaxation map showing the evolution versus reciprocal temperature of the relaxation times associated with the main α process (•) and with a secondary β process (◆). The behavior laws, respectively, of VTF and Arrhenius type are also shown. See also [31, 32].

parts of the main chain, resulting in local conformational changes as crankshaft-type rotations [33], rotations of side groups, or movements of a part of the side groups [34]. Several localized relaxations (or secondary relaxations) can be observed, and they are named after the greek letters (β, γ, δ, etc.) with increasing localization (decreasing relaxation time, thus increasing appearance frequency or decreasing appearance temperature). It should be noted that, since the discovery by Johari and Goldstein of the existence of a secondary β process even for some small rigid molecules [35], much discussion has arisen on the possibility of a peculiar β process type, called β of Johari–Goldstein type (β_{JG}), which is a universal property of the glass formers, which for polymers would involve the same movement as the main α process but are more restrained. It is interpreted in terms of islands of mobility or, in the framework of the coupling model [36], a genuine process precursor for the dynamic glass transition.

The secondary relaxations proceed by jumping the energy barrier E_a of a double-well potential. They are thermally activated, and the relaxation time obeys an Arrhenius law (Figure 8.13):

$$\tau(T) = \tau_0 \exp\left(\frac{E_a}{RT}\right) \tag{8.8}$$

where τ_0 is the time at infinite temperature. Typical values of the activation energy of secondary relaxations are between 20 and 50 kJ mol^{-1}, and τ_0 should be close to 10^{13} s. Higher values of τ_0 could be related to an entropic contribution that can be expressed, in the framework of Eyring's theory [37], as

$$\tau(T) = \frac{h}{k_B T} \exp\left(-\frac{\Delta S}{R}\right) \exp\left(\frac{\Delta H}{RT}\right) \tag{8.9}$$

where ΔH and ΔS are, respectively, the activation enthalpy and entropy of the process.

8.3.2.2 Main Relaxation Process

This process, sometimes called *dynamic glass transition*, involves motions of several repetitive units over a characteristic length of a few nanometers (2–10 nm), also called *segmental motions*. What characterizes the main α relaxation is that it is non-Arrhenian and non-exponential.

Indeed, as the temperature decreases from the molten state, the relaxation time, which is proportional to the viscosity through the Maxwell relation $\tau = \eta/G$, dramatically increases and its evolution departs from an Arrhenius behavior. This increase is so high that on a DSC experiment, due to the timescale, this movement is frozen at $T = T_g$, which corresponds to a viscosity of 10^{13} poise or a value of the relaxation time of the process close to 100 s. This evolution of τ is usually described by the so-called Vogel–Tamman–Fulcher (VTF) law [38–40] (Figure 8.13):

$$\tau(T) = \tau_0 \exp\left(\frac{C}{T - T_0}\right) \tag{8.10}$$

where τ_0, C, and T_0 are constants. This law predicts an infinite relaxation time at T_0. This departure from the Arrhenius behavior can be interpreted as follows:

- A decrease of the free volume around the moving unit, which allows its movement, till it becomes 0 at T_0 in the free volume theory of Cohen and Turbull (C is the reciprocal of the expansion coefficient of the free volume fraction [41–43]).
- An increase in cooperativity of the relaxing units. As the temperature and, thus, the volume decrease, neighbouring segments (of the same or different chains) have to move in a more and more cooperative way, which slows down the process (Adam–Gibbs model [44, 45], Donth model [46–48]).

The curvature of the evolution depends on the material, and a tool to describe it is to determine the fragility index m related to the slope of the evolution at $T = T_g$:

$$m = \left[\frac{\partial \log \tau}{\partial\left(T_g/T\right)}\right]_{T=T_g} \tag{8.11}$$

In the Angell framework [49], materials are classified as "strong liquids" when τ follows an Arrhenius behavior ($m = 16$) since the liquid vitrifies slowly, or "fragile liquids" when τ departs from the Arrhenius behavior ($m > 100$). The higher the value of m, the higher the slope (apparent activation energy), and the quicker the liquid vitrifies.

Dielectric spectroscopy is a suitable tool to follow the evolution of τ, as the accessible frequencies range from 10^{10} to 10^{-2} Hz. In mechanical spectroscopy, since the frequency range is smaller, master curves are constructed by making the assumption of a time/temperature superposition principle, that is, the shape of the dynamic function response does not change with temperature [50], and the $\log G^*(T, \log \omega)$ curves are shifted by a factor $\log a_T$:

$\log G^*(T, \log \omega) = \log G^*(T_{ref}, \log \omega + \log a_T)$, with $\log G^*(T_{ref}, \log \omega)$ the evolution of $\log G^*(\log \omega)$ recorded at the reference temperature T_{ref}. This leads to

$$\log a_T = \log \tau(T) - \log \tau(T_{ref}) \tag{8.12}$$

Williams, Landel, and Ferry proposed an empirical formula (WLF model) describing the temperature dependence of $\log a_T$ [51]:

$$\log a_T = -\frac{C_1(T - T_{ref})}{C_2 + T - T_{ref}} \tag{8.13}$$

where C_1 and C_2 are material-dependent constants. At $T_{ref} = T_g$, it is equivalent to the VTF law if the WLF parameters are

$$C_1 = \frac{C}{2.303(T_g - T_0)} \approx 14 - 18 \quad \text{and} \quad C_2 = T_g - T_0 \approx 30\,^\circ\text{C} - 70\,^\circ\text{C}$$

Another feature of the α process is that it is usually a non-exponential process. The simplest model to describe a relaxation process is the Debye model, which makes the assumption of a unique relaxation time. All entities move in the same way within the same time. However, this model usually fails to describe the α process. In a time-domain experiment, the evolution of the relaxation function $\phi(t)$ versus time is not a simple exponential decay characterized by τ:

$$\phi(t) = \exp\left(-\frac{t}{\tau}\right) \tag{8.14}$$

Instead, a stretched exponential decay is observed. It reveals a distribution of relaxation times, which emphasizes local dynamic heterogeneities. It is usually described by the Kohlrausch–Williams–Watts (KWW) relation [52, 53]:

$$\phi(t) = \exp\left[\left(-\frac{t}{\tau_{KWW}}\right)^{\beta_{KWW}}\right] \tag{8.15}$$

where β_{KWW} is the stretch exponent that characterizes the distribution of relaxation times ($0 < \beta_{KWW} < 1$), and τ_{KWW} is a mean relaxation time. Typical β_{KWW} values are on the order of approximately 0.5.

The α process is sensitive to the development of crystallinity. When the fraction of the amorphous part is reduced, the magnitude of the process decreases. But the crystalline parts can also reduce the mobility of the amorphous phase, thereby progressively increasing the relaxation time (shift of the α process to lower frequencies or higher temperatures). In some cases, as for poly(L-lactic acid) (PLA), an additional α process has been observed (Figure 8.14) [54, 55]. As crystallization occurs, the α process, due to motions in the amorphous interspherulitic region, disappears, while another one, α_{SC}, due to motions in the amorphous regions

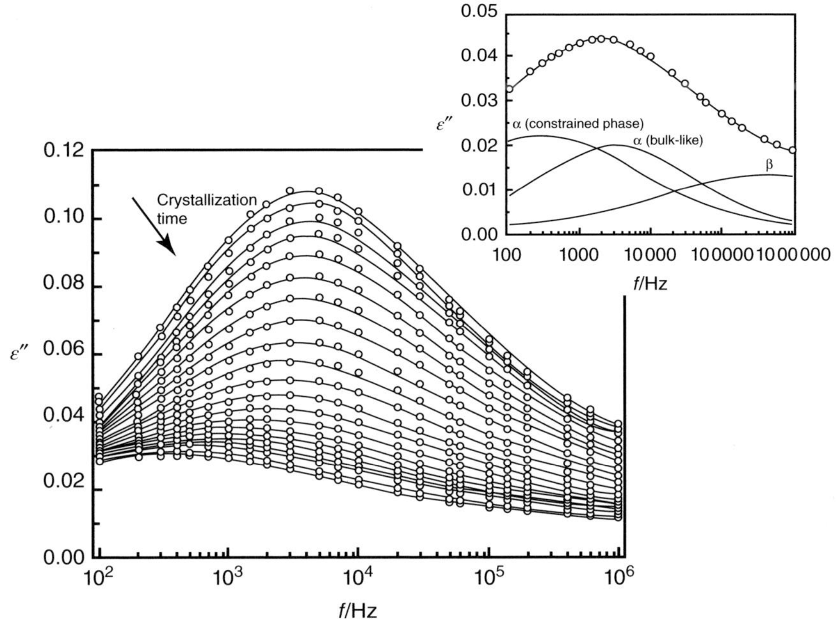

Figure 8.14 Evolution of the dielectric loss versus the frequency of the applied electric field upon isothermal crystallisation of PLA ($T = 80\,°C$). The inset represents ε'' recorded after 120 min and its fitting into three relaxations: a secondary process (β), the α processes assigned to the bulk amorphous phase, and the amorphous phase constrained by the crystallites. Adapted from Dionísio *et al.* [54]. Reproduced with permission of Wiley.

constrained by the crystalline lamellae (intraspherulitic), develops at lower frequencies (lower mobility).

8.3.2.3 **Global Chain Dynamics**

The viscous flow of the polymer occurs in the terminal flow region. Chains move relatively to each other, and the flow rate is linked to their disentangling time, which obviously depends on the molecular weight M. However, even though M influences the length of the rubbery-elastic plateau, the shape of $G'(\omega)$ and $G''(\omega)$ curves remains unchanged in the terminal flow relaxation (Figure 8.15). In particular, at very low frequencies, $G'(\omega)$ tends toward $J_e^{\,0}\eta_0^{\,2}\omega^2$ and $G''(\omega)$ toward $\eta_0\omega$, where $J_e^{\,0}$ and η_0 are, respectively, the recoverable shear compliance and the viscosity at zero shear rate. η_0 characterizes the viscous flow. It depends on the molecular weight according to the power law $\eta_0 = M^{\upsilon}$, where υ is a constant equal to 1 for $M < M_c$ (non-entangled polymers) and close to 3 for $M > M_c$ (entangled polymers) [56].

The mean relaxation time of the melt, τ, characterizing the dynamics of the global chain also depends on the molecular weight and also encounters a

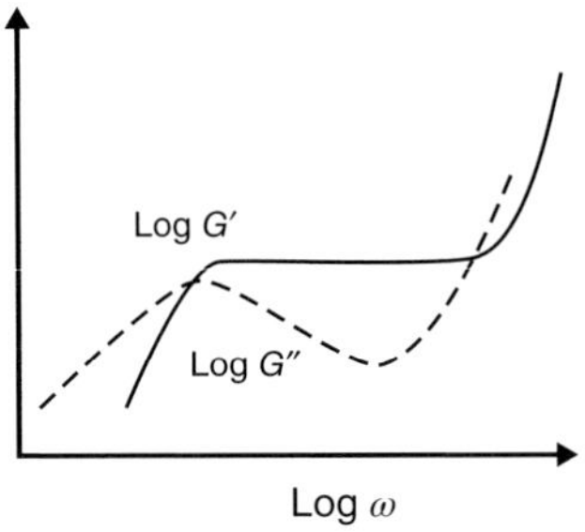

Figure 8.15 Schematic evolution of the real (G') and imaginary (G'') parts of the shear modulus with the pulsation of the applied mechanical field in the terminal flow region.

modification around M_c. For $M < M_c$, τ is proportional to M^2, and for $M > M_c$ it is, like η_0, proportional to $M^3 - M^{3.4}$.

For polymers that possess at least one component of the dipolar moment parallel to the chain backbone (polymers of A type [57, 58]), dynamic dielectric spectroscopy experiments reveal a relaxation mode appearing at lower frequencies (higher temperatures) than the α process: it is called *normal mode* and is due to this global chain dynamics (Figure 8.16). The dynamics of the global chain is generally described through the Rouse model [60, 61] for $M < M_c$ or the reptation model [62, 63] for $M > M_c$. In the case of non-entangled polymers, the associated relaxation time corresponds to the time needed for the motion to propagate from one end of the chain to the other end (Rouse mode with the longest relaxation time, i.e., the lowest order). For entangled polymers, the relaxation time of the normal

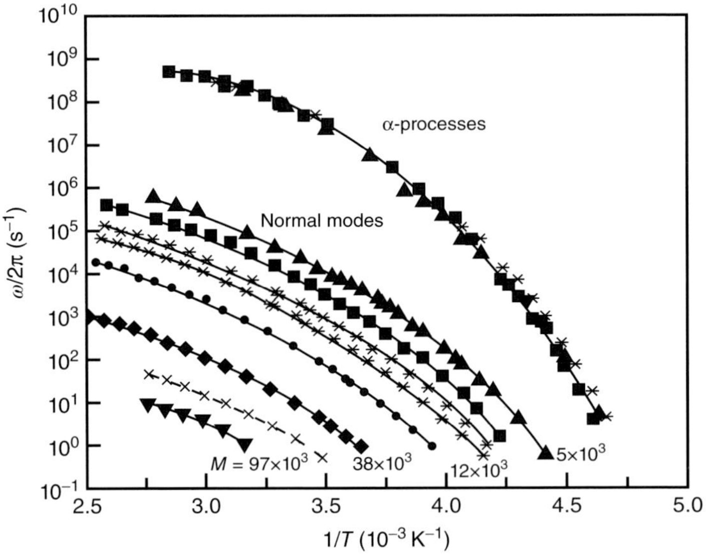

Figure 8.16 Relaxation map of the *cis*-polyisoprene obtained by dielectric spectroscopy showing the evolution of the α process and the normal mode for molecular weights ranging from 5×10^3 g mol^{-1} to 97×10^3 g mol^{-1}. Adapted from Strobl [2]. Reproduced with permission of Springer. Data from [59].

mode corresponds to the time needed for the chain to escape from the confinement tube created by the entanglements with other chains (disentangling time) (reptation mode with the longest relaxation time, i.e., the lowest order). Whatever the molecular weight, the relaxation time associated with the normal mode follows the VTF behavior as the α process.

8.4 Blends of Polymer and Small Molecules

The models allowing the description of the physical and chemical properties of blends have been initially developed for blends of small molecules. The complexity of polymers (architectures, number of possible conformations, etc.) leads to more and more complex models, which cannot be reviewed here. Therefore, in this part, we will focus on blends of polymers and small molecules, which is a case frequently found in the pharmaceutical products where drug/polymeric excipient amorphous formulations are used to improve the dissolution of poorly water-soluble drugs. It raises not only the question of miscibility but also the evolution of a liquid miscible blend when it is cooled: if crystallization is avoided, what are the effects of mixing on the overall mobility and, more precisely, on the glass transition? It also raises the question of the limit of solubility of the small molecules in the polymer.

8.4.1 Glass Transition of the Blends

In the case of a blend of two completely immiscible amorphous compounds, and if a sufficiently high cooling rate is applied, each phase becomes vitreous. The glassy blend is then characterized by two glass transition temperatures associated with each pure compound.

On the contrary, in the case of a blend of completely miscible compounds, the mixing at the molecular level of the different species leads to a homogeneous glass characterized by a single glass transition temperature. Assuming that no volume change occurs on mixing (regular solution), Gordon and Taylor proposed an expression for estimating the T_g of the blend [64]:

$$T_g = \frac{\omega_A T_{gA} + K\omega_B T_{gB}}{\omega_A + K\omega_B} \tag{8.16}$$

where T_{gA} and T_{gB} are the glass transition temperatures of the pure compounds, ω_A and ω_B are their weight fractions in the mixture, and K is a fitting parameter that accounts for the deviation from a linear evolution. K can be related to the ratio of the C_p jumps at T_g of pure compounds: $K = \Delta C_{pB}/\Delta C_{pA}$ [65, 66]. The example of PVP (poly(vinyl pyrrolidone))/indomethacin mixtures is shown in Figure 8.17 [67].

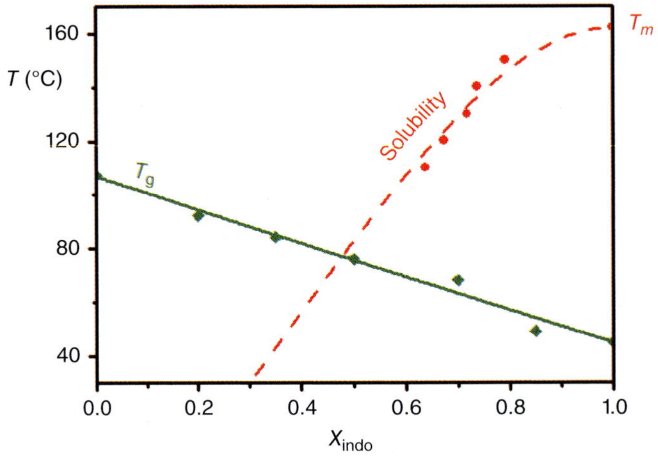

Figure 8.17 Blend of indomethacin and poly(vinyl pyrrolidone). Evolution of the glass transition temperature of the solid solution versus indomethacin weight fraction (◆: experimental points, solid line: fitted with the Gordon–Taylor law (Eq. 8.16)) and solubility limit (•: experimental points, dashed line: fitted with Eq. (8.22)). For more details, see [67].

When the interactions taking place between the two compounds are equal to those taking place between molecules of the same type, the evolution is linear (Figure 8.17). When the developed interactions between the two compounds are stronger (e.g., development of hydrogen bonds), the overall mobility is reduced and the glass transition of the blend is higher than that predicted by the linear evolution. On the contrary, if interactions are lower, the overall mobility is enhanced and the T_g of the blend is reduced. In most cases, when small molecules are added to a polymer, the glass transition of the blend lies between those of the pure compounds. In the framework of the free volume theory, it is usually explained by the fact that the small molecules inserted between polymeric chains increase the free volume and enhance the mobility of the chains, which shifts the main relaxation α to higher frequencies or lower temperatures. The T_g of "the polymer" is thus reduced (actually it is the T_g of the blend), and one says that the polymer is "plasticized" by the small molecules (plasticizers). From an opposite point of view, the mixing of small molecules with a polymer with a higher T_g allows increase of the T_g of "the small molecules" (actually it is the T_g of the blend): it is a commonly used process to stabilize APIs in their amorphous state.

It should be noted that the effect of the mixing on the main relaxation of a compound is not inevitably the same on the secondary relaxations. For example, water has often a plasticizing effect on the main relaxation, but it can have at the same time an antiplasticizing effect on local mobilities. It depends on the involved mobility and on the local interactions that water can develop or hinder (case, e.g., of poly(cellulose acetate) [68]). In the case of partial miscibility, several glass transition temperatures can be observed that are linked to the composition of each phases.

8.4.2
Mixing Free Energy of Blends of Polymers and Small Molecules

The miscibility of blends containing polymers is governed by the free energy of mixing: $\Delta G_{mix} = \Delta H_{mix} - T\Delta S_{mix}$. In the case of regular solutions of polymers with small molecules, Flory and Huggins have proposed an expression for ΔG_{mix} using a lattice model. This lattice is filled with small molecules (A) of volume v and macromolecules (B) divided into x volumes v (x is the molar volumes ratio) and no volume change on mixing is assumed [69]:

$$\Delta G_{mix} = RT\chi n_A \phi_B + RT(n_A \ln \phi_A + n_B \ln \phi_B) \quad (8.17)$$

where n_A and n_B are the number of moles of, respectively, small molecules A and polymer B; ϕ_A and ϕ_B are the volume fractions of A and B; and χ is the so-called Flory–Huggins parameter that originally accounts for the interaction energies between the species. Mixing occurs if $\Delta G_{mix} < 0$. Usually, as ΔS_{mix} is positive, the miscibility depends on the sign of ΔH_{mix}. If $\Delta H_{mix} < 0$, the blend is miscible; if $\Delta H_{mix} > 0$, there is a gap in the miscibility, which depends on the temperature, the molecular weight, and the composition. It should be noted that several variants have been developed/proposed to take into account changes of factors such as volume on mixing and polymer/polymer solution. Readers are referred to specific bibliography for more details [70].

8.4.3
Solubility Limit of Small Molecules in Polymers

In the case of miscible ($\Delta G_{mix} < 0$) regular solutions of small molecules A and a polymer B, the variation of the chemical potential of A in B upon mixing can be written as

$$\mu_A - G_A^L = RT\left[\ln \phi_A + (1 - \phi_A)\left(1 - \frac{1}{x}\right) + \chi(1 - \phi_A)^2\right] \quad (8.18)$$

where μ_A is the partial molar free energy of A (chemical potential of A in the solution) and G_A^L the molar free energy of the pure liquid of A. The activity of A in the solution, a_A, is then given by

$$\ln a_A = \frac{1}{RT}(\mu_A - G_A^L) = \ln \phi_A + (1 - \phi_A)\left(1 - \frac{1}{x}\right) + \chi(1 - \phi_A)^2 \quad (8.19)$$

The molecules of A will dissolve in the polymer as long as $\mu_A < G_A^C$, the molar free energy of pure crystallized A. The limit of solubility of A in the polymer B is reached when $\mu_A = G_A^C$. Then

$$\ln a_A = \frac{\mu_A - G_A^L}{RT} = \frac{G_A^C - G_A^L}{RT} = -\frac{\Delta G_{L-C}}{RT} \quad (8.20)$$

where $\Delta G_{L-C} = {}^{\Delta H_m (T_m - T)}/_{T_m}$ (with ΔH_m and T_m the melting enthalpy and melting temperature of A, respectively).

Replacing $\ln a_A$, this leads to the equilibrium condition

$$\frac{\Delta H_m}{R}\left(\frac{1}{T_m}-\frac{1}{T}\right)=\ln\phi_A+(1-\phi_A)\left(1-\frac{1}{x}\right)+\chi(1-\phi_A)^2 \tag{8.21}$$

The temperature at which phases are in equilibrium obeys

$$T=\left[\frac{T_m\Delta H_m}{\Delta H_m-RT_m\left[\ln\phi_A+(1-\phi_A)\left(1-\frac{1}{x}\right)+\chi(1-\phi_A)^2\right]}\right] \tag{8.22}$$

For a given ϕ_A content, above this temperature, all molecules of A dissolve in the polymer. Below this temperature, a part of A molecules crystallize (Figure 8.17).

Of course, the kinetics of the mixing and demixing processes depends on the overall mobility of the blend. Above the glass transition, the molecular weight-dependent viscosity of polymer slows down the diffusion of small molecules all the more when the temperature decreases. Below the glass transition, the mobility being extraordinarily reduced, these processes will become extremely slow and that is a way of stabilizing an amorphous solid solution of API/polymer [71]. However, it requires a good knowledge of the phases and states diagrams (Figure 8.17).

8.5 Effect of the Structural Properties of Pharmaceutical Polymers on Their Physical Behavior

From the theoretical discussion presented earlier in this chapter, it is clear that the polymer structure can have a significant impact on the degree of disorder and the physical behavior of the polymer. Polymers are widely used in the pharmaceutical industry from dosage form production through to packaging materials. The characteristics of a particular polymer's architecture, such as the chain length, branching, conformation, and chain flexibility, can have a significant impact on the behavior of the polymer when it is used in a pharmaceutical product. Several reviews regarding pharmaceutical polymers have been published, focusing on aspects such as polymer sources (natural, semisynthetic, synthetic), architecture (linear, branched, crosslinked, style of network), hydrophobicity (hydrophilic, hydrophobic), and their common uses (conventional dosage forms, controlled release dosage forms, packaging) [72, 73]. In terms of the *in vivo* performance of pharmaceutical polymers used in dosage forms, many studies have revealed that the MW, polydispersity, charge, and the hydrophilic–hydrophobic balance of the polymer can impact on the drug biodistribution, clearance, biological activity, and toxicity to a great extent [74, 75]. In this chapter, we take an unconventional viewpoint and discuss the classification of the commonly used pharmaceutical polymers based on both the polymer structure and their degree of disorder. We use this approach in an attempt to reveal the relationship between the disorder and physical property and behavior of pharmaceutical polymers, which may be used to inform early stage formulation development and excipient selection. It is

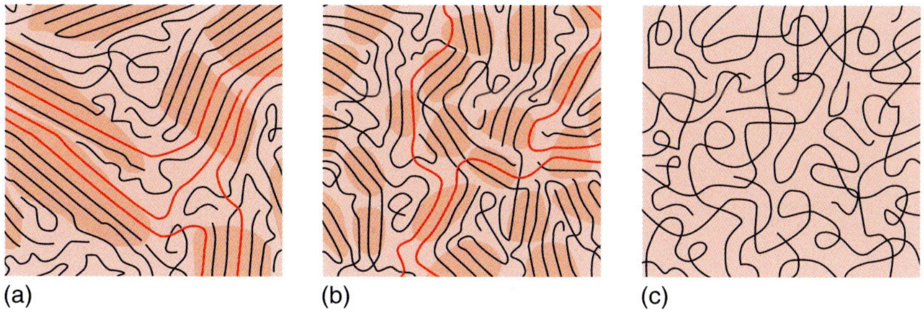

Figure 8.18 Illustration of the microstructure of a semicrystalline polymer with (a) low degree and (b) high degree of disordered aggregates and (c) being completely amorphous. The darker shadowed areas are ordered area and spaghetti-like amorphous area. Adapted from Noriega *et al.* [78]. Reproduced with permission of Nature Publishing.

important to point out that the discussion of the effect of molecular architecture on the performance of the polymer is highly application-dependent [73]. In this chapter, the discussions will be concerned only with solid-state pharmaceutical applications of the polymers and those in media following drug release. The uses of hydrogel-forming polymers and cyclodextrin-based polymers are discussed specifically elsewhere either in other chapters of this book or in other review papers [76, 77]. A list of the most commonly used GRAS (generally regarded as safe) pharmaceutical polymers in solid dosage forms is presented in Table 8.1 and further discussions on a few example polymers from the list are provided for our proof-of-principle analysis (Table 8.1).

8.5.1 Linear Polymers

8.5.1.1 Semicrystalline Polymers

As discussed earlier in this chapter, polymers with a linear backbone often have many degrees of conformational freedom and the possibility to form weak interactions between polymer chains and branches (short-range intermolecular aggregation). This leads to complex microstructures of these polymers in the solid state, as seen in Figure 8.18. Therefore, many linear polymers show the coexistence of amorphous and ordered phases, with different polymers exhibiting varying degrees of order [79]. When this type of polymer is used as a carrier of therapeutic agents, such as low-MW drugs, the degree of disorder of the polymer may be affected by the presence of the guest drug. In turn, the degree of disorder of the polymer structure may also impact on the physical state of the drug after being incorporated into the polymer matrix [78].

Poly(ethylene glycol) (PEG), which is commonly used as a pharmaceutical excipient, is a semicrystalline, water-soluble polymer. As a result of its unique long-circulation effect *in vivo*, it is also often used to conjugate with drugs and nanoparticles for targeted drug delivery and prolonged half-life [80].

Table 8.1 Structural and physical characteristics of example synthetic and natural pharmaceutical polymers.

Name	Architecture	Disorder	Phase transitions	Example applications
Examples of hydrophobic polymers				
Ammoniomethacrylate Copolymers (Eudragit® RL100/RLPO and RS100/RSPO)	Coplymer	Amorphous	T_g 55 °C	Controlled release coating
Poly(ethylacrylate–methylmethacrylate) (Eudragit® NE 30D and Eudragit® NE 40D)	Copolymer	Amorphous	T_g −8 °C	Controlled release coating and matrix
Poly(vinyl acetate) (PVAc) (Kollicoat® SR)	Linear	Amorphous	T_g 40–47 °C	Controlled release coating and hydrogel formation
Ethylcellulose	Linear (branched possible)	Semicrystalline or amorphous depending on the degree of substitution	T_g 130–133 °C T_m 165–173 °C	Controlled release coating
Poly(ethylene vinyl acetate) (EVA)	Copolymer	Semicrystalline	T_g −40 °C T_m is VA content-dependent (70–100 °C)	Controlled release matrix and coating for implants
Poly(ε-caprolactone) (PCL)	Linear	Semicrystalline	T_g −65 to −60 °C, T_m 49–65 °C	Biodegradable implants, controlled release, and tissue engineering

Examples of hydrophilic polymers				
Poly(ethylene glycol) (PEG)	Linear	Semicrystalline	T_g −53 °C, T_m 65 °C	Drug delivery matrix, drug–PEG conjugates, self-assembly and stabilization agents in solutions
Poly(vinyl alcohol) (PVA)	Linear	Semicrystalline	T_g 85 °C, $T_m > T_d$	Hydrogel formation, emulsion stabilizer, sustained release matrix
Poly(vinyl pyrrolidone) (PVP)	Linear	Amorphous	T_g 180–106 °C (MW-dependent [10])	Binder, film former, emulsion stabilizer gelling agents, matrix for solid dispersions
PVP/VA (60:40) copolymer	Block copolymer	Amorphous	T_g 102 °C	Matrix for solid dispersions, film former
Poly(acrylic acid) (PAA)	Linear	Amorphous	T_g 70 °C	Biodegradable and bioadhesive matrix
Poly(*N*-(2-hydroxypropyl) methacrylamide) (HPMA)	Linear copolymer	Amorphous	T_g 106 °C	Plasma expander, polymer–drug conjugate, targeted drug delivery

(*continued overleaf*)

Table 8.1 *(Continued)*

Name	Architecture	Disorder	Phase transitions	Example applications
Examples of amphiphilic polymers				
Hypromellose acetate succinate (HPMCAS)	Linear	Amorphous	T_g 120 °C	Solid dispersion matrix
PCL-PVAc-PEG (Soluplus®)	Graft copolymer	Amorphous	T_g 70 °C	Solid dispersion matrix
PEO-*b*-PPO-*b*-PEO (Poloxamer®)	Triblock copolymer	Semicrystalline	T_m 56 °C	Thermosensitive gel formation, solubilizing agent, stabilizer in solutions
Examples of natural polymers				
Hyaluronic acid (HA)	Linear (helical after hydration)	Amorphous	T_g not measureable due to the overlapping the decomposition	Gelation, tissue engineering, wound healing, sustained release matrix for macromolecules, targeted delivery
Chitosan derivatives	Linear	Amorphous	T_g 203 °C	Gene delivery, controlled drug delivery and tissue engineering
Cellulose ethers: Hydroxypropylmethyl cellulose (HPMC), hydroxypropyl cellulose (HPC), hydroxyethyl cellulose (HEC)	Linear	Semicrystalline	T_g HPC (105 °C), HPMC (170–198 °C), MC (184–197 °C), L-HPC (220 °C)	Solid dispersion matrix, film former, controlled release matrix or coating, gelation

Because of its hydrophilicity, it has been also used as a building block for many other commercially available amphiphilic pharmaceutical copolymers, such as Poloxamer and Soluplus. During the early development of the solid dispersion concept – which is the incorporation of a drug in a polymer or fat matrix and used widely as an effective formulation approach in the delivery of poorly soluble drugs – PEG was studied intensively as a carrier polymer. A large body of literature has documented the behavior of PEG–drug dispersions (prepared using the melt-mixing method since crystalline PEG has a low melting point 60–80 °C). It was noted that the crystallization kinetics of the guest drug was highly affected by the physicochemical properties of the pure API and the interaction between the API and PEG [81, 82]. For some amorphous drugs, the crystalline portion of PEG can act as a favorable heterogeneous nucleation site in the solid dispersion to promote the drug phase separation and recrystallization [78, 81, 82]. As seen in Figure 8.19, the crystallization behavior of PEG can also be affected by the presence of the API. In terms of the mechanism of the effect of drug on the crystallization rate of PEG, many research groups have attributed it to the drug–PEG interaction via hydrogen bonding. For example, in Zhu's study, the crystallization rate of PEG in the IBP (ibuprofen)/PEG solid dispersion at 40 °C was slower than in pure PEG, whereas the crystallization rate of PEG in the fenofibrate/PEG system was almost the same as that for pure PEG. Spectroscopic analysis confirmed that the different behavior of PEG in the two systems may be attributed to the different ability of the two APIs to interact with PEG through hydrogen bonding. Ibuprofen can form hydrogen bonds with PEG, while fenofibrate has limited ability to interact with PEG through hydrogen bonding since it lacks donor groups and PEG predominantly contains acceptor groups.

However, the degree of crystallinity of PEG was found to recover to the original level despite the presence of the guest drug (either as molecules or as solid

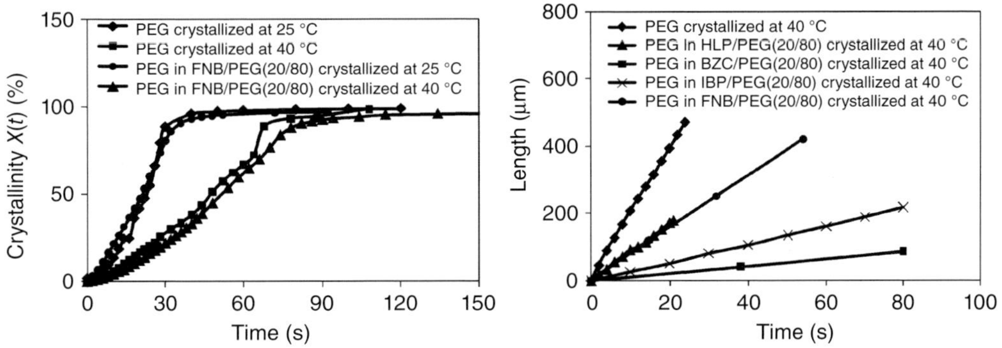

Figure 8.19 Effect of guest drugs (HLP, haloperidol, FNB, fenofibrate, BZC, benzocaine, and IBP, ibuprofen) on the crystallization of PEG. Adapted from Zhu *et al.* [80]. Reproduced with permission of American Chemical Society.

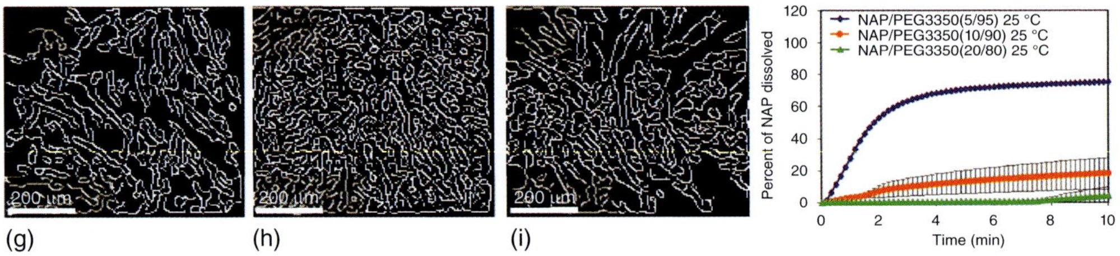

Figure 8.20 Second-harmonic generation (SHG) images of the edge detection of naproxen/PEG solid dispersions at 40 °C 5/95, 10/90, and 20/80 (% w/w), and the corresponding drug release profile. Adapted from Zhu *et al.* [81]. Reproduced with permission of American Chemical Society.

particles). This indicates that the amount of drug that can be "solubilized" in PEG-based molecular solid dispersions is largely located in the amorphous areas of the polymer. This may be used to explain the limited solubilization and stabilization capabilities of PEG for many low-MW drugs in the solid state (as molecular dispersions) in comparison to other fully amorphous polymers. However, PEG and PEG-containing polymer/lipid excipients can still improve the dissolution of some poorly water-soluble drugs [83, 84]. The mechanisms of this observation have been investigated. The hydrophilic nature of PEG was attributed to the dissolution enhancement as well as the high surface area of the recrystallized drug particles in the PEG matrix. This has been visualized using a novel imaging method called second-order nonlinear optical imaging of chiral crystals (SONICC), in which a reduced number of drug–crystal contact points in the PEG dispersion were observed in the dispersions with lower drug loading (Figure 8.20). The isolation of drug in systems with low drug loading can have more interaction with the dissolution media upon PEG dissolution, whereas a fairly continuous drug crystal network was formed in the dispersions with high drug loading, which may slow down the drug release, as seen in Figure 8.20. It has also been reported that increasing the molecular weight of PEG can affect the drug and polymer crystallization, which leads to differences in the drug release kinetics of the dispersions [85].

8.5.1.2 **Amorphous Polymers**

In comparison to semicrystalline polymers, amorphous polymers typically exhibit a much higher capability to inhibit the crystallization of APIs and produce an amorphous dispersion [86, 87]. Depending on the hydrophilicity of the polymer, amorphous linear polymers have been used for both solubility enhancement and controlled release. Many review papers have discussed the application and behavior of amorphous hydrophilic, hydrophobic, and natural polymers [88–90]. Here we specially discuss the potential impact of some key physical properties of these amorphous polymers on their drug loading (as dissolved drug molecules) and drug release performance. Here we use PVP, Eudragit E OP, and HPMC (hydroxypropylmethyl cellulose) as examples for hydrophilic, pH-dependent (random cationic copolymer), and natural linear amorphous polymers, respectively.

Among these three example polymers, Eudragit E PO is a linear copolymer with longer side chains, resulting in a lower T_g in comparison to PVP and HPMC, which are both linear amorphous polymers. Many physical properties of PVP and HPMC, such as T_g (as discussed in Section 8.2.1 of this chapter), melt shear viscosity (as discussed Section 8.3.2), aqueous viscosity, and hygroscopicity, are MW-dependent. As mentioned in Section 8.3.2 of this chapter, a power law can accurately describe the relationship between the melt viscosity and molecular weight of linear polymers [56]. From the structural point of view, PVP has highly exposed carbonyl groups (hydrogen-bonding acceptor), which leads to its excellent water solubility and low aqueous viscosity. More importantly, this also provides an excellent drug–polymer interaction opportunity for poorly water-soluble drugs that have hydrogen-bonding donor groups, promoting good miscibility of the drug and polymer as discussed in Section 8.4 of this chapter. In the literature, drug–polymer interactions have been widely acknowledged for their importance in providing high drug–polymer miscibility and the ability to form molecular solid dispersions with high physical stability. However, it also contributes to the high hygroscopicity of PVP, which often is a drawback when using PVP as the main carrier polymer for drug delivery.

In contrast, Eudragit E PO has weaker drug–polymer interactions with drugs having hydrogen donor groups, as measured by spectroscopic methods [91]. However, this does not reflect on the stabilization capability of this polymer for forming amorphous solid dispersions in comparison to PVP. Despite its low T_g of approximately 45 °C, Eudragit E PO still requires a much higher processing temperature than the T_g of the polymer when it is used for hot-melt extrusion. This is associated with the high melt shear viscosity of Eudragit E PO at low temperatures, which is suspected to at least partially contribute to the good stabilization effect of Eudragit-based amorphous solid dispersions. Both properties are related to the random entangled copolymer structure. As Eudragit E PO is a pH-sensitive polymer dissolving only at acidic pH, with low solution viscosity and lack of hydration and gelling behavior, the dissolution of Eudragit E PO-based dispersions can be rapid if the drug remains molecularly dispersed in the polymer [91, 92].

In terms of processibility during manufacturing, with much higher T_gs than Eudragit E, both PVP and HPMC are not suitable for hot-melt extrusion due to their high melt shear viscosities, unless significant amounts of excipients such as plasticizers are used in combination. However, both PVP and HPMC can be used in spray-drying, which involves the dissolution of the polymers and drug in suitable (aqueous or organic) solvents. As seen in Figure 8.21, structurally PVP can actively interact with the drugs with hydrogen-donor groups, which results in the high miscibility of many poorly soluble drugs in PVP [93, 94]. Therefore, based on structural analysis, PVP has good ability to form amorphous solid dispersions with poorly soluble drugs with hydrogen-donor groups. However, due to the high hygroscopicity of PVP, moisture uptake often poses as a major disadvantage of using this polymer solely for forming solid dispersions. After moisture uptake, PVP predominately bonds to water molecules, which leads to rapid phase separation and recrystallization of the originally dispersed drug [93, 94]. Additionally, the

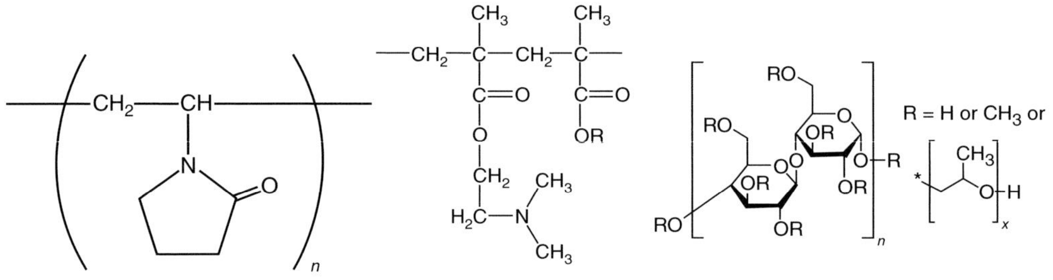

(a) (b) R = CH_3, C_4H_9 (c)

Figure 8.21 Monomer structures of the example polymers. (a) PVP. (b) Eudragit E PO. (c) HPMC.

low level of side chains in the structures of the polymers can contribute to a high level of chain entanglement during hydration of the polymer. This leads to the gelation behavior of the polymer during hydration. Therefore, in terms of drug release from PVP- or HPMC-based dispersions, although they are hydrophilic polymers, depending on the physical form of the formulation (tablets or small microparticles), the initial gelation that occurs during the initial wetting stage of the polymer base can often delay drug release from the polymeric matrices [95].

Despite the differences between the three example amorphous linear polymers, the miscibility of amorphous polymers with drugs is, in general, higher than with semicrystalline polymers. This may be largely due to the large amorphous area within the microstructure of the polymer. Even without strong polymer–drug interactions, the high viscosity of the polymers at room temperature is often sufficient to slow the motion of drug molecules entrapped in the polymer matrices and stabilize the amorphous dispersions over a pharmaceutically relevant shelf-life under suitable storage conditions. Therefore, amorphous polymers are often capable of solubilizing and stabilizing more drug in a molecular dispersions than semicrystalline polymers. However, this does not mean that the incorporation of drug molecules has no effect on the disordered structure of amorphous polymer. In recent studies, it has been reported that the size of the free-volume holes and the lifetimes of *ortho*-positronium atoms decreased with increase of the API (diclofenac sodium) concentration in Eudragit L 30D-55 [96]. The API distorted the original polymer structure, but the metastable structural relaxation totally restored the original structure after 3 weeks of storage at 17 °C and 65% RH [96].

8.5.2 Copolymers

In order to modify the physical and mechanical behavior of some homopolymers, copolymers are often synthesized by altering the chain stiffness, polarity, and architecture. As discussed earlier in Section 8.2.3 the complexity of the copolymer architecture can vary significantly and can impact on the physical behavior of the polymer. Copolymers can be random (random distribution of different types of chemical units), block (long segments of the same chemical unit), or graft (a chain of a type of chemical unit is grafted onto a linear chain composed by another type of chemical unit). The degree of disorder of copolymers is often expected to be higher than that of the homopolymer of each single monomer if the architecture of the polymer is significantly changed. PVP/VA 64 is an example of a random copolymer commonly used in pharmaceutical solid formulations. It is manufactured by free-radical polymerization of six parts of *N*-vinylpyrrolidone, which is a water-soluble monomer, and four parts of vinyl acetate, which is a lipophilic and water-insoluble monomer. Overall, PVP/VA 64 is still water-soluble, but with significantly reduced hygroscopicity in comparison to PVP homopolymers. As a result of the relatively bulky side groups, the polymer chain packing is largely disordered. Thus PVP/VA 64 is an amorphous polymer with a glass transition temperature of 106 °C, which is much lower than that of PVP homopolymers with similar molecular weight. PVP/VA 64 was originally invented as a binder for wet granulation and dry binder for direct compression. Because of the suitable glass transition temperature and good thermoplasticity, PVP/VA 64 was introduced as a polymeric carrier for hot-melt extrusion [91, 97]. Amorphous solid dispersions prepared using PVP/VA 64 have been reported to show significantly enhanced drug dissolution performance and improved physical stability in comparison to PVP-based solid dispersions [98].

Block copolymers are polymers that have two or more blocks or segments arranged in the main backbone polymer chain. Many block copolymers have self-assembly behavior in solvents because of the intrinsic affinity between the building blocks with the same physicochemical properties. The detailed features of self-assembled domains are sensitive to the architecture of the block copolymer. The self-assembly behavior of such copolymers can be easily modified by tuning the physicochemical properties of the polymer. Poloxamer is an example of a triblock copolymer of ethylene oxide (EO) and propylene oxide (PO), which has temperature-dependent self-assembling and thermogelling behavior in water. As seen in Figure 8.22, this copolymer has an alternating arrangement of EO and PO with the numbers of repeating units of *a* and *b*, respectively. Because of the linear arrangement with no side chain, the "streamlined" shape of the backbone facilitates chain packing. Poloxamer 407 has a molecular weight of approximately 12 600 (9840–14 600), and *a* and *b* are 95–105 and 54–60, respectively. It is a crystalline polymer with a T_m of 55 °C [99], and no glass transition of this polymer has been reported in the literature. It has applications in a wide range of formulations, such as gel formation, microemulsions, nanoparticles, and solid polymer blends [99]. Similar to PEG, the crystalline nature of the polymer reduces drug solubilization when it is used for molecular dispersions [100]. As a gel-forming material, Poloxamer also shows slow release of drug when used alone. When used in a solid dosage form, Poloxamer is often blended with other polymers to modulate the drug release rate [101].

Structurally, graft copolymers have an increased level of complexity. Most graft polymers are comb-type copolymers, with a high level of branching on the linear backbone. Therefore, most graft copolymers have a considerably higher degree of

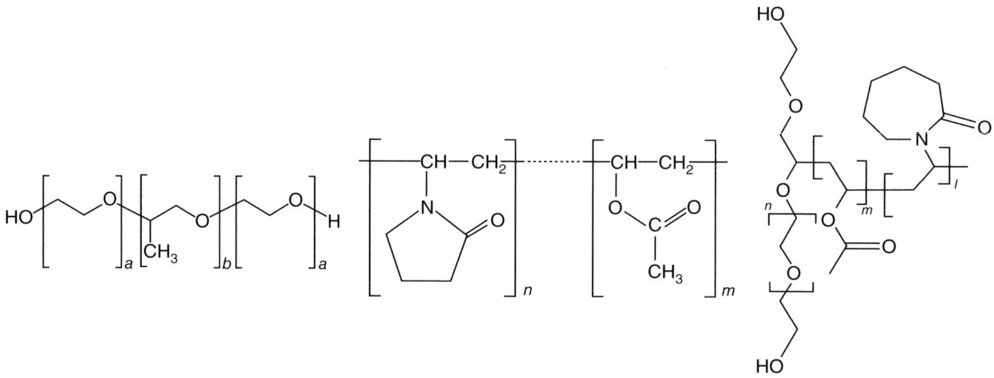

Figure 8.22 Monomer structures of the example copolymers Poloxamer®, PVP-VA, and Soluplus®.

disorder in comparison to random and block copolymers. PEG is the most popular side chain for grafted copolymer backbones because of its hydrophilicity and ability to form thermosensitive gels. Soluplus (poly(vinyl caprolactam)–poly(vinyl acetate)–poly(ethylene glycol) graft copolymer (PCL–PVAc–PEG)) is a new pharmaceutical excipient designed originally for preparing solid solutions of poorly water-soluble drugs by hot-melt extrusion technology. Soluplus is a water-soluble grafted copolymer with a T_g of approximately 70 °C. The relatively low T_g makes Soluplus suitable for thermal processing such as hot-melt extrusion. By forming solid dispersions with poorly soluble drugs using processes such as hot-melt extrusion, it has been reported to enhance the solubilization and dissolution of poorly water-soluble drugs [102]. For example, itraconazole, a highly insoluble drug, was formulated with Soluplus via hot-melt extrusion [103]. The *in vivo* assay revealed 6.9- and 11.6-fold increases in the AUC(0–t) and C(max) of ITZ/Soluplus in comparison to those of the pure drug and similar to the values obtained by the commercial itraconazole capsule (Sporanox®) [103].

8.6 Concluding Remarks

In this chapter, we discussed the theoretical basis of the disorder in the polymer structure and its impact on the physical behavior of pharmaceutical polymers. The different types of disorders were explained: structural disorders coming from the complex architectures of these giant molecules, which directly impact the physical state, and dynamic disorders, that is, mobilities in the amorphous state (supercooled liquid or glass). We showed that mobility can greatly influence the mixing/demixing or the dissolution processes of amorphous solid dispersions of drug/polymeric excipients by slowing down the kinetics. The link between the structural disorder and their *in vitro* performance as pharmaceutical excipients was also presented. From the data presented, it can be concluded that copolymers offer more versatile physicochemical and mechanical properties for pharmaceutical formulations with respect to drug–polymer interactions and modulation of the drug release. Amorphous polymers with a high degree of disorder, in general, show better drug solubilization and stabilization capability in the solid state in comparison to semicrystalline polymers. However, this disorder is often associated with a high level of chain entanglement during hydration, which in some cases can retard the drug release. Therefore, it is important to tailor the physical performance of the polymeric drug carriers by chemical modification or by simply blending with other polymers or copolymers to achieve the desired therapeutic outcome.

References

1. Sperling, L.H. (2006) *Introduction to Physical Polymer Science*, 4th edn, John Wiley & Sons, Inc., Hoboken, NJ.
2. Strobl, G. (1997) *The Physics of Polymers. Concepts for Understanding Their Structures and Behaviour*, 2nd edn, Springer-Verlag, Berlin, Heidelberg, New York.
3. McCrum, N.G., Read, B.E., and Williams, G. (1991) *Anelastic and Dielectric Effects in Polymeric Solids*, 2nd edn, Dover Publications, New York.
4. IUPAC (1974) Basic definitions of terms relating to polymers. *Pure Appl. Chem.*, **40**, 479–491.
5. Ring, W., Mita, I., Jenkins, A.D., and Bikales, N.M. (1985) Source-based nomenclature for copolymers. *Pure Appl. Chem.*, **57**, 1427–1440.
6. Kahovec, J., Kratochvil, P., Jenkins, A.D., Mita, I., Papisov, I.M., Sperling, L.H., and Stepto, R.F.T. (1997) Source-based nomenclature for non-linear macromolecules and macromolecular assemblies. *Pure Appl. Chem.*, **69**, 2511–2521.
7. Bunns, C.W. and Howells, E.R. (1954) Structures of molecules and crystals of fluorocarbons. *Nature*, **174**, 549–551.
8. Natta, G. and Corradini, P. (1960) General considerations on the structure of crystalline polyhydrocarbons. *Nuovo Cimento, Suppl. 10*, **15**, 9–39.
9. Bunn, C.W. (1948) Crystal structure of polyvinyl alcohol. *Nature*, **161**, 929–930.
10. Del Pilar Buera, M., Levi, G., and Karel, M. (1992) Glass transition in poly(vinylpyrrolidone): effect on molecular weight and diluents. *Biotechnol. Progr.*, **8**, 144–148.
11. Fox, T.G. and Flory, P.J. (1950) Second order transition temperatures and related properties of polystyrene. I. Influence of molecular weight. *J. Appl. Phys.*, **21**, 581–591.
12. Bunn, C.W. and Garner, E.V. (1947) The crystal structures of two polyamides ('Nylons'). *Proc. R. Soc. London, Ser. A*, **189**, 39–68.
13. Herrmann, K., Gerngross, O., and Abitz, W. (1930) Zur röntgenographischen Strukturforschung des Gelatinemicells. *Z. Phys. Chem. B.*, **10**, 371–394.
14. Flory, P.J. (1962) On the morphology of the crystalline state in polymers. *J. Am. Chem. Soc.*, **84**, 2857–2867.
15. Hoffman, J.D., Davis, G.T., and Lauritzen, J.I. Jr., (1976) in *Treatise on Solid State Chemistry*, vol. **3**, Chapter 7 (ed. N.B. Hannay), Plenum Press, New York, pp. 497–614.
16. Fischer, E.W. (1957) Stufen- und spiralförmiges Kristallwachstum bei Hochpolymeren. *Z. Naturforsch., A*, **12**, 753–754.
17. Keith, H.D. and Padden, F.J. Jr., (1996) Banding in polyethylene and other spherulites. *Macromolecules*, **29**, 7776–7786.
18. Keith, H.D., Padden, F.J. Jr., and Vadimsky, R.G. (1966) Intercrystalline links in polyethylene crystallized from

the molten state. *J. Polym. Sci. Part A-2*, **4**, 267–281.

19. Pennings, A.J., Langeveen, R., and de Vries, R.S. (1977) Hydrodynamically induced crystallization of polymers from solution. VII.: on the origin of persistent lamellar overgrowth of Shish-kebab fibers. *Colloid Polym. Sci.*, **255**, 532–542.
20. Pennings, A.J., van der Mark, J.M.A.A., and Kiel, A.M. (1970) Hydrodynamically induced crystallization of polymers from solution. III. Morphology. *Kolloid Z. Z. Polym.*, **237**, 336–358.
21. Dowling, L.M. and Sparrow, L.G. (1991) Sequences of wool keratin proteins: the CSIRO connection. *Trends Biochem. Sci.*, **16**, 115–118.
22. Prockop, D.J. and Fertala, A. (1998) The collagen fibril: the almost crystalline structure. *J. Struct. Biol.*, **122**, 111–118.
23. Hoffmann, J.D. and Lauritzen, J.I. (1961) Crystallization of bulk polymers with chain folding: theory of growth of lamellar spherulites. *J. Res. Nat. Bur. Stand.*, **65A**, 297–336.
24. Hoffman, J.D. and Miller, R.L. (1997) Kinetics of crystallization from the melt and chain folding in polyethylene fractions revisited: theory and experiment. *Polymer*, **38**, 3151–3212.
25. Armistead, J.P. and Hoffman, J.D. (2002) Direct evidence of regimes I, II, and III in linear polyethylene fractions as revealed by spherulite growth rates. *Macromolecules*, **35**, 3895–3913.
26. Chen, Y.-F. and Woo, E.M. (2008) Growth regimes and spherulites in thin-film poly(ε-caprolactone) with amorphous polymers. *Colloid Polym. Sci.*, **286**, 917–926.
27. Bates, F.S. and Fredrickson, G.H. (1990) Block copolymer thermodynamics: theory and experiment. *Annu. Rev. Phys. Chem.*, **41**, 525–557.
28. Salyer, I.O. and Kenyon, A.S. (1971) Structure and property relationships in ethylene-vinyl acetate copolymers. *J. Polym. Sci. Part A-1*, **9**, 3083–3103.
29. Zhang, Q., Lin, W., Yang, G., and Chen, Q. (2002) Studies on the phase structure of ethylene-vinyl acetate copolymers by solid-state ^{1}H and ^{13}C NMR. *J. Polym. Sci., Part B: Polym. Phys.*, **40**, 2199–2207.
30. Schultz, J.M. (1974) *Polymer Materials Science*, Prentice Hall, Englewood Cliffs, NJ.
31. Tyagi, M., Alegria, A., and Colmenero, J. (2005) Heterogeneous dynamics of poly(vinyl acetate) far above Tg: a combined study by dielectric spectroscopy and quasielastic neutron scattering. *J. Chem. Phys.*, **122**, 244909.
32. Tyagi, M., Alegria, A., and Colmenero, J. (2007) Broadband dielectric study of oligomer of poly(vinyl acetate): a detailed comparison of dynamics with its polymer analog. *Phys. Rev. E.*, **75**, 061805.
33. Schatzki, T.F. (1962) Statistical computation of distribution functions of dimensions of macromolecules. *J. Polym. Sci.*, **57**, 337–356.
34. Heijboer, J. (1978) Secondary loss peaks in glassy amorphous polymers, in *Molecular Basis of Transitions and Relaxations* (ed. D.J. Meier), Gordon and Breach, New York.
35. Johari, G.P. and Goldstein, M. (1970) Viscous liquids and the glass transition. II. Secondary relaxations in glasses of rigid molecules. *J. Chem. Phys.*, **53**, 2372–2388.
36. Ngai, K.L. and Paluch, M. (2004) Classification of secondary relaxation in glass-formers based on dynamic properties. *J. Chem. Phys.*, **120**, 857–873.
37. Eyring, H. (1935) The activated complex in chemical reactions. *J. Chem. Phys.*, **3**, 107–115.
38. Vogel, H. (1921) The law of the relationship between viscosity of liquids and the temperature. *Phys. Z.*, **22**, 645–646.
39. Fulcher, G.S. (1925) Analyis of recent measurements of the viscosity of glasses. *J. Am. Ceram. Soc.*, **8**, 339–355.
40. Tamman, G. and Hesse, W. (1926) Die Abhängigkeit der Viscosität von der Temperatur bie unterkühlten Flüssigkeiten. *Z. Anorg. Allg. Chem.*, **165**, 245–257.
41. Doolittle, A.K. (1951) Studies in Newtonian flow. II. The dependence of the viscosity of liquids on free space. *J. Appl. Phys.*, **22**, 1471–1475.

42. Cohen, M.H. and Turbull, D. (1959) Molecular transport in liquids and glasses. *J. Chem. Phys.*, **31**, 11641169.
43. Turbull, D. and Cohen, M.H. (1961) Free volume model of the amorphous phase: glass transition. *J. Chem. Phys.*, **34**, 120–125.
44. Gibbs, J.H. and Dimarzio, E.A. (1958) Nature of the glass transition and the glassy state. *J. Chem. Phys.*, **28**, 373–383.
45. Adam, G. and Gibbs, J.H. (1965) On the temperature dependence of cooperative relaxation properties in glass forming liquids. *J. Chem. Phys.*, **43**, 139–146.
46. Donth, E. (1982) The size of cooperatively rearranging regions at the glass transition. *J. Non-Cryst. Solids*, **53**, 325–330.
47. Donth, E. (1991) Characteristic length of glass transition. *J. Non-Cryst. Solids*, **131-133**, 204–206.
48. Donth, E. (1999) Phenomenological treatment of dynamic glass transition heterogeneity. *Acta Polym.*, **50**, 240–251.
49. Angell, C. (1995) Formation of glasses from liquids and biopolymers. *Science*, **267** (5206), 1924–1935.
50. Ferry, J.D. (1980) *Viscoelastic Properties of Polymers*, 3rd edn, John Wiley & Sons, Inc., New York.
51. Williams, M.L., Landel, R.F., and Ferry, J.D. (1955) The temperature dependence of relaxation mechanisms in amorphous polymers and other glass-forming liquids. *J. Am. Chem. Soc.*, 77, 3701–3707.
52. Kohlrausch, R. (1854) Theorie des elektrischen Rückstandes in der Leidener Flasche. *Poggendorff's Ann. Phys.*, **167**, 179–214.
53. Williams, G. and Watts, D.C. (1970) Non-symmetrical dielectric relaxation behaviour arising from a simple empirical decay function. *Trans. Faraday Soc.*, **66**, 80–85.
54. Dionísio, M., Viciosa, M.T., Wang, Y., and Mano, J.F. (2005) Glass transition dynamics of poly(L-lactic acid) during isothermal crystallisation monitored by real-time dielectric relaxation spectroscopy measurements. *Macromol. Rapid Commun.*, **26**, 1423–1427.
55. Brás, A.R., Malik, P., Dionísio, M., and Mano, J.F. (2008) Influence of crystallinity in molecular motions of poly(L-lactic acid) inverstigated by dielectric relaxation spectroscopy. *Macromolecules*, **41**, 6419–6430.
56. Colby, R.H., Fetters, L.J., and Graessley, W.W. (1987) The melt viscosity-molecular weight relationship for linear polymers. *Macromolecules*, **20** (9), 2226–2237.
57. Stockmayer, W. (1967) Dielectric dispersion in solutions of flexible polymers. *Pure Appl. Chem.*, **15**, 539–554.
58. Adachi, K. and Kotaka, T. (1993) Dielectric normal mode relaxation. *Prog. Polym. Sci.*, **18**, 585–622.
59. Boese, D. and Kremer, F. (1990) Molecular dynamics in bulk cis-polyisoprene as dielectric spectroscopy. *Macromolecules*, **23**, 829–835.
60. Rouse, P.E. (1953) A theory of the linear viscoelastic properties of dilute solutions of coiling polymers. *J. Chem. Phys.*, **21**, 1272–1280.
61. Bueche, F. (1954) Influence of rate of shear on the apparent viscosity of A-dilute polymer solutions, and B-bulk polymers. *J. Chem. Phys.*, **22**, 1570–1576.
62. de Gennes, P.G. (1971) Reptation of a polymer chain in the presence of fixed obstacles. *J. Chem. Phys.*, **55**, 572–579.
63. Doi, M. and Edwards, S.F. (1986) *The Theory of Polymer Dynamics*, Clarendon Press, Oxford.
64. Gordon, J.M. and Taylor, J. (1952) Ideal copolymers and the second-order transitions of synthetic rubbers. I. Non-crystalline copolymers. *J. Appl. Chem.*, **2**, 493–500.
65. Couchman, P.R. and Karasz, F.E. (1978) A classical thermodynamic discussion of the effect of composition on glass-transition temperatures. *Macromolecules*, **11**, 117–119.
66. Gordon, J.M., Rouse, G.B., Gibbs, J.H., and Risen, W.M. Jr., (1977) The composition dependence of glass transition properties. *J. Chem. Phys.*, **66**, 4971–4976.

67. Mahieu, A., Willart, J.-F., Dudognon, E., Danède, F., and Descamps, M. (2013) A new protocol to determine the solubility of drugs into polymer matrixes. *Mol. Pharm.*, **10**, 560–566.
68. Sousa, M., Brás, A.R., Isabel, H., Veiga, M., Castelo Ferreira, F., Norberta de Pinho, M., Correia, N.T., and Dionísio, M. (2010) Dynamical characterization of a cellulose acetate polysaccharide. *J. Phys. Chem. B*, **114**, 10939–10953.
69. Flory, P.J. (1953) *Principles of Polymer Chemistry*, Cornell University Press, Ithaca, NY.
70. Lipatov, Y.S. and Nesterov, A.E. (1997) *Thermodynamics of Polymer Blends*, vol. **1**, Technomic, Lancaster, Basel.
71. Marsac, P.J., Shamblin, S.L., and Taylor, L.S. (2006) Theoretical and practical approaches for prediction of drug-polymer miscibility and solubility. *Pharm. Res.*, **23**, 2417–2426.
72. Park, H. and Park, K. (1993) in *Polymers of Biological and Biomedical Significance*, vol. **540**, Chapter 1 (eds S.W. Shalaby, Y. Ikada, R. Langer, and J. Williams), American Chemical Society, pp. 2–15.
73. Qiu, L.Y. and Bae, Y.H. (2006) Polymer architecture and drug delivery. *Pharm. Res.*, **23** (1), 1–30.
74. Nori, A. and Kopeček, J. (2005) Intracellular targeting of polymer- bound drug drugs for cancer chemotherapy. *Adv. Drug Delivery Rev.*, **57**, 609–636.
75. Riebeseel, K., Biedermann, E., Löser, R., Breiter, N., Hanselmann, R., Mü lhaupt, R., Unger, C., and Kratz, F. (2002) Polyethyl- ene glycol conjugates of methotrexate varying in their molecular weight from MW 750 to MW 40000: synthesis, characterization, and structureYactivity relationships in vitro and in vivo. *Bioconjugate Chem.*, **13**, 773–785.
76. Kopeček, J. (2007) Hydrogel biomaterials: a smart future? *Biomaterials*, **28** (34), 5185–5192.
77. van de Manakker, F., Vermonden, T., van Nostrum, C.F., and Hennink, W.E. (2009) Cyclodextrin-based polymeric materials: synthesis, properties, and pharmaceutical/biomedical applications. *Biomacromolecules*, **10** (12), 3157–3175.
78. Noriega, R., Rivnay, J., Vandewal, K., Koch, F.P.V., Stingelin, N., Smith, P., Toney, M.F., and Salleo, A. (2013) A general relationship between disorder, aggregation and charge transport in conjugated polymers. *Nat. Mater.*, **12**, 1038–1044.
79. Knop, K., Hoogenboom, R., Fischer, D., and Schubert, U.S. (2010) Poly(ethylene glycol) in drug delivery: pros and cons as well as potential alternatives. *Angew. Chem. Int. Ed.*, **49**, 6288–6308.
80. Zhu, Q., Harris, M.T., and Taylor, L.S. (2012) Modification of crystallization behavior in drug/polyethylene glycol solid dispersions. *Mol. Pharm.*, **9** (3), 546–553.
81. Zhu, Q., Toth, S.J., Simpson, G.J., Hsu, H.-Y., Taylor, L.S., and Harris, M.T. (2013) Crystallization and dissolution behavior of naproxen/polyethylene glycol solid dispersions. *J. Phys. Chem. B*, **117** (5), 1494–1500.
82. Zhu, Q., Taylor, L.S., and Harris, M.T. (2010) Evaluation of the microstructure of semicrystalline solid dispersions. *Mol. Pharm.*, 7 (4), 1291–1300.
83. Zhu, Q., Harris, M.T., and Taylor, L.S. (2011) Time-resolved SAXS/WAXS study of the phase behavior and microstructural evolution of drug/PEG solid dispersions. *Mol. Pharm.*, **8**, 932–939.
84. Qi, S., Marchaud, D., and Craig, D.Q. (2010) An investigation into the mechanism of dissolution rate enhancement of poorly water-soluble drugs from spray chilled gelucire 50/13 microspheres. *J. Pharm. Sci.*, **99** (1), 262–274.
85. Ginésa, J.M., Arias, M.J., Moyanoa, J.R., and Sánchez-Sotob, P.J. (1996) Thermal investigation of crystallization of polyethylene glycols in solid dispersions containing oxazepam. *Int. J. Pharm.*, **143** (2), 247–253.
86. Leuner, C. and Dressman, J. (2000) Improving drug solubility for oral delivery using solid dispersions. *Eur. J. Pharm. Biopharm.*, **50**, 47–60.
87. Serajuddin, A.T.M. (1999) Solid dispersion of poorly water-soluble drugs:

early promises, subsequent problems, and recent break-throughs. *J. Pharm. Sci.*, **88**, 1058–1066.

88. Jones, D., Abu Diak, O., and Andrews, G. (2012) in *Fundamentals and Applications of Controlled Release Drug Delivery* (eds R. Siegel and M. Rathbone), pp. 345–376.
89. Kadajji, V.G. and Betageri, G.V. (2011) Water soluble polymers for pharmaceutical applications. *Polymers*, **3**, 1972–2009.
90. Reis, R.L., Neves, N.M., Mano, J.F., Gomes, M.E., Marques, A.P., and Azevedo, H.S. (2008) *Natural-Based Polymers for Biomedical Applications*, Elsevier, ISBN: 978-1-84569-264-3.
91. Yang, Z., Nollenberger, K., Albers, J., Craig, D., and Qi, S. (2013) Microstructure of an immiscible polymer blend and its stabilization effect on amorphous solid dispersions. *Mol. Pharm.*, **10** (7), 2767–2780.
92. Thakral, S., Thakral, N.K., and Majumdar, D.K. (2013) Eudragit: a technology evaluation. *Expert Opin. Drug Delivery*, **10** (1), 131–149.
93. Rumondor, A.C.F., Marsac, P.J., Stanford, L.A., and Taylor, L.S. (2009) Phase behavior of poly(vinylpyrrolidone) containing amorphous solid dispersions in the presence of moisture. *Mol. Pharm.*, **6** (5), 1492–1505.
94. Rumondor, A.C.F. and Taylor, L.S. (2010) Effects of polymer hygroscopicity on the phase behavior of amorphous solid dispersions in the presence of moisture. *Mol. Pharm.*, **7** (2), 477–490.
95. Fu, X.C., Wang, G.P., Liang, W.Q., and Chow, M.S. (2004) Prediction of drug release from HPMC matrices: effect of physicochemical properties of drug and polymer concentration. *J. Controlled Release*, **95** (2), 209–216.
96. Bölcskeia, É., Süveghb, K., Marekc, T., Regdon, G. Jr., and Pintye-Hódia, K. (2011) Testing of the structure of macromolecular polymer films containing solid active pharmaceutical ingredient (API) particles. *Radiat. Phys. Chem.*, **80** (7), 799–802.
97. Song, Y., Wang, L., Yang, P., Wenslow, R.M. Jr., Tan, B., Zhang, H., and Deng, Z. (2013) Physicochemical characterization of felodipine-kollidon VA64 amorphous solid dispersions prepared by hot-melt extrusion. *J. Pharm. Sci.*, **102** (6), 1915–1923.
98. Weuts, I., Kempen, D., Decorte, A., Verreck, G., Peeters, J., Brewster, M., and Van den Mooter, G. (2004) Phase behaviour analysis of solid dispersions of loperamide and two structurally related compounds with the polymers PVP-K30 and PVP-VA64. *Eur. J. Pharm. Sci.*, **22** (5), 375–385.
99. Dumortier, G., Grossiord, J.L., Agnely, F., and Chaumeil, J.C. (2006) A review of poloxamer 407 pharmaceutical and pharmacological characteristics. *Pharm. Res.*, **23** (12), 2709–2728.
100. Ali, W., Williams, A.C., and Rawlinson, C.F. (2010) Stochiometrically governed molecular interactions in drug: poloxamer solid dispersions. *Int. J. Pharm.*, **391** (1-2), 162–168.
101. Gonzalez, Y.M. and Ghaly, E.S. (2010) Modified drug release of poloxamer matrix by including water-soluble and water-insoluble polymer. *Drug Dev. Ind. Pharm.*, **36** (1), 64–71.
102. Linn, M., Collnot, E.-M., Djuric, D., Hempel, K., Fabian, E., Kolter, K., and Lehr, C.-M. (2012) Soluplus as an effective absorption enhancer of poorly soluble drugs in vitro and in vivo. *Eur. J. Pharm. Sci.*, **45**, 336–343.
103. Zhang, K., Yu, H., Luo, Q., Yang, S., Lin, X., Zhang, Y., Tian, B., and Tang, X. (2013) Increased dissolution and oral absorption of itraconazole/soluplus extrudate compared with itraconazole nanosuspension. *Eur. J. Pharm. Biopharm.*, **85** (3), 1285–1292.

9
Polymer Gels, Hydrogels, and Scaffolds – An Overview

Madeleine Djabourov and Kawthar Bouchemal

9.1
Introduction

The use of polymers in medicine dates back almost to the birth of the field of polymer science. Synthetic polymers were used in experimental surgical studies soon after their invention: nylon sutures were introduced in the early 1940s, followed in the mid-1940s by other synthetic polymers such as poly(methyl methacrylate), Dacron® polyester, and polyvinyl chloride, as reported at that time in medical journals [1, 2]. Later on, other polymers, such as Teflon®, high-density polypropylene, polyurethanes, and so on, found use in clinical medicine, as components of permanent prosthetic devices such as hip implants, artificial lenses, vascular grafts, catheters, and so on [3]. Whereas the original uses of polymers in surgery were primarily limited to replacements of connective tissues, other applications emerged with the development of new areas of molecular cell and developmental biology. New protein- and nucleic acid-based drugs, which cannot be taken in the classical form (pills), reinvigorated the search for new polymers for controlled drug delivery and gene therapy. In the new field of tissue engineering, polymers were designed to assist regeneration of three-dimensional tissues and organ structures and to be more integrated with biological demands. Engineering of tissues or organs to treat patients appeared as a promising therapeutic approach, which combines biomaterials, cells, and environmental factors, promoting tissue repair and/or functional restoration [4]. Biomaterials called *scaffolds* play an important role as extracellular matrices to enable the creation of the correct microenvironment and promotion of *in vitro* tissue development. Scaffolds are the porous solid materials that provide a three-dimensional framework for selective cell penetration and facilitate the formation of the new tissue. The scaffolds are designed to have optimal pore sizes depending on the cells that are expected to grow inside: for fibroblasts and hepatocytes, it should be around 20 μm; for skin regeneration, between 20 and 150 μm; and for bone regeneration, in the range of 100–250 μm [5, 6]. For smaller pore sizes (10–100 nm) and weak mechanical properties, biochemists and pharmacists designed other types of structures: these are "hydrogels," which

Disordered Pharmaceutical Materials, First Edition. Edited by Marc Descamps.

are mainly characterized by their high water content and soft and rubbery consistency, resembling living tissues. Apart from being implanted, hydrogels can also be injected. Hydrogels are "reversible" or "physical" gels when their structure (network) is stabilized by secondary forces (including ionic, H-bonding, or hydrophobic), and are "permanent" or "chemical" gels when they have covalently cross-linked networks [7].

The aqueous environment in hydrogels protects the cells as well as sensitive drugs incorporated into their network, and allows the transport of substances, for example, nutrients and byproducts from cell metabolism, inside and outside of the matrix. When required, hydrogels can also be derivatized with functional groups, which also promotes cell attachment and growth. Because of their weak mechanical properties, hydrogels were mainly investigated for soft tissue regeneration, but in the last few years interest has increased in testing hydrogels for hard tissue regeneration as well [8].

In a totally different field in other recent clinical investigations, gels turned out to be the most predominant forms for the delivery of microbicide products; for instance, gels are the most common dosage forms used for vaginal delivery. In such applications, the important parameters that come into play are the viscoelastic properties, stability, spreadability, and retention on mucosa (mucoadhesion) [9]. Gels formulated for this purpose are also called *semisolid dosage forms*. Besides, the physical and chemical characteristics of each microbicide drug candidate define the most suitable vehicles and dosage forms. The gels used in topical applications are often heat-set gels (gelled below body temperature) whose formation is driven by hydrophobic interactions of amphiphilic polymers.

A large number of natural and synthetic polymers and a variety of processes have been explored in the context of their biomedical and pharmaceutical applications. A growing number of publications highlight recent innovations in synthesizing new chemical structures and new manufacturing techniques, together with protocols for evaluating drug delivery and cell growth efficiencies [10]. Concurrent to such investigations, gels have been promoted in numerous other fields, for example, food preparations, personal products, technical applications, and so on. Nowadays, the terms *scaffolds* and *hydrogels* seem to be more specifically devoted to the formulations used in biomedical applications, but their precise definition is uncertain [11]. The term *hydrogel* was coined to denote polymer networks cross-linked by physical, chemical, or both types of links in aqueous media [12]. "Semisolid dosages" refer to the consistency of these gels, rather than to their composition, structure, or gel formation mechanism. Apparently, at the present time, the terminology used in the different contexts is not very clearly established.

In order to bridge the new areas that emerged, especially in biomaterials science, with the more traditional ones and with the fundamental knowledge in polymer materials, we review in this chapter various classes of networks or scaffolds and try to classify them in relation to the underlying mechanisms of polymer chain association: *scaffolds* seem to be mainly created with phase-separated (biphasic) systems, and *gels* are the networks more specifically built through molecular association. From the vast existing literature, it arises that the mechanisms that generate the microstructure in hydrogels combine molecular conformation changes, supramolecular interactions, and/or chemical cross-linking reactions, while in scaffolds they mostly involve phase transformations, foaming, and so on, or in some specific cases more sophisticated computationally designed structures.

In this review, we mainly concentrate on the physical aspects and, in particular, on the mechanisms underlying the assembly of molecules at the supramolecular scale in gels and scaffolds. The mechanisms of physical gelation are based on the subtle interplay between solubility and aggregation. The "sol state" is a true solution: polymers or particles are dissolved in the liquid. In aqueous solutions (binary systems), a combination of several factors, such as temperature, pH, polymer concentration, polymer molecular weight, and ionic strength, may decrease the ability to solubilize the polymers. In the absence of any chemical cross-linking, this would normally produce a precipitate, a phase-separated liquid (by sedimentation of two immiscible phases), a droplet morphology, or partial crystallization of the polymer. However, when a physical gel forms, it is observed that the polymer-rich domains or phases remain interconnected throughout the solvent. This important property is at the origin of the mechanical properties of gels. The sol–gel transition is a very special limit when the solution switches from a liquid state, where polymers are free to diffuse, to a solid-like state (otherwise called *semisolid*) in which the polymers entrap the solvent. The transition from one form to the other is defined as the sol–gel transition. It may occur through physical or chemical cross-linking. The fundamental characteristics of sol–gel transitions have been investigated by theoretical and experimental methods by physicists and physical chemists and are briefly recalled in this chapter.

Because of the complex processes that are needed in the preparation of the scaffolds or hydrogels, their final state is far from "equilibrium" and the resulting morphology/macromolecular assembly depends on the whole history. Therefore, it is difficult to clearly relate their inner structure simply, for instance, to the polymer composition, which in most cases also includes blends of several macromolecules. We shall concentrate in this chapter on the general properties that have been established in more fundamental experimental work for similar or simplified systems and try to elucidate the needs for the pharmaceutical/biomedical applications.

9.2 Gels and Hydrogels

Hydrogels, as already mentioned, are complex systems formulated in biomedical and pharmaceutical applications and belong to the large family of gels. We start by defining a gel, in a general context.

9.2.1
What Is a Gel?

The first definition of the colloidal state of matter was given by Graham [13] in the *Philosophical Transactions of the Royal Society*. Graham included in the category of colloidal materials (from the Greek $\kappa o\lambda\lambda\alpha$: glue) many substances such as hydrated silicic acid, starch, gelatin, albumen, gum, supersaturated solutions of inorganic particles or clay dispersions, and so on, all of which were known to be slowly diffusing substances. The molecular structure of these substances was not yet identified. While polymer synthesis was only at the beginning, in 1949 Hermans [14] proposed a more formal classification of gels. According to his initial definition

1) Gels should be coherent two-component systems formed by solid substances finely dispersed or dissolved in a solvent;
2) They should exhibit solid-like behavior;
3) The dispersed component and the solvent should extend continuously throughout the whole system (bicontinuous dispersions).

These criteria are valid even now. Since Hermans's definition, the polymeric nature of some colloids has been recognized, after the first experimental methods for measuring very high molecular weights (for rubber molecules, for instance) became available and the theoretical models for polymer conformations based on statistical mechanics developed by Staudinger were accepted (see, e.g., [15]).

Paul J. Flory proposed a more detailed classification, especially devoted to polymeric gels, which became since then the most actively investigated ones [16]. He set apart four different types of structures:

1) Well-ordered lamellar structures, including gel mesophases
2) Covalent polymeric networks that are completely disordered
3) Polymer networks formed through physical aggregation, predominantly disordered, but with regions of local order
4) Particular aggregated structures.

This classification still includes a large variety of molecules. This is why studying gels has been, from the beginning, a highly multidisciplinary activity. Gels were an aspect of colloid science: aggregation and association of small particles (proteins, mineral dispersions, etc.) can lead to physical and colloidal networks. Gels became part of the new polymer science, based on their increasing syntheses in organic chemistry. Thermodynamics is important: in chemical networks, the swelling behavior is related to polymer–solvent interactions. In physical gels, partial crystallization, phase separation, glass transition, conformational changes, and so on, as well as all solubility limitations, are also related to thermodynamics. Because of these multiple extensions of gel studies, it is important to review the development of this discipline, in light of the new areas to which it has been extended nowadays, and present the recent more fundamental progress.

First, we propose a simple classification of polymer gels according to the mechanisms of network formation or cross-linking: chemical gels, physical gels, and other types of networks that were found more recently.

9.2.2 Different Types of Gels

9.2.2.1 Chemical Gels

Chemical gels are irreversibly cross-linked networks [17]. Such gels can form either during the course of polymerization (e.g., a branched macromolecule forms from small molecular precursors during step addition polymerization) or by cross- or end-linking of previously existing macromolecules in solution (or in the melt) [18]. Chemical gels have been, from the beginning, a rich territory of organic chemists, who devised an impressive number of molecular configurations, currently used in all sorts of applications.

Synthetic hydrogels prepared in view of biomedical applications should have tailored and highly reproducible chemical characteristics, enabling the required functions, such as swelling, mechanical properties, or degradation properties, to be achieved [19, 20]. The networks can be modified with the incorporation of biofunctions, and their transport properties (e.g., for drug delivery) can also be adjusted by polymer chain lengths and cross-linking density [7, 8].

Cross-linking of the polymers can be carried out in the solid state or in solution with radiation (e.g., irradiation of poly(ethylene oxide) (PEO) in water) [21], photopolymerization [22], chemical cross-linking (e.g., collagen with glutaraldehyde or a bis-epoxide), using multifunctional reactive compounds (e.g., poly(ethylene glycol) PEG + diisocyanate), or by copolymerization of a monomer and a cross-linker in solution (e.g., HEMA (hydroxyethyl methacrylate) + EGDMA (ethylene glycol dimethacrylate)) [23].

9.2.2.2 Physical Gels

Different mechanisms giving rise to the formation of physical gels have been identified. They concern synthetic as well as biological polymers. We may summarize them as follows [24]:

a) *Conformational changes* of the polymer chains (e.g., coils to multiple helix transition) in particular for biopolymers (gelatin, polysaccharides). Examples are coil–triple helix transition in gelatin [25–28], double helixes in agarose [29–36], the aggregation of helixes in carrageenans [37–43], and the "egg box" structure found in alginates [44–48]. The first two cases are driven by lowering the temperature, and the last two are induced by temperature changes and/or specific ionic content.
b) *Denaturation* of globular proteins under conditions where the protein remains essentially globular induces *aggregation*, thereby producing colloidal type networks formed either by branched structures or linear ones (amyloid type fibers) [49–57]. This type of particulate networks includes also the casein networks formed in milk clotting or cheese making [58–62]. Such

aggregation is usually irreversible but shares some characteristics with the phase separation of model colloidal solutions [63, 64].

c) The *hydrophobic effect* occurring with amphiphilic polymers. The insertion of hydrophobic functional groups into a water-soluble (hydrophilic) polymer generates this effect, in which the hydrophobic entities tend to self-associate into micelle-like domains [65–71]. This is an important class of physical gels used in pharmaceutical topical applications [72].
d) *Synergy* between two different polymers occurs in some mixtures of polysaccharides and some specific synthetic polymers that do not gel as single components [73–80]. Such mechanisms have the features of both categories (a) (conformational change) and (e) (phase separation, see below).
e) *Phase separation.* This occurs mainly in binary polymer mixtures or in single polymer solutions when the solvent becomes poor, either at lower temperatures or with selected organic solvents. Sometimes, liquid–liquid phase separation is inhibited by the high viscosity or viscoelasticity of the mixtures, but this is not the only possible mechanism, since either glass transition or crystallization may interfere with the phase separation. Under these circumstances, the overall morphology of the mixture can change. The morphology of the networks created by phase separation in gels is similar to the scaffold architecture reported in Section 9.3. In these formulations, both terms *gels* and *scaffolds* should be acceptable. Freeze–thaw procedures can be included in this category [81–84].

9.2.2.3 Hydrogels

Most of the "hydrogels" used in biomedical applications are both chemically and physically cross-linked. We therefore include these new categories:

a) *Thermoreversible hydrogels* are synthesized from polymers exhibiting a lower critical solution temperature (LCST). The hydrogen bonding between the polymer's polar groups and water molecules leads to dissolution of the polymer at low temperature, whereas above the LCST dehydration of the hydrophobic groups induces the coil-to-globule transition of the polymer in very dilute solutions and partial aggregation in more concentrated ones. A model polymer often investigated is poly(*N*-isopropyl acrylamide) (poly-NIPAM). Poly-NIPAM in solution precipitates when heated above 32 °C. A list of synthetic or natural polymers that exhibit LCST has been prepared by Jeong *et al.* [85], which includes PEG, poly(propylene glycol) (PPG), methyl cellulose (MC), and others. The chemical cross-links are able to stabilize the interconnections between polymers in solution, thereby creating gels that can swell or shrink reversibly with the temperature, according to their LCST. Poly-NIPAM homopolymer and its copolymers were widely investigated for drug delivery [86] and tissue engineering because their LCST can be controlled by copolymerizing with other monomers with different hydrophobicity. Such hydrogels have adjustable swelling and deswelling properties versus temperature.
b) *Interpenetrating polymer networks:* More complex ternary systems have been designed to fulfill new requirements. Interpenetrating polymer networks

(IPNs) are "alloys" of cross-linked polymers in which at least one network is synthesized and/or cross-linked in the presence of the other. Entirely new properties are exhibited by IPNs, which are not observed in either of the single networks alone. IPNs can be generated in different scenarios: (i) one network is swollen and polymerized in the presence of the other and (ii) both of the network precursors are synthesized at the same time by independent routes. When one of the two polymers is non-cross-linked, a semi-IPN results, and when both the polymers are identical, a homo-IPN is obtained. Obviously, a very important parameter that must be considered in the design of IPNs is the miscibility of the polymers; in general, polymers do not mix well with each other, resulting in the phase separation of the blend, as already mentioned. However, because cross-linking provides a way to prevent demixing between two otherwise immiscible materials, many combinations of polymers could be explored as IPNs for a variety of applications. IPN hydrogels are typically produced first by synthesizing a hydrophilic polymer network, swelling it in a second aqueous monomer solution, and then polymerizing the latter, forming water-swollen meshes of two different polymers. The hydrophilicity/hydrophobicity and response to stimuli of IPNs can be controlled by the selection of the two polymers and by varying the composition. For example, (polyacrylamide and poly(acrylic acid) (PAAm/PAA) [87] establish cooperative hydrogen bonding between the two networks and exhibit large volume transitions *during heating*, while the non-ionic polymer has much less effect. Other examples of IPNs described in the literature include the poly(vinyl alcohol) (PVA)/PAA system developed by Gudeman and Peppas [88], the pH-sensitive chitosan/PAA systems developed by Lee *et al.* [89], and the temperature- responsive PVA/poly-NIPAM semi-IPNs synthesized by Zhang *et al.* [90]. The latter semi-IPN hydrogels, compared to the conventional poly-NIPAM hydrogels, exhibit significantly faster response and undergo full deswelling or have larger equilibrium swelling at room temperature.

c) *Double-network* (DN) hydrogels containing two interpenetrating networks is a new category of hydrogels. These gels have a high potential for use in artificial cartilage [91]. The mechanical properties of DN gels prepared from many different polymer pairs are enhanced compared to those of the individual components. These DN gels are synthesized from monomers in the presence of a cross-linker via a two-step network formation. The first step forms a tightly cross-linked network of polyelectrolyte. Then, the gel is immersed in an aqueous solution of a second monomer, with a low ratio of cross-linking agent, and a second polymerization is carried out. The second polymer is a neutral polymer, incorporated within a highly swollen polyelectrolyte network, and has a high relative molecular mass. Under an optimized composition, the DN gels, containing about 90 wt% water, possess elastic modulus of 0.1 – 1 MPa, failure tensile stress of 1 – 10 MPa, failure strain of 1000 – 2000%, and failure compressive stress of 20 – 60 MPa [92]. Although the DN structure has been found effective with many kinds of combinations, among all the polymer pairs studied so far, the one made of poly(2-acrylamido, 2-methyl, 1-propanesulfonic acid)

(PAMPS) polyelectrolyte and PAAm neutral polymer has outstanding properties. This property is observed when the molar concentration of the second network is 20–30 times that of the first network. The first network is tightly cross-linked, while the second network is loosely cross-linked, which requires also a very high molecular weight of the second polymer. The exceptional mechanical properties of these gels were explained by the effective local relaxation of the stress and dissipation of crack energy through a combination of the two networks, which exhibit different structures and densities [93]. It is interesting to note, however, that the enhanced properties are mainly observed with a very narrow, optimum proportion of the two components; therefore the particular mechanisms underlying this achievement are still to be elucidated.

d) *Hydrogels containing sliding cross-links* were produced for the first time by Okumura and Ito [94] with polyrotaxanes, which are polymer chains on which many cyclic molecules are threaded and which are trapped with bulky end groups. The authors produced a sliding, double ring, cross-linked hydrogel with two cyclodextrin molecules, each threaded on a different PEG chain and end-capped with a bulky group, such as adamantan. These cross-linked gels (Figure 9.1) possess freely movable sliding links, which results in outstanding mechanical properties: a high degree of swelling in water (400 times its initial weight) and a high stretching ratio without fracture (twice its length). These topological gels have "figure-of-eight" cross-links. The sliding double-ring cross-links in these gels apparently equalize the tension along the polymer chains ("pulley effect").

9.2.2.4 Hybrid Gels

Finally we include in our classification the emerging field of nanostructured IPNs containing organic and inorganic components.

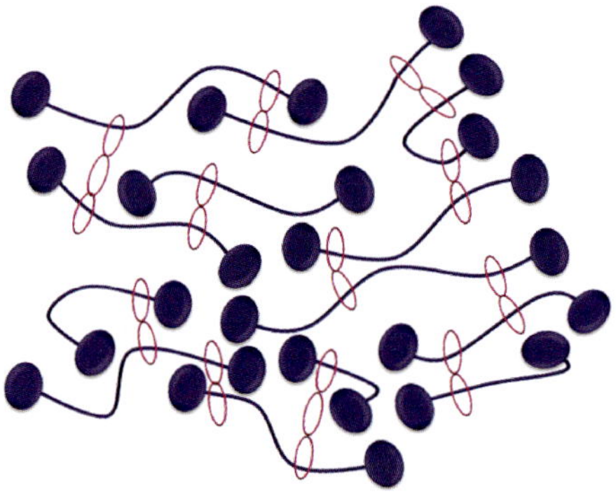

Figure 9.1 Schematic structure of topological sliding gels. α-Cyclodextrin molecules are threaded on PEG chains end-capped with bulky groups and then cross-linked, producing double or triple rings that can move freely along the PEG chain. (Adapted from Okumura and Ito [94]. Reproduced with permission of Wiley.)

A new area has opened up in the field of hybrid organic–inorganic materials. The formation of chemical bonds between organic and inorganic components allows molecular composites with novel properties to be produced. We take as an example hybrid networks containing clay particles. Clays are layered aluminosilicates (typically silica tetrahedra bonded to alumina octahedra) present in sheet-like structures with charge-compensating counter-ions (such as Li^+, Na^+, K^+, Ca^{++}) located in the interlayer space. The compatibility between organic polymers and inorganic hosts results in systems exhibiting the so-called *intercalated* or *exfoliated* morphologies. In intercalated structures, the organic component is inserted between the layers of the clay in such a way that the interlayer spacing is expanded. In exfoliated structures, the layers of the clay are completely separated and are randomly distributed throughout the organic matrix; the type of layer "delamination" determines the properties of the clay nanocomposites.

Haraguchi and Takehisa [95] were able to prepare a hybrid hydrogel based on poly-NIPAM, polymerized *in situ* in the presence of exfoliated, uniformly dispersed clay particles. The free-radical polymerization was initiated from the clay surface, without the use of an organic cross-linker, with the clay sheet itself acting as a large cross-linker. It was thought that the initiator was adsorbed onto the surface of the clay particles, and the monomer and catalyst were in the surrounding liquid. The properties of these gels are very different from those of the polymer networks with a conventional organic cross-linker. Compared to the latter, the change of mechanical properties of the hybrid gel is impressive. For example, in the stress–strain curve, the hybrid gel could reach a maximum elongation of 1000% before fracture, and its behavior was almost completely reversible, whereas the unswollen chemical gels were in general relatively brittle.

Organic–inorganic IPNs are an important class of hydrogels for hard tissue regeneration [8]. Nanocomposite hydrogels with an organic/inorganic semi-IPN were synthesized through *in situ* free-radical polymerization of NIPAM in the presence of an inorganic clay and linear PAA. The resulting nanocomposite gels showed mechanical and swelling/deswelling properties both with temperature and pH, and were greatly superior to those of other PNIPAM/PAA hydrogels. Among them, polymer/hectorite (Laponite) nanocomposite hydrogels have been widely studied because of their facile fabrication and tentative application in cell culture [96] without any chemical cross-linkers. The resultant hydrogels exhibited high mechanical properties together with high transparency. Polymer hydrogels with high mechanical properties have potential applications as articular cartilage, semilunar cartilage, tendon, and ligament.

9.2.3
General Properties of Polymer Networks

Chemical and physical polymer gels share in common a number of properties, which arise from the unique character of the linear flexible chains that build the network and from their cross-links.

9.2.3.1 Background on Entropic Elasticity

Synthetic polymers, which are long, linear macromolecules with hundreds of chemically bonded units, adopt a random coil conformation when dissolved in good solvents. Their conformation is similar to a 3D random-walk trajectory, which can be approximated by a Gaussian distribution of the end-to-end distance of the polymer chain in θ-solvents, and to the self-avoiding walk in good solvents [97]. The statistical distribution of random orientations of monomers in Gaussian coils implies that the average (or root-mean-square) radius of the chain $\overline{R}$ scales with the square root of the number of units:

$$\overline{R} = l\sqrt{N} \tag{9.1}$$

where l is the monomer or unit size, and N is the number of monomers or units. Taking $l = 1$ nm and $N = 100$, the average radius is $\overline{R} \sim 10$ nm, and with $N = 1000$, $\overline{R} \sim 32$ nm. When concentration of polymers increases above the so-called c^* concentration, in the semi-dilute regime, these long molecules tend to overlap and entangle (Figure 9.2). Under such conditions, by different means, such as UV radiation, chemical cross-linking, enzymatic actions, or physical processes as mentioned at the beginning of this section, the entangled chains may be linked to one another to create a 3D network. This is a gel. The solvent is then entrapped in the network, and the whole system behaves like a solid.

Cross-linked polymers above their glass transition temperature (also called elastomers in the polymer field) and gel networks (which entrap the solvent) are flexible solid-like materials that can be deformed by mechanical actions and recover their initial shape after the release of the mechanical force. This property, called rubber-like elasticity, is directly related to the coil conformation of the polymers. When the networks are stretched, the number of possible microscopic orientations that the monomers can adopt decreases, and therefore the entropy of the chain decreases. The free energy decreases and the polymer behaves like a spring, where the restoring force comes from entropy rather than the internal energy.

In rubber-like elasticity, when the force $\vec{F}$ acts on a single polymer with an end-to-end distance $\langle \vec{R} \rangle$, the polymer reacts like a spring whose stiffness is

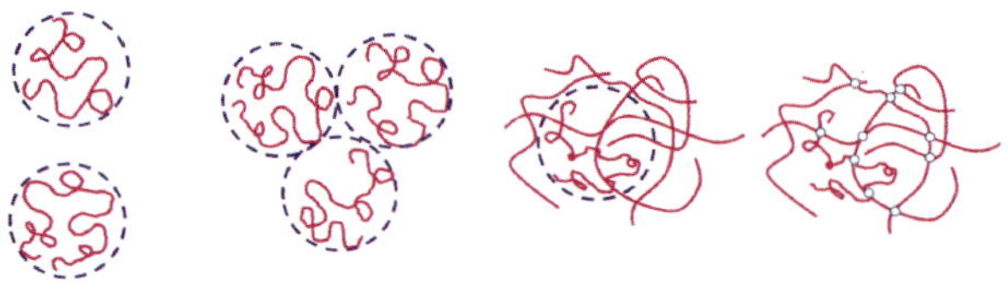

Figure 9.2 Polymer chains in solution at different concentrations. At $c < c^*$, in dilute solutions, chains have random coiled conformation in θ-solvent or are swollen (self-avoiding walks) in good solvents [97]. At $c = c^*$, polymer chains start to overlap. At $c > c^*$, solutions are semi-dilute and the chains are entangled (with viscoelastic properties). When they are cross-linked, they form a network and the solution becomes a gel.

given by

$$\vec{F} = -k_B T \frac{3\langle\vec{R}\rangle}{Nl^2} \tag{9.2}$$

The elastic coefficient of a single "ideal" chain, $3k_{\rm B}T/Nl^2$, is proportional to the temperature T, with $k_{\rm B}$ the Boltzman constant. The higher the temperature, the stiffer the spring. An extension of this molecular property to macroscopic networks made up of cross-linked flexible chains states that the shear modulus relating the stress and the strain under shear deformation (similar to the Young's modulus under elongation) is simply determined by

$$G = a\nu_{\rm e}k_{\rm B}T \tag{9.3}$$

where $\nu_{\rm e}$ is the number density of elastically active chains, and a is a prefactor close to unity. This very simple relation, however, encounters a real difficulty when one tries to relate the elastically active chains to the number density of the chains. Network topology is regarded as a major factor influencing the elasticity of networks, since the number of *elastically effective* junctions of the network is often very significantly different from the *total number* of junctions. The elastic properties are influenced by both the number of chemical junctions and their spatial distribution, which can be uniform, random, or clustered. The contribution of cross-links is difficult to establish, owing, for instance, to the complexity of the chemical reactions in physically entangled, semi-dilute solutions [98]. In addition, the "network defects" such as chain ends or closed loops do not generally contribute to connectivity and therefore to macroscopic properties. Chemical gels being swollen polymer networks, their elastic properties rely on the same analysis as that of the bulk elastomers: the shear moduli increase with the number density of elastically active chains.

9.2.3.2 Elasticity of Physical Gels

In physical gels, establishing the relation between cross-linking density and elasticity is an even more difficult task: the cross-links are not of a molecular size but include supramolecular assemblies with finite sizes and shapes (fiber-like, dense aggregates, etc.). If the cross-links are labile (i.e., hydrophobic interactions, ionic complexations, etc.), the storage modulus is frequency- (or time)-dependent and the solid-like behavior appears only if the frequency is above some critical value $f_{\rm c}$ (frequency at which the storage shear modulus is larger than the shear loss modulus; $G' \geq G''$). Such systems may flow or creep under a small strain when followed over a long time period. Many factors, such as the course of the thermal process, the nonequilibrium state, and the slow kinetics of reorganization of labile networks, are expected to induce substantial changes of the macroscopic properties, especially elasticity. Temperature dependence of the elastic moduli would be more complex than predicted by Eq. (9.3), as the number density of active junctions is directly related to the mechanism of gelation and obviously varies with temperature.

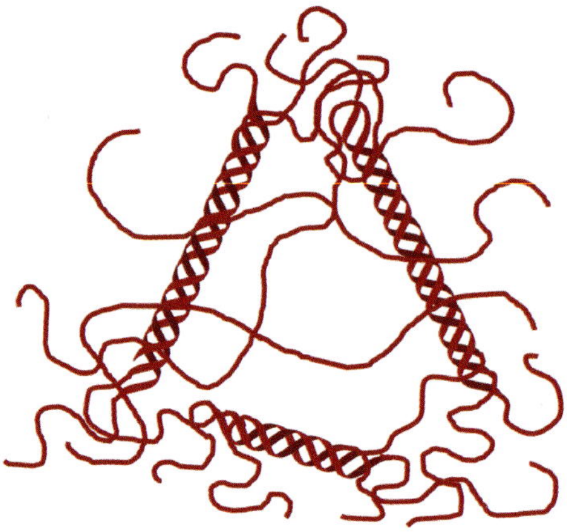

Figure 9.3 Gelatin gel structure, made of sequences of collagen-type triple helixes and random coils. The fraction of residues in helical conformation and the length of the sequences depend on many parameters such as temperature, concentration, molecular mass of the chains, time, and so on. The gel never reaches an equilibrium state.

Physical gels containing fiber-like structures have, in general, high shear storage modulus. Let us show this property for some well-known gels, namely gelatin, agarose, and gellan.

Gelatin and agarose are derived from natural polymers; gelatin is extracted from collagen (protein), particularly originating from mammals or fish, whereas agarose is a polysaccharide extracted from marine algae. In both cases, gelation is not driven by ionic interactions. Gelatin forms "cold-set" gels (below room temperature) when the concentration is high enough, above c^*, $c > \sim 1$ wt%, via a conformational transition from a disordered and relatively flexible coil conformation to an ordered, partly helical and rod-like structure, reminiscent of the collagen triple-helix conformation [28, 99]. At low temperatures, the network strands consist of both helical and coil-like portions (Figure 9.3). On heating, the gels "melt," and the individual molecular chains revert to the coil state (thermoreversible gels), and this process can be repeated many times. In agarose gels, it was found, by small-angle scattering techniques, that aggregation follows double helix formation upon cooling and consists in their hexagonal packing and in larger assemblies of the thin fibers [31] (Figure 9.4). The distribution of the aggregation number is likely to induce a thermal hysteresis, which is well established in these gels [36]. The two steps consisting of growth of double helixes and aggregation cannot be separated experimentally. At one level, these two gel networks consist of long, thin interconnected fibers.

A comparison is presented in Figure 9.5 for agarose and gelatin and reported within a large concentration range, up to 25 wt% for gelatin [100], from Clark *et al.* The plot highlights that agarose gels have a very high modulus at low concentrations compared to gelatin: for instance, at $c = 2$ wt%, the modulus is $\sim$100 kPa for agarose and 1 kPa for gelatin. The minimum concentration to obtain a gel is

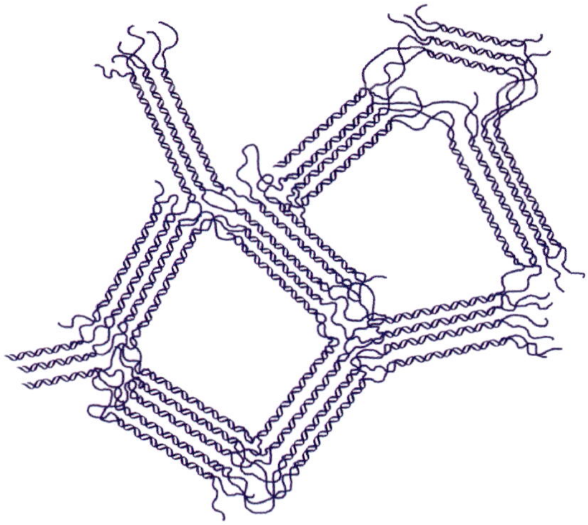

Figure 9.4 Agarose forms double helixes that aggregate avoidably when solutions are cooled. Gels are built by aggregation of the polysaccharide double helices. The aggregated strands form hexagonal packing and larger assemblies of thin fibers. It is believed that the polysaccharide chains adopt almost entirely the helical conformation, thus a very small amount of monomers is in the coil conformation. Structural and mechanical properties of agarose gels are stable with time at constant temperature.

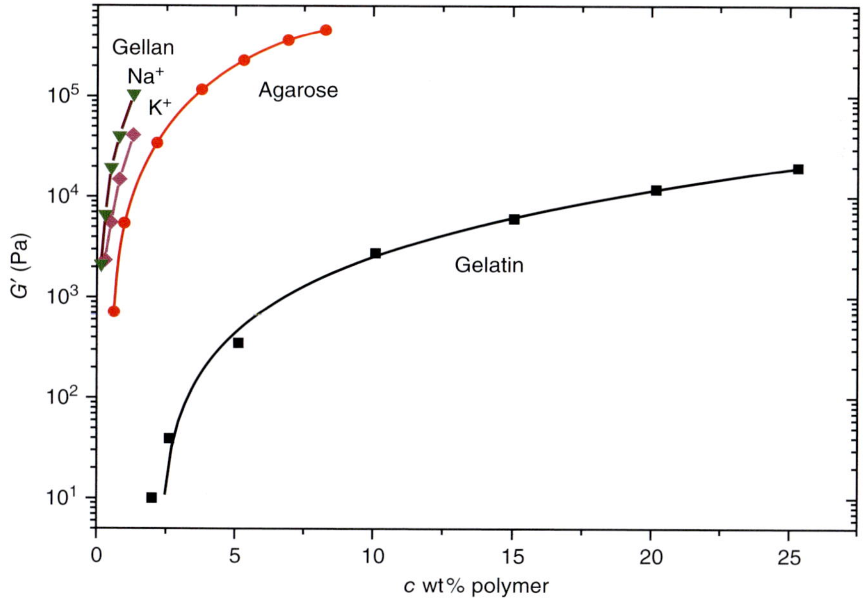

Figure 9.5 Comparison between shear storage moduli of fibril-type networks made of helixes in aqueous solutions: gelatin, agarose, and gellan with two types of counter-ions Na^+ and K^+.

about 0.2 wt% for agarose and 1–2 wt% for gelatin. It is not clear whether these concentrations are related to the overlapping concentrations c^* of the polymers or to the structure of the network. Indeed, differences in the fibrillar structure of agarose and gelatin are very important. The mesh size for agarose and gelatin gels at equal concentrations ($c \sim 2$ wt%) is, respectively, 100 nm and only a few nanometers, the latter containing an entangled network with both triple helixes and coils [101]. The fibrils in agarose gels are rigid and non-entangled. The rigidity of the agarose network thus comes from both (i) long and rigid rods with variable thicknesses, constituted by the aggregation of double helixes and (ii) junctions between the rods that are probably much less flexible than gelatin loops or coils. The non-helical content of agarose gels is probably very low; it is not discussed in the literature, but may be related to defects in the structure of the native polysaccharide.

For comparison, another helical type network is shown in Figure 9.5, which is the gellan gel. Gellan is a microbial polysaccharide. The data here shows the storage modulus derived from the Young's modulus at room temperature with the usual assumption that the shear modulus is one-third the Young's modulus [102]. The counter-ions have been carefully replaced from commercial samples, and the data shown contains either K^+ or Na^+ counter-ions with $M_w = 2.5 \times 10^5$ g mol^{-1}. The authors [102] explained also the role of cooling history on the gel properties.

It is evident from this comparison that very high shear moduli are observed in gellan compared to other types of helical networks. The gellan concentration did not exceed here 1.3 wt%, and the moduli are extremely high; the K^+ counter-ions give the highest moduli.

Helical type networks indeed show very high moduli (10–100 kPa) compared, for instance, to chemically cross-linked gels (0.1–1 kPa) at these low concentrations. Ionic effects give an additional rigidity to the helical aggregates in polyelectrolyte gels. The moduli for these gels remain constant (no creep) for long enough periods (longer than a few seconds), so that they can be considered as equilibrium or quasi-permanent.

Another interesting case that illustrates the strong influence of the local structures on the overall rheological properties is given by the gels obtained by denaturation of globular proteins, such as β-lactoglobulin (β-Lg), at different pH values. Differences in the heating protocol and ionic strength are likely to induce large differences for minimum concentration and the rate of increase of the shear modulus in relation to the morphology and flexibility of the aggregated structures, as shown in Figure 9.6. Aggregates studied by light and neutron scattering [103, 104] showed rod-like structures at pH 2, whereas globules and aggregates of globules appeared at pH 7. The rod-like structure was thought to occur when disulfide exchange (implied in globule formation) was inhibited. Increasing the salt concentration increases the level of branching (or rod flexibility) by screening out repulsive electrostatic interactions.

A number of studies have been carried out using TEM [105, 106], wide-angle X-ray diffraction, and Fourier transform infrared spectroscopy to try to understand which factors influence the relative propensity for fibrillar growth in the

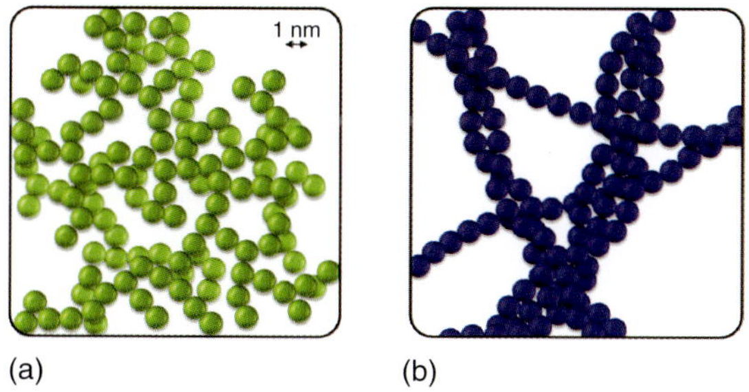

Figure 9.6 Globular protein gels. (a) Colloidal aggregation close to the isoelectric point. (b) Fibrillar aggregation of in amyloid-type gels at low pH.

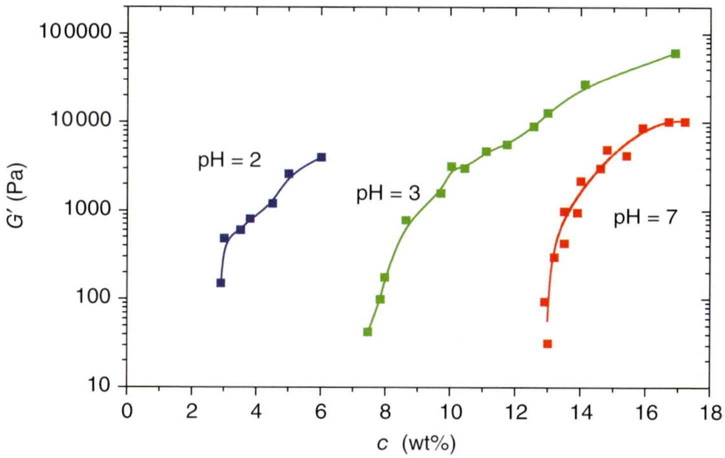

Figure 9.7 Comparison between the storage shear moduli of β-lactoglobulin (β-Lg) gels at different pHs. The colloidal gels may have a branched, compact, space filling structure at pH 7, while under some other conditions (pH 2) they can assemble into long fibrils. The minimum concentration and the shear moduli of gels are strongly related to the structure.

resultant gel [107]. Microscopy showed long linear aggregates forming in solutions at pH 2. Figure 9.7 shows a simple comparison of the shear moduli between three types of β-Lg heat-set gels at three different pH values: pH 2 from [108] and pH 3 and pH 7 from [107]. This figure illustrates the challenge in interpreting the rheological properties of these protein gels, clearly showing that globular protein concentration is not the unique parameter determining the amplitude of the shear modulus. Fiber-like aggregates are generated at low pH values, whereas more compact fractal-type structures shift the minimum concentration of the gel to much larger values.

The aggregates are more space-filling and compact at pH 7, while long amyloid-type fibers are likely to form at pH 2. At high pH values, the aggregated structures would be similar to those predicted in phase-separated colloidal fluids.

We shall illustrate, later on, a particular case where it was possible to relate the storage (elastic) modulus to its microscopic structure with different polymer concentrations and thermal treatments. This is the gelatin gel (Section 9.2.5.1). We also illustrate (Section 9.2.5.2) the interplay between physical and chemical cross-linking in this particular gel, which can be the starting point for understanding more complex hydrogels.

9.2.3.3 **Network Swelling**

When a chemically cross-linked gel is immersed in an excess of liquid and allowed to equilibrate, the size of the sample may increase, decrease, or remain constant. This effect depends strongly upon the thermodynamic quality of the liquid for the polymer and the degree of cross-linking. If a piece of gel is immersed in a "good solvent," the volume will increase from an initial value of V_0 to V. The swelling ratio q is defined as $q = V/V_0$. For very high degrees of swelling, this ratio can be defined in terms of the mass of gel. A chemically cross-linked gel will tend to swell to an equilibrium degree depending upon its composition, the amount of cross-linking, the nature of the solvent, the temperature, and so on. In a poor solvent, the polymer will appear to shrink.

Swelling of a gel is assumed to be governed by the quality of the solvent involving osmotic and, in addition, for polyelectrolyte systems, ionic contributions. The tendency to swell is limited by the elastic restoring force from rubber-elasticity theory (for a recent review, see [109]). For achieving a high degree of swelling, the retractive force must be small, and so the degree of cross-linking must itself be not far from the minimum amount necessary to interconnect the polymers.

The total swelling pressure Π can be written as the sum of several independent contributions:

$$\Pi = \sum_{i=1,4} \pi_i \tag{9.4}$$

Starting with the free energy for swelling, π_1 is the excluded volume (or mixing term) from the Flory–Huggins theory with an associated polymer–solvent "quality" parameter (Flory parameter χ). Then, π_2 of opposite sign reflects the change in configurational free energy with swelling (the rubber-elasticity term). Further terms include π_3, which is a measure of the difference in osmotic pressure between the gel and the solution. For a polyelectrolyte gel, this includes the Donnan contribution from the mixing of ions with the solvent, where the gel acts like an osmotic semipermeable membrane. For instance, if the polymer chains have a net negative charge, there will be a net accumulation of cations inside the gel. Finally, the term π_4 comes from the free energy of electrostatic interactions. For gels in non-polar solvents, only the first two terms are required. The osmotic swelling force depends on the factor $(\chi - 0.5)$ reflecting solvent quality ($\chi = 0.5$ is the θ-solvent). As already mentioned, the retractive force, in turn, is proportional to the number density of cross-links, so that the net elastic force is higher (the shear modulus is higher) at high degrees of cross-linking and the equilibrium degree of swelling is lower. Swelling equilibrium is then accomplished for polyelectrolyte gels when the chemical potentials of the solvent in the gel and in the surrounding solvent are the same.

A polyelectrolyte gel will swell to a large extent in pure water or in low-ionic-strength electrolyte solutions to reduce charge–charge interactions along the chain contour. On the contrary, at high electrolyte concentrations the gel will tend to deswell. For ionic gels containing weakly acidic pendent groups, the equilibrium degree of swelling increases as the pH of the external solution increases, while the degree of swelling increases as the pH decreases for gels containing weakly basic pendent groups.

Numerous properties (e.g., ionic content, ionization equilibrium considerations, nature of counter-ions, and nature of the polymer) contribute to the swelling of ionic hydrogels, and these have been extensively studied. Examples of some commonly studied ionic polymers include PAA, PAAm, poly(methacrylic acid), poly(diethylaminoethyl methacrylate), and poly(di-methylaminoethyl methacrylate) [110].

Since the delivery of small molecules is based on the control of the pore sizes and of possible specific interactions between the loaded molecules and the matrix, it is important to evaluate the pore size distribution with independent methods.

Labeled molecular probes (such as fluorescein-labeled dextrans) of a range of molecular weights or molecular sizes can be used to probe pore sizes and their interconnections in hydrogels [7]. Mercury intrusion porosimetry, BET (Brunauer, Emmett, and Teller) measurements, scanning electron microscopy (SEM), and thermoporometry are difficult to perform [111]. They require dried gels or microspheres, and because upon drying the pores collapse, these techniques are not suitable for the determination of the pore sizes of hydrated samples. Kim and Chu [112] succeeded to avoid the collapse of the gel by cryofixation of dextran hydrogels, but quantification of the results by SEM remains difficult. To evaluate the average pore sizes of dextran microspheres, Stenekes *et al.* [111] proposed the determination of the size cut-offs for the uptake of different proteins. The microspheres were prepared by cross-linking an aqueous solution of methacrylated dextran (dex-MA) emulsified in a continuous PEG phase. The release of myoglobin, ovalbumin, bovine serum albumin (BSA), and immunoglobulin G (IgG) from microspheres with various degrees of substitution (DS: the number of MA groups per 100 dextran glucopyranosyl monomer units) and equilibrium water contents between 71% and 62% w/w was studied. Hydrodynamic diameters of myoglobin, ovalbumin, BSA, and IgG are, respectively, 4.2, 5.6, 7.2, and 10.9 nm. It had been previously demonstrated that the proteins were not involved in the cross-linking reaction. The release profiles of the proteins from the different formulations show that a plateau is reached in a short period (about 5 h) because of the large surface area of the microspheres and the short diffusion distances (microsphere diameter between 5 and 15 μm). Because these microspheres are nondegradable, the protein that is not released is entrapped inside the matrix. It was found that the fraction of entrapped protein is larger when a larger protein is incorporated, and for a given protein a higher percentage of protein is entrapped in microspheres with a higher degree of substitution. From

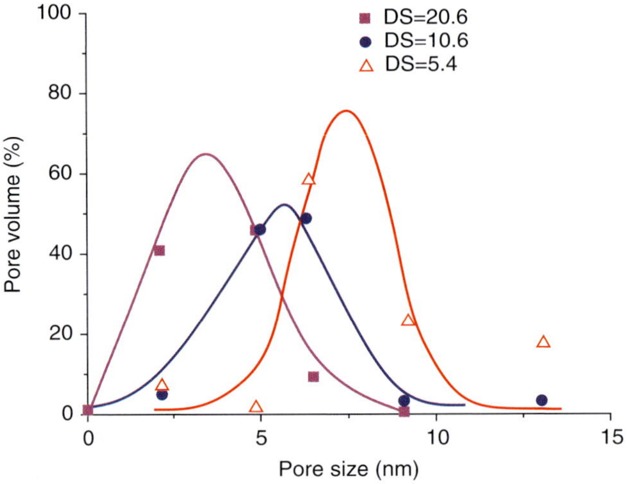

Figure 9.8 Pore size distribution constructed from protein release data of cross-linked dextran microspheres with various degrees of substitution (■ DS = 20.6, • DS = 10.6, △ DS = 5.4). The maximum of pore size distribution depends on the degree of substitution of dextran. The initial water content in the hydrogel is 50 wt%. (Adapted from Stenekes *et al.* [111]. Reproduced with permission of American Chemical Society.)

the release curves of proteins with different sizes, for each formulation a pore size distribution plot was constructed. The average pore sizes for the different hydrogels, taken as the maxima in Figure 9.8, were, respectively, 7.5, 5.7, and 3.6 nm for increasing DS. These results correspond fairly well with the average mesh size of the network calculated from micromanipulation experiments, where the authors evaluated the compression modulus of the microspheres using the classical rubber elasticity.

The swelling of physical gels is much more complex (e.g., alginate beads with calcium ionic cross-linking, PVA cryogels). Obviously, because the physical links are labile junctions, they can redistribute progressively under osmotic pressure, and the junctions may partially dissolve, releasing some polymer chains into solution. The ultimate state of swelling of physical networks may lead to dispersions of polymer aggregates or simply to a polymer solution.

Mesh size of physical gels can be measured by scattering techniques: the cooperative diffusion coefficient determines the hydrodynamic correlation length versus concentration in semi-dilute solutions and gels. Small-angle neutron scattering determines the static correlation length between chains (e.g., gelatin [26, 27], agarose [36]); the values are close to each other but not identical.

The release of a macromolecular drug from a hydrogel will be controlled not only by the pore volume fraction, pore size, and pore interconnections but also by the interactions between the drug and the polymer network. The re-partition of the drug inside the networks is another key factor that should be mentioned: a protein drug that has a net opposite charge to that of the polymer may concentrate toward the surface of the gels, whereas if it bears a net charge that is the same as that of the gel it may be excluded. Such noncovalent interactions, together with the mesh size after swelling and degradation of the gel network (see below), control the overall time of drug delivery.

9.2.3.4 Swelling and Biodegradation

Swollen hydrogels are aimed to release drugs in a controlled rate. Two types of mechanism can be distinguished for the release of drugs: diffusion-controlled, as explained before, and degradation-controlled. For diffusion-controlled release, the mesh size of the gel has to be larger than the hydrodynamic diameter of the molecule that would diffuse through the network. The equilibrium mesh size of swollen gels is related to their cross-linking density, solvent quality (pH, ionic content), temperature, and so on, but the swelling time is also controlled by the overall size of the gel. Since in pharmaceutical and medical applications what matters is the time dependence of the swelling, more than the equilibrium swelling degree, the macroscopic size of the matrix is important and microparticles are preferred to bulk gels. The synthesis of microgels with well-defined overall size distributions is a major challenge. As demonstrated before, the release of proteins from nondegrading macroscopic gels and microgels is indeed governed by Fickian diffusion laws when the protein is smaller than the mesh size of the gel [113]. Biodegradation of the polymers is an additional way of swelling or dispersing the networks, which can be tailored for both physically and chemically cross-linked gels. This unique feature has not been considered in other disciplines where gels are used and is typical to biomedical applications. The drug release in this case should be dependent on the degradation rate of the matrix instead of its mesh size.

While nondegradable hydrogels have been used in a number of medical applications (e.g., contact lenses, wound dressings), the clinical use of degradable hydrogels has only recently been explored. Current applications include membranes for the prevention of surgical adhesions and implantable drug delivery systems. The degradability of hydrogels is based on the hydrolysis of either the polymer backbone or the cross-links. The former can be achieved by introducing degradable units into the polymer and by bringing the polymer network into contact with a suitable enzyme [114].The degradation rate could be tailored by the gel characteristics and the amount of encapsulated enzyme. Model systems that degrade by hydrolysis of the polymer backbone consist, for instance, of dex-MA hydrogels, which are degraded by dextranase. When dextranase is entrapped within the hydrogel, bulk degradation occurs. The alternative model would be hydrogels that degrade by hydrolysis of the cross-links, such as dextran hydroxyethyl methacrylate (dex-HEMA): the hydrolyzable carbonate ester introduced between the methacrylate group and the dextran produces hydrolyzable cross-links [115]. The hydrolysis of the carbonate esters is base-catalyzed above pH 4 and acid-catalyzed below pH 4. Under physiological conditions (pH 7 and 37 °C), the degradation time of dex-HEMA hydrogels can be tailored by the initial water content and the degree of substitution DS, and ranges from a few days up to months in macroscopic gels.

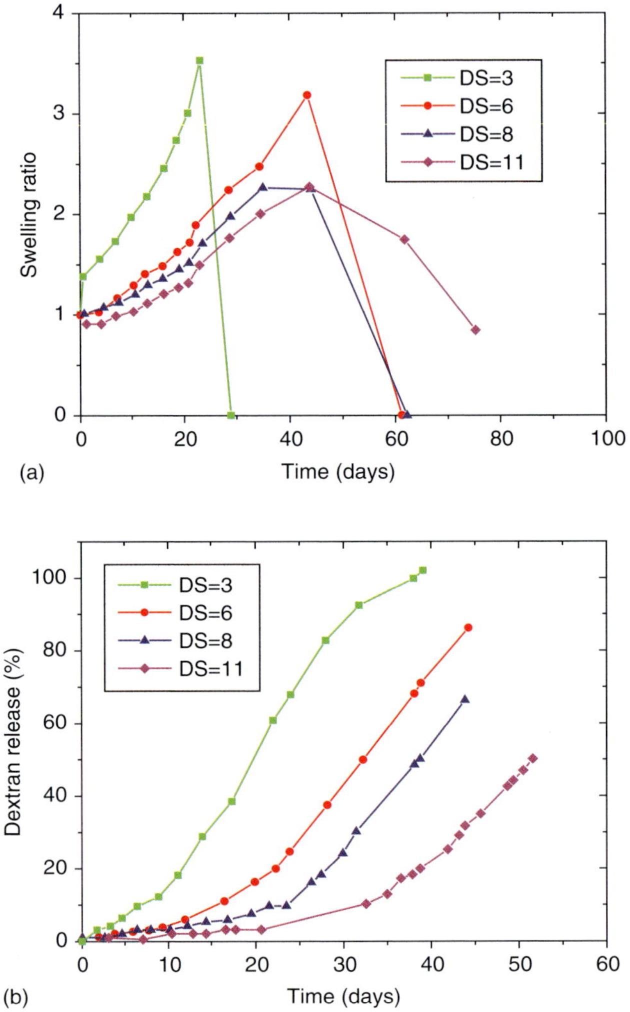

Figure 9.9 (a) Swelling ratio versus time of macroscopic dex-HEMA hydrogels with water content 70 wt% and different degrees of substitution. (b) Cumulative release of dextran in hydrogels with water content 70 wt% and different DS values. (Adapted from Franssen *et al.* [113]. Reproduced with permission of Elsevier.)

The example in Figure 9.9 illustrates the interplay between swelling and degradation rates driven by the degree of substitution DS (or the cross-linking degree) in macroscopic networks. The swelling takes place first, as it starts as soon as the gels are immersed in the excess of water in physiological conditions (Figure 9.9a), and the ultimate swelling degree is much larger for a low DS, as was mentioned earlier, whereas in Figure 9.9b the release of dextran occurs after a delay, which also depends on the DS: the delay varies from 5 days at DS = 3 and 20 days for DS = 11. Therefore, it is seen that the swelling continues during the period preceding the full dispersion of the networks; in the meantime, cross-links are progressively hydrolyzed. Meyvis *et al.* [114] tried to correlate the changes of shear moduli, swelling degree, and degradation rates of these gels, and showed that the three methods are complementary but the results of a specific analysis cannot be easily used to predict the results of the others. They discussed in particular the influence of the functionally at the junctions of the elastically active network chains: because of hydrolysis, the functionally of the junctions decreases, which means that the particular topology of the cross-links changes with time. When microspheres of dex-HEMA were synthesized (10 μm diameter) and let to swell in the same conditions, the timescales were completely different due to size effects: swelling and degradation times were accelerated and resulted in a complete dissolution of particles and dispersion of the polymers.

9.2.4 The Sol–Gel Transition

There is a particular aspect of gels that interests many chemists, physicists, and theoreticians: this is the very special limit when the liquid solution turns into a solid-like state and the polymers entrap the solvent. The transition from one form to the other may happen quite suddenly and is defined as the sol–gel transition or gel point. What is the best way to determine accurately the "gel point"? Can we have a single method for all systems, or can we choose the most convenient way to describe our own particular system? There are various ways of defining the gel point from both theoretical and experimental viewpoints.

It was Paul Flory who first developed a quantitative theory for the linear and nonlinear end-linking reactions of small molecular weight species. He was able to derive the equations relating both the number- and weight-average molecular mass to the degree of conversion p in these reactions. Work by Flory and Stockmayer [116–118] identified that, unlike the linear systems, for nonlinear reactions, when there are more than two potential reactive groups, very high molecular weights are achieved, and for a symmetric system the Flory–Stockmayer theory predicts that M_w will diverge to infinity following the relation

$$DP_w = \frac{1+p}{1-p(f-1)} \tag{9.5}$$

where f is the functionality of the cross-links. The denominator of this equation becomes zero when the product $p(f-1)=1$, and this marks the so-called Flory–Stockmayer gel point, so that p_c, the critical degree of conversion, is given by

$$p_c = \frac{1}{f-1} \tag{9.6}$$

where the molecular weight becomes infinite. In most experiments, however, it was found that measured values of p_c were greater than those predicted, and it was realized that the main reason for this was that, besides intermolecular bonds, a small proportion of intramolecular reactions takes place forming cycles.

This is one of the major limitations to this approach. New approaches came from physicists in the mid-1970s. The analogy between gelation and percolation was suggested simultaneously by Gennes [97] and Stauffer [119], and provided a general framework encompassing the phenomenon independently of microscopic details.

The percolation model is one of the theoretical techniques that deal with disordered systems. Percolation deals with dramatic changes of the connectivity of systems of infinite size around the so-called percolation threshold. The term *percolation* was coined by the mathematician Hammersley [120] to a statistical geometric model, which reminded him of the passage of a fluid through a network of channels in which a proportion of the channels randomly distributed were blocked, as symbolized by a traditional Italian espresso pot. Another popular image is illustrated by the "vandalized grid". A two-dimensional grid, with regular square lattice is shown in Figure 9.10a. The opposite sides of the resistor grid are connected to a voltage source, and an electric current flows in the circuit. The saboteur starts cutting the grid in random places, and the current drops progressively as more and more paths are disconnected. This is shown in Figure 9.10b, and continues until the moment when a certain critical fraction of bonds are cut. At the stage determined as $p = p_c$, where p is the fraction of uncut (connected) bonds, the electric current stops. This is the percolation threshold. No current flows below the percolation threshold.

Cutting the bonds above the threshold weakens the mechanical strength of the grid and the transport properties.

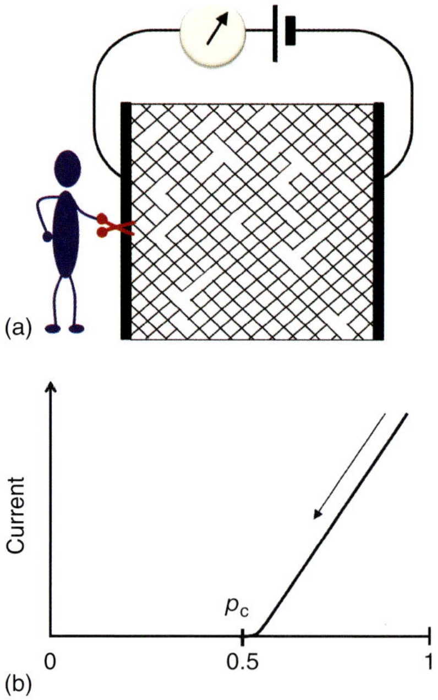

Figure 9.10 Electric current flowing in the network composed of a large number of resistors forming a square lattice grid. Some connections are cut randomly. (a) Vandalized grid. (b) Electric current deceases with the fraction of open connections. There is la critical value, $p_c = 0.5$ in this case, at which the current stops flowing. There is no percolating pathway between the two sides of the grid below p_c. The critical value depends on the lattice geometry and space dimension.

The percolation transition have been simulated repeatedly on 2D and 3D lattices: the simulations are generally performed by increasing the connections or by filling at random pre-existing sites on regular lattices. As the fraction of connections or filled sites increases, they form larger and larger clusters, whose geometrical size increases. At some stage, one infinite cluster appears, coexisting with smaller ones. This stage is called the *percolation threshold*. The value of p_c depends on the lattice geometry and the space dimensionality.

De Gennes and Stauffer recognized the analogy between permanent cross-link formation in a chemical reaction and bond formation in numerical simulations: the extent of the reaction, p, would be the probability of bond formation; the branched macromolecules would be the clusters of connected sites; the cross-link's functionality, f, is the coordination number of the sites. Then, the gel point is equivalent to the percolation threshold. The infinite cluster is the macroscopic polymer network spanning all over the solvent.

Similar to second-order phase transitions, thoroughly investigated by physicists in thermodynamics, it was shown by computer simulations and theoretical arguments that in the close vicinity of the percolation threshold some properties of the clusters would follow universal laws. Among these are the average size of the clusters and the electrical conductivity, versus the relative distance to the threshold, as determined by $|p-p_c| \rightarrow 0$.

As a consequence of this analysis, under certain conditions it is predicted that the viscosity of the polymer solutions containing highly branched macromolecules (sol state) tends to be infinite when approaching the threshold from below p_c, following a power law versus the distance to threshold with an exponent called k:

$$\eta \sim (p_c - p)^{-k} \quad p < p_c \quad p_c - p \rightarrow 0 \tag{9.7}$$

whereas at the threshold the elastic modulus (storage or Young's modulus) would start from zero and increase with a different power law, with the exponent called t:

$$E \sim (p - p_c)^{-t} \quad p > p_c \quad p - p_c \rightarrow 0 \tag{9.8}$$

The experimental evidence for these predictions was more complex to establish than was expected. However, the framework provided by the percolation models has generated very active research efforts in view of elucidating the unique features of the sol–gel transition and predicting the parameters that could control the formation of the gels.

When dealing with physical gels, because the cross-linking mechanisms are more difficult to establish, it was anticipated that the analogy with the percolation transition should not be evident. It was, however, found quite unexpectedly that gelatin gelation could be interpreted in the light of the percolation model, provided the mechanisms of gelation were properly analyzed (see below).

In the mid-1980s, Winter and Chambon [121] investigating the end-linking reaction of poly(dimethylsiloxane) PDMS chains were able to stop the chemical reaction at intermediate times of conversion and observed that the spectra of the shear storage G' and the loss G'' moduli of their system, at some stage of the conversion, exhibited a power law behavior, over the entire measurable angular frequency (ω) domain. They then proposed to define the gel point as a particular stage at which the frequency spectrum of the storage and loss moduli follow the same power law with exponent $n < 1$ over the infinite frequency domain:

$$G' \approx \omega^n \quad \text{and} \quad G'' \approx \omega^n \quad 0 < \omega < \infty \tag{9.9}$$

In this case, the loss angle $\delta(\omega)$ is independent of the frequency and is defined by

$$\tan\delta(\omega) = \frac{G''(\omega)}{G'(\omega)} = \tan\frac{n\pi}{2} \quad n < 1 \tag{9.10}$$

By generalizing their analysis, they envisaged three cases to determine the gel point with their criteria according to the value of the exponent n:

$$n < {}^1\!/_2, \quad G'' < G'$$
$$n > {}^1\!/_2, \quad G'' > G'$$
$$n = {}^1\!/_2, \quad G'' = G'$$

In the particular case where $n = 1/2$, $G' = G''$, the gel point is easily located by measurements at any frequency. This was the case, indeed, when Winter and Chambon investigated the samples with appropriately balanced stoichiometry. However, cases reported in the literature in a number of physical or chemical gels show an exponent $n > 1/2$, which is stage at which the loss modulus is larger than the storage modulus and, therefore, the usual definitions of the gel as a solid-like material would not apply. Therefore, there is no straightforward relation between the Winter–Chambon criteria of gelation and the appearance of solid-like behavior of the solution. The frequency dependence of the storage modulus is not taken into account in the percolation model; the rheological properties should be measured at the lowest possible frequency in order to detect the network formation at its early stages. In any case, the kinetics of gelation has to be slowed down as much as possible to allow time-resolved measurements at low frequencies.

9.2.5
Selected Examples of Gels

9.2.5.1 Gelatin Gels

When a gelatin solution is cooled below room temperature, the protein coils undergo a coil-to-helix transition and a polymer network begins to be formed. The helixes are reminiscent of the triple-helical native structure of tropocollagen. They are randomly oriented and do not further assemble with higher organization levels as native collagen fibers do. The polypeptide chains only partly recover their native conformation, and no real equilibrium is ever reached even after many days. It is therefore very important to relate the fraction of residues in helical conformation to the gel elasticity. Although many factors can take part in the renaturation process in solution (molecular weight of gelatin, temperature, amino acid composition/source of gelatin (e.g., mammalian fish), acid or basic process of extraction setting the iso-electric point, time, etc.), it was found experimentally that the helix concentration is the unique parameter that controls the storage

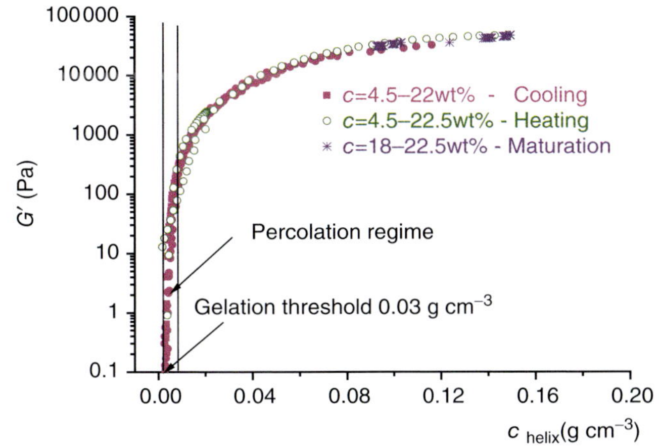

Figure 9.11 Correlation between the storage modulus G' and helix concentration c_{helix} in gelatine gels during various thermal protocols (cooling, heating, and maturation in a thermal range between 5 and 40 °C) and for different protein concentrations. The thermal protocols were identical for both types of measurements. (Adapted from Elharfaoui *et al.* [122]. Reproduced with permission of Wiley.)

(elastic) modulus of gels. The helical fraction can be measured independently by optical rotation or microcalorimetry, whereas the storage and loss moduli are determined by rheological means, with identical experimental protocols. All experiments performed on gelatin could be gathered on a single curve, which relates the storage modulus G' to the mass concentration of triple helixes, c_{helix} (g helixes per cubic centimeter), formed in solution, whatever the conditions in which these triple helixes have been renatured, with no adjustable parameter. The thermal protocols were identical in both types of experiments.

The result is shown in Figure 9.11, and different data could be superposed on this unique curve; the storage modulus is controlled solely by the helix concentration. It is seen that the gelation threshold is obtained at a very low helix concentration $c_{helix} = 0.003\,\mathrm{g\,cm^{-3}}$, and beyond the threshold value the elasticity increases steeply (log scale for the storage modulus G'). Very close to the gel point, the theoretical predictions based on the percolation models are achieved for elasticity (shear modulus) versus helix concentration. This example also shows that, despite the many different experimental conditions and samples, there is an underlying rule for elasticity versus cross-linking density, which was anticipated for disordered systems with percolation models, very close to the gel point. In gelatin gels, this correlation extends over the whole range of concentrations.

9.2.5.2 Gelatin Hydrogels

Rheological and the structural properties of gelatin hydrogels, which are both physically and chemically cross-linked, were investigated using the cross-linker bisvinylsulfonemethyl (BVSM). The vinyl double bonds of the BVSM link to the amine groups of the protein to form the covalent network. In order to prepare the hydrogels with both types of cross-links, three different routes can be adopted, from different thermal treatments:

1) The chemical network is allowed to form first (at high temperature), followed by the physical gelation (by cooling).
2) Physical gel and chemical gel are formed simultaneously (at intermediate temperatures).
3) Physical gelation takes place first (quench at low temperature), followed by chemical gelation (going back to high temperatures).

If the gelatin and reactant concentrations are kept constant, it is shown [123, 124] that the elasticity is strongly modified by changing the sequence in which gels could form, with only the thermal protocols. It was deduced that the cross-linker would preferentially react with the helical structures when these are already formed by physical gelation, and a heterogeneous network giving a higher modulus (by a factor of 10 for the storage moduli) is formed when the helixes are melted out (Figure 9.12). It was found, when both physical and chemical cross-links are present, that the contribution to the shear modulus due to physical cross-links (the helical structure) is clearly the dominant one. This should be the case for most hydrogels prepared with polysaccharide or collagen biopolymers.

9.2.5.3 Poloxamer Gels

There is a very important class of associating polymers that exhibit large thermal effects resulting from the hydrophobic interactions between long sequences of copolymers that promote gelation. Thermogelling systems, based on neutral triblock copolymers with hydrophobic cores, such as PEO_n-PPO_m-PEO_n (PEO and poly(oxypropylene) PPO), are well known in pharmaceutical applications. These

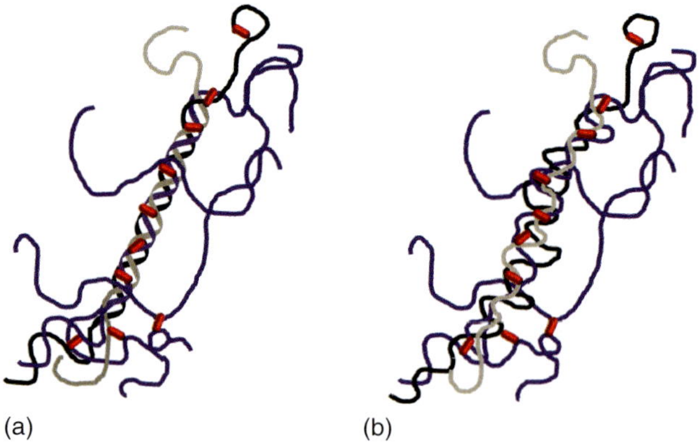

Figure 9.12 (a) Schematic structure of gelatine hydrogels cross-linked at low temperature with BVSM. The chemical cross-links may preferentially form on the existing helical structure (b) After heating, the hydrogen bonds of the triple helixes are broken, but the chains remain cross-linked and the structure behaves like a rigid sequence. (Adapted from Hellio-Serughetti and Djabourov [124]. Reproduced with permission of American Chemical Society.)

copolymers called Pluronics® exhibit temperature-dependent micellization and gelation in aqueous solutions. Because of the marked change in water solubility of the central PPO block, these copolymers form various aggregates depending on the degree of polymerization of each block. A number of triblock copolymers of this family have been shown to aggregate in the form of micelles with a core dominated by PPO and a corona dominated by hydrated PEO blocks [72]. Pluronics® with a high PPO content form spherical micelles at low concentrations and low temperatures: the primary factor in the micellization process of Pluronics® is PPO. Numerous studies deal with rather dilute solutions ($c \sim 1\%$). Since we are concerned here with the gelation phenomenon, solutions with larger polymer concentrations are considered.

Gelation of this type of polymers is basically different from those described with percolation models. Pham Trong *et al.* [125] analyzed the correlation between micelle formation and rheological changes of solutions of F127® ($n = 98$ and $m = 67$) with increasing temperature. Microcalorimetry is a very sensitive method for detecting micellization: the signal is endothermic during heating and perfectly reversible on cooling.

Figure 9.13 shows both the rheological measurements and the heat flow measured during the heating ramp of a solution of F127® at a concentration $c = 20$ wt%. At the end of the enthalpic peak, a small endothermic peak is observed, which corresponds to the crystallization temperature of the micelles. The figure shows the correlation between enthalpy and rheological measurements. Three domains appear, which correspond to three microscopic structures. In region 1 ($T < 10\,°C$), the polymers are well dispersed, and the overall viscosity decreases as temperature increases, meaning that the polymer is less swollen, its hydrodynamic volume is decreased, and so the solvent becomes less good. In region 2 ($10 < T < 22\,°C$), progressive micellization takes place, the molecules start to assemble, and the viscosity increases, with no measurable storage modulus. The micelles repel each other forming grafted brushes on colloidal particles ("hairy grains") or a suspension of hard spheres. Region 3 ($T > 22\,°C$) corresponds to a colloidal crystal of spherical micelles [126].

For this system, gelation is the transition from a colloidal suspension of micelles, which behave as hard spheres, to a colloidal crystal formed of these micelles. For lower polymer concentrations, the crystallization peak is much smaller; possibly, the solution is partially structured into ordered domains, and rheological measurements show a less steep temperature variation. In the latter case, the solution becomes a viscoelastic liquid. This is also the case for mixtures of Pluronics such as F68® and F127® (Zhang *et al.* [127]). The shear moduli of the crystalline phase show little frequency dependence. The origin of the elasticity in the colloidal crystal phase can be attributed either to the interfacial tension between water and PPO, by analogy with an emulsion system [128], or to the entanglement of PEO chains [129], while the loss modulus could be related to friction between micelles. There is no quantitative estimation of these contributions, and this is still an open field for investigation.

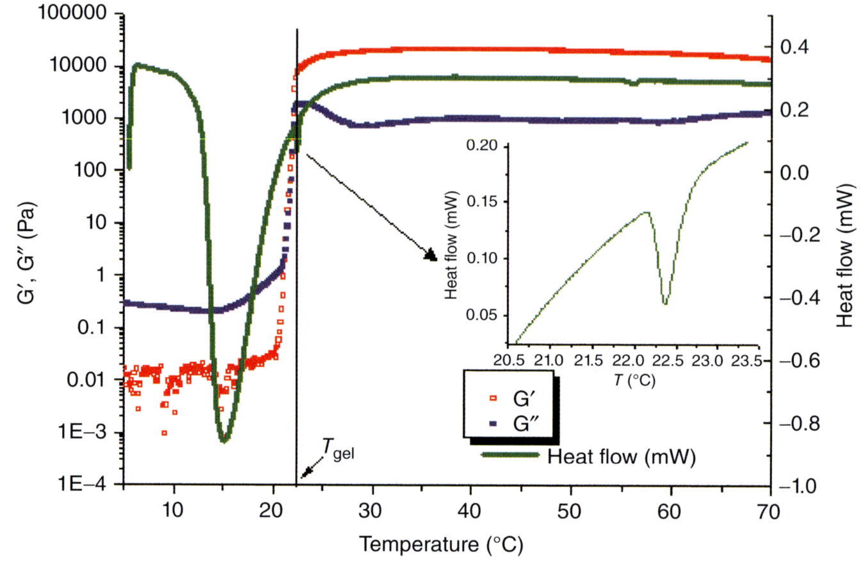

Figure 9.13 Heat flow and rheological measurements during a heating ramp of F127® aqueous solution at concentration $c = 20$ wt%. The gelation temperature T_{gel} coincides with the formation of a colloidal crystal. The solution containing micelles is viscous, but not viscoelastic, as the storage modulus G' is very low. The micelles behave as repulsive hard spheres. (Adapted from Pham Trong *et al.* [125]. Reproduced with permission of Elsevier.)

After this short overview of the complex systems named gels and hydrogels, we analyze a different class of materials, namely scaffolds, and compare them with the gels.

9.3 Scaffolds

Polymer scaffolds mimic many roles of the natural extracellular matrices that regulate the functions of cells and allow the diffusion of nutrients, metabolites, and growth factors [130]. For tissue engineering, specific cells must be isolated from a small tissue biopsy from the patient and incorporated into the polymer scaffold: the surgeon makes incisions that enable placement of the polymer/cell assembly. As already mentioned, fabrication of scaffolds needs a process for generating highly interconnected networks of pores, with size distributions ranging from 50 to 150 μm. Scaffolds ideally should disappear after the tissue is restored and be eliminated within the time frame of tissue rebuilding.

On the whole, the design of scaffolds for tissue engineering should answer several demands (according to Sokolsky-Papkov *et al.* [131]).

1) They should have *mechanical properties* matching those of the tissue at the implantation site or properties that are sufficient to protect the cells from damage (e.g., under compressive or tensile forces).
2) They must possess acceptable *biocompatibility* and *low toxicity.*
3) *Cells* or *proteins* should *attach* on scaffold surface.
4) The scaffold should have a highly *porous structure.*
5) They should mimic the native extracellular matrix, providing signals that *promote cellular development* and morphogenesis.
6) The polymers should be *biodegradable* and capable of being eliminated. The space initially occupied by the scaffolds should be replaced by the grown tissue.
7) They must have the ability to control the *release of growth factors* over the period of tissue regeneration.

There is a significant challenge in the design and manufacture of scaffolds: the list of requirements is even longer [131]. Scaffolds should possess additional requirements regarding the *drug release*:

- Sufficient *loading capacity*: defined as the amount of drug that can be mixed into the scaffold
- *Homogeneous drug distribution*: the drug needs to be dispersed homogenously throughout the scaffold
- *Adequate binding affinity*: how tightly the drug should bind the scaffold to allow release
- *Controlled release kinetics*: which allows the appropriate dose of drug to reach the cells over a given period
- *High stability*: drugs need to maintain their structure and activity over a prolonged period at physiological temperature.

In this long list of physical characteristics of the scaffolds (points 1 and 4), chemical composition of the polymers (points 2, 3, and 6), and biological functions (points 5 and 7) should all be appropriate. Pore size distribution, which is an important factor for selective cell culture, should be tailored. The pore sizes needed for cell growth are much larger than in hydrogels. The therapeutic and biological exchanges should be validated by *in vitro* experimentation. Obviously, materials selected for the preparation of scaffolds for cells growth depend on the application [130].

In general, phase transformations in polymer solutions can lead to coarse network formation under specific circumstances [132]:

1) When the phase transformation is incomplete: for instance, a crystalline polymer fraction co-exists with an amorphous part
2) When the newly formed macroscopic phase does not fully develop because of kinetic constraints
3) When connectivity is achieved either between individual molecules (gels) or by the phase morphology, not specifically related to the phase transformation itself.

Common approaches in the creation of biocompatible scaffold matrices are those in which the polymer is precipitated by a freeze-drying process (cryogels), where phase separation of liquid/liquid type takes place by gas expansion, particle embedding, or electrospinning methods.

Some of these methods and their typical porous structures are briefly presented below.

9.3.1 Thermally Induced Phase Separation (TIPS) with Solvent Crystallization: Cryogels

A well-known example is PVA. PVA is a water-soluble synthetic polymer with a high degree of hydrolysis (98–99%) and is considered to be atactic. Physical cryogels prepared by repeated freezing and thawing cycles of an aqueous solution of PVA have captured the attention of both academic and industrial researchers ever since the pioneering work of Inoue [133] and Peppas [134] because of the potential applications of such gels in many fields. Because they are biocompatible, PVA hydrogels are suitable for a variety of biomedical and pharmaceutical applications (artificial tissues, contact lenses, controlled-release devices for drug delivery). Freeze/thaw PVA hydrogels show high mechanical strength and good elastic properties since they are able to support large deformations upon stretching and to recover their original shape and dimensions on releasing this tension.

PVA is dissolved in water at high temperature, but when cooled to room temperature the solutions do not gel and remain transparent for more than a month (c ~10 w/w%) [135]. The elastic properties of cryogel films are obtained only by subjecting the aqueous solutions of the polymer to several repeated freeze/thaw cycles, consisting of several hours (20 h) of freezing steps at −22 °C followed by shorter (4 h) thawing steps at room temperature. The elasticity and tenacity of the gel increase with the number of cycles. Also, with increasing freeze–thaw cycles, the original cryogels become more turbid and almost nontransparent above three cycles, suggesting the development of a micro phase-separated structure [81]. In parallel with changes of elasticity, crystalline reflections become more distinct. Traces of crystal growth of ice are left as pores in the cryogel, as suggested by scanning electron micrographs of the xerogels (dry gels). Pores, of ~10 μm, are oriented along a direction nearly normal to the freezer plate. Under freezing, the ice crystals grow along the direction of temperature gradient, the PVA-rich solution phases being segregated around them, and the gelation proceeds in the segregated solution phases, which then forms the continuous porous gel skeleton [81]. Repeated freezing stages are nothing but repeated recrystallizations and cycling, a kind of refinement process for ice crystals. Therefore, it is expected that polymer chains will be rejected from the growing ice crystals, as an "impurity," whenever the PVA–water system is frozen, giving rise to the porous structures. It seems that the pore size (or the size of ice crystals) increases when lowering the PVA concentration. This trend is consistent with the mechanism in which diffusion plays a major role because on lowering the concentration of PVA, it will become easier to exclude chains from the ice crystal growth front and a larger ice crystal size is then expected. Therefore, network formation can be attributed to kinetically frustrated crystallization. The average crystallite size remains small, in the range of a few nanometers.

Hassan and Pepppas [82] examined the long-term (i.e., over several months) morphology changes of PVA on *swelling* in water at 37 °C as a function of the preparation conditions. They observed that, during swelling, a significant fraction of PVA chains dissolved in the solution. PVA is not degradable in most physiological situations. Therefore, these gels are mostly likely useful as a long-term or permanent scaffold (regeneration of artificial articular cartilage, hybrid-type artificial pancreas, and bone-like apatite formation). Oligopeptide sequences have been introduced onto the surface of PVA gels to enhance cellular interaction [130].

Recently, the cryogel method was developed for the preparation of polyester scaffolds in an organic solvent (poly(lactic acid) PLA and 1,4-dioxane as solvent). The freezing point of the pure solvent is close to 12 °C, which decreases by several degrees with increase in polymer concentration (cryoscopic effect). The glass transition temperature of the bulk polymer is between 50 and 65 °C depending on its molecular weight. The process reported by Schugen *et al.* [136] consists of first preparing a homogeneous solution, and then cooling it rapidly to liquid nitrogen temperature. In the intermediate range of polymer concentration (5–10 wt%), a bicontinuous phase-separated material gives rise to a network of interconnected pores after sublimation. Then the freeze-drying step proceeds at 0 °C during several days, allowing sublimation of the solvent. Then the solid sample (with no solvent left) is let slowly to warm to room temperature; this temperature does not exceed the glass transition temperature of the amorphous PLA and no shrinkage of the porous structure occurs upon solvent removal. The pore size distribution is centered on a diameter of 10 μm. These scaffolds support solvent removal while the hydrogel would have been totally collapsed.

9.3.2
Spinodal Decomposition in Ternary Systems: Polymer/Solvent/Non-Solvent

Preparation of open macroporous scaffolds, ranging in size from 20 to 300 μm, was worked out on the concept of arresting the morphology of the *spinodal phase separation* stage during *liquid–liquid demixing* of a poly(glycolic acid) PLA, or copolymers poly(lactide-*co*-glycolide) of PLA and PGA (PLGA) dioxane/water ternary systems. These polymers have a long history in medical applications as degradable synthetic suture materials [137] and are considered safe by the Food and Drug Administration. They can be dissolved in binary 1,4-dioxane/water mixtures with selected compositions and polymer concentrations up to 12 wt%. The cloud point temperatures of these mixtures were found to depend on the polymer concentration, solvent composition, and polymer molecular weight. A dramatic increase of the cloud point temperature with water content indicated that liquid–liquid demixing depends mainly on the content of non-solvent liquid in the mixture. Solutions are prepared first and heated about 15 °C above the cloud point temperatures, rapidly quenched at different temperatures below

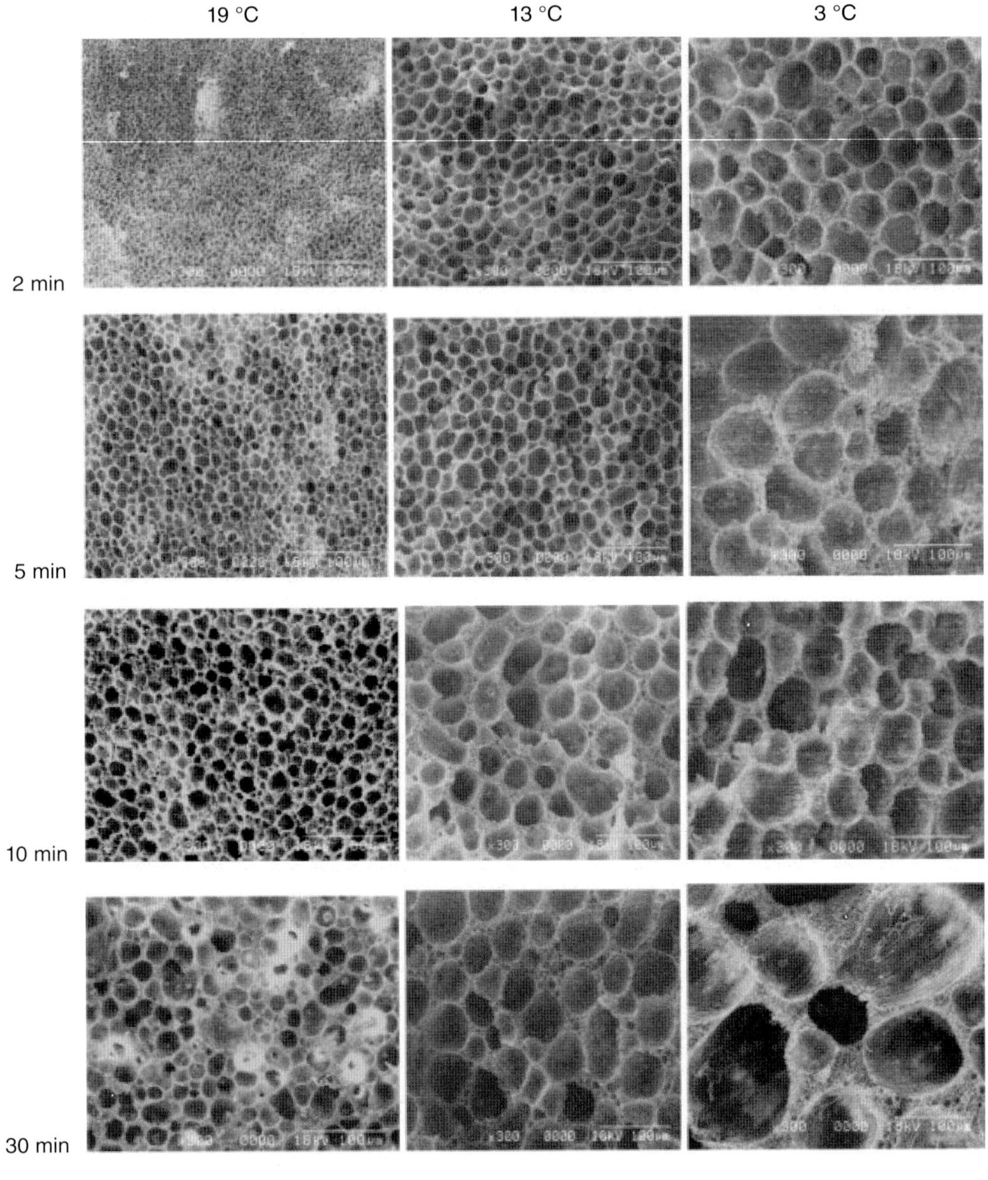

Figure 9.14 SEM micrographs of scaffolds prepared by quenching a PLGA solution ($c = 9$ wt%) in dioxane/water mixture at three different temperatures as a function of time. The scale bar is 100 μm. (Adapted from Hua *et al.* 2003 [5]. Reproduced with permission of Elsevier.)

cloud point temperatures, and then kept for various periods of time to anneal the phase-separated structures. Then the samples were frozen under vacuum and the solvent allowed to sublimate (freeze drying) to obtain the scaffolds before SEM observations.

In Figure 9.14, open porous structures are observed during the early stage (2 min) of the phase separation. The porous structure is controlled by the quenching temperature and the annealing time. Quenching temperatures close to the cloud point (19 °C, first column) generate fine porous structures that grow with time, but remain under 20 μm. With larger quenching depths (13 °C, second column), the initial pore size (measured at the same time) is larger because the kinetics of phase separation is more rapid. The liquid–liquid phase separation in the unstable region proceeds rapidly and enters the coarsening phase via a droplet coalescence mechanism in which the thermodynamic driving force comes from the interfacial tension. After 30 min of aging at 3 °C (third column), the pores develop into large, closed, and irregular structures with sizes of more than 150 μm.

Deeper quenching temperatures were reported by Hua *et al.* [5], which allow solvent crystallization; water or dioxane shows separate melting peaks, generating a more complex porous structure via the nucleation and growth of the two types of crystals. This investigation illustrates the intricate interplay between polymer molecular weight, solution concentration, phase diagram, and processing conditions for preparation of scaffolds with controlled porosity by thermodynamic phase separation.

9.3.3
Spinodal Decomposition in Organic–Inorganic Composite Scaffolds

Biodegradable polyesters such as PLA, PGA, and PLGA have been widely used for the preparation of 3D scaffolds for bone tissue engineering applications because of their good biocompatibility. Most of them, however, have poor cell affinity, which limits their application in tissue engineering, and are too flexible and too weak to meet the mechanical strength requirement. Composite scaffolds with PLGA and nanosized bioceramic nano-hydroxyapatite (NHA) dissolved in a mixed solvent of distilled water and 1,4-dioxane could be fabricated by the thermally induced phase separation (TIPS) technique. The dispersion of the bioceramic was performed by ultrasonication, and the polymer was added subsequently. The effect of solvent composition, polymer concentration, coarsening time, and temperature, as well as the NHA content, on the microstructure and properties of scaffolds was investigated [138]. Optimum PLGA/NHA ratio was found to be 95 : 5 with 10% (w/v) PLGA in the mixed solvent. The protocols of scaffold fabrication are similar to those of Hua *et al.* [5], but the structure of the foam changes considerably with NHA content.

Figure 9.15 shows the SEM images of scaffolds coarsened for 4 h at different temperatures prepared from 10% (w/v) PLGA/NHA/dioxane/water mixtures at different quenching temperatures. They were freeze-dried before SEM observations. A

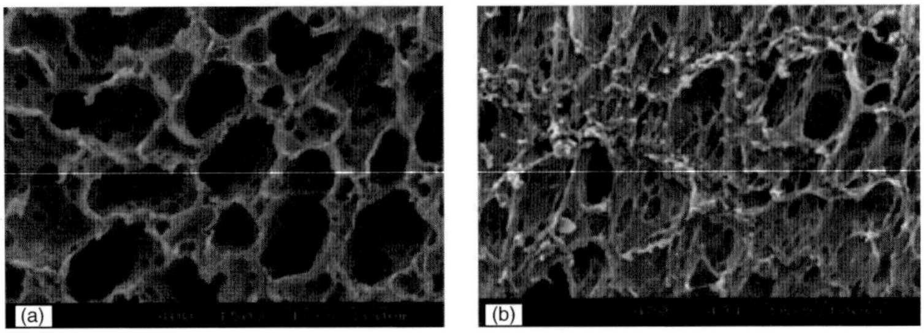

Figure 9.15 SEM micrographs of PLGA/NHA scaffolds prepared in mixed dioxane/water solvent by quenching at (a) 9 °C and (b) 5 °C. (Adapted from Huang *et al.* [138]. Reproduced with permission of Sage Publications.)

macroporous, uniform "foam" (>50 μm) is observed by thermal quenching at 9 °C, whereas the scaffold quenched at 5 °C shows irregular and small pores, which is the opposite of what is seen in Figure 9.14 with no NHA. From these images, it was deduced that NHA particles disturb the coalescence of the droplets during the coarsening process and, therefore, reduce significantly the pore sizes.

The benefit expected is that NHA would greatly increase the mechanical properties and water absorption ability: the compression strength is increased by a factor of 3, water absorption by a factor of 3 or 4, while volume fraction of pores (void fraction) is decreased from greater than 90% to 60% when the concentration of the bioceramic reached 25%. Cell adhesion and proliferation in the scaffold increased with the NHA content. These results suggest that PLGA/NHA composite scaffolds may serve as an excellent 3D substrate for cell attachment and migration in bone tissue engineering.

9.3.4
Supercritical Processing

Supercritical fluids SCFs have densities and solvating properties similar to those of liquids, while diffusion and viscosity properties are those of a gas. Because of its low toxicity, ease of use, and cost, CO_2 is the most common candidate. The technique normally involves the *dissolution* of the polymer (and growth factors) in supercritical CO_2, followed by a rapid expansion of the mixture into a low temperature and pressure environment, which leads to scaffold formation. The benefits of using these process methodologies are due to the ability to form growth-factor-loaded scaffolds by avoiding the use of organic solvents and harsh process conditions [131].

Another technique utilizing the SCF involves using CO_2 to *plasticize* the polymer by lowering the glass transition temperature of the polymer. This technique involves the application of the SCF under pressure into the polymer and growth factor mixture until the mixture is saturated with CO_2 and the polymer is devitrified. This is followed by depressurization through a nozzle, leading to the formation of highly porous scaffolds as the gas comes out of the liquefied polymer. This process has the advantages that the starting material does not have to be soluble in the SCF and that no organic solvents are required during processing [139]. Upon the release of the pressure, pores of typical diameters between 50 and 400 μm are formed [131]. Mooney *et al.* [140] first showed the possibility of forming scaffolds with PLA or PLGA using CO_2.

9.3.5
Other Scaffold Manufacturing Techniques

Several other techniques been reported to produce scaffolds, suggesting that each method should be selected according to the final requirements. They combine phase changes of the polymer solutions with different processing methods [131].

Porous matrices can be prepared by *embedding soluble ingredients* (porogens) into PLA solutions. Salt particles, paraffin microspheres, and emulsion particles of defined sizes are examples of porogens. This technique involves the pouring of the polymer solution into a bed of salt particles with a defined size. Phase separation (precipitation of the polymer) by evaporation of the solvent under vacuum, followed by leaching of the salt particles in distilled water, leads to the formation of a highly porous scaffold with well-defined pores. The polymer solidifies around the porogen particles. Organic solvents are used with paraffin particles [141]. The manufactured scaffolds using leaching techniques show an internal architecture with a designed pore lattice (>100 μm) globally interconnected, coexisting with size-distributed local voids (10–300 μm size).

Fiber-based scaffolds have been recently prepared by electrospinning. Electrospinning is the process of electrostatic spraying that occurs when electrical forces at the surface of the polymer solution overcome the surface tension, forming a polymer solution jet. Electrospinning is a useful technology for the generation of fine fibers, ranging from 15 nm to 10 μm or greater. The fibers are collected in a nonwoven mesh characterized by high surface area/unit mass. A large fraction of the volume of such nonwoven meshes is in the form of interconnected pores [142]. Solvent volatility is an important factor. This method uses organic solvents and has negative influence on the cell culture. Subsequent treatment with vacuum and heating to remove residual solvent well below 100 ppm is necessary to completely eliminate downstream biological effects. Development of this processes, allowing the fabrication of structures with porosity and topography that mimics *in vivo* conditions, is still under investigation.

9.4
Conclusion

These highly disordered networks – gels, hydrogels, or scaffolds – display a vast range of physical and chemical properties and are capable of responding to numerous requirements of the pharmaceutical applications, and have gone so far as to mimic biological systems. Gels, in general, are also present in everyday life, as one can find them commonly on the shelves of food or personal products stores. The fundamental knowledge on polymer gels has progressed in the last decades through elaborate theoretical approaches, new experimental facilities, and model chemical reactions. By exploring deeper using the available experimental tools and by combining several methods, it has become possible to analyze the process of gelation and the resulting macroscopic properties of the gels, by relating, for instance, their structure to their elasticity, although this challenge still remains a complex one. Hydrogels and scaffolds that have been designed for specific biomedical applications were processed in such a way as to meet a long list of demands. Each molecular composition and processing route is able to generate a different system and to tailor its properties, which are far from the equilibrium state. The available scientific literature highlights the intricate relation between the chemical composition of these systems and the process of fabrication, and especially evaluates the ability of the systems to meet the criteria for biological/pharmaceutical functions. Selecting and modeling reference systems could be a goal for a deeper understanding or better prediction of their major properties, but creating new materials is still a vital issue. The aim of this review was to bring together the expertise from different fields and to try to generate new innovative routes and better explain the existing ones.

References

1. Blaine, G. (1946) The uses of plastics in surgery. *Lancet*, **2**, 525–528.
2. Ingraham, F.D., Alexander, E., and Matson, D.D. (1947) Synthetic plastic materials in surgery. *N. Engl. J. Med.*, **236**, 362–368.
3. Griffith, L.G. (2000) Polymeric biomaterials. *Acta Mater.*, **48**, 263–277.
4. Malafaya, P.B., Silva, G.A., and Reis, R.L. (2007) Natural–origin polymers as carriers and scaffolds for biomolecules and cell delivery in tissue engineering applications. *Adv. Drug Delivery Rev.*, **59**, 207–233.
5. Hua, F.J., Park, T.G., and Lee, D.S. (2003) A facile preparation of highly interconnected macroporous poly(d,l-lactic acid-co-glycolic acid) (PLGA) scaffolds by liquid–liquid phase separation of a PLGA–dioxane–water ternary system. *Polymer*, **44**, 1911–1920.
6. Spector, M., Michno, M.J., Smarook, W.H., and Kwiatkowski, G.T. (1978) A high-modulus polymer for porous orthopedic implants: biomechanical compatibility of porous implants. *J. Biomed. Mater. Res.*, **12**, 665–677.
7. Hoffman, A.S. (2002) Hydrogels for biomedical applications. *Adv. Drug Delivery Rev.*, **54**, 3–12, http://www.sciencedirect.com/science/article/pii/S0169409X12002700 (accessed 24 November 2015).
8. Guarino, V., Gloria, A., Raucci, M.G., and Ambrosio, L. (2012) Hydrogel-based platforms for the regeneration of osteochondral tissue and intervertebral disc. *Polymers*, **4**, 1590–1612.
9. Rohan, L.C. and Sassi, A.B. (2009) Vaginal drug delivery systems for HIV prevention. *AAPS J.*, **11**, 78–87.
10. Krsko, P. and Libera, M. (2005) Biointeractive hydrogels. *Mater. Today*, **8**, 36–44.
11. Graham, N.B. and McNeill, M.E. (1984) Hydrogels for controlled drug delivery. *Biomaterials*, **5**, 27–36.
12. Van Vlierberghe, S., Dubruel, P., and Schacht, E. (2011) Biopolymer-based hydrogels as scaffolds for tissue engineering applications: a review. *Biomacromolecules*, **12**, 1387–1408.

13. Graham, T. (1861) Liquid diffusion applied to analysis. *Philos. Trans. R. Soc. London*, **151**, 183–224.
14. Hermans, P.H. (1949) in *Colloid Science*, vol. **2** (ed H.R. Kruyt), Elsevier Publishing Company Inc., pp. 483–651.
15. Mülhaupt, R. (2004) Hermann Staudinger and the origin of macromolecular chemistry. *Angew. Chem. Int. Ed.*, **43**, 1054–1063.
16. Flory, P.J. (1974) Introduction. *Faraday Discuss.*, **57**, 7–18.
17. Stepto, R.F.T. (1998) *Polymer Networks: Principles of Their Formation, Structure and Properties*, Chapman & Hall.
18. Iza, M., Stoianovici, G., Viora, L., Grossiord, J.L., and Couarraze, G. (1998) Hydrogels of poly(ethylene glycol): mechanical characterization and release of a model drug. *J. Controlled Release*, **52**, 41–51.
19. Nair, L.S. and Laurencin, C.T. (2007) Biodegradable polymers as biomaterials. *Prog. Polym. Sci.*, **32**, 762–798.
20. Tessmar, J.K. and Göpferich, A.M. (2007) Customized PEG-derived copolymers for tissue-engineering applications. *Macromol. Biosci.*, **7**, 23–39.
21. Kofinas, P., Athanassiou, V., and Merrill, E.W. (1996) Hydrogels prepared by electron irradiation of poly(ethylene oxide) in water solution: unexpected dependence of cross-link density and protein diffusion coefficients on initial PEO molecular weight. *Biomaterials*, **17**, 1547–1550.
22. Nguyen, K.T. and West, J.L. (2002) Photopolymerizable hydrogels for tissue engineering applications. *Biomaterials*, **23**, 4307–4314.
23. Elouali, F.Z. and Maschke, U. (2011) Kinetics and equilibrium swelling properties of hydrophilic polymethacrylate networks. *Macromol. Symp.*, **303**, 71–77.
24. Djabourov, M., Nishinari, K., and Ross-Murphy, S.B. (2013) *Physical Gels from Biological and Synthetic Polymers*, Cambridge University Press.
25. Djabourov, M., Leblond, J., and Papon, P. (1988) Gelation of gelatin aqueous solutions. I. Structural investigation. *J. Phys. Fr.*, **49**, 319–332.
26. Herning, T., Djabourov, M., Leblond, J., and Takerkart, G. (1991) Conformation of gelatin chains in aqueous solutions: 2. A quasi-elastic light scattering study. *Polymer*, **32**, 3211–3217.
27. Pezron, I., Djabourov, M., and Leblond, J. (1991) Conformation of gelatin chains in aqueous solutions: 1. A light and small-angle neutron scattering study. *Polymer*, **32**, 3201–3210.
28. Joly-Duhamel, C., Hellio, D., and Djabourov, M. (2002) All gelatin networks: 1. Biodiversity and physical chemistry. *Langmuir*, **18**, 7208–7217.
29. Arnott, S. *et al.* (1974) The agarose double helix and its function in agarose gel structure. *J. Mol. Biol.*, **90**, 269–272.
30. Norton, I.T., Goodall, D.M., Austen, K.R.J., Morris, E.R., and Rees, D.A. (1986) Dynamics of molecular organization in agarose sulphate. *Biopolymers*, **25**, 1009–1029.
31. Djabourov, M., Clark, A.H., Rowlands, D.W., and Ross-Murphy, S.B. (1989) Small-angle X-ray scattering characterization of agarose sols and gels. *Macromolecules*, **22**, 180–188.
32. Watase, M., Nishinari, K., Clark, A.H., and Ross-Murphy, S.B. (1989) Differential scanning calorimetry, rheology, X-ray, and NMR of very concentrated agarose gels. *Macromolecules*, **22**, 1196–1201.
33. Bulone, D. and San Biagio, P.L. (1991) Microgel regions in dilute agarose solutions: the notion of non-gelling concentration, and the role of spinodal demixing. *Chem. Phys. Lett.*, **179**, 339–343.
34. Schafer, S.E. and Stevens, E.S. (1995) A reexamination of the double-helix model for agarose gels using optical rotation. *Biopolymers*, **36**, 103–108.
35. San Biagio, P.L., Bulone, D., Emanuele, A., Palma-Vittorelli, M.B., and Palma, M.U. (1996) Spontaneous symmetry-breaking pathways: time-resolved study of agarose gelation. *Food Hydrocolloids*, **10**, 91–97.
36. Aymard, P. *et al.* (2001) Influence of thermal history on the structural and

mechanical properties of agarose gels. *Biopolymers*, **59**, 131–144.
37. Smidsrød, O., Andresen, I., Grasdalen, H., Larsen, B., and Painter, T. (1980) Evidence for a salt-promoted 'freeze-out' of linkage conformations in carrageenans as a prerequisite for gel-formation. *Carbohydr. Res.*, **80**, C11–C16.
38. Morris, V.J. and Belton, P.S. (1980) Specific ion effects in ι-carrageente gels. *J. Chem. Soc., Chem. Commun.*, 983–984. doi: 10.1039/C39800000983
39. Norton, I.T., Morris, E.R., and Rees, D.A. (1984) Lyotropic effects of simple anions on the conformation and interactions of kappa-carrageenan. *Carbohydr. Res.*, **134**, 89–101.
40. Zhang, W., Piculell, L., and Nilsson, S. (1992) Effects of specific anion binding on the helix-coil transition of lower charged carrageenans. NMR data and conformational equilibria analyzed within the Poisson-Boltzmann cell model. *Macromolecules*, **25**, 6165–6172.
41. Viebke, C., Borgström, J., Carlsson, I., Piculell, L., and Williams, P. (1998) A differential scanning calorimetry study of κ-carrageenan in the NaCl/NaI/CsI/CsCl systems and analysis by poisson–boltzmann calculations. *Macromolecules*, **31**, 1833–1841.
42. Ikeda, S., Morris, V.J., and Nishinari, K. (2001) Microstructure of aggregated and nonaggregated κ-carrageenan helices visualized by atomic force microscopy. *Biomacromolecules*, **2**, 1331–1337.
43. Ikeda, S. and Nishinari, K. (2001) 'Weak gel'-type rheological properties of aqueous dispersions of nonaggregated κ-carrageenan helices. *J. Agric. Food Chem.*, **49**, 4436–4441.
44. Draget, K.I., Skjåk Bræk, G., and Smidsrød, O. (1994) Alginic acid gels: the effect of alginate chemical composition and molecular weight. *Carbohydr. Polym.*, **25**, 31–38.
45. Stokke, B.T. *et al.* (2000) Small-angle X-ray scattering and rheological characterization of alginate gels. 1. Ca–alginate gels. *Macromolecules*, **33**, 1853–1863.
46. Campa, C. *et al.* (2004) Biochemical analysis of the processive mechanism for epimerization of alginate by mannuronan C-5 epimerase AlgE4. *Biochem. J.*, **381**, 155–164.
47. Donati, I., Holtan, S., Mørch, Y.A., Borgogna, M., and Dentini, M. (2005) New hypothesis on the role of alternating sequences in calcium–alginate gels. *Biomacromolecules*, **6**, 1031–1040.
48. Fang, Y. *et al.* (2007) Multiple steps and critical behaviors of the binding of calcium to alginate. *J. Phys. Chem. B*, **111**, 2456–2462.
49. De Kruif, K.G. *et al.* (1995) Gelation of proteins from milk. *Faraday Discuss. Chem. Soc.*, **101**, 185–200.
50. Clark, A.H. (1998) *Functional Properties of Food Macromolecules*, 2nd edn, Aspen Publishers, pp. 77–142.
51. Gosal, W.S., Clark, A.H., Pudney, P.D.A., and Ross-Murphy, S.B. (2002) Novel amyloid fibrillar networks derived from a globular protein: β-lactoglobulin. *Langmuir*, **18**, 7174–7181.
52. Bromley, E.H.C., Krebs, M.R.H., and Donald, A.M. (2005) Aggregation across the length-scales in beta-lactoglobulin. *Faraday Discuss.*, **128**, 13–27.
53. Nicolai, T. and Durand, D. (2007) Protein aggregation and gel formation studied with scattering methods and computer simulations. *Curr. Opin. Colloid Interface Sci.*, **12**, 23–28.
54. Hughes, V.A. and Dunstan, D.E. (2009) *Modern Biopolymer Science*, Elsevier, pp. 559–594.
55. Chiti, F. and Dobson, C.M. (2009) Amyloid formation by globular proteins under native conditions. *Nat. Chem. Biol.*, **5**, 15–22.
56. Krebs, M.R.H., Domike, K.R., and Donald, A.M. (2009) Protein aggregation: more than just fibrils. *Biochem. Soc. Trans.*, **37**, 682–686.
57. Van der Linden, E. and Foegeding, E.A. (2009) *Modern Biopolymer Science*, Elsevier, pp. 29–91.
58. Bremer, L.G., van Vliet, T., and Walstra, P. (1989) Theoretical and experimental-study of the fractal nature

of the structure of casein gels. *J. Chem. Soc., Faraday Trans. 1*, **85**, 3359–3372.
59. Roefs, S.P.F.M., De Groot-Mostert, A.E.A., and Van Vliet, T. (1990) Structure of acid casein gels 1. Formation and model of gel network. *Colloids Surf.*, **50**, 141–159.
60. Roefs, S.P.F.M. and Van Vliet, T. (1990) Structure of acid casein gels 2. Dynamic measurements and type of interaction forces. *Colloids Surf.*, **50**, 161–175.
61. Mellema, M., Heesakkers, J.W.M., van Opheusden, J.H.J., and van Vliet, T. (2000) Structure and scaling behavior of aging rennet-induced casein gels examined by confocal microscopy and permeametry. *Langmuir*, **16**, 6847–6854.
62. Mellema, M., Walstra, P., van Opheusden, J.H.J., and van Vliet, T. (2002) Effects of structural rearrangements on the rheology of rennet-induced casein particle gels. *Adv. Colloid Interface Sci.*, **98**, 25–50.
63. Lodge, J.F.M. and Heyes, D.M. (1997) Brownian dynamics simulations of Lennard-Jones gas/liquid phase separation and its relevance to gel formation. *Faraday Trans.*, **93**, 437–448.
64. Lodge, J.F.M. and Heyes, D.M. (1998) Structural evolution of phase-separating model colloidal liquids by Brownian dynamics computer simulation. *J. Chem. Phys.*, **109**, 7567.
65. Nyström, B., Thuresson, K., and Lindman, B. (1995) Rheological and dynamic light-scattering studies on aqueous solutions of a hydrophobically modified nonionic cellulose ether and its unmodified analog. *Langmuir*, **11**, 1994.
66. Durand, A., Hourdet, D., and Lafuma, F. (2000) Thermoassociative graft copolymers: NMR investigation and comparison with rheological behaviour. *J. Phys. Chem. B*, **104**, 9371.
67. Shedge, A.S. *et al.* (2005) Hydrophobically modified poly(acrylic acid) using 3-pentadecylcyclohexylamine: synthesis and rheology. *Macromol. Chem. Phys.*, **206**, 464.
68. Beheshti, N., Zhu, K., Kjoniksen, A.L., and Nyström, B. (2008) Interaction behaviors in aqueous solutions of negatively and positively charged hydrophobically modified hydroxyethylcellulose in the presence of an anionic surfactant. *Colloids Surf., A*, **328**, 79.
69. Nyström, B., Kjoniksen, A.-L., Beheshti, N., Zhu, K., and Knudsen, K.D. (2009) Rheological and structural aspects on association of hydrophobically modified polysaccharides. *Soft Matter*, **5**, 1328.
70. Desbrieres, J., Hirrien, M., and Rinaudo, M. (1998) A calorimetric study of methylcellulose gelation. *Carbohydr. Polym.*, **37**, 145.
71. Desbrieres, J., Hirrien, M., and Ross-Murphy, S.B. (2000) Thermogelation of methylcellulose: rheological considerations. *Polymer*, **41**, 2451.
72. Alexandridis, P., Holzwarth, J.F., and Hatton, T.A. (1994) Micellization of poly(ethylene oxide)-poly(propylene oxide)-poly(ethylene oxide) triblock copolymers in aqueous solutions: thermodynamics of copolymer association. *Macromolecules*, **27**, 2414.
73. Cairns, P., Miles, M.J., Morris, V.J., and Brownsey, G.J. (1987) X-ray fibre-diffraction studies of synergistic, binary polysaccharide gels. *Carbohydr. Res.*, **160**, 411–423.
74. Masakuni, T. (1991) Synergistic interaction between xanthan and tara-bean gum. *Carbohydr. Polym.*, **16**, 239–252.
75. Goycoolea, F.M., Richardson, R.K., Morris, E.R., and Gidley, M.J. (1995) Stoichiometry and conformation of xanthan in synergistic gelation with locust bean gum or konjac glucomannan – evidence for heterotypic binding. *Macromolecules*, **28**, 8308–8320.
76. Miyoshi, E., Takaya, T., Williams, P.A., and Nishinari, K. (1996) Effects of sodium chloride and calcium chloride on the interaction between gellan gum and konjac glucomannan. *J. Agric. Food Chem.*, **44**, 2486–2495.
77. Ridout, M.J., Cairns, P., Brownsey, G.J., and Morris, V.J. (1998) Evidence for intermolecular binding between deacetylated acetan and the glucomannan konjac mannan. *Carbohydr. Res.*, **309**, 375–379.

78. Pai, V.B. and Khan, S.A. (2002) Gelation and rheology of xanthan/enzyme-modified guar blends. *Carbohydr. Polym.*, **49**, 207–216.
79. Nitta, Y. *et al.* (2003) Synergistic gel formation of xyloglucan/gellan mixtures as sudied by rheology, DSC, and circular dichroism. *Biomacromolecules*, **4**, 1654–1660.
80. Kim, B.-S., Takemasa, M., and Nishinari, K. (2006) Synergistic interaction of xyloglucan and xanthan investigated by rheology, differential scanning calorimetry, and NMR. *Biomacromolecules*, **7**, 1223–1230.
81. Yokoyama, F., Masada, I., Shimamura, K., Ikawa, T., and Monobe, K. (1986) Morphology and structure of highly elastic poly(vinyl alcohol) hydrogel prepared by repeated freezing-and-melting. *Colloid Polym. Sci.*, **264**, 595–601.
82. Hassan, C.M. and Peppas, N.A. (2000) Structure and morphology of freeze/thawed PVA hydrogels. *Macromolecules*, **33**, 2472–2479.
83. Doyle, J.P., Giannouli, P., Martin, E.J., Brooks, M., and Morris, E.R. (2006) Effect of sugars, galactose content and chainlength on freeze-thaw gelation of galactomannans. *Carbohydr. Polym.*, **64**, 391–401.
84. Lazaridou, A. and Biliaderis, C.G. (2004) Cryogelation of cereal beta-glucans: structure and molecular size effects. *Food Hydrocolloids*, **18**, 933–947.
85. Jeong, B., Kim, S.W., and Bae, Y.H. (2002) Thermosensitive sol–gel reversible hydrogels. *Adv. Drug Delivery Rev.*, **54**, 37–51.
86. Ramkissoon-Ganorkar, C., Liu, F., Baudyš, M., and Kim, S.W. (1999) Modulating insulin-release profile from pH/thermosensitive polymeric beads through polymer molecular weight. *J. Controlled Release*, **59**, 287–298.
87. Ilmain, F., Tanaka, T., and Kokufuta, E. (1991) Volume transition in a gel driven by hydrogen bonding. *Nature*, **349**, 400–401.
88. Gudeman, L.F. and Peppas, N.A. (1995) pH-sensitive membranes from poly(vinyl alcohol)/poly(acrylic acid) interpenetrating networks. *J. Membr. Sci.*, **107**, 239–248.
89. Lee, J.W. *et al.* (1999) Synthesis and characteristics of interpenetrating polymer network hydrogel composed of chitosan and poly (acrylic acid). *J. Appl. Polym. Sci.*, **73**, 113–120.
90. Zhang, J.-T., Cheng, S.-X., and Zhuo, R.-X. (2003) Poly(vinyl alcohol)/poly(N-isopropylacrylamide) semi-interpenetrating polymer network hydrogels with rapid response to temperature changes. *Colloid. Polym. Sci.*, **281**, 580–583.
91. Gong, J.P., Katsuyama, Y., Kurokawa, T., and Osada, Y. (2003) Double-network hydrogels with extremely high mechanical strength. *Adv. Mater.*, **15**, 1155–1158.
92. Gong, J.P. (2010) Why are double network hydrogels so tough? *Soft Matter*, **6**, 2583.
93. Kopecek, J. (2009) Hydrogels: from soft contact lenses and implants to self-assembled nanomaterials. *J. Polym. Sci., Part A: Polym. Chem.*, **47**, 5929–5946.
94. Okumura, Y. and Ito, K. (2001) The polyrotaxane gel: a topological gel by figure-of-eight cross-links. *Adv. Mater.*, **13**, 485–487.
95. Haraguchi, K. and Takehisa, T. (2002) Nanocomposite hydrogels: a unique organic-inorganic network structure with extraordinary mechanical, optical, and swelling/de-swelling properties. *Adv. Mater.*, **14**, 1120.
96. Hu, X. *et al.* (2010) Preferential adsorption of poly(ethylene glycol) on hectorite clay and effects on poly(N-isopropylacrylamide)/hectorite nanocomposite hydrogels. *Langmuir*, **26**, 4233–4238.
97. Gennes, P.G.D. (1979) *Scaling Concepts in Polymer Physics*, Cornell University Press.
98. Abete, T. *et al.* (2006) Kinetics of bond formation in cross-linked gelatin gels. *J. Chem. Phys.*, **125**, 174903-1–174903-7.
99. Joly-Duhamel, C., Hellio, D., Ajdari, A., and Djabourov, M. (2002) All–gelatin networks: 2. The master curve for elasticity. *Langmuir*, **18**, 7158–7166.
100. Clark, A.H., Richardson, R.K., Ross-Murphy, S.B., and Stubbs, J.M.

(1983) Structural and mechanical properties of agar/gelatin co-gels. Small-deformation studies. *Macromolecules*, **16**, 1367–1374.
101. Carn, F. *et al.* (2012) Biopolymer folding driven nanoparticle reorganization in bio-nanocomposites. *Soft Matter*, **8**, 2930.
102. Milas, M. and Rinaudo, M. (1996) The gellan sol-gel transition. *Carbohydr. Polym.*, **30**, 177–184.
103. Aymard, P., Durand, D., and Nicolai, T. (1996) The effect of temperature and ionic strength on the dimerisation of β-lactoglobulin. *Int. J. Biol. Macromol.*, **19**, 213–221.
104. Aymard, P., Gimel, J.C., Nicolai, T., and Durand, D. (1996) Experimental evidence for a two-step process in the aggregation of β-lactoglobulin at pH 7. *J. Chim. Phys. Phys.- Chim. Biol.*, **93**, 987–997.
105. Langton, M. and Hermansson, A.M. (1992) Fine-stranded and particulate gels of β-lactoglobulin and whey-protein at varying pH. *Food Hydrocolloids*, **5**, 523–539.
106. Stading, M., Langton, M., and Hermansson, A.M. (1992) Inhomogeneous fine-stranded β-lactoglobulin gels. *Food Hydrocolloids*, **6**, 455–470.
107. Kavanagh, G.M., Clark, A.H., Gosal, W.S., and Ross-Murphy, S.B. (2000) Heat-induced gelation of β-lactoglobulin/α-lactalbumin blends at pH3 and pH7. *Macromolecules*, **33**, 7029–7037.
108. Veerman, C., Ruis, H., Sagis, L.M.C., and van der Linden, E. (2002) Effect of electrostatic interactions on the percolation concentration of fibrillar beta-lactoglobulin gels. *Biomacromolecules*, **3**, 869–873.
109. Hong, W., Zhao, X., and Suo, Z. (2010) Large deformation and electrochemistry of polyelectrolyte gels. *J. Mech. Phys. Solids*, **58**, 558–577.
110. Peppas, N.A., Hilt, J.Z., Khademhosseini, A., and Langer, R. (2006) Hydrogels in biology and medicine: from molecular principles to bionanotechnology. *Adv. Mater.*, **18**, 1345–1360.
111. Stenekes, R.J.H. *et al.* (2000) Pore sizes in hydrated dextran microspheres. *Biomacromolecules*, **1**, 696–703.
112. Kim, S. and Chu, C. (2000) Pore structure analysis of swollen dextran-methacrylate hydrogels by SEM and mercury intrusion porosimetry. *J. Biomed. Mater. Res.*, **53**, 258–266.
113. Franssen, O., Vandervennet, L., Roders, P., and Hennink, W.E. (1999) Degradable dextran hydrogels: controlled release of a model protein from cylinders and microspheres. *J. Controlled Release*, **60**, 211–221.
114. Meyvis, T.K.L., De Smedt, S.C., Demeester, J., and Hennink, W.E. (2000) Influence of the degradation mechanism of hydrogels on their elastic and swelling properties during degradation. *Macromolecules*, **33**, 4717–4725.
115. Van Dijk-Wolthuis, W.N.E., Hoogeboom, J.A.M., van Steenbergen, M.J., Tsang, S.K.Y., and Hennink, W.E. (1997) Degradation and release behavior of dextran-based hydrogels. *Macromolecules*, **30**, 4639–4645.
116. Flory, P.J. (1941) Molecular size distribution in three dimensional polymers. I. Gelation. *J. Am. Chem. Soc.*, **63**, 3083–3090.
117. Flory, P.J. (1942) Constitution of three-dimensional polymers and the theory of gelation. *J. Phys. Chem.*, **46**, 132–140.
118. Stockmayer, W.H. (1943) Theory of molecular size distribution and gel formation in branched-chain polymers. *J. Chem. Phys.*, **11**, 45–55.
119. Stauffer, D. (1976) Gelation in concentrated critically branched polymer solutions. Percolation scaling theory of intramolecular bond cycles. *J. Chem. Soc., Faraday Trans.*, **72**, 1354.
120. Hammersley, J.M. (1957) Percolation processes. II. The connective constant. *Proc. Cambridge Philos. Soc.*, **53**, 642.
121. Winter, H.H. and Chambon, F.J. (1986) Analysis of linear viscoelasticity of a crosslinking polymer at the gel point. *J. Rheol.*, **30**, 367.
122. Elharfaoui, N., Djabourov, M., and Babel, W. (2007) Molecular weight influence on gelatin gels: structure,

enthalpy and rheology. *Macromol. Symp.*, **256**, 149–157.
123. Hellio-Serughetti, D. and Djabourov, M. (2006) Gelatin hydrogels cross-linked with bis(vinylsulfonyl)methane (BVSM): 1. The chemical networks. *Langmuir*, **22**, 8509–8515.
124. Hellio-Serughetti, D. and Djabourov, M. (2006) Gelatin hydrogels cross-linked with bisvinyl sulfonemethyl. 2. The physical and chemical networks. *Langmuir*, **22**, 8516–8522.
125. Pham Trong, L.C., Djabourov, M., and Ponton, A. (2008) Mechanisms of micellization and rheology of PEO–PPO–PEO triblock copolymers with various architectures. *J. Colloid Interface Sci.*, **328**, 278.
126. Artzner, F. *et al.* (2007) Interactions between poloxamers in aqueous solutions: micellization and gelation studied by differential scanning calorimetry, small angle X-ray scattering, and rheology. *Langmuir*, **23**, 5085.
127. Zhang, M., Djabourov, M., Bourgaux, C., and Bouchemal, K. (2013) Nanostructured fluids from pluronic® mixtures. *Int. J. Pharm.*, **454**, 599–610.
128. Hvidt, S., Jørgensen, E.B., Schillén, K., and Brown, W. (1994) *J. Phys. Chem.*, **98**, 12320.
129. Lau, B.K., Wang, Q., Sun, W., and Li, L. (2004) Micellization to gelation of a triblock copolymer in water: thermoreversibility and scaling. *J. Polym. Sci., Part B: Polym. Phys.*, **42**, 2014.
130. Lee, K.Y. and Mooney, D.J. (2001) Hydrogels for tissue engineering. *Chem. Rev.*, **101**, 1869–1880.
131. Sokolsky-Papkov, M., Agashi, K., Olaye, A., Shakesheff, K., and Domb, A.J. (2007) Polymer carriers for drug delivery in tissue engineering. *Adv. Drug Delivery Rev.*, **59**, 187–206.
132. Keller, A. (1995) Introductory lecture. Aspects of polymer gels. *Faraday Discuss.*, **101**, 1–49.
133. Inoue, T. (xxxx) Poly(vinyl alcohol) gels.
134. Peppas, N.A. (1975) Turbidimetric studies of aqueous poly(vinyl alcohol) solutions. *Makromol. Chem.*, **176**, 3433–3440.
135. Ricciardi, R. *et al.* (2005) Structural organization of poly(vinyl alcohol) hydrogels obtained by freezing and thawing techniques: a SANS study. *Chem. Mater.*, **17**, 1183–1189.
136. Schugens, C., Maquet, V., Grandfils, C., Jérôme, R., and Teyssie, P. (1996) Biodegradable and macroporous polylactide implants for cell transplantation: 1. Preparation of macroporous polylactide supports by solid-liquid phase separation. *Polymer*, **37**, 1027–1038.
137. Tessmar, J.K. and Göpferich, A.M. (2007) Matrices and scaffolds for protein delivery in tissue engineering. *Adv. Drug Delivery Rev.*, **59**, 274–291.
138. Huang, Y.X., Ren, J., Chen, C., Ren, T.B., and Zhou, X.Y. (2007) Preparation and properties of poly(lactide-co-glycolide) (PLGA)/nano-hydroxyapatite (NHA) scaffolds by thermally induced phase separation and rabbit MSCs culture on scaffolds. *J. Biomater. Appl.*, **22**, 409–432.
139. Quirk, R.A., France, R.M., Shakesheff, K.M., and Howdle, S.M. (2004) Supercritical fluid technologies and tissue engineering scaffolds. *Curr. Opin. Solid State Mater. Sci.*, **8**, 313–321.
140. Mooney, D.J., Baldwin, D.F., Suh, N.P., Vacanti, J.P., and Langer, R. (1996) Novel approach to fabricate porous sponges of poly(D,L-lactic-co-glycolic acid) without the use of organic solvents. *Biomaterials*, **17**, 1417–1422.
141. Taboas, J.M., Maddox, R.D., Krebsbach, P.H., and Hollister, S.J. (2003) Indirect solid free form fabrication of local and global porous, biomimetic and composite 3D polymer-ceramic scaffolds. *Biomaterials*, **24**, 181–194.
142. Lannutti, J., Reneker, D., Ma, T., Tomasko, D., and Farson, D. (2007) Electrospinning for tissue engineering scaffolds. *Mater. Sci. Eng., C*, **27**, 504–509.

10
Use of the Pair Distribution Function Analysis in the Context of Pharmaceutical Materials

Pierre Bordet and Pauline Martinetto

10.1
Introduction

Determining the structural arrangement of small molecule materials, and more specifically of compounds of pharmaceutical interest, has been a source of constant progress in the understanding of the mechanisms governing their effects. Until recently, almost all structural studies were carried out by using single-crystal X-ray diffraction, which is now a mature technique allowing solving and refining structures of very high complexity. However, in the last decade, the use of X-ray powder diffraction has been constantly growing in the area of small-molecule crystallography [1]. The use of powder diffraction was in part motivated by the requirement of pharmaceutical industry to determine the structures of the numerous polymorphs occurring for each type of active molecule. Polymorphism is known to be a general feature of molecular compounds because of the relative weakness of intermolecular bonding in such crystals, and the existence of a large number of different, almost iso-energetic, structural arrangements is expected [2]. Most importantly, different polymorphs may display different physicochemical properties, especially in terms of solubility, curing efficiency, and so on. However, not necessarily all of these polymorphic forms can be prepared as single crystals suitable for X-ray diffraction studies. Crystal growth may require thermodynamic and/or kinetic conditions, type of solvents, pH, and so on, which can be incompatible with the existence of some specific polymorphs. In cases were no single crystals of suitable size can be prepared, X-ray powder diffraction has to be used. Indeed, the need for structure analysis of small-molecule compounds has been a strong incentive for the development of *ab initio* structure solution from X-ray powder diffraction data. For example, this has led to the emergence of innovative algorithms for the direct space approach of structure solution that are well suited to working with molecules, such as simulated annealing and Monte Carlo-based techniques, genetic algorithm, and so on, coupled to a rigid block approach that allows limiting the number of free parameters to those that are chemically meaningful. Although it cannot be considered yet a routine technique, *ab initio* structure solution from powder data for small-molecule

Disordered Pharmaceutical Materials, First Edition. Edited by Marc Descamps.

compounds is undergoing steady progress and is now widely used based on data from synchrotron as well as laboratory sources. Another motivation for the use of powder diffraction is the possibility to study the material *in situ* as a function of an external parameter, such as temperature or pressure, or mixed with excipients as a fully prepared drug. From the same data, by analyzing the diffraction peak shapes, quantitative information can also be obtained on the sample microstructure, for example, the size of the structurally coherent domains or the presence of micro-strains.

In recent years, the search for new drugs has made more and more intense use of various techniques for the preparation of active pharmaceutical ingredients (APIs) and excipients in the form of nanoparticles of even as amorphous compounds, in order to improve the solubility and/or bioavailability of drugs. Among them, physical techniques, such as high-energy milling, spray-drying, dehydration of hydrates, and so on, are widely used and they all may have a strong impact of the structural as well as thermodynamic and physicochemical properties of materials [3]. In general, strong modifications of the compounds can be observed under such treatments: amorphization, polymorphic transitions, recrystallization, changes in aging properties, solubility, and so on. Most of the time, the consequences on the powder diffraction diagram are drastically unfavorable for the application of standard crystallography methods: decrease of the Bragg diffracted intensity and increase of amorphous humps until complete disappearance of the former, marked broadening of the Bragg peaks due to reduction of the size of nanoparticles or of coherent crystalline domains, and broadening due to the effects of strain. Therefore, the somewhat featureless powder diffraction diagram obtained becomes essentially useless for detailed structural and microstructural studies. Its main use will be for a phenomenological estimate of the sample crystallinity from the intensity ratio between the remaining Bragg signal and the amorphous humps. When amorphization takes place, or when the nanoparticle size becomes very small (i.e., a few nanometers), even identification of the compound from its diffraction diagram becomes unfeasible. In practice, both cases (amorphous or nanoparticle), although physically and structurally quite different, are almost unidentifiable from the sole standard diffraction diagram.

We will show here that pair distribution function (PDF) analysis can provide quantitative structural and microstructural information for molecular compounds regardless of their crystalline/amorphous state, and therefore may become a useful tool for pharmaceutical research in the very near future.

10.2 What Is the PDF?

Historically, as described in the books by Guinier [4] or Warren [5], the PDF analysis was used to characterize the local order in amorphous compounds using data from wide-angle X-ray scattering. Indeed, this function is somehow a histogram of interatomic distances in a compound, regardless of its crystalline or amorphous state, and yields the probability of finding an atom at a distance r from another atom.

Experimentally, it is obtained by Fourier transform of a powder diffraction diagram. For a long time, the main limitation was the possibility of obtaining diffraction data up to large $Q=4\pi\sin(\theta)/\lambda$, to limit the effects of cut-off in the Fourier transform. This requires working with short wavelengths less accessible to laboratory instruments, and has long limited the use of the PDF to the field of amorphous materials and liquids. More recently, the PDF analysis took a new development thanks to its usefulness in the study of local structural effects and nanomaterials to the emergence of instruments well suited for PDF measurements at synchrotron and neutron sources and to the development of user-friendly software to process and analyze the data.

Depending on the authors and the communities, there are different definitions of the PDF. We will not go into the details here, and the interested reader can refer to the article by Keen [6], where the relationships between different definitions are explained. We will here follow the formalism developed by Egami and Billinge [7].

The PDF $G(r)$ yields the probability of finding a pair of atoms separated by a distance r. It can be calculated from a structural model describing the distribution of atoms in a sample according to the formula:

$$G(r) = \frac{1}{r}\sum_{i}\sum_{j}\left[\frac{b_i b_j}{\langle b\rangle^2}\delta(r - r_{ij})\right] - 4\pi r\rho_0 \tag{10.1}$$

where the summation is throughout the atoms contained in the structural model, separated by the distance r_{ij}. b_i is the scattering power of atom i (its Fermi length for neutrons, its form factor for X-rays or electrons), $\langle b\rangle$ is the average scattering power of the sample, and ρ_0 is the numerical density of compound, that is, the number of atoms per unit volume. This function will have peaks for values of r corresponding to interatomic distance in the model, the intensity of these peaks being proportional to the product of the scattering factors of the atoms forming the pair. All the contributions of all pairs of atoms add up in the PDF. However, it is possible to calculate partial PDFs by taking into account selected atoms only (Figure 10.1). It is worth noting that Eq. (10.1) is strictly valid only for neutron scattering for which the Fermi length b is independent of Q in reciprocal space so that $\langle b\rangle$ can be calculated unambiguously. For X-rays and electrons, the form factors are Q-dependent and the shape of $F(Q)$ depends on the element considered so that Eq. (10.1) is only an approximation [8]. However, in practice the error introduced is small enough for quantitative analysis of X-ray PDFs to remain valid. Since PDF analysis of pharmaceuticals with a neutron source is complicated by the strong incoherent scattering of hydrogen, we will limit ourselves here to the case of X-rays.

The position of the peaks of the PDF reflects the distribution of distances in the material. In principle, the contribution of each pair consists of a δ-function. For a completely random distribution of atoms, the PDF will then be flat and equal to 0

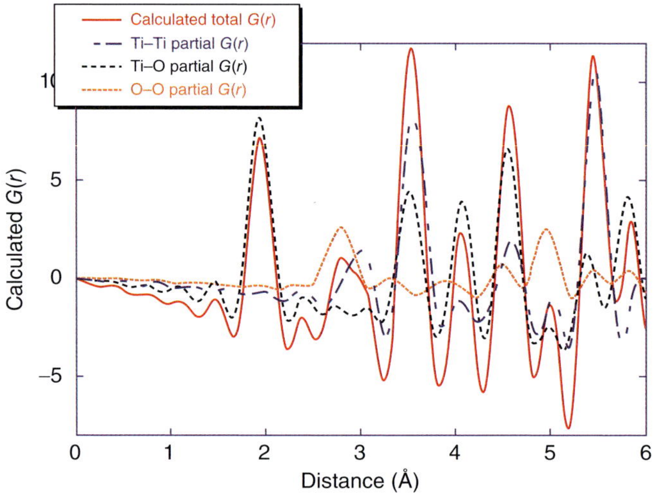

Figure 10.1 Calculated $G(r)$ for TiO_2 rutile, showing the total $G(r)$ and partial $G(r)$ for the Ti–Ti, Ti–O, and O–O couple of atomic species.

for any r. The second term $-4\pi r\rho_0$ in Eq. (10.1) is a normalization term introduced to compensate for the increase of the number of atomic pairs with r, given the numerical density ρ_0. Therefore, the frequent (i.e., more than average) presence or absence of interatomic distances appears as fluctuations of the PDF about zero. For short distances corresponding to close atomic neighbors, isolated peaks can be observed (e.g., at ~1.9 or 3.5 Å in Figure 10.1). The width of the peaks depends on the distribution of the distances around their average value, either due to thermal vibrations or a statistical distribution (disorder).

For a crystalline solid, the distribution of interatomic distance is defined by the coordinates of atoms, the atomic displacement parameters, and the periodicity of the unit cell. The PDF will yield a peak at each of these distances, the width of the peak being related to the atomic displacement parameters of the atoms forming the pair and its surface to the product of their scattering powers. In principle, the PDF will show such peaks up to the largest interatomic distances in the crystal. For example, Figure 10.2 displays the experimental PDF of LaB_6 obtained from high-resolution synchrotron powder diffraction up to 1000 Å. Obviously, structural information can be retrieved from the PDF. For a well-ordered and crystallized solid, it is in fact equivalent to a 1D projection of the Patterson function. In that case, standard crystallographic techniques can be applied and are generally more convenient than the PDF description to study the structure.

In the case of nanoparticles, the sets of coherently scattering atoms are limited to the volume of the individual particles, and therefore no peak in the PDF will

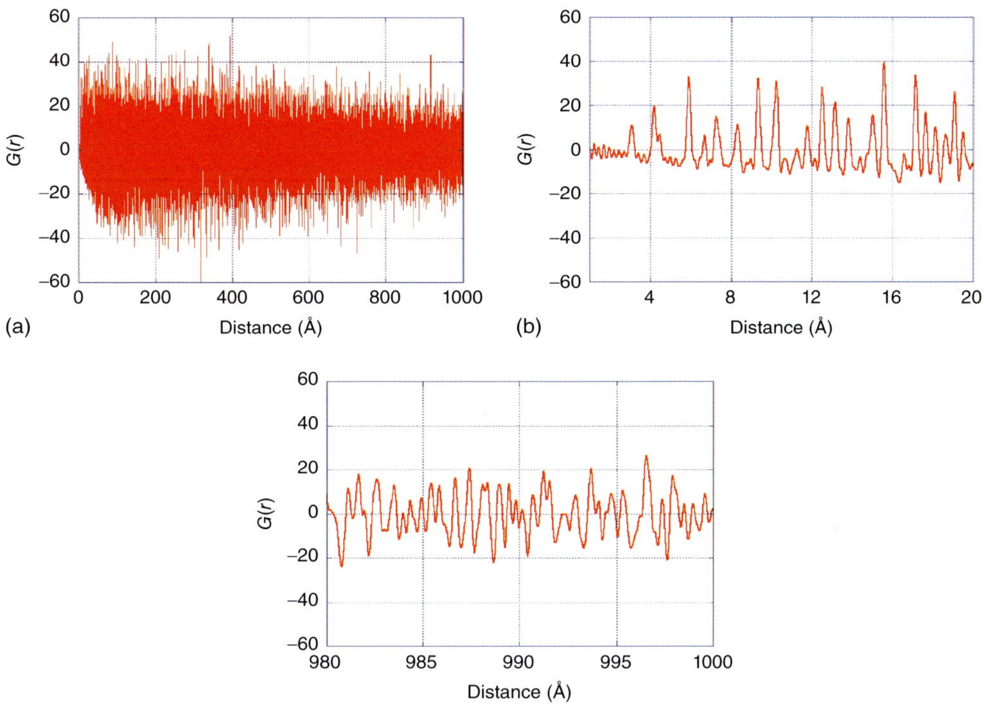

Figure 10.2 Experimental PDF for LaB_6 obtained up to 1000 Å from data collected at the ESRF ID31 beam line, $E = 35$ keV. Below are shown the details of the PDF between 1–20 Å (a) and 980–1000 Å (b).

be observed above the longest interatomic distance within the particles. Above this distance, only randomly distributed interatomic distances between atoms belonging to different particles will exist, and the PDF will vanish as a result of the normalization term of Eq. (10.1). The PDF will thus provide a direct estimate of the particle size, plus the distribution of interatomic distances inside the particles. The result will be similar for amorphous compounds: the PDF will vanish with the loss of structural coherence induced by disorder. The case of titanium oxide nanoparticles of a diameter less than 1 nm is exemplified in Figure 10.3 for PDF data obtained using a laboratory diffractometer in the Bragg–Brentano geometry and Mo Kα radiation. A particle size of approximately 8–9 Å can be readily deduced by simple observation of the PDF; a model of atomic arrangement inside the particle can be checked and fitted to the data to provide a quantitative estimate of the interatomic distances and, ultimately, of the structural arrangement.

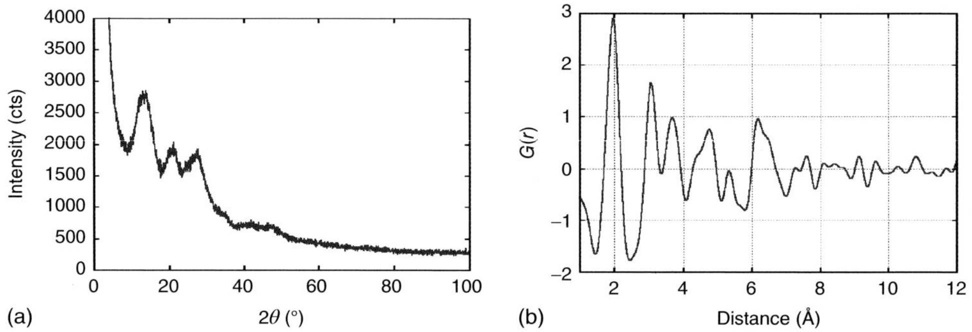

Figure 10.3 (a) Part of the diffraction pattern for nanocrystalline titanium oxide measured up to $2\theta = 150°$ on a laboratory diffractometer with the Bragg–Brentano geometry and Mo Kα radiation. (b) The corresponding PDF.

10.3
How to Measure the PDF

Experimentally, the PDF is obtained from a powder diffraction pattern by realizing the Fourier transform as

$$G(r) = 4\pi r[\rho(r) - \rho_0] = \frac{2}{\pi}\int_0^{\infty} Q[S(Q) - 1]\sin(Qr)dQ \tag{10.2}$$

where $\rho(r)$ is the microscopic pair density, and $S(Q)$ is the total structure function, that is, the normalized coherent scattered intensity. It is worth noting that the whole diffraction pattern is used to obtain $G(r)$, not only the Bragg peak intensities as for a classical crystallographic analysis. Therefore, the contributions of disorder, defects, and so on, which contribute to the pattern through diffuse scattering outside of the Bragg peaks, will also be included in $G(r)$. Obtaining $S(Q)$ from diffraction data requires performing a set of corrections to eliminate all incoherent (Compton) or inelastic (e.g., fluorescence) contributions, the signal coming from the environment (sample holder, cryostat, air scattering, etc.), as well as the effects of absorption, multiple scattering, and so on. These corrections require additional measurements of these contributions (from, e.g., the empty sample holder). The self-scattering contribution is then subtracted, and the pattern is normalized by $\langle b\rangle^2$, resulting in $S(Q)$. Then the function $F(Q) = Q(S(Q) - 1)$ is Fourier transformed according to Eq. (10.2). User-friendly programs to calculate the PDF from experimental patterns are now available for neutrons as well as X-rays, and the procedure has become almost routine.

The calculation of the Fourier transform on a bounded domain will have consequences on the experimental PDF, as it will be necessary to take into account when modeling by the function $G(r)$. These effects include the presence of termination wiggles, broadening of the peaks of the PDF, and a decrease of their intensity at long distances. In principle, these effects can be taken into account in the calculation of the theoretical PDF. However, if one is interested in fine structural details on a wide range of distances, it is necessary to minimize them. For that, we must

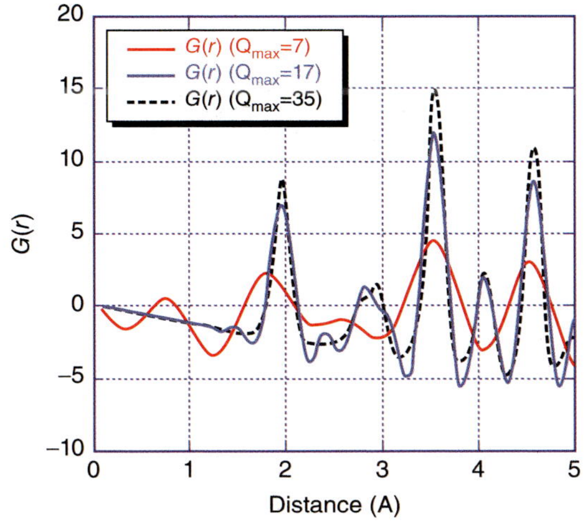

Figure 10.4 Comparison of theoretical PDFs of rutile calculated for increasing values of Q_{max}.

measure the powder diffraction pattern up to high values of Q, ideally well above the disappearance of a distinct scattering signal due to thermal and/or static disorder. Figure 10.4 shows a comparison between the PDFs simulated for values of Q_{max} corresponding to the Cu Kα ($\approx$7 Å^{-1}) and Mo Kα ($\approx$17 Å^{-1}) radiations and for a value of Q_{max} significantly higher (35 Å^{-1}) corresponding to a wavelength of 0.35 Å reachable at a synchrotron instrument. It is clear that the increase in the value of Q_{max} provides access to much more detailed structural information. In principle, PDFs from patterns collected only up to only low Q_{max} (e.g., using Cu Kα radiation) may be used [9]; however, the PDFs obtained are relatively featureless and dominated by spurious effects from the Fourier transform or poor normalization, which can lead to misinterpretations of the data. It is thus more favorable to use shorter wavelength radiations for PDF analysis [10]. Note that the angular resolution of the diffractometer is less restrictive, though too low a resolution in the reciprocal space leads to a rapid decrease of the PDF intensity with distance in the direct space.

Because of the decrease versus Q of the scattering power in the case of X-rays, the normalization of $S(Q)$ by dividing with $\langle b \rangle^2$ will significantly amplify the data at large Q, which may be very noisy. It is therefore important to ensure sufficiently high counting statistics in the high-Q range; otherwise, a high noise level at high Q for $F(Q) = Q(S(Q) - 1)$ will introduce a spurious signal in the Fourier transform and therefore the PDF.

To obtain a powder diffraction pattern usable for PDF analysis, one must combine a high Q_{max} (meaning small wavelength or high-energy radiation) and high counting statistics on a wide Q range. Both conditions can be fulfilled at synchrotron sources. Two main kind of setups can be used at synchrotron sources: high-resolution powder diffraction using a parallel beam and analyzer crystals (e.g., ID31 at ESRF, CRISTAL at SOLEIL, etc.). This has the advantage of eliminating the inelastic scattering and thus simplifies the corrections, but leads to long measurement times (several hours). The high Q_{max} and ultimate angular resolution of such an instrument allows obtaining PDFs up to several hundreds of angstroems (Figure 10.2). The alternate setup relies on the access to very high energies (~100 keV) at which a flat-plate 2D detector allows collection of data to high Q in a short time (e.g., at ESRF beam lines ID11, ID15, etc.) [11]. In this case, although the spatial resolution may be poorer, *in situ* studies, investigations of phase transitions, solidification, and so on, become feasible.

Nevertheless, good-quality PDF measurements are also possible with modern laboratory X-ray diffraction instruments. In order to reach high Q_{max}, it is necessary to use the Mo Kα ($\lambda = 0.7107$ Å, $Q_{max} = 17$ Å^{-1}) or better Ag Kα ($\lambda = 0.5608$ Å, $Q_{max} = 21.7$ Å^{-1}) radiation and measure up to the highest accessible angles (>150°). The diffractometer must be set to optimize the intensity, with relaxed monochromatization and a 1D detector, which must remain efficient for the wavelength used. Either the Bragg–Brentano or Debye–Scherrer geometry can be used. In the first case, a set of variable slits will help increasing the intensity at high angles. For the latter case, a 2D detector can also be used. This can be done, for instance, by using a single-crystal diffractometer equipped with a charge-coupled device (CCD) camera. Diffraction images taken at increasing values of 2θ are converted to a $I(2\theta)$ diffraction pattern, which is used to compute the PDF (Figure 10.5). In any case, the collection time for measuring a PDF with a modern laboratory X-ray diffractometer is of the order of 1 day, depending on the instrument performance and the sample's scattering power.

10.4
Modeling of the PDF

The main feature of the PDF is to provide a representation of the structure of a compound through the distribution of interatomic distances, and this in a variety of scales ranging from the local order (some angstroems) to an average structure (a few tens of nanometers). It then constitutes a true multiscale structure investigation tool and will be fruitfully applicable in cases where these local and average structures differ, either because of the coherent domain/grain size (nanocrystalline and amorphous compounds) or because of localized defects or structural disorder (static and dynamic), which cannot be taken into account efficiently by conventional crystallographic techniques. We have seen that the PDF can be obtained experimentally from total scattering measurements and can also be calculated provided some kind of distribution of atoms is given. As usual, in order to validate and improve the structural model, it will be necessary to search for the configuration yielding the best agreement between the observed and calculated PDF.

Two major types of methods are used to model the PDF. The first is based on numerical minimization techniques of the reverse Monte Carlo type

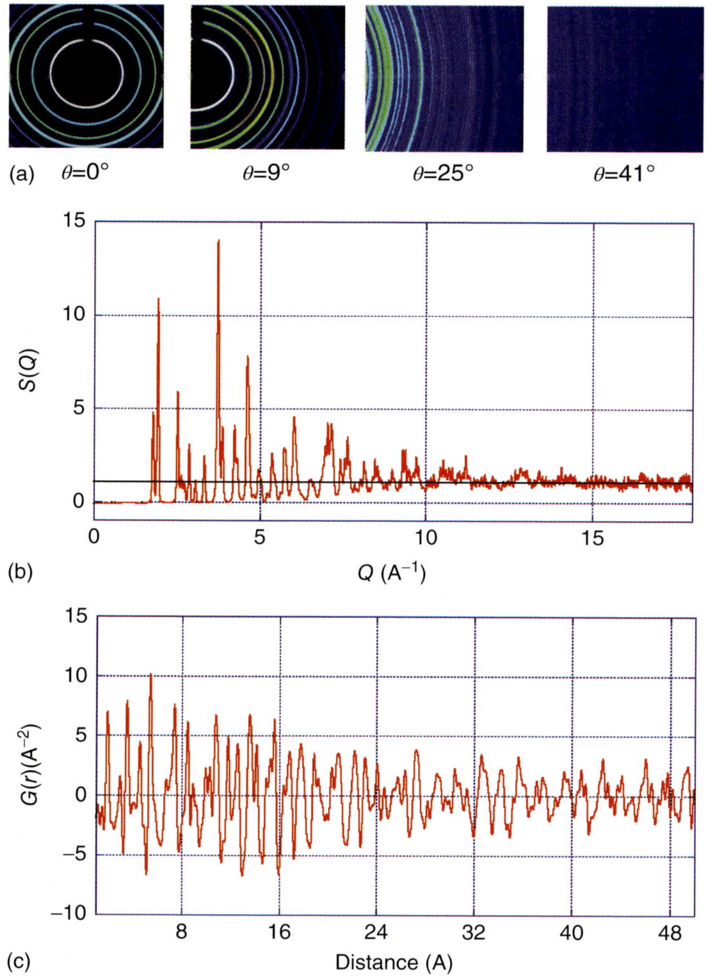

Figure 10.5 Measurement of the PDF of a mixture of TiO_2 oxides using a laboratory single-crystal diffractometer with a CCD camera. (a) Selection of diffraction images taken at increasing θ angles. (b) $S(Q)$ scattering function obtained after integration and summation of images. (c) The corresponding PDF.

(e.g., program RMCProfile [12]). The structural model is built by distributing more or less randomly atoms in a (large) box and is optimized by using a reverse Monte Carlo algorithm minimizing the difference between the observed and calculated PDFs [13]. It therefore requires large computing resources when the problem size increases, and its chances of success are heavily dependent on the choice of the physicochemical constraints applied to the model (distance/angle limits, coordination imposed, bond valence sum, etc.). It requires dedicated tools for statistical analysis of the resulting structure. This method is particularly well suited to amorphous compounds and liquids.

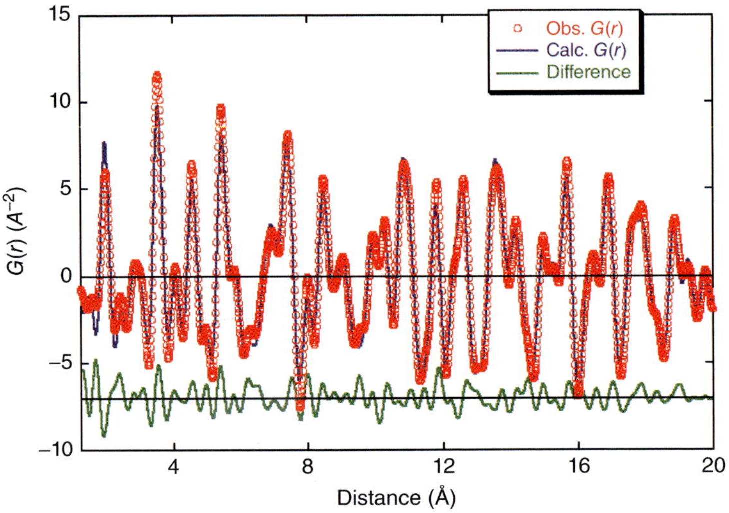

Figure 10.6 Refinement using PDFGui of the PDF of rutile from data measured up to $2\theta = 150°$ on a laboratory diffractometer with the Bragg–Brentano geometry and Mo Kα radiation.

The second type of technique (e.g., program PDFGui [14]) consists in a least squares profile refinement of the experimental PDF with a periodic structural model, a sort of Rietveld refinement in direct space (Figure 10.6). This method is faster and better suited to compounds that are close to a crystalline state. Symmetry constraints can be imposed to all or part of the structure, and refinements can be applied to different ranges of distances, to compare the local and average structure. The effects of various instrumental parameters (Q_{max}, spatial resolution, etc.) and sample-dependent effects (atomic displacement correlations, spherical nanoparticle size, etc.) can be applied. The model can also comprise several different phases with individual scale factors, allowing quantitative analysis to be performed regardless of the crystalline state of the different phases.

10.5 Applications of PDF Analysis to Molecular and Pharmaceutical Compounds

Most recent developments of PDF analysis have been applied to inorganic and often simple compounds. In principle, any of these applications is also valid for small-molecule compounds. However, these materials are generally much more complex, with low symmetry structures, and are constituted of weakly diffracting atomic species. Also, the lack of implementation of practical rigid-block molecular description in PDF refinement programs has hindered its use in the structural analysis of small-molecule materials, where the number of structural parameters becomes rapidly too large. Most applications of the PDF in the case of pharmaceuticals have therefore relied on direct inspection and comparison between

PDF patterns or with simulations. However, PDF can still be a helpful tool for the structure determination of ill-ordered materials, as a structural model obtained by other means can be easily tested against an experimental PDF. This approach was used by Schmidt *et al.* [15] in their study of the α and β phases of Pigment Yellow 213 ($C_{23}H_{21}N_5O_9$), the α phase being long-range-ordered and the β phase nanocrystalline. Having determined the structure of the α phase using classical crystallographic methods, the authors used PDF analysis to show that the local arrangement of the β phase was similar to that of the α phase.

As discussed above, one of the obvious applications of PDF in the field of pharmaceutics is the investigation of nanocrystalline or amorphous compounds. Bates *et al.* [9] stressed the usefulness of the technique to analyze these kinds of samples from their X-ray diffraction patterns. For example, they reported a quantitative description of the local structure evolution during various stages of grinding in the case of indomethacin. The same team also investigated the mixture of amorphous APIs, using PDF as a complement to traditional differential scanning calorimetry (DSC) measurements.

More recently, Billinge *et al.* [16] demonstrated the usefulness of PDF to investigate nanocrystalline and amorphous molecular compounds. Using carbamazepine and indomethacin as test materials, they showed the possibility to unambiguously identify a melt-quenched molecular compound from its PDF signature, provided XRPD data were recorded in a convenient way, that is, up to sufficiently high Q_{max} (20 Å^{-1} in their case) [10]. They were also able to discriminate between different polymorphs of a given API in the amorphous state and proposed the use of PDF as a powerful technique for fingerprinting molecular compounds. Along the same line, PDF analysis was used by Nollenberger *et al.* [17] to investigate the local structural order in mixed felodipine amorphous melt extrudates, providing insights into the control of dissolution rates of the API.

We already noted that PDF could be a powerful technique for quantitative analysis, irrespective of the crystallization state of the various phases in a mixture. This is in contrast with the Rietveld method, where, in general, only crystalline phases can be quantified. Using a complex procedure and the use of additional measurements with a standard added to the sample, the proportion of the "amorphous part" can be estimated by Rietveld refinement, but without specifying the nature of this "amorphous part" or even the number of phases it contains. In principle, such a drawback does not exist in the case of PDF analysis, and all phases in a mixture can be quantified as long as they can be identified. This specificity may also be useful in the case of pharmaceutical materials. For example, it was recently used for the study of recrystallization effects of cryo-milled sulfamerazine [18]. In that case, the composition of cryo-milled nanocrystalline powders containing sulfamerazine form 1, form 2, and an amorphous phase could be determined using PDF analysis. It was shown that the composition depended on the cryo-milling protocol, which is obviously an important issue for the use of this technique in producing drugs in a perfectly controlled way.

More examples of (potential) applications demonstrate that it is indeed the case.

10.5.1
Polymorphs of *rac*-Modafinil

Seven polymorphic forms of *rac*-modafinil $C_{15}H_{15}NSO_2$ (2-(diphenylméthyl)sul finyl)-*N*-acétamide), a wake-promoting agent, have been obtained up to now. Among them, form I is monoclinic ($P2_1/a$, $Z' = 2$, $a = 14.502$ Å, $b = 9.678$ Å, $c = 20.844$ Å, $\beta = 110.17°$) while form III is orthorhombic ($Pna2_1$, $Z' = 2$, $a = 14.510$ Å, $b = 9.710$ Å, $c = 19.569$ Å) [19]. Both structures are known and differ essentially by the stacking of a slab along the common *a* axis, similar to a periodic twinning or regular stacking fault arrangement. Both polymorphs thus present very similar structures and are quasi-isoenergetic. There PDFs were measured at the D2AM beam line of the ESRF in the Debye–Scherrer geometry using 22 keV radiation up to $2\theta = 120°$ with an XPAD pixel 2D detector [20]. Figure 10.7 [21] shows the $F(Q) = Q(S(Q) - 1)$ data and the resulting PDF up to 50 Å calculated with $Q_{max} = 16$ Å^{-1} [22]. Peaks corresponding to interatomic distances are clearly observable in the whole data of range. On comparing the PDFs of polymorphic forms I and III in Figure 10.8, it appears that both are nearly identical up to approximately 20 Å, above which the two PDFs begin to differ strongly. Interestingly, this distance is larger than the longest intramolecular bonds in *rac*-modafinil, but similar to the thickness of the common structural slab, the *c* parameter of form III. It defines the largest distance below which the two structures are essentially identical. This shows that the PDF can be used by direct observation to identify the minimum dimension of structural blocks common to different polymorphs. If differences start to appear at distances below the shortest intermolecular distance, it can be readily inferred that the conformation of the molecule is not conserved between the polymorphic forms, which is very important information for structure solution.

We have attempted a structure refinement of the PDF of both polymorphs using the PDFGui software. Figure 10.9 shows the result for form III up to 30 Å ($R = 12\%$). Similar agreement is found for form I. Although the refinement looks

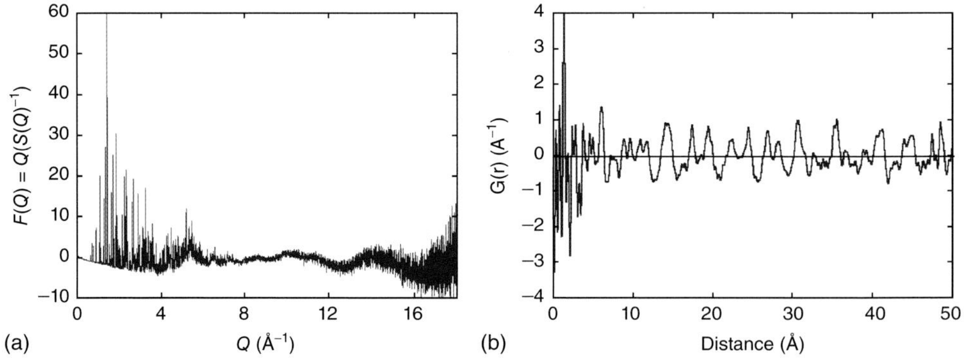

Figure 10.7 Experimental *F(Q)* (a) and *G(r)* (b) for form I of *rac*-modafinil from data measured at the D2AM beam line of ESRF.

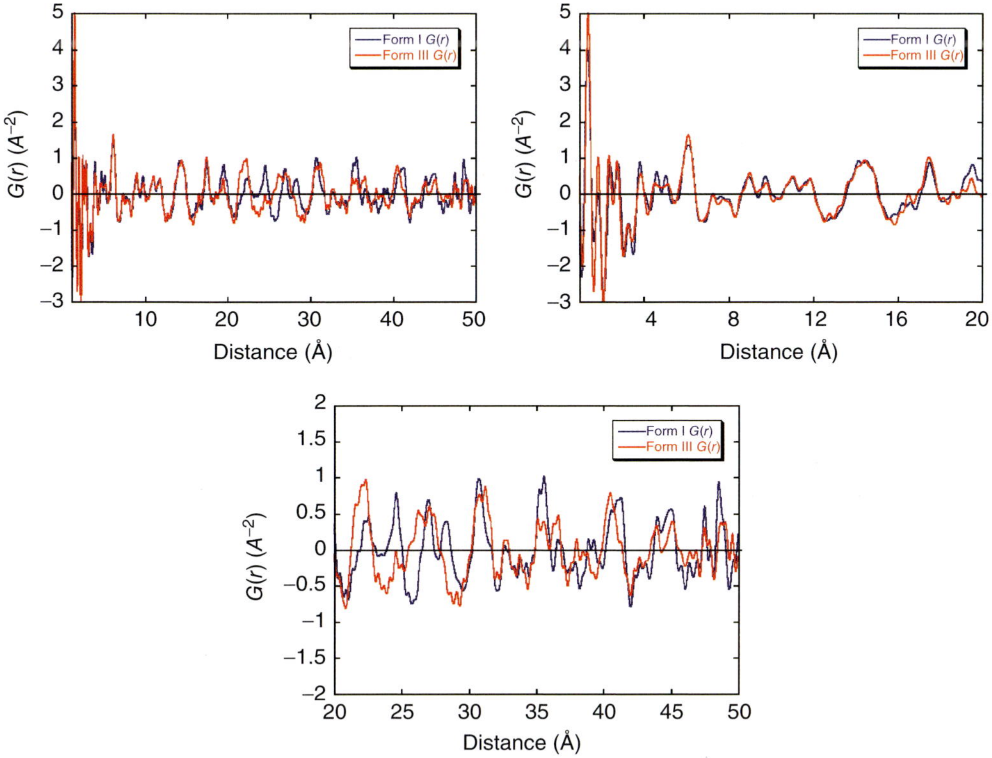

Figure 10.8 Comparison between the PDFs of forms I and III of *rac*-modafinil.

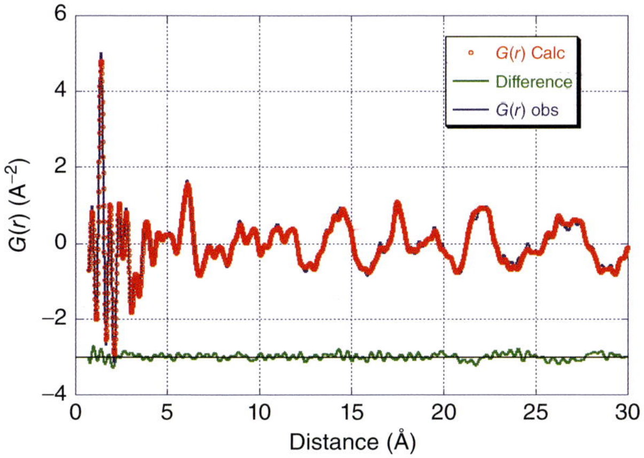

Figure 10.9 Structure refinement using the PDF of *rac*-modafinil form III using PDFGui.

quite good, one must be aware that the only constraints were those imposed by space group symmetry. Although all H atoms were fixed and a single overall isotropic a.d.p. was refined, 135 structural parameters were released. This number is too large for a sound and stable least-squares minimization, and additional constraints between parameters must be introduced. A software for PDF refinement, including rigid block molecular description, is expected to appear very soon (A. Bytchkov and J. Rodriguez-Carvajal, 2013, personal communication). This will be a crucial step for the use of PDF in the context of pharmaceutical compounds.

10.5.2
Effects of Grinding on Hydrochlorothiazide

Hydrochlorothiazide (HCT) ($C_7H_8Cl_1N_3O_4S_2$) is known as a diuretic widely used in the clinical treatment of a number of disorders. A sample of polymorph I (monoclinic, $P2_1$, $Z=2$, $a=7.34$ Å, $b=8.508$ Å, $c=9.964$ Å, $\beta=111.33°$ [23]) was submitted to high-energy milling, and three samples (a pristine sample, one milled for 15 min, and one for 4 h) were studied at the CRISTAL beamline of the SOLEIL synchrotron in high-resolution parallel beam geometry with an X-ray energy of 22.8 keV [24]. The corresponding powder patterns and PDFs are shown in Figures 10.10 and 10.11. After 15 min, the powder pattern shows a strong broadening of the Bragg peaks, and with some diffuse scattering present in the background. However, the compound is still identifiable, and its structure could be refined by the Rietveld method including a rigid-body description of

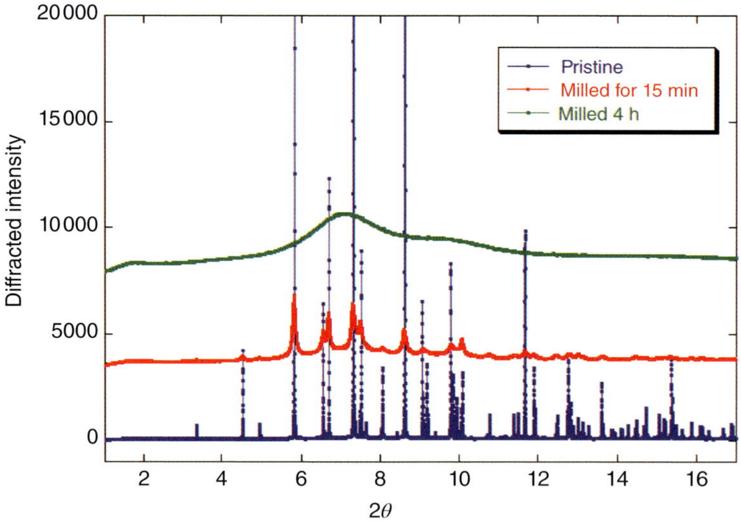

Figure 10.10 Comparison of the diffraction patterns of hydrochlorothiazide: pristine (bottom), milled for 15 min (middle), and milled for 4 h (top). The curves have been shifted for better readability.

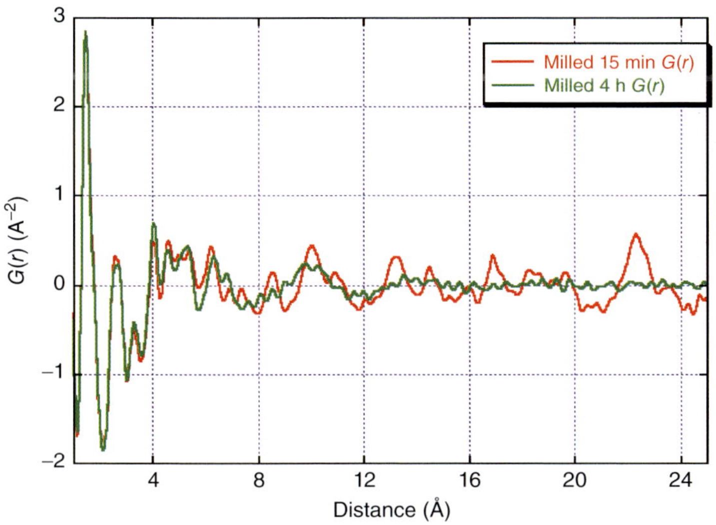

Figure 10.11 Comparison of the PDFs of hydrochlorothiazide milled for 15 min and for 4 h.

the molecule. After 4 h of grinding, the compound appears as fully amorphized and its powder pattern seems to be of little help for any structural investigation. However, the PDF clearly demonstrates that the molecular structure is preserved, since it remains almost unchanged for the two samples up to approximately 7 Å, which corresponds to the longest intramolecular distances for this compound. Weak humps at longer distances (~10 and 14 Å) seem to indicate the persistence of correlations with neighboring molecules. The PDF completely vanishes above approximately 15 Å, indicating that such correlations do not propagate farther than the nearest molecular neighbor.

Such results demonstrate that PDF analysis can be used to continuously and quantitatively to monitor the evolution of the structure and microstructure during an amorphization process. Moreover, provided a rigid-block description is available, structure refinement using the PDF should be an efficient way to obtain quantitative structural information even for compounds with a coherent domain size as small as a few nanometers.

10.6 Conclusion

With the trend toward production of drugs in the form of nano or amorphous materials, new structure investigation tools have become necessary to determine the crystalline arrangement and microstructure of badly diffracting samples. PDF analysis is especially well suited as such a tool. Instruments are now widely available at synchrotrons around the world to provide high-quality PDF data, but laboratory diffractometers equipped with short wavelength sources can also be used successfully. Efficient and user-friendly software is available to analyze the data. The lack of a software allowing refinement of the PDF using rigid-block molecular models is still a drawback for small-molecule compounds, but this problem is expected to be solved very soon. PDF analysis provides a way of refining (and even maybe solving) the structures of materials regardless of their crystallinity, and provides at the same time information on the size as well as the internal structure of nanoparticles. There is no doubt that the use of PDF analysis will soon increase very rapidly in the field of small-molecule and pharmaceutical science, just as it has become now for inorganic compounds.

Acknowledgments

The authors wish to express their gratitude to I.E. Grey and I. Madsen for fruitful discussions and experimental facilities and N. Boudet and J-F. Bérar at the D2AM-ESRF beamline, E. Elkaïm at the CRISTAL-SOLEIL beamline, G. Coquerel, M. Descamps, and coworkers for collaboration and samples.

References

1. Shankland, K. and David, W.I.F. (2001) *Structure Determination from Powder Diffraction Data*, vol. **13**, IUCr Commission on Powder Diffraction, Oxford Science Publications, pp. 252–282.
2. (a) Threlfall, T.L. (1995) *Analyst*, **120**, 2435; (b) Hilfiker, R. (ed) (2006) *Polymorphism in the Pharmaceutical Industry*, Weinheim, Wiley-VCH Verlag GmbH.
3. (a) Brittain, H.G. and Fiese, E.F. (1999) in *Polymorphism in Pharmaceutical Solids*, vol. **95** (ed H.G. Brittain), Marcel Dekker, New York, pp. 331–361; (b) Willart, J.-F. and Descamps, M. (2008) *Mol. Pharm.*, **5**, 905.
4. Guinier, A. (1963) *X-Ray Diffraction in Crystals, Imperfect Crystals, and Amorphous Bodies*, W.H. Freeman, San Francisco, CA.
5. Warren, B.E. (1969) *X-Ray Diffraction*, Addison-Wesley Publishing Co..
6. Keen, D.A. (2001) *J. Appl. Crystallogr.*, **34**, 172–177.
7. Egami, T. and Billinge, S.J.L. (2012) *Underneath the Bragg Peaks: Structural Analysis of Complex Materials*, 2nd edn, Pergamon Press, Oxford.
8. Masson, O. and Thomas, P. (2012) *J. Appl. Crystallogr.*, **46**, 461.
9. Bates, S., Zografi, G., Engers, D., Morris, K., Crowley, K., and Newman, A. (2006) *Pharm. Res.*, **23** (10), 2333.
10. Dykhne, T., Taylor, R., Florence, A., and Billinge, S.J.L. (2011) *Pharm. Res.*, **28**, 1041.
11. Chupas, P.J., Qiu, X., Hanson, J.C., Lee, P.L., Grey, C.P., and Billinge, S.J.L. (2003) *J. Appl. Crystallogr.*, **36**, 1342.
12. Keen, D.A., Tucker, M.G., and Dove, M.T. (2005) *J. Phys.: Condens. Matter*, **17**, S15.
13. McGreevy, R.L. (2001) *J. Phys.: Condens. Matter*, **13**, R877.
14. Farrow, C.L., Juhas, P., Liu, J.W., Bryndin, D., Božin, E.S., Bloch, J., Proffen, T., and Billinge, S.J.L. (2007) *J. Phys.: Condens. Matter*, **19**, 335219.
15. Schmidt, M.U., Brühne, S., Wolf, A.K., Rech, A., Brüning, J., Alig, E., Fink, L., Buchsbaum, C., Glinnemann, J., van de Streek, J., Gozzo, F., Brunelli, M., Stowasser, F., Gorelik, T., Mugnaioli, E., and Kolb, U. (2009) *Acta Crystallogr., Sect. B: Struct. Sci.*, **B65**, 189.

16. Billinge, S.J.L., Dykhne, T., Juhas, P., Bozin, E., Taylor, R., Florence, A.J., and Shankland, K. (2010) *CrystEngComm*, **12**, 1366.

17. Nollenberger, K., Gryczke, A., Meier, C., Dressman, J., Schmidt, M.U., and Brühne, S. (2009) *J. Pharm. Sci.*, **98** (4), 1476.

18. Davis, T., Johnson, M., and Billinge, S.J.L. (2013) *Cryst. Growth Des.*, **13**, 4239.

19. Pauchet, M., Morelli, T., Coste, S., Malandain, J.-J., and Coquerel, G. (2006) *Cryst. Growth Des.*, **6** (8), 1881.

20. Boudet, N., Berar, J.-F., Blanquart, L., Breugon, P., Caillot, B., Clemens, J.-C., Koudobine, I., Delpierre, P., Mouget, C., Potheau, R., and Valin, I. (2003) *Nucl. Instrum. Methods Phys. Res., Sect. A*, **510** (1), 41.

21. Martinetto, P. *et al.* (2006) to be published.

22. Juhás, P., Davis, T., Farrow, C.L., and Billinge, S.J.L. (2013) *J. Appl. Crystallogr.*, **46**, 560.

23. Leech, C.K., Fabbiani, F.P.A., Shankland, K., David, W.I.F., and Ibberson, R. (2008) *Acta Crystallogr., Sect. B: Struct. Sci.*, **64**, 101.

24. Pagnoux, W. *et al.* (2006) to be published.

11
Application of Broadband Dielectric Spectroscopy to Study Molecular Mobility in Pharmaceutical Systems

Katarzyna Grzybowska, Karolina Adrjanowicz, and Marian Paluch

11.1
Introduction to Broadband Dielectric Spectroscopy

11.1.1
Broadband Dielectric Spectroscopy Technique (10^{-5} to 10^{11} Hz)

Broadband dielectric spectroscopy (BDS) is a very useful method for investigating the molecular dynamics of various materials because it enables monitoring the molecular mobility as reflected in the relaxation processes over a wide frequency range up to 16 decades, at different temperatures and pressures. BDS finds application in many fields of science and technology, including investigations of pharmaceutical systems [1, 2]. This spectroscopic method is successfully employed in the studies of (i) the molecular dynamics of liquids, liquid crystals, glasses, disordered crystals, (ii) charge transport in ionic liquids, semiconductors, organic crystals, ceramics, and so on, (iii) structural material properties such as phase compositions, and (iv) nonlinear electrical and optical effects. BDS is also a very useful tool to monitor chemical reactions (including polymerization, tautomerization of drugs, mutarotation of sugars) and phase transitions, for example, crystallization or vitrification.

The essence of dielectric spectroscopy is the investigation of the effects caused by an external electric field applied to the sample and the measurements of the quantities that characterize the response of the material (polar dielectrics) to the applied alternating electric field $\mathbf{E}(\omega)$. In the frequency range $f = \omega/2\pi$ from 10^{-5} to 10^{11} Hz of the electric field $\mathbf{E}(\omega)$, the following phenomena can occur in the material [1]:

- dielectric dispersion $\varepsilon'(\omega)$ and absorption $\varepsilon''(\omega)$ caused by dipole relaxation arising from the reorientational motion of molecular dipoles, which are represented by the complex dielectric permittivity $\varepsilon^*(\omega) = \varepsilon'(\omega) - i\varepsilon''(\omega)$;
- electrical conduction arising from the translational motions of electric charges (ions, electrons), which can be described by the complex conductivity $\sigma^*(\omega) = \sigma'(\omega) + i\sigma''(\omega)$ (related to the complex permittivity by the equation $\sigma^*(\omega) =$

Disordered Pharmaceutical Materials, First Edition. Edited by Marc Descamps.

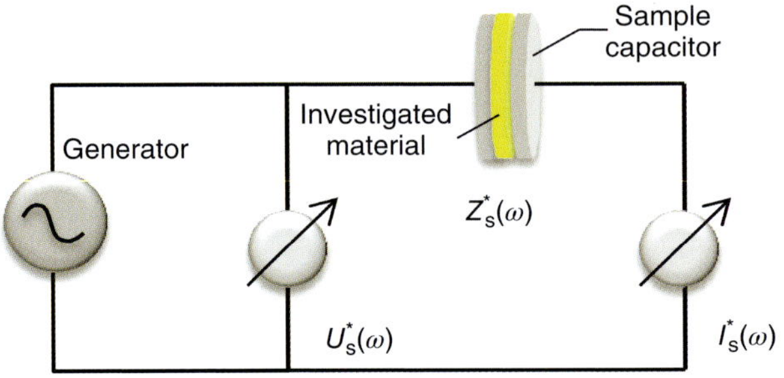

Figure 11.1 Diagram of a typical impedance analyzer illustrating the principle of dielectric measurements in the frequency domain. Figure taken from [3].

$i\omega\varepsilon_0\varepsilon^*(\omega)$, where ε_0 is the vacuum permittivity) or the complex electrical modulus $M^*(\omega) = M'(\omega) + iM''(\omega)$ (related to the complex dielectric permittivity as $M^*(\omega) = 1/\varepsilon^*(\omega)$).

The characteristic complex quantities such as $\varepsilon^*(\omega)$, $\sigma^*(\omega)$, and $M^*(\omega)$ can be derived from the complex electrical impedance $Z^*(\omega)$, which is measured by means of the BDS technique. In a dielectric experiment, the sample to be investigated is placed in a measurement capacitor, to which an alternating voltage $U^*(\omega)$, typically a sine wave voltage, is applied by a generator. Then, an impedance analyzer (see Figure 11.1) determines the complex impedance of the sample $Z^*_s(\omega) = U^*_s/I^*_s$ by measuring the amplitudes and the phases of the voltage $U^*_s(\omega)$ between the plates of the sample capacitor as well as the sample current $I^*_s(\omega)$ [4].

Since the complex dielectric permittivity $\varepsilon^*(\omega)$ of the sample can be found from the ratio of the complex capacitance of the capacitor filled with the investigated material $C^*_s(\omega)$ and the capacitance of the empty capacitor C_0, that is, $\varepsilon^*(\omega) = C^*_s(\omega)/C_0$, and $C^*_s(\omega)$ is related to the sample impedance $C^*_s(\omega) = -i/\omega Z^*_s(\omega)$, the dielectric function $\varepsilon^*(\omega)$ of the material can be established from the sample impedance $Z^*_s(\omega)$ measurements in the following way:

$$\varepsilon^*(\omega) = \varepsilon'(\omega) - i\varepsilon''(\omega) = -\frac{i}{\omega Z^*_s C_0} = -\frac{i}{\omega \frac{U^*_s}{I^*_s} C_0}$$

where the capacitance of the empty capacitor C_0 depends on its geometry (e.g., $C_0 = \varepsilon_0 S/d$ for a parallel capacitor, where d is the distance between the capacitor plates and S is the area of one plate).

It should be noted that the extremely wide frequency range (10^{-5} to 10^{11} Hz) cannot be covered by only one dielectric spectroscopy measurement technique. Therefore, if we need to examine the electrical/dielectric properties of the materials over such a broad frequency rage, we have to employ a few experimental techniques such as frequency-response analysis, AC-bridge methods, coaxial-line reflectometry, and network analysis (coaxial-line transmission), which would enable us to conduct the dielectric study in the frequency domains shown in Figure 11.2. The measurement techniques are described in detail in [1, 2, 4].

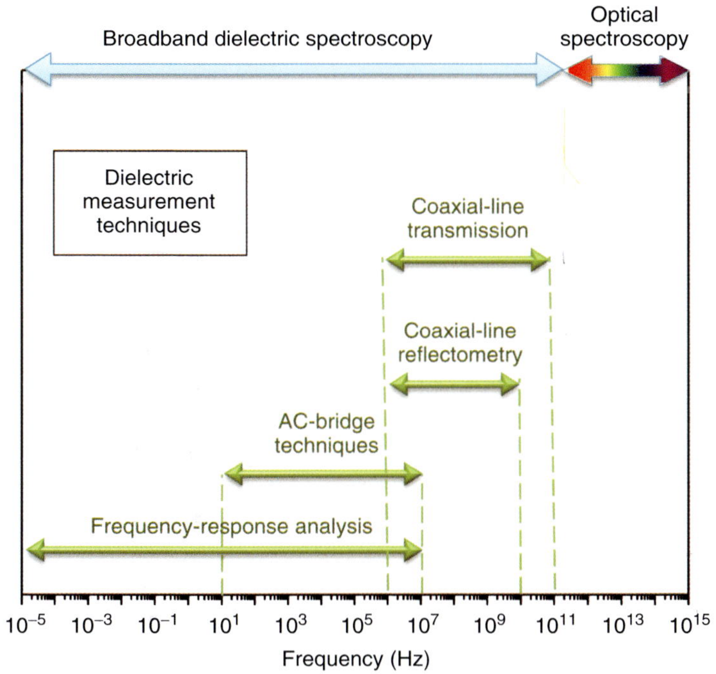

Figure 11.2 Dielectric measurement techniques exploited in the frequency range from 10^{-5} to 10^{11} Hz. The broadband dielectric spectroscopy frequency range reaches the low-frequency limit of optical spectroscopy.

Nowadays, using a commercial apparatus we can perform accurate dielectric measurements of the various types of liquid and solid materials not only over a broad frequency range but also over wide temperature and pressure ranges [5–11]. The investigations involving, besides temperature T, also other thermodynamic variables such as pressure P or density $\rho(T, P)$ provide many important findings, for instance, in the field of the glass transition physics. From the point of view of pharmaceutical applications, compression can have a significant effect on the properties of drugs (i.e., during the tableting process) or on the physical stability of amorphous drugs.

11.1.2 Principles and Models of Dielectric Relaxation

11.1.2.1 Debye Relaxation

The fundamental model that enables us to understand the nature of dielectric dispersion and absorption induced by an electric field in polar materials is the Debye model of dipole relaxation. To find the Debye equation for the complex dielectric permittivity ε^* of an investigated polar material, we consider the capacitor containing by the polar sample, to which a rectangular impulse of the electric field $E(t)$ is applied (Figure 11.3).

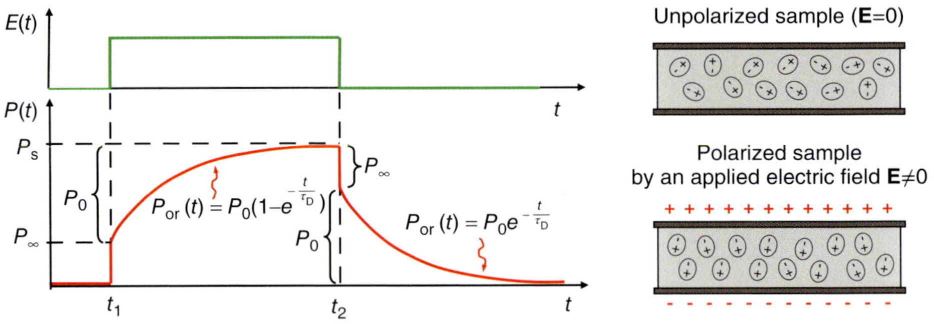

Figure 11.3 Changes in the polarization of a polar sample as a function of time, *P*(*t*), due to an applied rectangular impulse of the electric field *E*(*t*). On the right side, dipole orientations in the sample placed between capacitor electrodes when the electric field **E**(*t*) = 0 and **E**(*t*) ≠ 0 are shown.

After the electric field is switched on at the moment $t = t_1$, the deformation polarization P_∞ (which involves the electron P_e and atomic P_a polarizations, which are related to the elastic deformations of electron shells and interatomic bonds, respectively) suddenly increases. This kind of polarization depends on the external electric field according to the following equation:

$$P_\infty = P_e + P_a = \varepsilon_0(\varepsilon_\infty - 1)E \tag{11.1}$$

where ε_0 is the dielectric permittivity of vacuum, and ε_∞ is the dielectric permittivity of the material in the high-frequency limit.

However, the orientation polarization $P_{or}(t)$, which is related to rotational motions of suppressed polar molecules, increases much more slowly than P_∞, due to intermolecular friction, and reaches a maximum value P_s at $t = t_2$. The saturated orientation polarization is given by

$$P_s = \varepsilon_0(\varepsilon_s - 1)E \tag{11.2}$$

where ε_s is the static dielectric permittivity of the material. From Eqs. (11.1) and (11.2), one can obtain the saturated orientation polarization as

$$P_0 = P_s - P_\infty = \varepsilon_0(\varepsilon_s - \varepsilon_\infty)E \tag{11.3}$$

After the electric field is switched off at $t = t_2$, the deformation polarization P_∞ immediately decreases, whereas the orientation polarization $P_{or}(t)$ requires a sufficiently long time to completely decay, which is needed for the permanent dipoles to relax to equilibrium.

The phenomenon of the increase and decrease in the orientation polarization P_{or} with time is called *dielectric relaxation*. The increase in $P_{or}(t)$ can be described by the following differential equation:

$$\tau_D \frac{dP_{or}(t)}{dt} = P_0 - P_{or}(t) \tag{11.4}$$

where τ_D is a characteristic relaxation time, which is the same for all molecules in the material.

A solution of Eq. (11.4) is given by

$$P_{or}(t) = P_0[1 - \exp(-t/\tau_D)] \tag{11.5}$$

which involves the exponential relaxation function proposed by Debye

$$\phi(t) = \exp(-t/\tau_D) \tag{11.6}$$

to describe the response of the sample to the applied electric field. One can note that the relaxation function (Eq. (11.6)) describes also the orientation polarization decay after the electric field is switched off, because Eq. (11.4) with $P_0 = 0$ can be solved by the following function:

$$P_{or}(t) = P_0 \exp(-t/\tau_D) \tag{11.7}$$

Since the sine function can represented by a sequence of rectangular of impulses with varying amplitudes, one can assume that Eq. (11.4) is also valid if we apply a an electric field of the form $E^*(t) = E_0 e^{i\omega t}$. Exploiting Eqs. (11.3) and (11.4), we obtain the following differential equation describing changes of orientation polarization $P_{or}(t)$ to the applied AC electric field $E^*(t)$:

$$\tau_D \frac{dP^*_{or}(t)}{dt} = \varepsilon_0(\varepsilon_s - \varepsilon_\infty)E_0 e^{i\omega t} - P^*_{or}(t) \tag{11.8}$$

which can be solved as

$$P^*_{or}(t) = \frac{\varepsilon_0(\varepsilon_s - \varepsilon_\infty)}{1 + i\omega\tau_D} E_0 e^{i\omega t} \tag{11.9}$$

Then, taking into account Eqs. (11.1) and (11.7), the total polarization $P^*(t)$ is

$$P^*(t) = P^*_\infty(t) + P^*_{or}(t) = \varepsilon_0(\varepsilon_\infty - 1)E^*(t) + \frac{\varepsilon_0(\varepsilon_s - \varepsilon_\infty)}{1 + i\omega\tau_D} E^*(t) \tag{11.10}$$

From the fundamental relations between vectors of electric induction (electric displacement) **D**, polarization **P**, and electric field **E**, we have

$$\mathbf{D} = \varepsilon_0 \varepsilon \mathbf{E} \tag{11.11a}$$

$$\mathbf{D} = \varepsilon_0 \mathbf{E} + \mathbf{P} \tag{11.11b}$$

The complex dielectric permittivity can be expressed by the equation

$$\varepsilon^* = 1 + P^*(t)/\varepsilon_0 E^*(t) \tag{11.12}$$

Substitution of the complex total polarization $P^*(t)$ for Eq. (11.8) in the above equation results in the Debye equation

$$\varepsilon^*(\omega) = \varepsilon'(\omega) - i\varepsilon''(\omega) = \varepsilon_\infty + \frac{\varepsilon_s - \varepsilon_\infty}{1 + i\omega\tau_D} \tag{11.13}$$

which is the basic model of dipole relaxation.

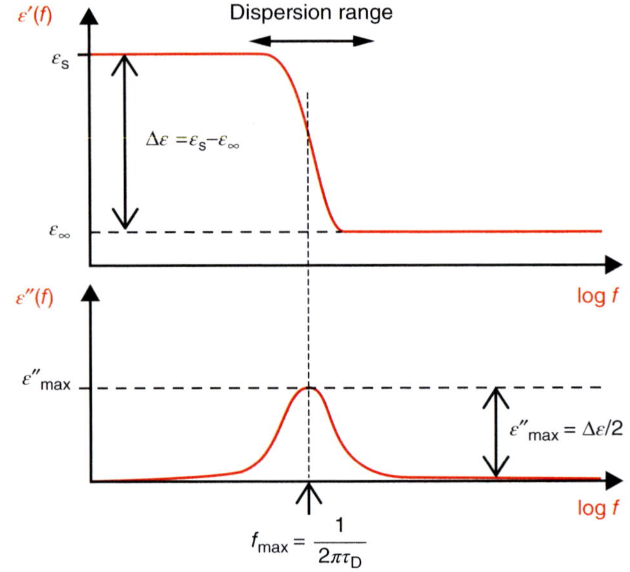

Figure 11.4 Schematic frequency dependences of dispersion $\varepsilon'(f)$ and absorption $\varepsilon''(f)$ for Debye dielectric relaxation.

According to the Debye equation, the dielectric dispersion and absorption, respectively, represented by real and imaginary parts of the complex permittivity, are described in the following way:

$$\varepsilon'(\omega) = \varepsilon_\infty + \frac{\varepsilon_s - \varepsilon_\infty}{1 + (\omega\tau_D)^2} \tag{11.14}$$

$$\varepsilon''(\omega) = (\varepsilon_s - \varepsilon_\infty)\frac{\omega\tau_D}{1 + (\omega\tau_D)^2} \tag{11.15}$$

They are illustrated as a function of frequency $f = \omega/2\pi$ in Figure 11.6.

From Figure 11.4, one can see that the dielectric dispersion $\varepsilon'(f)$ decreases nearly in step with increasing frequency, while the dielectric absorption (dielectric loss) $\varepsilon''(f)$ achieves a maximum ε''_{max} at the frequency f_{max} related to the characteristic relaxation time of the reorienting dipoles, τ_D, given by

$$\tau_D = 1/2\pi f_{max} \tag{11.16}$$

The dielectric strength $\Delta\varepsilon = \varepsilon_s - \varepsilon_\infty$ is a measure of the magnitude of the relaxation process. According to the Onsager model, $\Delta\varepsilon \sim N\mu^2$, which means that the dielectric strength depends on the number of relaxing dipoles per unit of volume, N, which are characterized by the permanent dipole moment μ.

The loss peak of the Debye relaxation is symmetric and narrow. Its full width at half-maximum (FWHM) equals 1.14 decade.

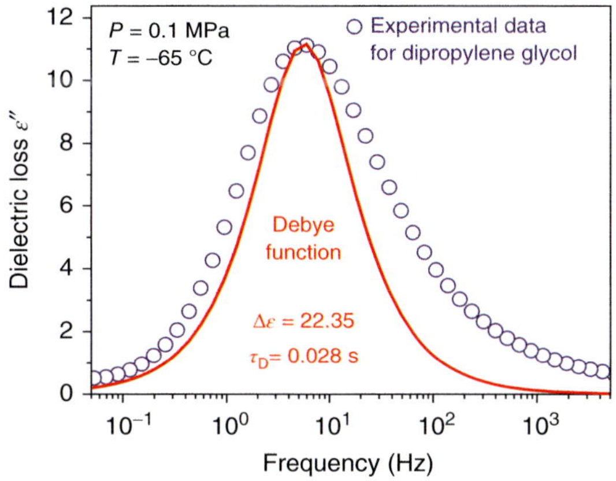

Figure 11.5 Structural relaxation process of dipropylene glycol. The experimental data cannot be described by the Debye model (Eq. (11.13)).

11.1.2.2 **Non-Debye Relaxation Models**

For most materials, the measured dielectric loss peaks are characterized by a much larger FWHM (even up to six decades) in comparison with that of the Debye relaxation process. Additionally, the shape of dielectric absorption peak is usually asymmetric (Figure 11.5). It means that relaxation processes occurring in such materials are too complex to be described by the Debye model (Eq. (11.13)), that is, by the simple exponential function (Eq. (11.6)). In other words, the molecular motions of all dipoles in the sample cannot be characterized by a single relaxation time τ_D (Eq. (11.16)).

Such a broadening of relaxation process can be related to the heterogeneity of the environment nearest to the reorienting molecules. According to the concept of heterogeneous relaxation, molecular motions can be regarded as a superposition of single Debye relaxation processes with different relaxation times τ, which can be described by a continuous, normalized function of the distribution of relaxation times $G(\tau)$ [1, 12]:

$$\varepsilon^*(\omega) = \varepsilon_\infty + \Delta\varepsilon \int_0^\infty \frac{G(\tau)}{1 + i\omega\tau} d\tau, \quad \text{where} \quad \int_0^\infty G(\tau) d\tau = 1 \tag{11.17}$$

Then, we can formulate a non-Debye relaxation function as follows:

$$\Phi(t) = \int_0^\infty e^{-(t/\tau)} G(\tau) d\tau \tag{11.18}$$

whereas the average relaxation time $<\tau>$ can be defined by the integral

$$\langle\tau\rangle = \int_{\text{process}} \tau G(\tau) d\tau \tag{11.19}$$

For the single Debye process, the distribution function $G(\tau)$ is given by the Dirac distribution $\delta(\tau - \tau_D)$, and then Eq. (11.19) yields $<\tau> = \tau_D$.

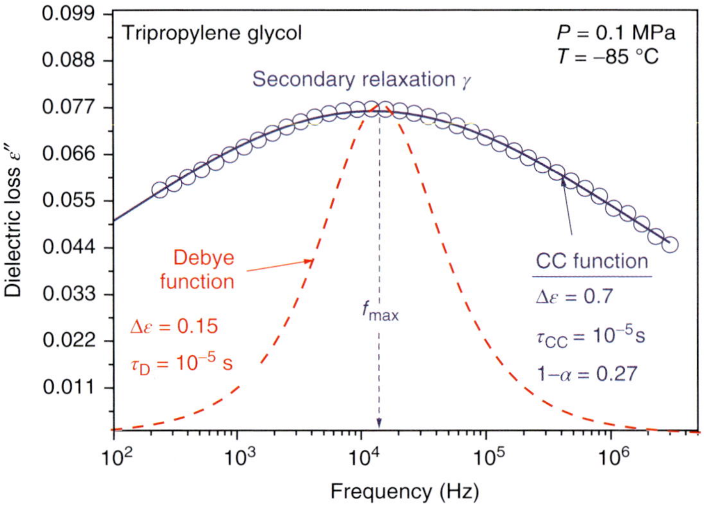

Figure 11.6 Example of fitting the broad and symmetric secondary γ-relaxation loss peak for tripropylene glycol by using the imaginary part of the CC equation (Eq. (11.20)). The experimental data cannot be described by the Debye model (Eq. (11.13)).

Since the distribution functions $G(\tau)$ for non-Debye relaxation processes are difficult to establish analytically, different empirical equations are used to describe the experimental dielectric spectra, such as the following:

- Cole–Cole (CC)
- Cole–Davidson (CD),
- Havriliak–Negami (HN)
- Kohlrausch–Williams–Watts (KWW) function originally proposed to describe the non-Debye relaxation behavior in the time domain, which is based on a modified Debye relaxation function, but also used in the frequency domain.

These models are used to fit the relaxation processes in the frequency domain. They are modifications of the Debye model with some empirical parameters that characterize the relaxation time distributions.

Cole–Cole function

A broad and symmetric relaxation peak (see Figure 11.6) can be described by the CC function [13]

$$\varepsilon^*(\omega) = \varepsilon_\infty + \Delta\varepsilon \frac{1}{1 + (i\omega\tau_{CC})^{1-\alpha}}, \quad 0 \le \alpha < 1 \tag{11.20}$$

where α is a symmetric broadening parameter. One can see that the CC equation reduces to the Debye equation if $\alpha = 0$ (Eq. (11.13)). Similar to the Debye model, the CC process is characterized by the relaxation time τ_{CC}, which can be determined from the frequency at which the maximum dielectric loss occurs, $\tau_{CC} = 1/2\pi f_{max}$, and the average relaxation time $<\tau> = \tau_{CC}$.

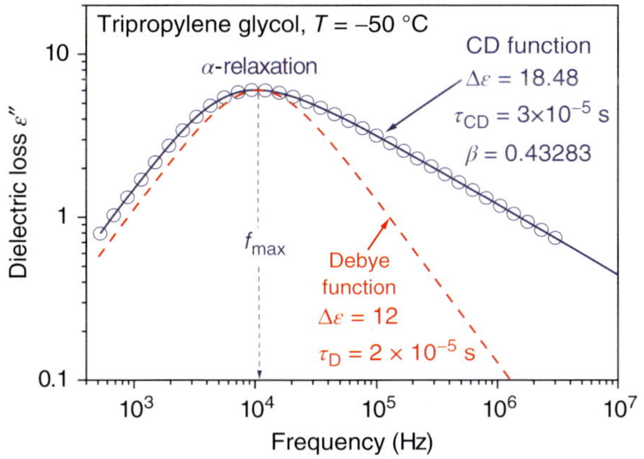

Figure 11.7 Example of fitting the asymmetric structural α-relaxation loss peak for tripropylene glycol by using the imaginary part of the CD equation (Eq. (11.21)). The experimental data cannot be described by the Debye model (Eq. (11.13)).

Cole–Davidson function

Many experimental results, especially for low molecular weight liquids, show that the dielectric loss peaks $\varepsilon''(f)$ are often narrow but asymmetric due to the broadening of its high-frequency side. Such dielectric processes can be described (see Figure 11.7) by the CD function [14, 15]

$$\varepsilon^*(\omega) = \varepsilon_\infty + \Delta\varepsilon = \frac{1}{(1 + i\omega\tau_{CD})^\beta}, \quad 0 < \beta \leq 1 \tag{11.21}$$

where β is an asymmetric broadening parameter for frequencies $f > 1/2\pi\tau_{CD}$, where τ_{CD} is the relaxation time of the CD function. The CD equation becomes the Debye equation (Eq. (11.13)) if $\beta = 1$.

It is worth noting that in the case of the asymmetric CD function, the value of its characteristic relaxation time τ_{CD} is not equal to the relaxation time τ determined from the frequency f_{max} at which maximum dielectric loss occurs. One can find that τ depends on both τ_{CD} and the shape parameter β [1].

$$\tau = \tau_{CD} \operatorname{ctg}\left(\frac{\pi}{2\beta + 2}\right) \tag{11.22}$$

whereas the average relaxation time $<\tau> = \beta\tau_{CD}$ in the CD model [16].

Havriliak–Negami Function

A generalization of the CC and CD models was formulated by Havriliak and Negami [17, 18] (HN function), who proposed the following empirical function to describe broadened and asymmetric dielectric processes (Figure 11.8):

$$\varepsilon^*(\omega) = \varepsilon_\infty + \Delta\varepsilon\frac{1}{[1 + (i\omega\tau_{HN})^{1-\alpha}]^\beta}, \quad 0 \leq \alpha < 1 \text{ and } 0 < \beta \leq 1 \tag{11.23}$$

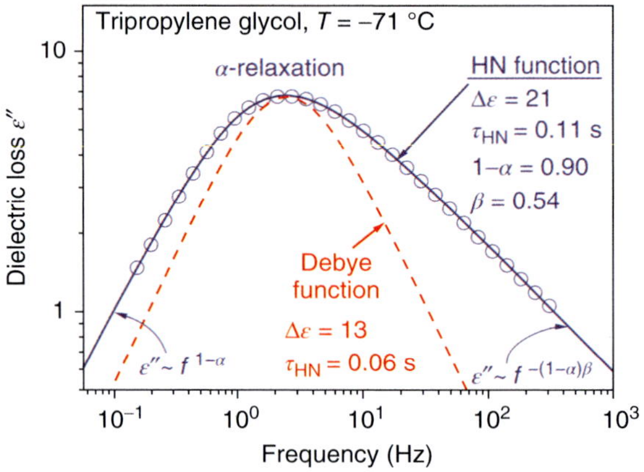

Figure 11.8 Example of fitting the broad and asymmetric structural α-relaxation loss peak for tripropylene glycol by using the imaginary part (Eq. (11.24b)) of the HN equation (Eq. (11.23)). The experimental data cannot be described by the Debye model (Eq. (11.13)).

where the real and imaginary parts of the complex HN function can be represented as

$$\varepsilon'(\omega) = \varepsilon_\infty + \Delta\varepsilon \frac{\cos(\beta\phi)}{\left\{1 + 2\left(\omega\tau_{\mathrm{HN}}\right)^{1-\alpha} \sin\left(\pi\alpha/2\right) + (\omega\tau_{\mathrm{HN}})^{2(1-\alpha)}\right\}^{\beta/2}} \tag{11.24a}$$

$$\varepsilon''(\omega) = \Delta\varepsilon \frac{\sin(\beta\phi)}{\left\{1 + 2\left(\omega\tau_{\mathrm{HN}}\right)^{1-\alpha} \sin\left(\pi\alpha/2\right) + (\omega\tau_{\mathrm{HN}})^{2(1-\alpha)}\right\}^{\beta/2}} \tag{11.24b}$$

with

$$\phi = \mathrm{arctg}\left[\frac{\left(\omega\tau_{\mathrm{HN}}\right)^{1-\alpha} \cos\frac{1}{2}\pi\alpha}{1 + (\omega\tau_{\mathrm{HN}})^{1-\alpha} \sin\frac{1}{2}\pi\alpha}\right]$$

The meaning of the HN parameters (α and β) can be elucidated by considering the following limiting cases: the parameter $(1-\alpha)$ is the slope of the dependence $\log\varepsilon''$ ($\log f$) in the low-frequency limit $f \ll f_{\mathrm{max}}$, while the product of the shape parameters $-(1-\alpha)\beta$ is the slope of the dependence $\log\varepsilon''$ ($\log f$) in the high-frequency limit $f \gg f_{\mathrm{max}}$.

The relaxation time τ established from the frequency f_{max} at which maximum dielectric loss occurs is expressed by the relaxation time τ_{HN} and shape parameters of the HN function in a more complicated way [1]:

$$\tau = \tau_{\mathrm{HN}}\left[\sin\left(\frac{(1-\alpha)\,\pi}{2\beta+2}\right)\right]^{-(1/(1-\alpha))}\left[\sin\left(\frac{(1-\alpha)\,\beta\pi}{2\beta+2}\right)\right]^{1/(1-\alpha)} \tag{11.25}$$

The average relaxation time $<\tau>$ cannot be derived analytically in the case of the HN function [16].

The dielectric spectra of various materials measured in the broad frequency range at a given temperature and pressure often reveal more than one dielectric relaxation process. Moreover, in the low-frequency limit, a DC conduction contribution to the experimental absorption function $\varepsilon''(f)$ is observed, which originates from the translational motions of electric charges (ions and electrons).

Then the entire dielectric spectra can be analyzed as a superposition of the individual relaxation processes (each of them described by a separate HN function) and the DC conduction term:

$$\varepsilon^*(\omega) = \varepsilon'(\omega) - i\varepsilon''(\omega) = -i\left(\frac{\sigma}{\varepsilon_0\omega}\right)^N + \varepsilon_\infty + \sum_k \frac{\Delta\varepsilon_k}{[1 + (i\omega\tau_k)^{1-\alpha_k}]^{\beta_k}} \quad (11.26)$$

where σ is the DC conductivity, N most frequently equals 1, and k is the number of dielectric processes identified in a dielectric spectrum.

Kohlrausch–Williams–Watts Function (KWW)

The non-Debye relaxation processes in the time domain can be described by the empirical exponential function proposed by Kohlraush, Williams, and Watts (KWW function) or the stretched exponential law [19], given by

$$\phi(t) = \exp\left[-\left(\frac{t}{\tau_{\mathrm{KWW}}}\right)^{\beta_{\mathrm{KWW}}}\right] \quad (11.27)$$

where τ_{KWW} is the characteristic KWW relaxation time, whereas the exponent $0 < \beta_{\mathrm{KWW}} \leq 1$ is a measure of the deviation of the time-dependent relaxation function from the Debye exponential decay. If $\beta_{\mathrm{KWW}} = 1$, the KWW function reduces to the Debye relaxation function (Eq. (11.6)), and the KWW relaxation time $\tau_{\mathrm{KWW}} = \tau_{\mathrm{D}}$ (Figure 11.9). For many glass-forming liquids, in the vicinity of the glass transition temperature T_g, the values of the stretching parameter are in the range $0.3 < \beta_{\mathrm{KWW}} < 0.7$ [20].

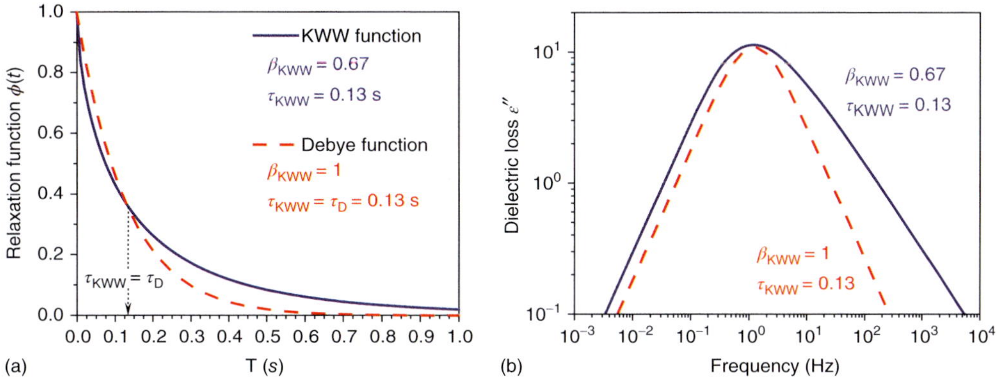

Figure 11.9 Comparison of example Debye (red line) and KWW (blue line) characterized by the same relaxation times $\tau_{\mathrm{D}} = \tau_{\mathrm{KWW}}$ in (a) the time domain and (b) the frequency domain (b).

The KWW function can be also used in the frequency domain. In order to do that, we can employ the Laplace transformation L. Then the dielectric permittivity $\varepsilon*(\omega)$ can be express as follows:

$$\frac{\varepsilon*(\omega)-\varepsilon_{\infty}}{\Delta\varepsilon} = L\left[-\frac{d\phi(t)}{dt}\right] = \int_0^{\infty} e^{-i\omega t}\left[-\frac{d\phi(t)}{dt}\right]dt \tag{11.28}$$

where $\phi(t)$ is the KWW function. The above transformation cannot be, in general, found analytically. Therefore, it is numerically evaluated by the asymptotic series expansion [12]

$$\frac{\varepsilon^*(\omega)-\varepsilon_{\infty}}{\Delta\varepsilon} = \sum_{m=1}^{\infty}\frac{(-1)^{m-1}\Gamma(m\beta_{\mathrm{KWW}}+1)}{(\omega\tau_{\mathrm{KWW}})^{n\beta_{\mathrm{KWW}}}\Gamma(m+1)}e^{\left(im\beta_{\mathrm{KWW}}\frac{\pi}{2}\right)} \tag{11.29}$$

which exploits the special function Γ.

Examples of the KWW function in the time and frequency domains are plotted in Figure 11.9a,b, respectively.

As can be seen in Figure 11.9a, the KWW function with the parameter $\beta_{\mathrm{KWW}} < 1$ decays faster within the time range $t=0$ to $t=\tau_{\mathrm{KWW}}$, and it decreases more slowly at $t>\tau_{\mathrm{KWW}}$ in comparison to the Debye function. This time behavior of the KWW function results in a broadening of the KWW loss peak in the frequency domain (Figure 11.9b). The shape of KWW function in the frequency domain has similar features as those of the CD function.

For the KWW function in the frequency domain, one can evaluate the relaxation time τ determined from the frequency $f_{\max}$ at which maximum dielectric loss occurs, using the parameters τ_{KWW} and β_{KWW} [21].

$$\tau \approx \tau_{\mathrm{KWW}}\, e^{-0.60607(\beta_{\mathrm{KWW}}-1)} \tag{11.30}$$

The average relaxation time for the KWW function is as follows [1]:

$$<\tau>= \frac{\tau_{\mathrm{KWW}}}{\beta_{\mathrm{KWW}}}\Gamma\left(\frac{1}{\beta_{\mathrm{KWW}}}\right)$$

11.1.3 Identification of Relaxation Processes in Dielectric Spectra

In the dielectric spectra of various materials, one can observe several relaxation processes, depending on molecular structure of examined systems as well as thermodynamic conditions (temperature and pressure). As an example, we identify relaxation processes in the dielectric spectra obtained for the popular glass formers such as propylene glycols (PPGs) of different molecular weights (Figure 11.10), which are widely used in pharmaceutical applications.

11.1.3.1 Structural Relaxation α

A common feature of the glass-forming liquids is the structural relaxation α, which can be observed at temperatures higher than the glass transition temperature ($T > T_g$) in the liquid state. The dielectric loss peak of the structural relaxation is usually broad and asymmetric, and its magnitude is significantly

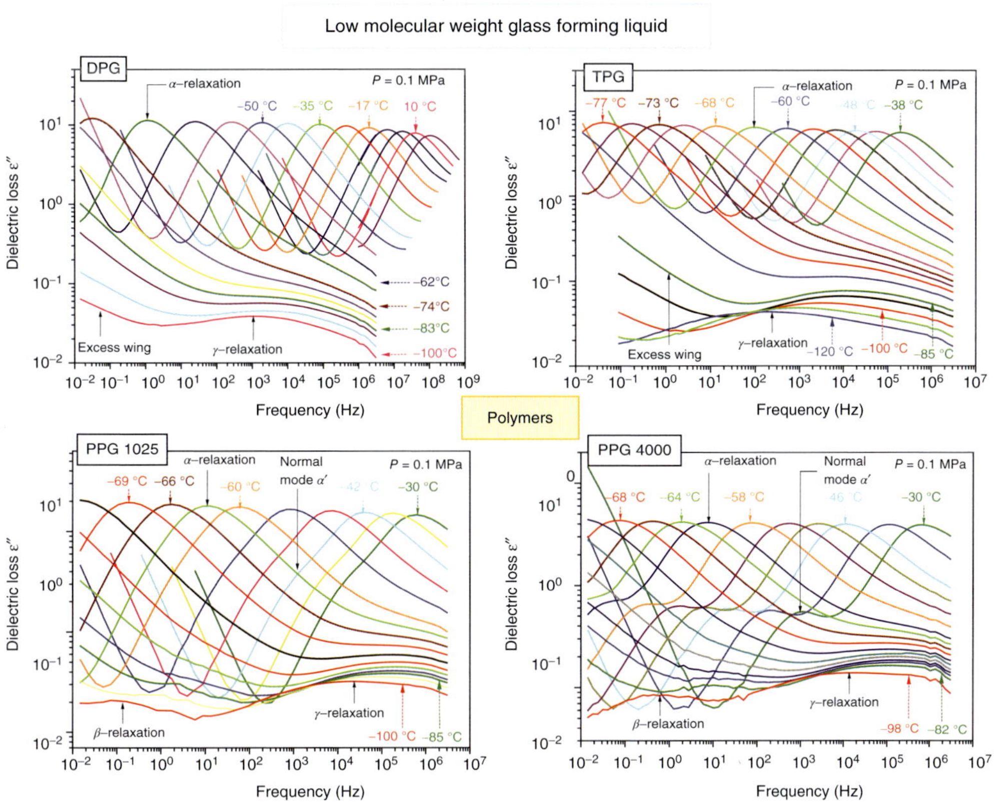

Figure 11.10 Selected dielectric spectra for low molecular weight glass formers DPG and TPG and the polymers PPG 1025 and PPG 4000, measured at ambient pressure and at different temperatures.

larger than that for other relaxation processes. In the case of low molecular weight materials such as the oligomers dimer propylene glycol (DPG) and the trimer propylene glycol (TPG), the α-relaxation reflects reorientation of whole molecules, which leads to the changes of the local structure. For polymers, the structural α-relaxation is related to the some segmental motions in the polymer chains (segmental relaxation), which leads to conformation changes in the macromolecules. As can be seen in Figure 11.10, when the temperature decreases, the structural relaxation peaks are shifted toward lower frequencies, which indicates that the molecular motions associated with α-relaxation become slower. At sufficiently low temperatures, the structure of liquid becomes "frozen" and the system reaches the glassy state. Then, the molecular mobility reflected in the structural relaxation becomes too slow to be experimentally measured (the detection of α-relaxation in the glassy state requires extremely time-consuming measurements). Such a rapid slowing down in the structural relaxation due to

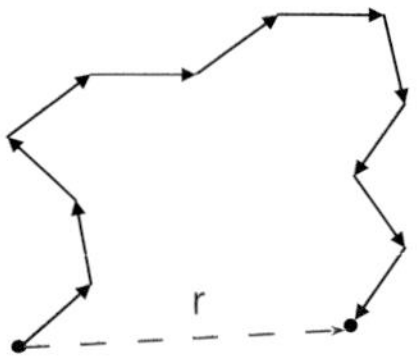

Figure 11.11 In the case of PPG, the normal mode originates from the sum of the segmental dipole moments parallel to the polymer chain backbone [22], which is proportional to the end-to-end vector **r**.

isobaric cooling or isothermal squeezing of the liquid is the main characteristic of its transition to the glassy state.

11.1.3.2 Relaxation α' (Normal Mode)

In the case of polymer melts, for example, PPG 1025 or PPG 4000, one can identify the low-frequency relaxation α', the so-called the *normal mode.* As can be seen in Figure 11.10, the normal mode is observed at frequencies lower than those at which the α-process occurs, which means that the timescale of molecular motions reflected in the normal mode is larger than that of the structural relaxation. The α′-process shifts toward lower frequencies with decreasing temperature, similar to the α-relaxation. It is strongly dependent of the molecular weight.

The molecular origin of the normal mode is the motions of entire polymer chain. According to the Stockmayer theory [22], the normal mode reflects the reorientation of the total dipole moment of the entire polymer chain (i.e., the sum of segmental dipole moments parallel to the polymer chain), which is proportional to the distance vector **r** between the ends of the polymer chain (see Figure 11.11).

11.1.3.3 Secondary Relaxations and Excess Wing

In the supercooled liquid, the structural α-relaxation is the dominant process, which reflects cooperative motions of many molecules and is responsible for the liquid–glass transition. In the glassy state, the α-relaxation process becomes very slow, and, usually, only secondary relaxations, called β, γ, δ, and so on, are observed in the dielectric spectra. Therefore, in fact, the secondary relaxations (of either inter or intramolecular origin) reflecting fast local motions provide us information on the molecular dynamics in the glassy state. Among the different secondary relaxations, those reflecting motions of the whole molecule (intermolecular secondary relaxation) are of particular interest, because they, also called Johari–Goldstein (JG) relaxations, are regarded as precursors of the molecular mobility of the cooperative α-relaxation.

As can be seen in Figure 11.10, in the dielectric spectra obtained in the glassy state of polypropylene glycols, we can distinguish two secondary relaxations, the slower β- and the faster γ-relaxation. The γ-process is observed for all materials, whereas well-separated β-relaxation can be observed only for polymers. In the case of low molecular weight liquids, that is, DPG and TPG, dielectric loss maxima of the β-process are hidden by the dominating structural relaxation. Therefore, a well-developed β-peak cannot be observed; instead, an additional power law to the α-relaxation (a so-called *excess wing*) appears, which is, in fact, a high-frequency flank of the β-process. The molecular origin of the excess wing in glass-forming

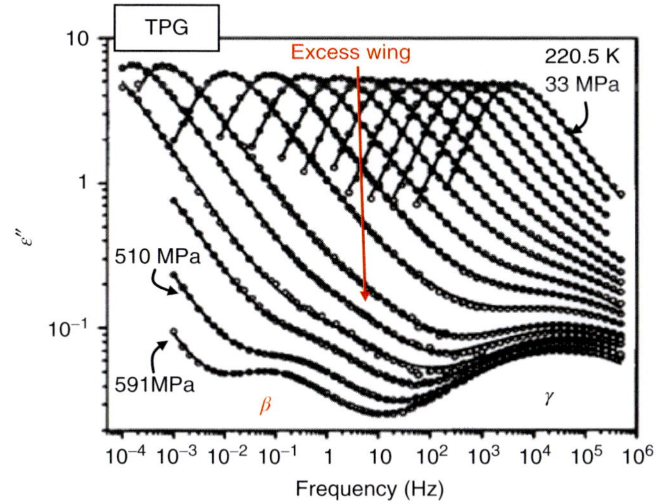

Figure 11.12 Dielectric loss spectra for TPG measured at a constant temperature −53 °C and at different pressures (from right to left) 33.4, 61.9, 93.0, 120.7, 150.0, 180.2, 209.3, 237.5, 268.6, 297.2, 331.3, 373.4, 415.3, 447.2, 463.7, 510.2, and 591.3 MPa. (Adapted from Casalini and Roland [24]. Reproduced with permission of American Physical of Society.)

liquids is one of the most discussed aspects of the glass transition dynamics. Johari and Pathmanathan [23] were the first to suggest that the excess wing is the β-process hidden underneath the α-peak. This interpretation has been confirmed by the high-pressure dielectric measurements on TPG [24]. As can be seen in Figure 11.12, when the pressure increases, the α-peak quickly shifts toward lower frequencies, while the γ-process remains practically independent of pressure (the frequency f_{max} at which maximum dielectric loss occurs does not change). Consequently, the excess wing becomes more and more prominent, and at $P = 0.5$ GPa we can distinguish a well-separated β-process in the dielectric spectra of TPG.

In order to precisely determine the relaxation times and shape parameters characterizing all the individual relaxation processes observed in the dielectric spectra, it is necessary to carry out a numerical fitting of the whole spectra according to Eq. (11.26).

Examples of the approximation of dielectric spectra for PPG as a superposition of the individual relaxation processes are shown in Figure 11.13. The α-peaks are usually broad and asymmetric, thus they can be described in the frequency domain by means of the HN (Eq. (11.23)) or CD approaches (Eq. (11.21)). The CC function (Eq. (11.20)) is often used to describe the broad and symmetric normal modes as well as the secondary relaxation processes. In fact, the β-process of PPG has been described successfully by the CC function, but the faster secondary relaxation γ-peak in this material is broad and asymmetric (which suggests that the molecular mobility reflected in the γ-relaxation is more complex than in standard glass-forming liquids) and has to be approximated by the HN function.

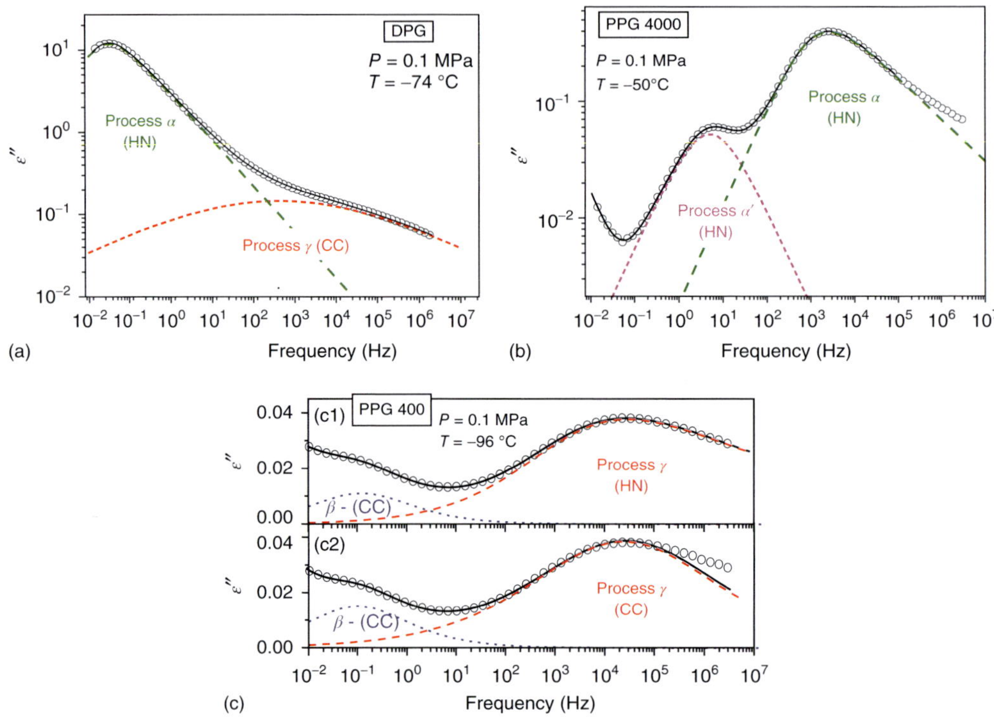

Figure 11.13 Examples for decomposition of the experimental dielectric loss into a sum of elementary relaxation processes (a) α and β for DPG at $T = -74\,°C$, (b) α' and α for PPG 400 at $T = -50\,°C$, and (c1,c2) β and γ for PPG 400 at $T = -96\,°C$. Solid black lines indicate the fitting curves of the entire spectra.

11.2 Molecular Dynamics in Amorphous Pharmaceutical Systems

11.2.1 Molecular Mobility in the Liquid and Glassy States of Drugs – Physical Factors that Can Govern Crystallization

Since amorphous drugs are usually thermodynamically unstable and may undergo recrystallization during the processing, storage, and use of the product, the main challenge in working with amorphous form of drugs is improving their physical stability as well as proper identification and understanding of the physical factors that govern crystallization from the glassy state. Probably the most important factor determining the stability of amorphous drugs is their molecular mobility [25–30]. For the purpose of determining molecular mobility, it is helpful to study the relaxation processes in the supercooled liquid and glassy states.

A useful method to determine the timescales of molecular motions of pharmaceuticals in the glassy and liquid states is BDS (described in the Section 11.1.1), which enables measurements of relaxation times over a wide frequency range of up to 16 decades at different temperatures and pressures. As already mentioned, glass-forming materials usually exhibit several relaxation processes of different nature in the dielectric relaxation spectra, which can be distinguished by their properties including their dependence on temperature and pressure.

11.2.1.1 Molecular Mobility in Supercooled Liquids (Structural α-Relaxation, Fragility)

The dominant relaxation process in the supercooled liquid state observed at temperatures T higher than the glass transition temperature T_g is the structural α-relaxation. It reflects the cooperative and correlated motions of many molecules together and is responsible for the liquid–glass transition.

During vitrification, which can be achieved by sufficiently fast cooling of a liquid to avoid its crystallization, the α-process slows down drastically (see Figure 11.14). Such an extremely rapid increase in the structural relaxation time τ_α of the liquid approaching the glassy state is the hallmark of the liquid–glass transition. The structural relaxation time τ_α and viscosity η of supercooled liquids vary between the melting temperature T_m and the glass transition temperature T_g by 15 orders of magnitude. It is usually assumed that the glass transition occurs at $\tau_\alpha(T_g) \approx 100$ s and $\eta(T_g) \approx 10^{13}$ P.

To quantitatively describe the temperature dependence of α-relaxation times in the supercooled liquid, the empirical Vogel–Fulcher–Tamman (VFT) equation is

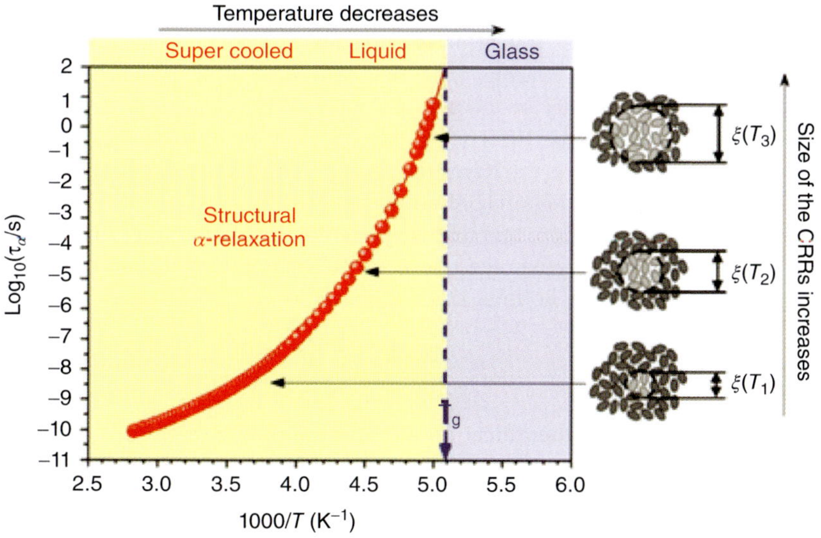

Figure 11.14 Non-Arrhenius temperature dependence of α-relaxation times for a glass-forming liquid together with schematic illustration of increasing cooperatively rearranging regions during cooling (ξ is the so-called cooperativity length).

most commonly used [31–34]

$$\tau_\alpha = \tau_\infty \exp\left(\frac{DT_0}{T - T_0}\right) \quad (11.31)$$

where τ_∞, T_0, and D are fitting parameters.

The nonlinear increase in $\tau_\alpha(T)$ with decreasing temperature is often explained by the increase in cooperativity of molecular motions near the glass transition. In a supercooled liquid, motions of particular molecules are highly limited (because of the limited free volume) and depend on the motions of neighboring molecules. Rearrangement of a given molecule in such a system is possible only if some of the neighboring molecules also move. Adam and Gibbs [35] introduced the concept of a *cooperatively rearranging region* (CRR), which denotes a group formed by some molecules that can rearrange, as a result of thermal fluctuations, to another configuration independently of its neighbors (there is no interaction between different CRRs). A reason for the rapid slowdown of molecular dynamics of a supercooled liquid with temperature is the growth of the average size ξ of CRR, which is illustrated in Figure 11.14. The Adam–Gibbs (AG) model relates the thermodynamic properties of supercooled liquids to their molecular dynamics by the assumption that the size of CRR depends on the configurational entropy, $S_c = S^{liq} - S^{cryst}$, and suggests describing the temperature dependences of structural relaxation times by the following equation:

$$\log \tau_\alpha = \log \tau_{AG} + \frac{C}{TS_c} \quad (11.32)$$

where C is a constant.

According to this model, the slowdown in the molecular dynamics of supercooled liquids near the glass transition is caused by the decrease in the number of possible system configurations (the decrease in S_c). It is worth noting that the fundamental model proposed by Adam and Gibbs implies the VFT equation (Eq. (11.31)). Since the configurational entropy can be expressed as $S_c = a(T-T_K)/T$, where T_K is the Kauzamnn temperature, the AG equation leads to the VFT equation, with the interpretation that $T_0 = T_K$.

One of the important parameters that characterize the temperature function of structural relaxation is the dynamic *isobaric fragility* parameter m (also called the *steepness index*), defined by Böhmer *et al.* [20] as follows:

$$m \equiv \left.\frac{d \log \tau_\alpha}{d\left(T_g/T\right)}\right|_{T=T_g} = \frac{D(T_0/T_g)}{(1-(T_0/T_g))^2 \ln(10)} \quad (11.33)$$

Using the parameter m, we can classify supercooled liquids into two types [36]: "strong" if the temperature dependence of τ_α is close to Arrhenius behavior in the plot of $\log \tau_\alpha$ versus the scaled temperature T_g/T, and "fragile" if the temperature dependence of $\log \tau_\alpha$ deviates significantly from Arrhenius law but can be expressed by means of the VFT equation (see Figure 11.15). "Strong" glass formers are usually characterized by $m \leq 30$, whereas the "fragile" ones have $m \geq 100$. If

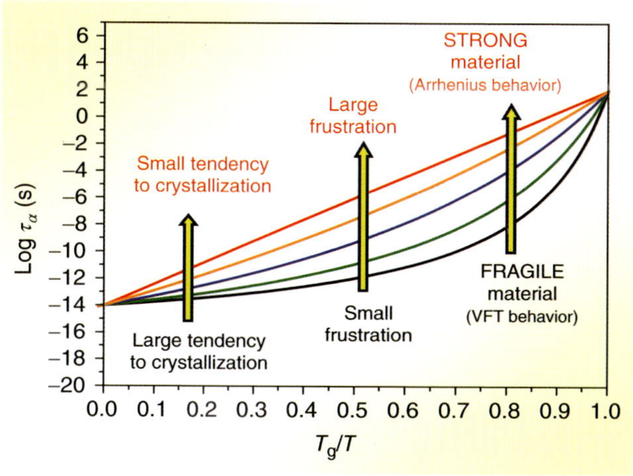

Figure 11.15 Illustration of the Tanaka concept of frustration against crystallization. The fragile system crystallizes more easily because its frustration against crystallization is weaker than in a strong material.

m falls within the range $30 < m < 100$, the liquid is classified as intermediate glass formers.

The concept of fragility is of interest in current research fields such as the formulation of amorphous drugs and food preservation, because it is considered a key factor that correlates with the glass-forming ability and physical stability of amorphous systems [25, 29, 30, 37–39]. This is because the parameter m is related to the average degree of molecular mobility reflected in structural relaxation near the glass transition. "Fragile" glassformers have molecular mobility that varies rapidly with temperature near T_g in contrast to that occurring in "strong" liquids. This difference has been considered by some as the reason why strong liquids are more physically stable than fragile liquids.

The correspondence between the large fragility and the ease of crystallization can be better realized within the framework of the two-order-parameter (TOP) model proposed by Tanaka [40]. According to this model, a liquid near the glass transition tends to order into the equilibrium crystal (long-range ordering), but it does not crystallize because of frustration (i.e., locally favored short-range ordering). Frustration effects increase the free-energy barrier for nucleation and act as impurities against crystallization and facilitate vitrification. Therefore, a fragile system crystallizes more easily than strong glass formers, because its frustration against crystallization is weaker, which is illustrated in Figure 11.15. This hypothesis was later supported by simulations [41].

Two decades ago, based on the analysis of many experimental data of glass formers measured at ambient pressure, Böhmer *et al.* [20] suggested a linear, empirical correlation between the isobaric fragility and the stretch exponent β_{KWW} of the KWW function (Eq. (11.27)):

$$m_{corr} = 250 \pm 30 - 320\beta_{KWW} \tag{11.34}$$

According to this relation, fragile materials with large values of m should be characterized by broad structural relaxation peaks near T_g, that is, by a high level of non-exponentiality of dielectric relaxation response or a small value of β_{KWW}, which means large deviation from the Debye relaxation behavior. On the other hand, strong glass formers (small m) should exhibit narrow relaxation peaks or large values of β_{KWW} in the vicinity of T_g. Faster modes of molecular motions within the spectrum of relaxation times can be responsible for nucleation in the glassy state, and therefore a more fragile glass former (of smaller β_{KWW}) would be more susceptible to nucleation [42]. Thus, the exponent β_{KWW}, which is often considered as a measure of the degree of cooperativity, length scale, or dynamic heterogeneity of the structural relaxation [5, 20], can be also regarded as an alternative parameter to the fragility index m for the purpose of correlating or predicting the physical stability of amorphous pharmaceuticals. However, it should be noted that there are a few materials [43] that do not satisfy the correlation suggested by Böhmer *et al.* In the case of pharmaceutical systems, such an exception to this rule has been found, for example, for the anti-inflammatory drug celecoxib (CEL) [30], which is characterized by a large value of the fragility index ($m = 110$, calculated at $\tau_\alpha = 100$ s) as well as by narrow peaks of α-relaxation near T_g (i.e., the exponent β_{KWW} describing the width of the structural relaxation peak is equal to 0.67). Assuming that the Böhmer correlation is valid for CEL, it has been predicted that the fragility equals only $m_{P\_corr} = 36 \pm 30$. However, the real value of fragility is much larger ($m = 110$). Thus, CEL does not satisfy the Böhmer correlation. Very recently, it has been suggested [44] that the parameter β_{KWW} is not related to the dynamic heterogeneity near the glass transition, which may explain the found exceptions for the relation between m and β_{KWW}.

Fragility manifests not only kinetically in the temperature dependence of log τ_α (i.e., different degrees of departures from the Arrhenius behavior) but also in the thermal response of glass formers. Contrary to strong materials, fragile liquids usually show large changes in thermal response (e.g., in the heat capacity C_p) near the glass transition. Fragile liquids are expected to have larger values of configurational heat capacities C_{pconf}, resulting from their configurational entropy changing rapidly with temperature, whereas strong liquids have small C_{pconf} because their configurational entropy changes only slowly with T [36, 45].

Many attempts have been made to define a thermodynamic measure of fragility and predict the parameter m using calorimetric methods as well as to find a proper correlation between the dynamic fragility and thermodynamic fragility [46–48].

Angell [36] proposed that a thermodynamic measure of fragility can be a ratio of the heat capacities in the liquid and crystalline states, that is, C_p^l/C_p^c, evaluated at T_g. In the case of materials that do not exhibit any crystalline form (as is the case for many polymers), the ratio C_p^l/C_p^c is replaced with the ratio C_p^l/C_p^g in the liquid and glassy states [46, 49]. Huang and McKenna [46] compared the dynamic and thermodynamic fragilities of many compounds from different classes of materials and showed that, in general, there is no positive correlation between them (which was the commonly accepted picture). They obtained (i) an increase in m with increasing C_p^l/C_p^c for inorganic glass formers, (ii) a decrease in m with increasing C_p^l/C_p^g for polymers, and (iii) a very weak correlation between m and C_p^l/C_p^c for organic and H-bonding small molecules.

Another thermodynamic parameter, γ_{C_p}, that is often used to characterize the fragility of pharmaceutical glass formers is defined as follows [25, 27, 29, 37]:

$$\gamma_{C_p} = \left.\frac{C_p^l - C_p^g}{C_p^l - C_p^c}\right|_{T=T_g} = \left.\frac{\Delta C_P}{C_{p\,\mathrm{conf}}}\right|_{T=T_g} \tag{11.35}$$

The value of γ_{C_p} can vary between 0 (for extremely fragile systems) and 1 (for extremely strong materials).

An empirical relation between dynamic and thermodynamic fragility has been proposed by Wang and Angell [47, 48, 50]. Using the experimental data for 42 materials, they established the correlation

$$m = 56\frac{T_g \Delta C_P}{\Delta H_m} \tag{11.36}$$

which involves a jump in the heat capacity ΔC_P at T_g and the enthalpy of fusion ΔH_m.

Based on the random first-order transition theory, Lubchenko and Wolynes [51] derived a similar correlation:

$$m = 34.7\frac{T_m \Delta C_P}{\Delta H_m} \tag{11.37}$$

where T_m is the temperature of melting. The correlation was successfully tested for 44 substances belonging to different material classes [47].

However, because of the complex nature of the glass transition, this correlation is not ideally satisfied, and there are exceptions to this prediction (see Table 11.1).

As suggested by Cangialosi *et al.* [52], the reason for the discrepancy between dynamic and thermodynamic fragility might be associated with the residual excess entropy at T_0. At this temperature, according to the AG equation, no α-relaxation-related excess entropy is expected. However, as it turns out, many materials have nonzero excess entropy at T_0, which represents all motions not related to the structural relaxation (i.e., secondary relaxations) and arising from the internal

Table 11.1 Dynamic and thermodynamic fragilities for several materials.

Material	*m* from Wang and Angell (Eq. (11.36))	*m* from Lubchenko and Wolynes (Eq. (11.37))	*m* from dielectric measurements
Adenosine	83	71	90
Thymidyne	37	33	83
Uridine	82	66	89
Celecoxib	83	67	110
Sildenafil	85	73	85
Telmisartan	80	67	87

degrees of freedom. Consequently, thermodynamic fragility is composed of contributions from primary and secondary processes, but only the former is relevant to the dynamic fragility and must be extracted for a credible comparison of the thermodynamic and dynamic fragilities [53]. Consequently, the fact that thermodynamic fragility suffers from the effect of secondary relaxations might be a direct reason why correlation between thermodynamic and dynamic fragilities does not work well for nucleosides.

Therefore, further studies and experiments are necessary to verify the correlation between dynamic and thermodynamic fragility and to establish the most general one.

11.2.1.2 Molecular Mobility in the Glassy State (Secondary Relaxation Processes)

As already mentioned, in the glassy state the α-relaxation process becomes very slow, and usually only secondary relaxations, called β, γ, and so on, are observed in the dielectric spectra, as illustrated in Figure 11.16. Therefore, the secondary relaxations, reflecting fast local motions, provide us information on the molecular dynamics in the glassy state. The temperature dependence of the secondary relaxation times (τ_β, τ_γ, etc.) in the glassy state is commonly described by the Arrhenius equation:

$$\tau(T) = \tau_\infty \exp\left(\frac{\Delta E}{k_B T}\right) \tag{11.38}$$

where τ_∞ is a pre-exponential factor, ΔE is the energy barrier for a given secondary process, and k_B is Boltzmann constant. However, it often happens that evaluation of secondary relaxation times in the liquid state is very difficult because of the

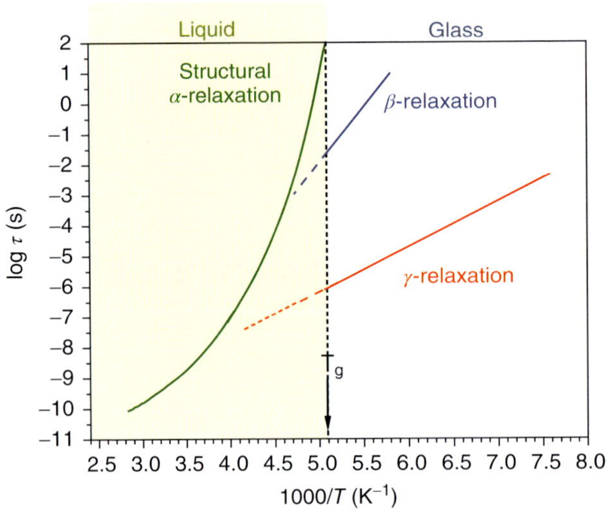

Figure 11.16 Diagram of typical temperature dependences of structural and secondary relaxations observed near the liquid–glass transition.

strong coupling between the structural and secondary processes. Then, the temperature dependence of the secondary relaxation times in the liquid state can be more complex [54] than those described by a single Arrhenius equation.

Among the different secondary relaxations, one that has an intermolecular origin, called JG relaxation, is of particular interest. It is assumed that the JG β-relaxation is related to local reorientations of the whole molecules. There are two conceptions of the molecular origin of the JG relaxation. (i) In a *heterogeneous scenario* [55, 56], it is assumed that JG reflects local reorientations of the molecules only in the so-called *mobility islands*, which are the isolated regions of lower density in the glass, where molecules are loosely packed within the glassy matrix. (ii) In the alternative *homogeneous scenario* [57], it is regarded that the JG β-relaxation originates from small angular reorientations of all molecules, which are not limited to any isolated mobility islands. These motions have smaller amplitudes than the structural α-process.

It should be stressed that the JG process, irrespective of the heterogeneous or homogeneous model, reflects local reorientations of entire molecules both in the case of rigid or flexible molecules. This intermolecular secondary relaxation is postulated to be a universal feature of all glass formers and is considered as the precursor of structural relaxation (such local molecular motions lead to the α-relaxation).

Besides the intermolecular secondary process, one can distinguish also *intramolecular* secondary relaxations in the dielectric spectra at higher frequencies than those at which the JG β-relaxation occurs. The non-JG secondary processes are considered to originate from intramolecular reorientations of some parts of the molecules.

Ngai and Paluch [58], based on the extended coupling model (CM) [59], proposed a criterion to classify secondary relaxations according to their inter or intermolecular character. In the CM model, it is assumed that there is a characteristic microscopic time t_c, independent of thermodynamic conditions, that distinguishes domains of noncooperative and cooperative molecular motions. At $t < t_c$, molecules relax independently with the characteristic primitive (noncooperative) relaxation time τ_p, according to the Debye relaxation function $\phi(t) = \exp(-t/\tau_p)$; whereas at $t > t_c$, the molecules interact in a cooperative manner with the characteristic relaxation time τ_α, according to the KWW function $\phi(t) = \exp[-(t/\tau_p)^{\beta_{KWW}}]$. Thus, the primitive relaxation time τ_p can be derived from the continuity condition at $t = t_c$ as follows:

$$\tau_p = t_c^{1-\beta_{KWW}} \tau_\alpha^{\beta_{KWW}} \tag{11.39}$$

where $t_c \approx 2$ ps for small molecular and polymeric glass formers, which has been found from quasielastic neutron scattering [60, 61]. Since the JG relaxation is regarded as the precursor of structural relaxation, its relaxation time τ_{JG} should correspond well to the primitive relaxation time τ_p of the CM: that is

$$\tau_{JG} \approx \tau_p \tag{11.40}$$

Thus, based on Eq. (11.40), one may classify the secondary relaxation observed in dielectric spectra as the JG β-process if its relaxation time corresponds well to the primitive relaxation time τ_p; otherwise, the secondary relaxation is a non-JG process.

Many researchers argue that nucleation in the glassy state may results from local molecular motions, reflected in the experiment as secondary relaxations [26, 42, 62]. Studies of organic molecular glasses have shown that these localized motions, but not the long-range diffusion of the α-relaxation process, determine the nucleation and crystallization rates in the glassy state [42, 62–65]. The study of the JG relaxation is of fundamental importance because this process occurs in almost all glass-forming systems and the investigation of its relation with structural relaxation should allow a deeper understanding of the glass transition phenomenon and provide answer to the question of how the secondary processes influence the crystallization of amorphous drugs.

11.2.2 Physical Stability of Drugs in the Liquid and Glassy States

11.2.2.1 Prediction of Physical Stability of Amorphous Drugs from Molecular Mobility Studies (At $T < T_g$)

A very important merit of amorphous drugs is their significantly higher water solubility and, consequently, better bioavailability. Unfortunately, they are usually physically unstable and may crystallize during storage. Thus, often drugs on the market will not be amorphous, thus posing a big problem. Therefore, an important challenge in pharmaceutical studies is ensuring sufficient glass stability to guarantee a long shelf-life of the product. From the practical point of view, amorphous pharmaceuticals that are stable near room temperature T_{RT} are of great interest. The knowledge of storage conditions that will ensure long-term stability of the amorphous drugs can be obtained from investigations of the timescales of molecular motions in the glassy state, which are reflected in both structural and secondary relaxations. As already mentioned, secondary relaxations in the glassy state can be directly measured, whereas such measurements of the structural relaxation are almost impossible below T_g because of the exceedingly long timescale. Thus, the temperature dependence of α-relaxation time below T_g, especially its VFT or Arrhenius-like character, is unidentified. However, there are a few methods based on dielectric measurements for the prediction of the timescale of global molecular mobility in the glassy state.

A commonly used method for evaluating τ_α below T_g is based on the AG model [35] (Eq. (11.32)) extended [27, 29, 45] to the glassy state. According to the extended AG approach, the temperature dependence of structural relaxation times can be evaluated from the following formula:

$$\tau_\alpha(T, T_f) = \tau_\infty \exp\left(\frac{DT_0}{T\left(1 - T_0/T_f\right)}\right) \quad (11.41)$$

where τ_∞, D, T_0 are the fitting parameters found from the VFT equation (Eq. (11.31)) for the dependence $\tau_\alpha(T)$ in the liquid state. T_f in Eq. (11.41) is the fictive temperature, which can be estimated by using the thermodynamic parameter γ_{Cp} (Eq. (11.42)) in the following way:

$$\frac{1}{T_f} = \frac{\gamma_{Cp}}{T_g} + \frac{1-\gamma_{Cp}}{T} \tag{11.42}$$

The fictive temperature is a function of temperature in terms of the heat capacities of the crystal C_p^c, of glass C_p^g, and of the supercooled liquid C_p^l, and is bounded the Kauzmann temperature T_K and the real glass transition temperature T_g, ($T_K < T_f < T_g$). According to the definition of the parameter γ_{C_p}, its value can vary between 0 (for extremely fragile systems) and 1 (for extremely strong materials). It means that the fictive temperature T_f defined by Eq. (11.42) is constant and $T_f = T_g$ if $\gamma_{C_p} = 1$. Then Eq. (11.41) becomes the Arrhenius equation. On the other hand, if $\gamma_{C_p} = 0$, then $T_f = T$ and Eq. (11.41) results in the VFT equation. Thus, the extended AG model enables us to predict the temperature evolution of structural relaxation times in the glassy state, which ranges between the Arrhenian and VFT behavior. As an example, we can see in Figure 11.17 the extended AG model prediction for telmisartan with $\gamma_{C_p} = 0.75$ [66].

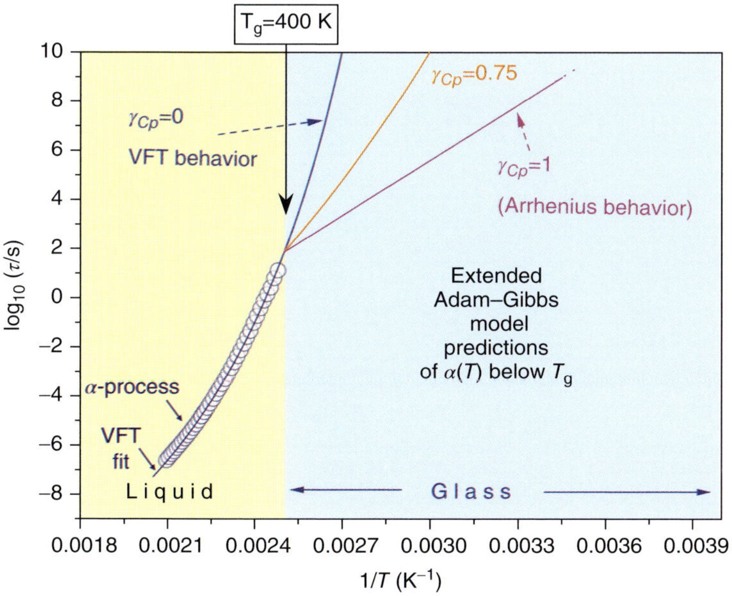

Figure 11.17 Extended AG prediction for τ_α below T_g (orange line) for telmisartan (with γ_{C_p}=0.75) compared with two extreme cases of Eq. (11.41) with γ_{C_p}=0 and γ_{C_p}=1, which are the VFT (blue line) and Arrhenius (magenta line) equations, respectively. The blue circles denote the experimental temperature dependence of structural relaxation times in the liquid state, whereas the blue line indicates its VFT approximation. (Adapted from Adrjanowicz *et al.* [66]. Reproduced with permission of Elsevier.)

To predict the structural relaxation time in the glassy state by using the extended AG model, we need to know the heat capacity data besides the structural relaxation times in the liquid state. However, there are other methods for predicting $\tau_\alpha(T)$ in glass widely described in the literature, which are based only on the experimental dielectric relaxation data. Structural relaxation times in glass can be predicted by constructing a master plot for the structural relaxation in glass [30, 66], by using physical aging of the secondary relaxation [66–68], or by using the extended CM for the JG secondary relaxation [60, 69, 70].

Is There any Correlation Between Physical Stability and Molecular Mobility in the Glassy State? – The Case of Telmisartan

By comparing the predicted timescales of structural relaxation for telmisartan in the glassy state (Figure 11.18), we can obtain similar results by constructing the master plot for α-relaxation ($\tau_\alpha^{\text{master}}$), performing physical aging of the secondary JG relaxation (τ_{ag}), and using the extended CM model (τ_α^{CM}). However, the temperature dependence of α-relaxation times below T_g predicted from the extended AG model corresponds to the other predictions only in the vicinity of the glass

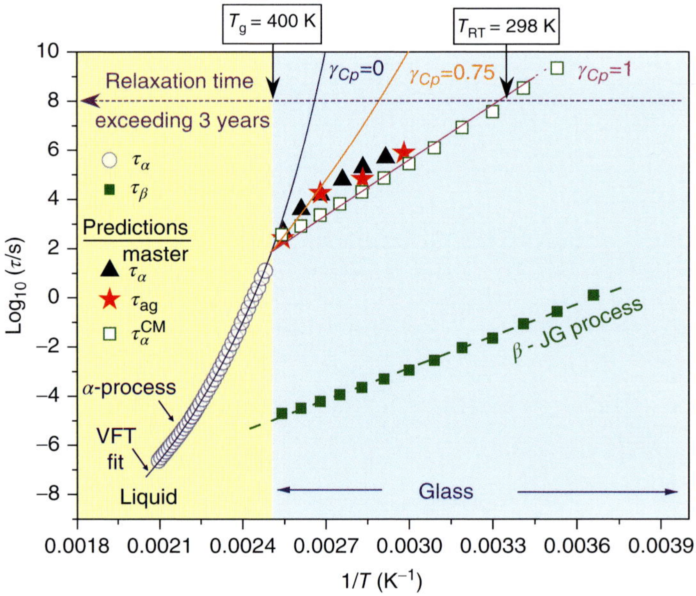

Figure 11.18 Predictions for structural relaxation τ_α below T_g for telmisartan obtained from different methods: the masterplot (solid up-triangles), aging of the secondary JG relaxation (red stars), the extended CM model (open squares), and the extended AG model (orange line). The blue circles denote the experimental temperature dependence of structural relaxation times in the liquid state, whereas the blue line indicates its VFT approximation. The olive squares indicate the experimental temperature dependence of secondary JG β-relaxation times, while the olive line represents its Arrhenius approximation. (Adapted from Adrjanowicz *et al.* [66]. Reproduced with permission of Elsevier.)

transition; deep in the glassy state, it considerably deviates (increases) from the other predicted dependences with decreasing temperature. Such a discrepancy between the predicted α-relaxation times can suggest a limitation on the extended AG method to predict molecular dynamics of only freshly formed glasses.

It should be stressed that, if we take into account even the predictions that provide the shortest τ_α in the glassy state, the timescale of global molecular motions (reflected in the structural relaxation) at typical storage at room temperature ($T_{RT} = 293$ K) and even human body temperature will exceed 3 years, which is a typical desirable shelf-life for a drug. This observation very well correlates with the high physical stability of the amorphous drug telmisartan. It was confirmed experimentally by XRD measurements that the glassy state of the drug is still physically stable after 2.5 years of storage at ambient conditions (which makes the amorphous telmisartan prepared by vitrification a promising candidate for further pharmaceutical investigations). Thus, the extremely slow structural relaxation at T_{RT} can determine the high stability of the amorphous drug, whereas the faster molecular mobility in the glassy telmisartan, as reflected in the secondary relaxations, does not destroy the high stability of the glassy telmisartan during its storage.

11.2.2.2 Relation Between Cold Crystallization of Liquid and Molecular Mobility

Various amorphous drugs show different tendencies for recrystallization. Some of them (e.g., telmisartan) are very good glass formers and, in general, they are resistant to recrystallization; others recrystallize from the supercooled liquid state but their glassy forms are physically stable during their storage (e.g., sildenafil); and yet others easily recrystallize both in the glassy and liquid states (e.g., CEL). Therefore, it is very important to investigate the crystallization phenomenon of amorphous pharmaceuticals near their glass transitions (both in the supercooled liquid and glassy states). Since molecular mobility is considered to play a key role in the crystallization phenomenon, a better understanding of the nucleation and crystal growth mechanisms, usually investigated by experiments of crystallization kinetics, can be gained at the molecular level by finding correlations between molecular mobility and crystallization.

Crystallization Kinetics Study of Supercooled Liquid by BDS

BDS is a powerful method for detecting and quantifying recrystallization processes occurring in drugs because it is a very sensitive technique to investigate the changes in molecular dynamics during the crystallization process in terms of a reduction in the molecular mobility, as reflected in the structural relaxation, which corresponds to a drop in the dielectric strength $\Delta\varepsilon_\alpha$ of the structural α-process. The dielectric strength $\Delta\varepsilon_\alpha$ depends on the number N of relaxing dipoles per unit of volume, which are characterized by the permanent dipole moment μ (e.g., $\Delta\varepsilon_\alpha \sim N\mu^2$ in the Onsager model). As the degree of crystallization increases, the number of relaxing dipoles decreases as a result of their immobilization, which is observed as a drop in $\Delta\varepsilon_\alpha$. Thus, we can investigate the non-isothermal and isothermal crystallization of materials by analyzing the changes in $\Delta\varepsilon_\alpha$ as a function of temperature or time.

As an example, we consider the non-isothermal and isothermal crystallization of supercooled sildenafil, which is a representative of drugs that easily recrystallize in the supercooled liquid state and reveal a resistance to recrystallization during their storage in the glassy state [71].

Non-Isothermal Crystallization By analyzing the structural relaxation, we can detect the cold crystallization of sildenafil under non-isothermal conditions. Figure 11.19 presents the dielectric spectra of supercooled sildenafil obtained on non-isothermally heating the sample from the glassy to the liquid state at atmospheric pressure. As can be seen in Figure 11.19a, the α-process peaks move toward higher frequencies on heating up to $T_{cryst} = 360$ K, which indicates

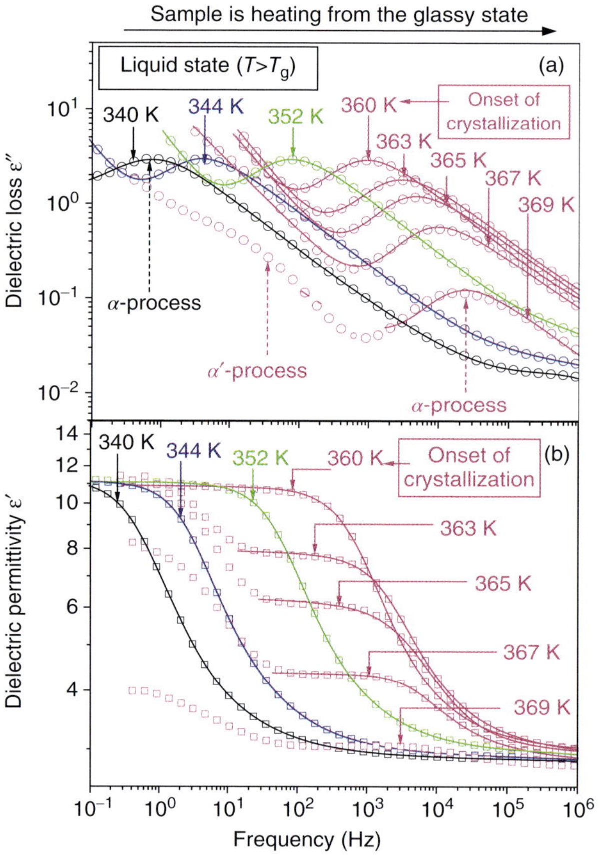

Figure 11.19 (a) Imaginary and (b) real part of the complex dielectric permittivity of sildenafil at different temperatures at $T > T_g$. (Adapted from Kołodziejczyk *et al.* [71]. Reproduced with permission of American Chemical Society.)

the increase in molecular mobility of the system. At $T > 360$ K, amplitudes of both the structural relaxation loss peaks (Figure 11.19a) and static dielectric permittivity of the α-process (Figure 11.19b) begin to decrease rapidly with increasing temperature. Such a sudden drop in the dielectric strength of the α-process is caused by the onset of the sample recrystallization on heating and reflects the increasing degree of crystallinity. It is worth noting that a new relaxation process (α') emerges in the low-frequency flank of the α-peak at high crystallization degrees of sildenafil. This α'relaxation can be attributed to conformational mobility originating from the remaining amorphous fraction adjacent to the crystal surfaces [72].

Isothermal Crystallization BDS can also be successfully used to study the isothermal crystallization kinetics of supercooled drugs. In such measurements, the spectra of the complex dielectric permittivity of the examined sample, $\varepsilon^*(\omega) = \varepsilon'(\omega) - i\varepsilon''(\omega)$, are recorded under isothermal conditions (at a given crystallization temperature T_{cr} within the range $T_g < T_{cr} < T_m$) at specified time intervals throughout the crystallization process. For example, the representative frequency dependences of the real (ε') and the imaginary (ε'') part of the complex dielectric permittivity measured for supercooled sildenafil during the course of its crystallization at $T_{cr} = 350$ K are presented in Figure 11.20.

It is clearly seen that both the static dielectric permittivity and amplitudes of structural relaxation loss peaks begin to rapidly decrease with time after an induction time of crystallization during which the changes in the spectra are not observed. As already mentioned, such a dramatic drop in the dielectric strength of the structural relaxation $\Delta\varepsilon_\alpha$ is typical for the crystallization process because an increase in the crystalline volume fraction leads to a reduction in the number of reorienting dipoles contributing to the α-relaxation. The increase in the crystallization degree with time can be analyzed from the normalized real permittivity ε'_N by using the following formula:

$$\varepsilon'_N(t) = \frac{\varepsilon'(0) - \varepsilon'(t)}{\varepsilon'(0) - \varepsilon'(\infty)} \tag{11.43}$$

where $\varepsilon'(0)$ is the static dielectric permittivity at the beginning of the crystallization, $\varepsilon'(\infty)$ is the long-time limiting value, and $\varepsilon'(t)$ is the value at a given time t of crystallization.

The time dependences of normalized real permittivity ε'_N under isothermal conditions can be described by different models of crystallization kinetics, among which the Avrami model [73, 74] is the most commonly used. According to this model, which is modified by adding the induction crystallization time t_0, we can analyze the time evolution of ε'_N using the following equation:

$$\varepsilon'_N(t) = 1 - \exp(-K(t - t_0)^n) \tag{11.44}$$

where $K = k^n$ is a crystallization rate constant, which depends on the crystallization temperature and geometry of sample, and n is the Avrami exponent, which is related to the time dependence of the nucleation rate and the dimensionality of

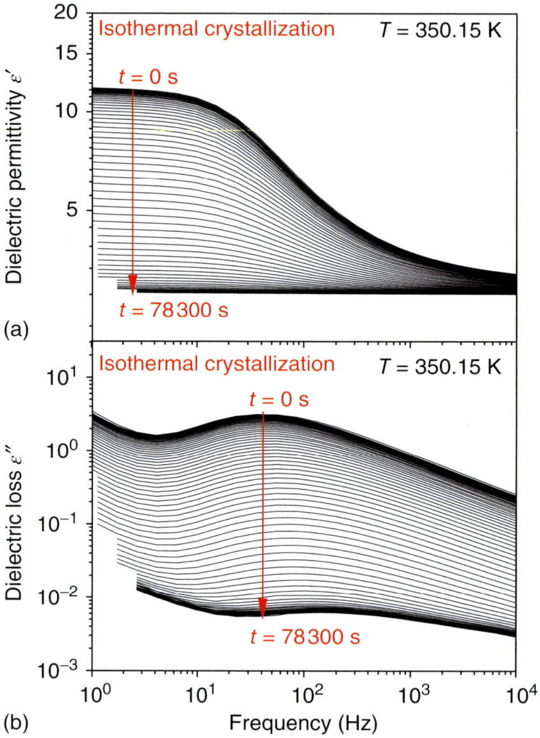

Figure 11.20 Dielectric spectra of (a) the real and (b) imaginary part of the complex dielectric permittivity collected during isothermal crystallization at $T = 350$ K. (Adapted from Kołodziejczyk *et al.* [71]. Reproduced with permission of American Chemical Society.)

the crystallization. It is worth noting that Eq. (11.44) with $t_0 = 0$ implies another equation

$$\log(-\ln(1 - \varepsilon'_N)) = \log K + n\log(t) \tag{11.45}$$

which shows that the dependence of $\log(-\ln(1 - \varepsilon'_N))$ on $\log(t)$ should be linear. This underlies the criterion of applicability of the Avrami model, which can be easily verified by constructing the so-called Avrami plot, which is the plot of $\log(-\ln(1 - \varepsilon'_N))$ versus $\log(t)$.

The Avrami plot is not linear for the entire time range. Its nonlinear behavior, especially at the beginning of the crystallization process, can make the induction time evaluated from the Avrami model unreliable. This problem is removed in the model proposed by Avramov *et al.* [75]:

$$\varepsilon'_N(t) = 1 - \exp\left(-\left(\frac{t - t_0}{\tau_{cr}}\right)^n\right) \tag{11.46}$$

where τ_{cr} is a characteristic time for the overall isothermal crystallization. The parameters n and t_0 have the same meaning as those in the Avrami model (Eq. (11.44)), whereas τ_{cr} is related to the Avrami parameters as $\tau_{cr} = K^{-1/n}$.

This method allows us to avoid significant estimation errors of t_0 and estimate correctly the crystallization induction time t_0 by constructing the so-called Avrami–Avramov plot, which combines the plots of the normalized real permittivity ε'_N and its derivative $d\varepsilon'_N(t)/d\ln t$ versus $\ln t$. Note that the derivative $d\varepsilon'_N(t)/d\ln(t-t_0) = n((t-t_0)/\tau_{cr})^n \exp(-((t-t_0)/\tau_{cr})^n)$ achieves the maximum value $(\varepsilon'_N)'_{max} = n/e$ at $t - t_0 = \tau_{cr}$. Thus, preliminarily assuming that $t_0 = 0$, we can easily find $\ln\tau_{cr}$ for $(\varepsilon'_N)'_{max}$ from the Avrami–Avramov plot, and the parameter n can be simply calculated as $n = (\varepsilon'_N)'_{max}/0.368$. It should be noted that $\varepsilon'_N(\tau_{cr}) = 1 - 1/e \approx 0.63$. However, we usually obtain from experimental dependences that $\varepsilon'_N(\tau_{cr}) < 0.63$. It means that the induction time $t_0 > 0$. In this case, we can estimate the proper value of t_0 by finding t from the condition $\varepsilon'_N = 0.63$ and then calculating $t_0 = t - \tau_{cr}$.

In the case of sildenafil, the isothermal cold crystallization has been investigated at selected temperatures ($T_{cr} = 348$, 350, 353, 356, and 358 K) in the range of the supercooled liquid. The time evolution of the normalized real permittivity $\varepsilon'_N(t)$ for this drug, which reflects the increase in the degree of crystallization with time, is presented for each crystallization temperature in Figure 11.21a. Although the fitting of $\varepsilon'_N(t)$ to the Avrami model (Eq. (11.44)) is quite good, to find values of the Avrami model parameters n and k, we construct the Avrami plots (Eq. (11.45)) for each temperature of crystallization in the time ranges in which the linear character of the obtained dependences is maintained. Moreover, to evaluate reliable values of t_0, we construct the Avrami–Avramov plots for each T_{cr} (see an example in Figure 11.21b). From the analyses for the cold isothermal crystallization of sildenafil, the value of the Avrami exponent n changes with decreasing crystallization temperatures from 2.60 to 3.50 and from 2.57 to 3.60, based on the Avrami model and the Avramov model, respectively. This result suggests that sildenafil can form coexistent rather three-dimensional crystallites from instantaneous and sporadic nucleation [76, 77]. The obtained values of the crystallization rate constant k can be used to determine the activation energy for crystallization, E_a, according to the Arrhenius law:

$$\log k = \log k_0 - \frac{E_a}{RT}\log e \tag{11.47}$$

where k is the crystallization rate expressed in inverse time units, which is related to the Avrami parameters as $k = K^{1/n}$, R is the gas constant, and k_0 and E_a are fitting parameters. Equation (11.47) can be also used with the characteristic crystallization time τ_{cr}, which is a parameter of Eq. (11.46), because $k = 1/\tau_{cr}$. The plots based on Eq. (11.47) are shown in the lower inset to Figure 11.32 and the inset to Figure 11.21b for values of the parameters k and τ_{cr} found from the Avrami model and the Avramov model, respectively, which yield very similar values of the activation energies for overall crystallization of sildenafil: $E_a = 175 \pm 15$ and, $169 \pm 10\,\mathrm{kJ\,mol^{-1}}$. These activation energies as well as that found using the induction time t_0 instead of the k in the Arrhenius law (Eq. (11.47)) can be compared with the activation energy for the structural relaxation to consider a potential correlation between the timescales of the crystallization process and the structural α-process, which is discussed further.

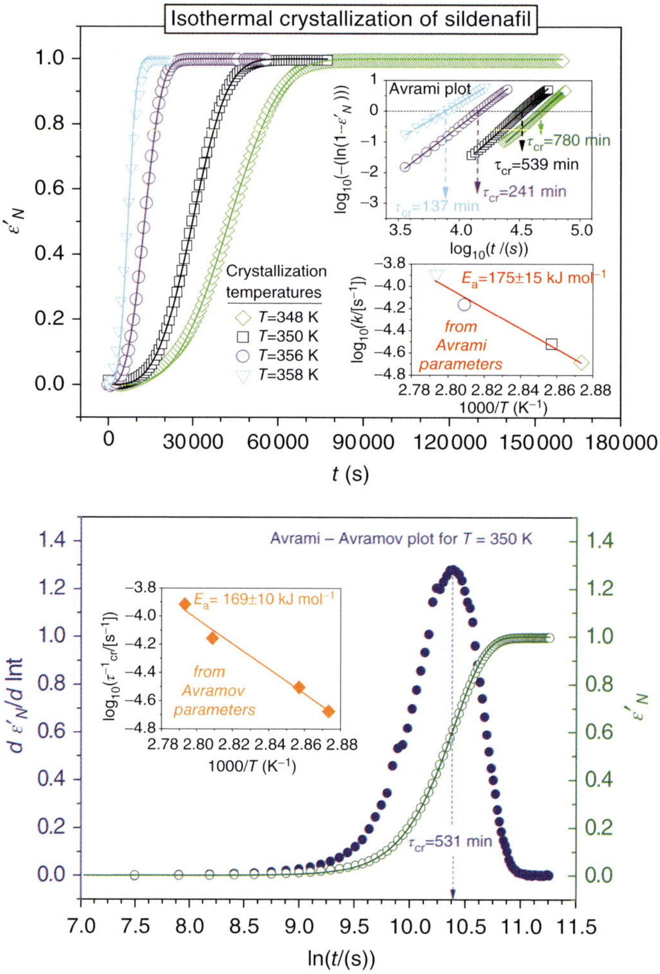

Figure 11.21 (a) Time dependence of normalized real permittivity ε'_N. Solid lines represent Avrami fits in terms of Eq. (11.44). The upper inset shows the Avrami plots in terms of Eq. (11.45) for each crystallization temperature in the time ranges in which the dependences are linear. The crystallization times τ_{cr}, graphically determined at $\log(-\ln(1-\varepsilon'_N)) = 0$, which corresponds to $\varepsilon'_N = (\varepsilon'_N)'_{max} = 0.63$ according to the Avramov model, are the same as those calculated from the Avrami parameters K and n. The lower inset presents the temperature dependence of the logarithm of the crystallization rate k related to the Avrami parameters as $k = K^{1/n}$. The solid line in the lower inset denotes the linear fit to Eq. (11.47). (b) Example of the Avrami–Avramov plot. The evolution of normalized real permittivity ε'_N (open green circles) and its first derivative versus the natural logarithm of the time (closed blue circles). The solid line is the best fit to Eq. (11.46). The inset presents the temperature dependence of the logarithm of the inverse crystallization time from the Avramov model. The solid line in the inset denotes the linear fit to Eq. (11.47) with $k = 1/\tau_{cr}$. (Adapted from Kołodziejczyk *et al.* [71]. Reproduced with permission of American Chemical Society.)

Timescales of Crystallization and Structural Relaxation

A general issue that should be addressed in crystallization studies is a possible relation between molecular mobility and crystallization kinetics. As already mentioned, the molecular dynamics of supercooled liquids monitored by BDS is characterized by α-relaxation. Therefore, it is important to verify whether molecular motions reflected in the structural relaxation control the cold crystallization of supercooled liquids. To find a possible correlation between α-relaxation dynamics and isothermal cold crystallization kinetics for a given glass former, it has been suggested to investigate the relations of the structural relaxation time τ_α with the characteristic crystallization time τ_{cr} or/and the induction time t_0, which can be found, for instance, by constructing the Avrami–Avramov plot (see Figure 11.21) based on the Avramov model (Eq. (11.46)). By exploring the dependences $\log \tau_{cr}$ ($\log \tau_\alpha$) or $\log t_0$ ($\log \tau_\alpha$), their slopes S are analyzed. Such a slope S, called a coupling coefficient, is regarded as a measure of correlation between τ_{cr} and τ_α or t_0 and τ_α, where the induction time t_0 determined from the Avramov model is an estimate of the lag time for nucleation. This kind of analysis was originally suggested for the crystallization onset time and the dielectric relaxation time (i.e., for the correlation between t_0 and τ_α) to investigate whether the nucleation process is controlled by the molecular mobility reflected in the structural relaxation process [78–80]. However, the correlation between τ_{cr} and τ_α has been also recently studied [71, 81] to answer the question whether the molecular mobility reflected in the structural relaxation can be straightforwardly exploited to explain the diffusion-controlled mechanism of the overall crystallization kinetics. In these studies, the coupling coefficient $S = 1$ has been interpreted to indicate that the crystallization is controlled by diffusion on the assumption that diffusivity and viscosity (or structural relaxation time) are inversely proportional according to the Stokes–Einstein equation. Otherwise, if the value of S is <1, the crystallization of the drug is not controlled solely by diffusion, or the relation between diffusivity and dielectric structural relaxation time is more complex than given by the simple Debye–Stokes and Stokes–Einstein equations.

As can be seen in Figure 11.22 for sildenafil, there is a large difference in the timescales of the isothermal crystallization and the structural relaxation. Nevertheless, this simple observation should be supplemented with the analysis of the activation energies, which shows that the activation energy of isothermal cold crystallization ($E_a = 169\,\mathrm{kJ\,mol^{-1}}$) found from the Arrhenius law is less than that of structural relaxation ($E_\alpha = 350\,\mathrm{kJ\,mol^{-1}}$), which also can be established from the Arrhenius law in the considered crystallization temperature range (see Figure 11.22). The obtained ratio of the activation energies, that is, $S = 0.49$ for sildenafil, defines the slope of the $\log \tau_{cr}$ versus $\log\tau_\alpha$ curve (see the inset to Figure 11.22). For comparison, a deviation from $S = 1$ has been also reported by Dantuluri *et al.* [82] for the dependence $\log \tau_{cr}$ ($\log \tau_\alpha$) for CEL ($S \approx 0.7$), which can be interpreted such that the cold crystallization of sildenafil is less influenced by diffusion reflected in the dielectric structural relaxation than that of CEL.

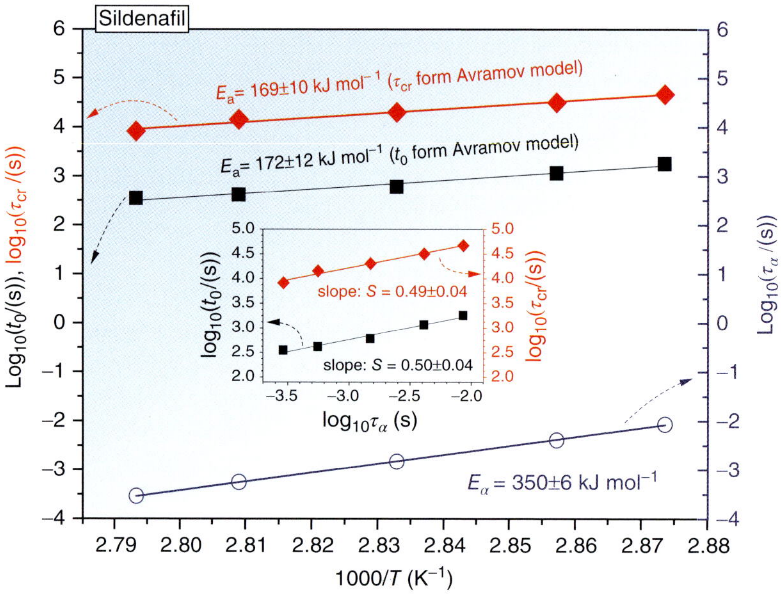

Figure 11.22 Comparison of temperature dependences of the characteristic isothermal crystallization times τ_{cr} found from the Avramov model of crystallization kinetics (red diamonds), the induction times t_0 determined from the Avramov model (black squares), and structural relaxation times τ_α (open blue circles) at the temperatures at which the isothermal crystallization was performed. The solid lines indicate linear fits to the Arrhenius law. The inset shows log–log plots τ_{cr} and t_0 versus τ_α within the crystallization temperature range. The solid lines in the inset denote linear fits. (Adapted from Kołodziejczyk *et al.* [71]. Reproduced with permission of American Chemical Society.)

Using the induction time t_0 determined from the Avramov model as an estimate of the lag time for nucleation, we examine the correlation between t_0 and the structural relaxation time τ_α to determine the corresponding coupling coefficient S. As a result (see Figure 11.22), we find nearly the same values of the activation energy ($E_a = 172\,\text{kJ mol}^{-1}$) and the coupling coefficient $S = 0.50$ for the induction time as those established for the characteristic crystallization time τ_{cr}. It confirms that the diffusion reflected in the dielectric structural relaxation is not a dominant factor that controls the cold crystallization of sildenafil. Moreover, it is worth noting that the linear dependences of log τ_{cr} and log t_0 on the inverse temperature can be superimposed by vertical shifting, which is a consequence of the very similar values of the activation energies found from these dependences. It means that τ_{cr} is linearly proportional to t_0 in the case of sildenafil. A similar proportionality has been also established by Avramov *et al.* for other materials [75].

Since the analysis based on the coupling coefficient S does not provide any definite answer concerning the mechanisms that govern the recrystallization of supercooled liquids, especially if $S < 1$, it has been suggested to investigate a potential correlation of the characteristic overall crystallization time or/and the lag time for nucleation with the average number of the dynamically correlated molecules estimated from the maximum of the four-point dynamic susceptibility function [83, 84], which is a promising research trend that requires, however, further studies.

11.2.2.3 Crystallization in the Glassy State

From the scientific point of view, in the study of the stability of amorphous drugs, the most interesting are those characterized by a high tendency to crystallize both in liquid and glassy states. It is believed that the molecular mobility governs the recrystallization of amorphous materials. Therefore, by studying the relaxation behaviors of drugs in the liquid and glassy states by using the BDS, one should try to find answers to the following questions:

- Which kind of molecular mobility (reflected in the structural or secondary relaxations) can be responsible for the amorphous drugs' recrystallization?
- Is there any correlation between the dynamic fragility and the large tendency of amorphous drugs to crystallize?
- Can we enhance the physical stability of amorphous drugs by changing their molecular mobility?

These nontrivial problems can be discussed using as example the amorphous CEL, prepared by quench-cooling the melt of the crystalline form, which is highly unstable and recrystallizes at temperatures below its glass transition temperature T_g (during storage of the glass in isothermal conditions) as well as on heating the drug above T_g in the supercooled liquid temperature range [30]. It should be noted that there are not many measurements of the recrystallization of amorphous materials (at $T < T_g$), because such experiments are usually highly time consuming. Amorphous CEL is a good candidate to study devitrification because we have established that it recrystallizes very fast at temperatures far below T_g.

Isothermal crystallization studies of the amorphous form of CEL (at room temperature $T = 293$ K which is of 35 K lower than T_g) have been performed by means of X-ray diffraction, which is a very sensitive and foolproof method for solid-state characterization. Diffraction patterns for the initial crystalline form of CEL and its amorphous counterpart are presented in Figure 11.23a,b, respectively.

It is clearly seen in Figure 11.23(b1–b7) that the area of sharp Bragg peaks increases with time, indicating that the degree of crystallization (D_c) of the initial fully amorphous CEL increases. It is based on the observation that some white crystalline spots occur and grow in the initially transparent amorphous sample. The degree of crystallization D_c for each diffraction pattern has been evaluated and plotted as a function of storage time t_S (see Figure 11.23c). The shelf-life of amorphous CEL, defined as the storage time after which 90% of the drug remains still amorphous, has been evaluated to be only 40 h. According to this definition, amorphous CEL loses its efficacy only 40 h after its preparation, which clearly shows that the drug is highly unstable. By calculating the derivative of D_c with respect to t_S, we estimated that the maximum rate of recrystallization is achieved at $t_{s\_vmax} \approx 100$ h of storage (see the inset to Figure 11.23). It is worth noting that

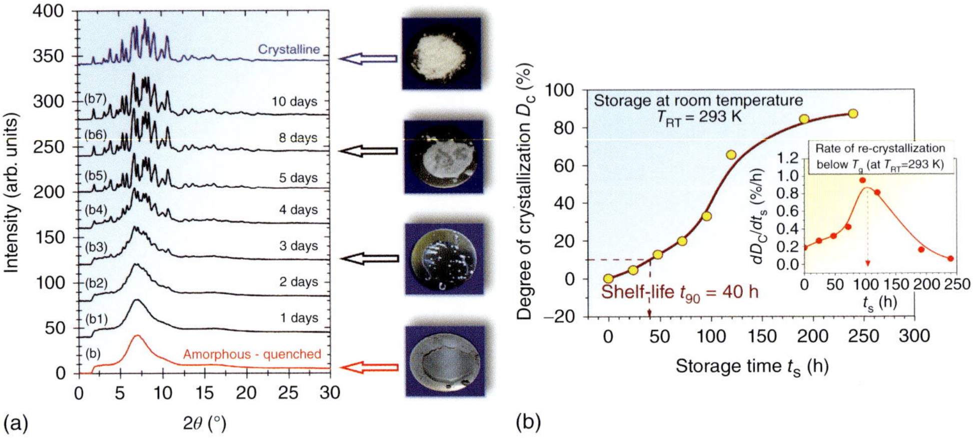

Figure 11.23 X-ray diffraction patterns for various solid-state forms of celecoxib performed at room temperature $T_{RT} = 293$ K (35 K below its T_g). (a) Initial crystalline form celecoxib. (b) Amorphous celecoxib prepared by quench-cooling of the melt of crystal (a). (b1–b7) Gradual isothermal recrystallization of the amorphous form (b) observed at specified time periods from 1 to 10 days. (c) The relative degree of isothermal recrystallization D_C of amorphous celecoxib as a function of storage time t_S at $T_{RT} = 293$ K. The inset presents the rate of celecoxib recrystallization evaluated as a derivative of D_C in terms of t_S. Solid lines are guides for the eyes. (Adapted from Grzybowska *et al.* [30]. Reproduced with permission of American Chemical Society.)

this estimation corresponds to the Avrami–Avramov plot based on Eq. (11.46) and the found t_{s_vmax} is the characteristic crystallization time τ_{cr}.

An important way to determine which kind of molecular mobility can be responsible for the devitrification of the drug is to compare the estimated characteristic time τ_{cr} for the recrystallization process at room temperature T_{RT} with the timescales of molecular mobility reflected in the dielectric structural and secondary relaxations in the glassy state of CEL at T_{RT}.

First, one should analyze the dielectric spectra obtained for the liquid and glassy CEL. As can be seen in Figure 11.24c, the glassy state of CEL is characterized by a large molecular mobility reflected in three secondary relaxations: the slowest β-process, a faster γ-process, and the fastest δ-process (each of them described by using the CC equation (Eq. (11.20)), which correlates with the low stability of the amorphous drug below T_g. Among them, the most important one seems to be that of intermolecular origin (the JG secondary relaxation) because it has been suggested [42, 62–65] that this type of secondary relaxation may control the rate of crystal growth below T_g. On the basis of extended CM (Eq. (11.40) with Eq. (11.39)), the β-process is classified as the intermolecular JG relaxation, and thus this process can play a potential role in the devitrification of the drug.

From the analysis of dielectric spectra obtained for CEL above its glass transition temperature T_g (see Figure 11.24a), it turns out that the supercooled liquid is unstable and crystallizes during its heating at $T > T_g$. The peaks due to

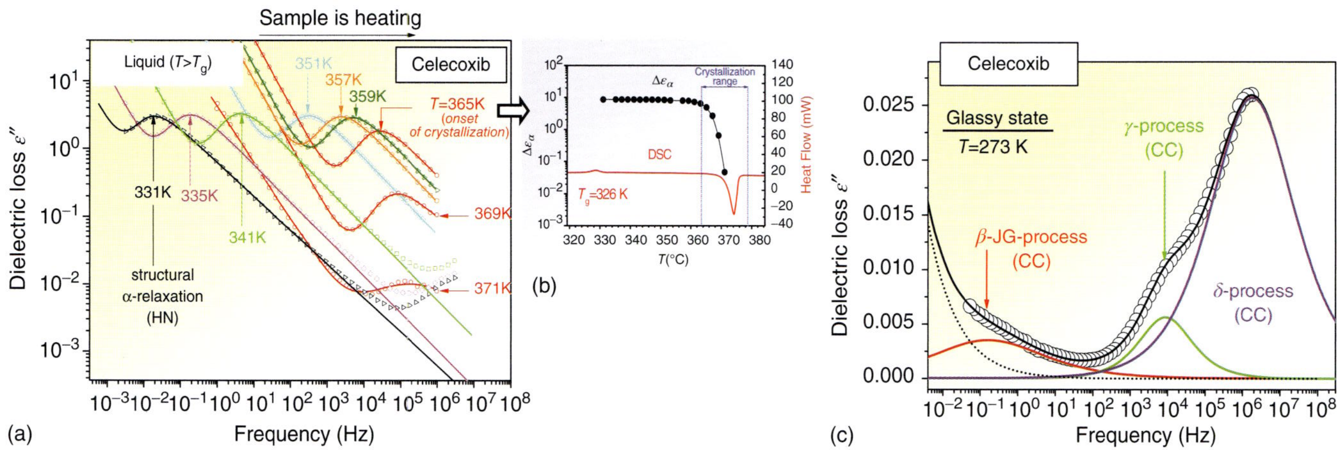

Figure 11.24 Dielectric loss spectra for celecoxib obtained (a) above and (c) below T_g as well as examples of the fitting procedure of the selected dielectric loss spectra obtained both in the liquid and glassy states of the drug in terms of Eq. (11.26). (b) Temperature dependence of dielectric strength of structural relaxation during the non-isothermal crystallization of supercooled celecoxib compared with the corresponding DSC thermogram. (Adapted from Grzybowska *et al.* [30]. Reproduced with permission of American Chemical Society.)

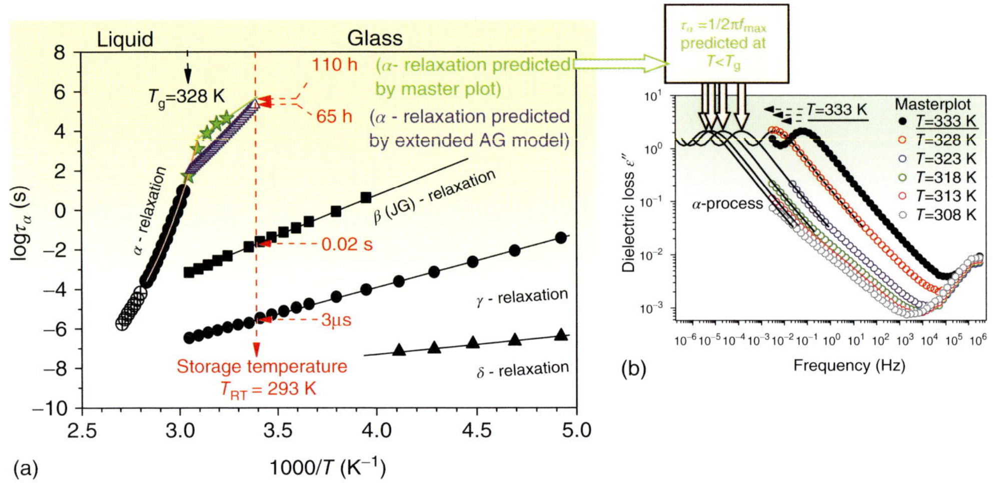

Figure 11.25 The experimental relaxation map of celecoxib (panel (a)): temperature dependence of structural relaxation times (solid circles) was fitted to VFT equation (solid orange line, Eq. (11.31)). Crossed circles show α-relaxation times above T_g in the crystallization range. Temperature dependences of secondary relaxations: β, γ, and δ were fitted to the Arrhenius equation (black lines, Eq. (11.38)). Moreover, the prediction of structural relaxation times in the glassy state and timescales of molecular motions at the temperature of storage of amorphous celecoxib ($T = 293$ K) are shown. Open blue triangles denote the predicted τ_α ($T < T_g$) from the extended AG model (Eq. (11.41)) with $\gamma_{C_p} = 0.875$ [37], whereas the solid green stars indicate τ_α ($T < T_g$) predicted on the basis of the master plot (shown in panel (b)). (Adapted from Grzybowska *et al.* [30]. Reproduced with permission of American Chemical Society.)

the α-process move toward higher frequencies on heating, and its temperature dependence of dielectric strength $\Delta\varepsilon_\alpha$ is nearly constant up to about $T = 365$ K (see Figure 11.24b). Above this temperature, one can observe a sudden drop in $\Delta\varepsilon_\alpha(T)$, which is caused by the onset of sample crystallization. As can be seen, this result is in a good agreement with calorimetric measurements (the red line in Figure 11.24b is the differential scanning calorimetry (DSC) thermogram, which exhibits the cold crystallization exotherm).

From the best fits of entire dielectric spectra (see Figure 11.24a,c) obtained both in the liquid and glassy states of CEL, which assume a superposition of the individual relaxation processes according to Eq. (11.26), the temperature dependences of relaxation times for all dielectric processes are determined (see Figure 11.25a). The temperature dependences of secondary relaxation times are described by the Arrhenius law (Eq. (11.38)), whereas the temperature dependences of α-relaxation times follow the VFT equation (Eq. (11.31)).

In order to estimate the α-relaxations times in the glassy state, two prediction methods have been exploited. Taking into account the parameter VFT evaluated for supercooled liquid as well as the thermodynamic parameters (fictive temperature T_f and γ_{C_p}), the α-relaxations times below T_g are generated in terms of the extended AG model (Eq. (11.41)). To support the obtained result, another prediction method based on the construction of *master plot* is also used (the clearly visible α-peak in the liquid state measured at $T = 333$ K is shifted horizontally to lower frequencies to overlap spectra collected in the glassy state (see Figure 11.25b). As can be seen in Figure 11.25a, the predicted temperature dependences of α-relaxations times by constructing the master plot (solid green stars) and the extended AG model prediction (open blue triangles) are roughly similar, and they become quite close (110 and 65 h, respectively) at room temperature T_{RT}.

The predicted timescale of structural relaxation at $T_{RT} = 293$ K is in a very good agreement with the time of storage of amorphous CEL at which the maximum rate of recrystallization occurs of amorphous CEL ($t_{s\_vmax} \approx 100$), which is the characteristic crystallization time τ_{cr}. It indicates that α-relaxation can play an important role in the devitrification of CEL. Nevertheless, the molecular mobility reflected in secondary relaxations, characterized by much faster molecular motions (i.e., $\tau_\beta = 0.02$ s, $\tau_\gamma = 3$ μs at T_{RT}), can also affect the recrystallization of the amorphous drug. Thus, a role of either structural relaxation or secondary relaxations cannot be excluded in the recrystallization of amorphous CEL.

11.2.2.4 Enhancement of the Physical Stability of Amorphous Drugs by Preparing Amorphous Mixtures

The last decade has seen a growing interest in exploiting amorphous mixtures to improve the physical stability of amorphous drugs. The basic idea in stabilizing the amorphous systems consists in the preparation of the mixtures of a drug with excipients having a higher T_g than that of the drug. As a result, the prepared mixture has a higher T_g compared to that of the pure drug, thereby reducing its molecular mobility and tendency to devitrify. In a typical case, the mixing of two materials results in the glass transition temperature of their binary mixture, which can be described by the Gordon–Taylor (GT) equation [85]:

$$T_g(x) = \frac{xT_{g1} + K(1-x)T_{g2}}{x + K(1-x)} \tag{11.48}$$

where T_{g1} and T_{g2} are the glass transition temperatures of the two components, and K is a fitting parameter which is approximately equal to the ratio of changes in heat capacity ($\Delta C_{p2}/\Delta C_{p1}$) at T_g [86]. The GT equation implies that the values of T_g for binary mixtures change monotonically with the composition of the mixtures. Thus, if we mix one drug with another component, the glass transition temperature (T_{g2}) of which is higher than that of the drug (T_{g1}), Eq. (11.48) predicts that the glass transition temperature of the mixture T_g will be higher than that of the pure drug. The commonly used excipients in mixtures with drugs, which result in the higher T_g of the mixture than that of the pure drug, are macromolecules such as polyvinylpyrrolidone (PVP) [87, 88], hydroxypropyl methylcellulose (HPMC), and so on.

Here we discuss an interesting case of the composition based on CEL and octaacetylmaltose (acMAL), the high physical stability of which cannot be related to the change in T_g predicted by the GT equation but finds its origin in the suppression of some molecular motions caused by specific interactions between molecules of CEL and acMAL [89]. It should be noted that both the components are characterized by nearly the same glass transition temperatures (331 K for CEL and 332 K for acMAL), and therefore the enhancement of stability by the addition of acMAL to CEL will not be related with significant changes in T_g. However, in contrast to CEL, acMAL reveals no tendency for cold crystallization during its heating in the wide temperature range – it is a desirable feature of a crystallization inhibitor.

Physical stability investigations have been performed for binary amorphous mixtures of CEL with different contents of acMAL (the chemical structures of the molecules are shown in Figure 11.26) prepared by a simple quench-cooling of the molten phase of the mixture. The isothermal X-ray diffraction measurements of

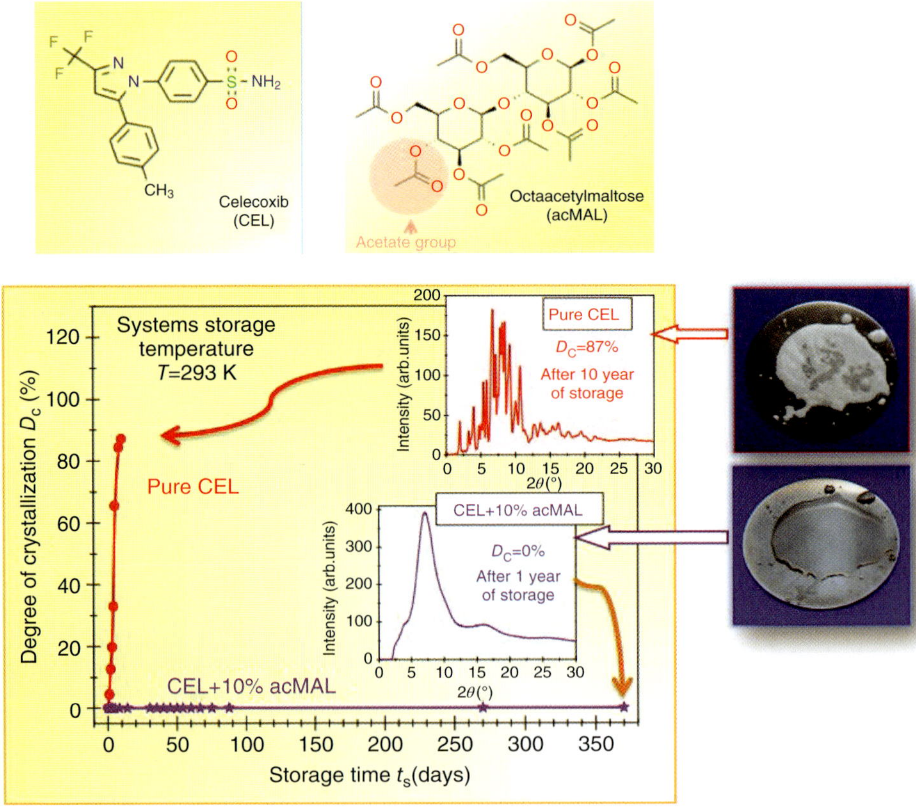

Figure 11.26 Relative degree of isothermal recrystallization D_C of amorphous CEL and CEL + 10% acMAL as a function of storage time t_S at room temperature (T_{RT} = 293 K). The insets present the diffraction patterns for pure CEL and the CEL + 10% acMAL system after 10 and 360 days of storage, respectively. Solid lines are only guides for the eyes. (Adapted from Grzybowska *et al.* [89]. Reproduced with permission of American Chemical Society.)

amorphous binary mixtures of CEL/acMAL have shown that the system with only 10 wt% of acMAL did not revealed any tendency for crystallization for >1 year of storage at room temperature T_{RT} (T_{RT} = 293 K, which is 30 K more than the T_g of the system). As can be seen in Figure 11.26, the degree of crystallization (D_c) of the mixture during its storage continuously tends to zero, while that of the pure amorphous CEL reaches nearly 90% after only 10 days of storage under the same conditions. Thus, one can claim that acMAL is a good inhibitor of the amorphous drug's recrystallization if the storage temperature is lower than the T_g of the binary system.

However, the small amount of acMAL in the binary system (i.e., 10 wt% of acMAL), which stabilizes the glassy state of CEL, does not protect the CEL/acMAL binary system against cold crystallization above T_g during its heating in the supercooled liquid state. From dielectric spectra obtained for binary mixtures with various amount of acMAL (10, 30, and 50 wt%) in the wide temperature range, the temperature dependences of dielectric strength $\Delta\varepsilon_\alpha(T)$ of the structural relaxation have been determined (see Figure 11.27). For pure CEL and binary mixtures with 10 and 30 wt% of acMAL, a characteristic rapid drop in dielectric strength of the structural relaxation was observed at some temperature, which indicates the onset of crystallization of the system. It is clearly seen that, with increasing amounts of acMAL in the mixture, the system recrystallizes at higher temperatures and gradually becomes more stable. Supercooled liquid crystallization is fully suppressed if the weight concentration of acMAL in the mixture with CEL is larger than 30 wt%.

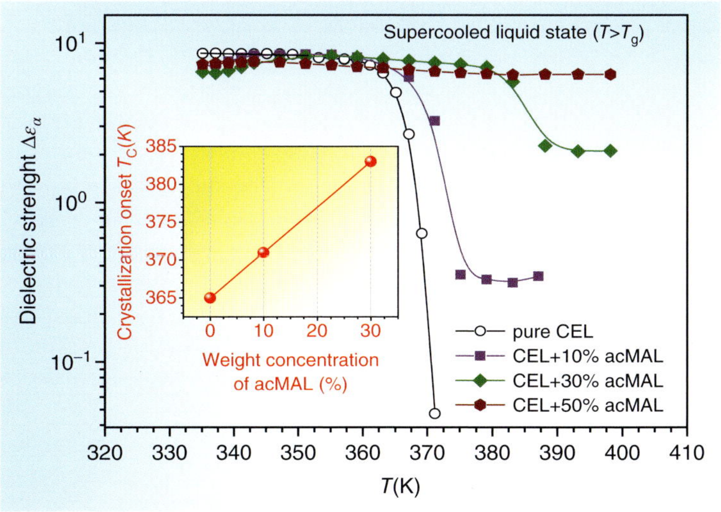

Figure 11.27 Temperature dependence of the dielectric strength $\Delta\varepsilon_\alpha$ of α-relaxation for supercooled mixtures of CEL/acMAL with 0, 10, 30, and 50 wt% acMAL content. The inset shows the changes of the crystallization onset temperatures when adding acMAL to the mixtures. Solid lines are guides for the eyes. (Adapted from Grzybowska *et al.* [89]. Reproduced with permission of American Chemical Society).

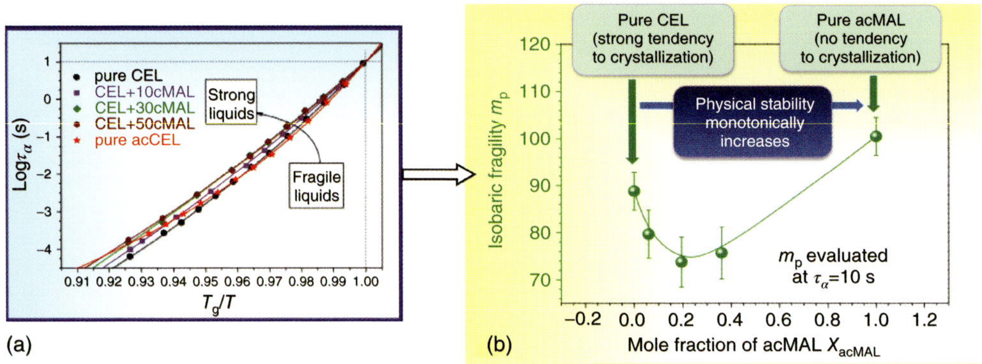

Figure 11.28 (a) Dependence of structural relaxation times for the CEL/acMAL binary systems on the inverse of temperature scaled by T_g at which the α-relaxation time of each material is 10 s (the so-called the Angell plot). The solid lines indicate fits of the experimental data to the Vogel–Fulcher–Tammann (VFT) formula. (b) Dependences of isobaric fragility m for the CEL/acMAL binary systems on the mole fraction of acMAL in the mixture. Values of m_p have been evaluated at $\tau_\alpha = 10$ s. Panel (b) is (Adapted from Grzybowska *et al.* [89]. Reproduced with permission of American Chemical Society.)

As already mentioned, the concept of fragility (m) is considered a key factor that correlates with the glass-forming ability and physical stability of amorphous systems, so we analyze whether a correlation exists between the dynamic fragility (Eq. (11.33)) and the weaker tendency of binary mixtures to crystallize with increasing amount of acMAL in the CEL/acMAL mixture. From temperature dependences of structural relaxation times for all examined systems (see Figure 11.28a), the values of the isobaric fragility m were evaluated for the systems (at $\tau_\alpha = 10$ s) and plotted against the molar fraction of acMAL (see Figure 11.28b). One can see that pure CEL (experimentally the most unstable system) and acMAL (experimentally the most stable system) are the most fragile liquids, whereas their mixtures belong to the stronger glass formers. The initial drop in fragility with increasing amounts of acMAL in the mixture could suggest that the fragility correlates with better physical stability of the binary mixtures with larger amounts of acMAL in the mixture. However, this dependence m (x_{acMAL}) reaches a minimum and then starts increasing, while physical stability of the binary mixtures monotonically increases with molar fraction of acMAL (see the inset to Figure 11.27). Thus, the stability of mixtures, which continuously increases with the addition of acMAL, cannot be rather correlated with the mixture's fragility. It is a consequence of the fact that pure acMAL does not reveal any tendency to crystallize although its fragility is even higher than that for the pure CEL, which recrystallizes easily.

The successful enhancement of amorphous CEL's stability (during its storage at T_{RT}) by adding only 10 wt% acMAL raises the following important questions: (i) How does acMAL modify the relaxation dynamics of CEL below T_g? (ii) What is the molecular mechanism of the stabilization?

On comparing the selected dielectric spectra presented in Figure 11.29a obtained at two temperatures below T_g for two systems, namely pure CEL (black points) and binary mixtures of CEL with 10% acMAL (violet points), one can conclude the following:

1) 1. 10 wt% of acMAL in the mixture suppresses the β- (JG) relaxation, which is often considered responsible for nucleation and crystal growth in the solid structure of amorphous pharmaceuticals. In contrast to pure CEL, for the binary system we cannot distinguish the β-process from the spectra. It indicates that the β-relaxation becomes slower in the mixture than in pure CEL and shifts toward lower frequency. It can be one of the important reasons for the stabilization of the amorphous form of CEL in CEL/acMAL binary systems. Such a suppression of these fast small-angle reorientations of entire CEL molecules by acMAL can reduce the ability of the amorphous CEL to form critical nuclei and devitrify the system.
2) 2. 10 wt% of acMAL in the binary mixture causes the freezing of some molecular motions reflected in the γ-process. After adding acMAL to CEL, one can observe a significant (or even complete) decrease in the dielectric strength of the γ-process.
3) 3. 10% acMAL in the mixture does not influence only the δ-relaxation, which originates from pure CEL; δ-relaxation in the mixture is characterized by the same relaxation times as those observed for pure CEL.

Thus, the molecular mobility of the mixture of CEL/10% acMAL in the glassy state is found to be much more limited than in the case of pure CEL. The limitation on molecular mobility in the glassy state of binary mixture is also seen in the relaxation maps (see Figure 11.29b). For pure CEL, we observe three secondary processes (black points), namely β (JG), γ, and δ, with the following values of the activation energy (evaluated using Eq. (11.38)): 80, 50, and 20 kJ mol^{-1}, respectively. However, the mixture with 10% acMAL reveals only one secondary process (violet points) with the same relaxation time and the same activation energy (20 kJ mol^{-1} also found in terms of Eq. (11.38)) as for pure CEL. It indicates that acMAL does not influence the δ-process. This analysis shows that the suppressed molecular mobility reflected in β (JG)- and γ-relaxations can be responsible for preventing the devitrification of the drug.

To gain better insight into the molecular mechanism that governs the stabilization of amorphous CEL in the acMAL matrix, we need to know the molecular origin of the secondary process of CEL.

To explore this problem, density functional theory (DFT) calculations were performed for single molecules of CEL and acMAL as well as for complexes of two molecules CEL/CEL and CEL/acMAL [89]. It has been found (Figure 11.30a) that the activation energy for rotation of the phenyl ring with the methyl group (Ph-CH_3), which does not reveal any tendency to form external hydrogen bonds with other molecules of CEL, is ~17 kJ mol^{-1} and well corresponds to value of the activation energy of the δ-process (20 kJ mol^{-1}).

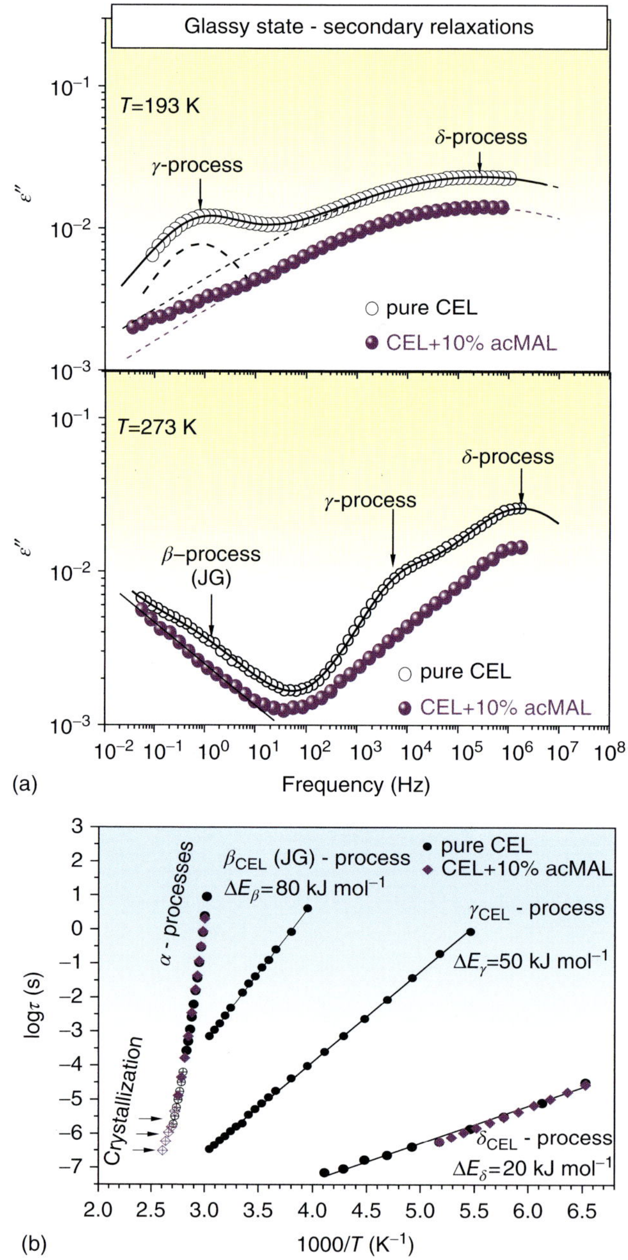

Figure 11.29 (a) Comparison of the dielectric spectra of the pure CEL and the CEL/10% acMAL mixture measured at $T = 193$ and 273 K. Solid lines indicate fits of the entire dielectric spectra based on the superposition of Cole–Cole functions, which describe the individual secondary processes (dotted lines). (b) Comparison of relaxation maps for the pure CEL, its mixture CEL/10% acMAL. (Adapted from Grzybowska *et al.* [89]. Reproduced with permission of American Chemical Society.)

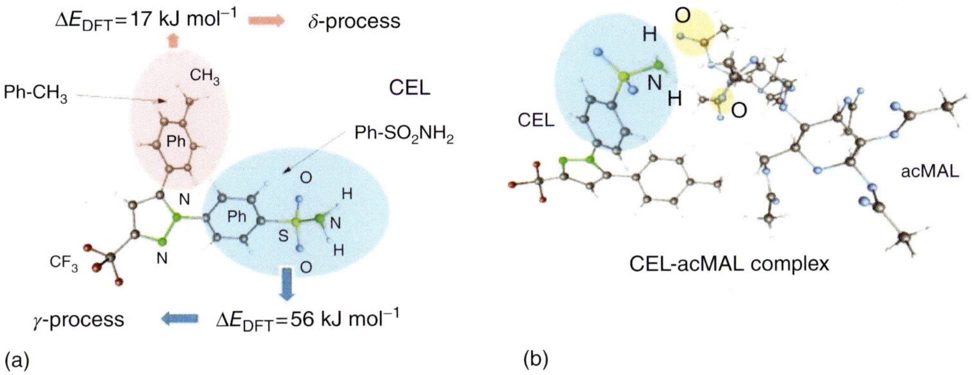

Figure 11.30 (a) Optimized structure of the single CEL molecule with activation energies for rotations of the phenyl ring with the methyl group (Ph-CH_3) and the phenyl ring with the sulfonamide group (Ph-$SONH_2$) evaluated from DFT calculations, (b) Optimized structure of the CEL/acMAL complex. Two major hydrogen bonds are marked for each structure. Protons from sulfonamin group are donors, while the oxygens in case of acMAL and nitrogen and fluorine in case of second CEL molecule are acceptors. (Adapted from Grzybowska *et al.* [89]. Reproduced with permission of American Chemical Society.)

Moreover, taking into account that the phenyl ring with the sulfonamide group (Ph-$SONH_2$) is characterized by a large tendency to form H-bonds with other molecules of CEL, it has been evaluated that the activation energy for rotation of this group equals 56 kJ mol^{-1}, which corresponds well with the value of activation energy of the γ-process (50 kJ mol^{-1}).

acMAL as a pure compound is unable to form hydrogen bonds with other acMAL molecules. However, if acMAL is mixed with CEL, the acMAL molecule can create H-bonds with CEL. From the DFT calculation of the complex of two molecules of CEL and acMAL (see Figure 11.30b), it has been found that H-bonds can be formed between the –C=O group of acMAL and the –N–H group of CEL, and a single acMAL molecule can bind three CEL molecules. The rotation of the phenyl ring with the sulfonamide group in CEL (which probably is reflected in the γ-relaxation in pure CEL) can be suppressed by two hydrogen bonds formed between –Ph-$SONH_2$ group and two –C=O groups in acMAL (while in the case of two molecules of CEL, this group can form only one H-bond). This may be a reason for the rapid decrease in the dielectric strength of the γ-process after adding acMAL to CEL.

In summary, one can conclude that acMAL is a good inhibitor of recrystallization of amorphous CEL below T_g, because acMAL molecules effectively form hydrogen-bond networks with CEL molecules and prevent these molecules from recrystallization from the glassy state. A similar molecular mechanism of the amorphous CEL stability enhancement has been reported by Gupta *et al.* [87] for the CEL/PVP binary system, where hydrogen bonds are formed between the –NH_2 group of CEL and the –C=O group of PVP. However, PVP has a strong antiplasticizing effect on amorphous CEL and significantly increases the T_g of

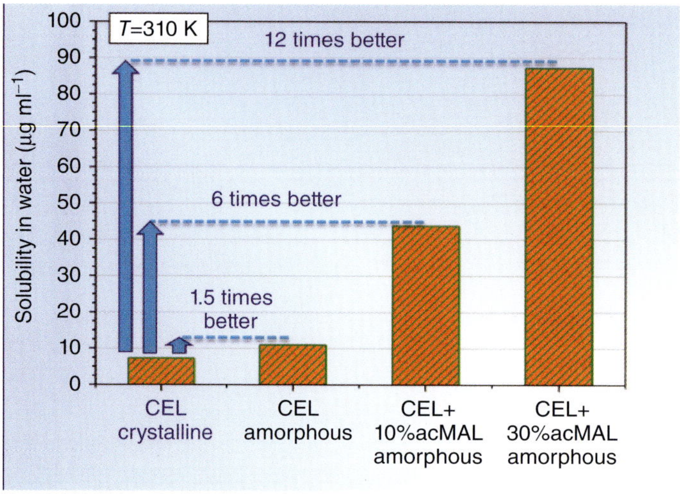

Figure 11.31 Comparison of aqueous solubility of crystalline and amorphous CEL as well as the amorphous CEL/acMAL binary mixtures with 10 and 30 wt% of acMAL at 310 K. (Adapted from Grzybowska *et al.* [89]. Reproduced with permission of American Chemical Society.)

the system, whereas the T_g of the CEL/acMAL binary system is lower than that for pure CEL.

Finally, it should be emphasized that this first application of a sugar with acetate groups to stabilize the amorphous drug has not only improved its physical stability but also enhanced its water solubility. As can be seen in Figure 11.31, amorphous CEL is much more water soluble in the amorphous acMAL matrix than itself. The water solubility of pure amorphous CEL is only 1.5 times more than its crystalline counterpart, whereas the water solubility of CEL in the binary amorphous mixtures with 10% and 30% of acMAL is 6 and 12 times, respectively, better than that for crystalline CEL. Thus, these results for binary mixtures are very promising from the therapeutic point of view.

11.3 Molecular Mobility and Dielectric Response in Partially Ordered Pharmaceutical Systems

Despite intensive studies on modern drug delivery systems, solid dosage forms that contains well-defined crystalline form of the active pharmaceutical ingredient remain still the most commonly applied route of drug administration. This is a very rational procedure, because crystalline materials are generally physically and chemically stable upon processing and storage, which is mostly related to the three-dimensional order in their structure. The aspects related to the crystalline state formation and the so-called crystal engineering are becoming more and more recognized in the pharmaceutical field [90, 91]. In addition, an integral part of solid pharmaceutical characterization constitutes polymorphic transformations, which are a source of different crystalline states of the same molecule exhibiting different physiochemical and biological properties [92, 93]. It has been demonstrated that in some cases polymorphs can be predicted theoretically [94]. However, one should also be aware that polymorphic transformation might occur upon processing and storage, which emphasizes the importance of careful studies on their levels before and during formulation levels [95, 96]. Studies on polymorphism are not an easy task because of the considerable number of pharmaceutical substances that are estimated to be polymorphic. In fact, the ability to isolate, differentiate, and characterize individual polymorphs is a major challenge to the pharmaceutical industry.

When an ordinary polymorphic transformation takes place, no rotational degrees of freedom are allowed. However, in Nature, even a more frequent polymorphic transition occurs when an ordered crystalline state (with fixed positions and orientations) transforms to orientationally disordered crystalline state [97]. A disordered crystal with reorientational motions and positional order of molecules is termed as a “rotator phase” or a “plastic crystal” (typically soft and waxy to the touch). When a plastic crystal is cooled down, depending on the cooling rate an orientationally ordered crystalline phase or orientationally disordered crystalline state (also termed as glassy crystal) can be formed. Mesophases with partial or complete lack of positional order but having orientational order of the constitutent molecules are known more commonly as “liquid crystals”. A schematic illustration of the difference between a crystal, various crystalline mesophases, and a liquid is given in Figure 11.32.

Partially ordered systems can be also found among pharmaceutical compounds, giving an alternative option to improve solubility, bioavailability, as well as formulation procedures of drug substances [99, 100]. However, in the routinely performed characterization of the pharmaceutically important drug substances, which focuses mainly on the crystalline and amorphous states, partially ordered systems might sometimes go unnoticed. This is because pharmaceuticals are often classified as crystalline or amorphous based on the results extracted from X-ray diffraction as well as calorimetric studies. However, some mesophases do not possess long-range order, so a conventional X-ray diffraction pattern looks much like that of a very typical amorphous solid. In addition, in DSC, partially ordered

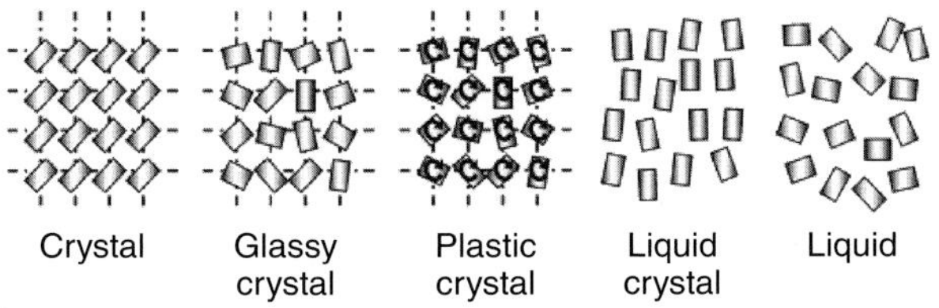

Figure 11.32 Schematic illustration of various phases that might occur in condensed matter. (Adapted from Dionisio *et al.* [98]. Reproduced with permission of American Chemical Society).

systems might reveal a liquid-like behavior and a glass-transition event, which only brings about more confusion regarding their real nature. Therefore, routinely performed analyses of solid pharmaceutical substances should be supplemented by additional methods that are able to distinguish amorphous solids from partially ordered crystalline phases. Definitely, a very helpful technique can be dielectric spectroscopy, which allows the characterization of mobility in partially ordered systems. In the traditional sense, this three-dimensional order means fixing molecules in repeatable three-dimensional arrays, where their orientational and positional degrees of freedom are blocked. Therefore, polar molecules of a typical crystal will not show any variation of the dielectric susceptibility when an external electric field is applied. However, for partially ordered systems, orientational freedom is allowed, which means that their dielectric signal becomes activated, and rotational mobility in the crystalline phases can be probed by the dielectric technique [101, 102].

The usefulness of the dielectric technique in studies of partially ordered pharmaceutical systems can be clearly demonstrated in the example of plastic and crystalline states of caffeine studied by Descamps and coworkers [103]. This very important substance from the food and pharmaceutical perspectives has in fact two polymorphic forms. The high-temperature form I is highly disordered, and always accompanies (as an unwanted impurity) the room-temperature stable form II, causing instability problems [104]. As from X-ray diffraction measurements it was not possible to extract any conclusive outcome on the nature of form I, dielectric measurements were carried out in order to answer the crucial question whether its molecular disorder is due to static or dynamic reasons. From analyzing the dielectric strength of the loss peaks of ε'' (T) recorded in the isochronal mode (Figure 11.33) for caffeine forms I and II, one can get the very important

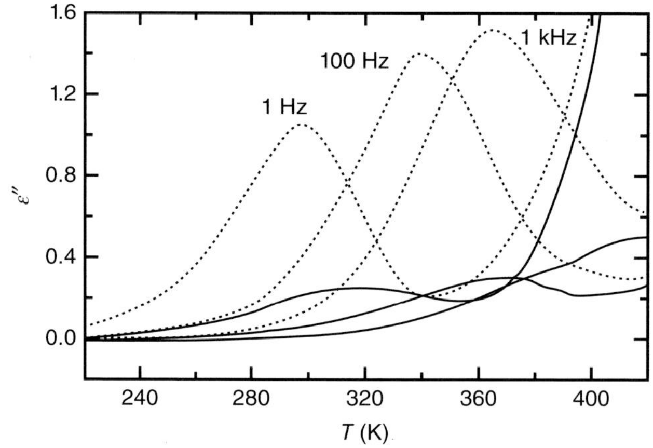

Figure 11.33 Temperature dependence of the imaginary part of the complex dielectric function at fixed frequencies. Solid lines refer to dielectric relaxation in form II whereas dotted line is for form I of caffeine. (Adapted from Descamps *et al.* [103]. Reproduced with permission of American Chemical Society.).

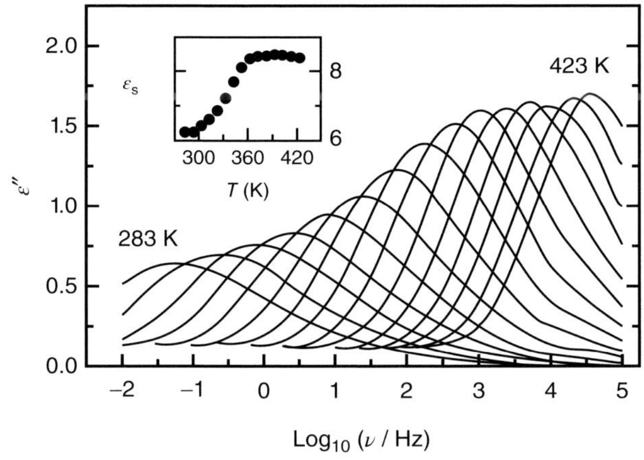

Figure 11.34 Dielectric loss spectra recorded of caffeine (form I) recorded between 424 and 283 K (in steps of 10 K). The inset shows the corresponding temperature evolution of static permittivity increment. (Adapted from Descamps *et al.* [103]. Reproduced with permission of American Chemical Society).

information that form I shows definitely much higher dynamic dipolar mobility compared to form II.

If dielectric measurements are performed over a wide range of temperatures and frequencies, it is possible to detect and characterize thoroughly the dipolar rotational motions that appear in the disordered crystalline form. As illustrated in Figure 11.34, the plastic crystal of caffeine can be supercooled and, with decreasing temperature, a well-defined dielectric loss peak shifts toward lower frequencies. Its distribution of relaxation times is greater than that expected for exponential Debye decay, and significant broadening of the relaxation peak is observed with decreasing temperature. By analyzing the temperature dependence of the characteristic relaxation times, it was found that the dynamic rotation that appears in form I of the caffeine molecule has an Arrhenius temperature behavior, and in the temperature range not very far from room temperature (260 K), it enters the orientationally disordered glassy state. In contrast to typical glassy materials, where both orientational and positional disorder are frozen, in partially ordered systems, below the glass transition (then termed as a glassy crystal) reorientational motions are arrested, but molecules still possess positional order.

By looking more closely at Figure 11.34, Decamps and coworkers noticed that the amplitude of the relaxation peak observed in the plastic phase systematically decreased. This observation is also supported by very a pronounced decease of the static ε_s with temperature, as seen in the inset of Figure 11.34. This can be related to antiparallel organization of dipoles when temperature decreases (Kirkwood factor decreases). However, it was also demonstrated that the polymorphic transformation from the metastable form I to the more stable form II could be monitored in real time during the dielectric spectroscopy experiment (Figure 11.35).

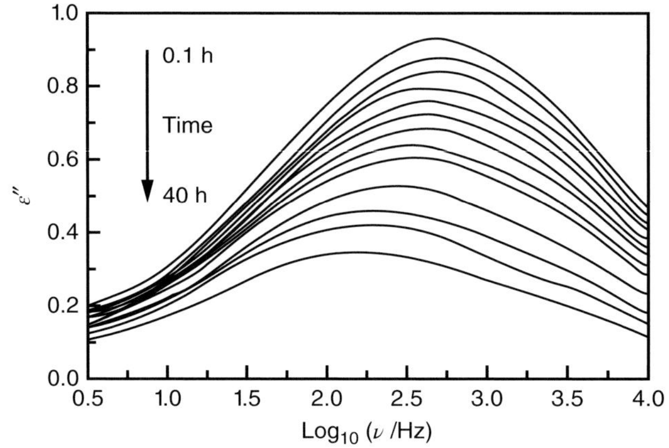

Figure 11.35 Time evolution of the imaginary part of complex dielectric permittivity during isothermal transformation from form I to form II of caffeine at 363 K. (Adapted from Descamps *et al.* [103]. Reproduced with permission of American Chemical Society.)

In that case, a decrease in the relaxation peak intensity is due to the lowering of the number of active rotating dipoles contributing to the relaxation process as a result of transformation to the less mobile but still dynamically disordered form II. The case study of caffeine clearly demonstrates the application of dielectric spectroscopy to study the plastic and glassy crystal states of pharmaceutical substances, providing remarkable information considering their apparent complexity, which basically originates from phase transformations and orientational mobility. This knowledge is essential to address instability problems of drug products and to work out appropriate storage and processing conditions.

Among different pharmaceutical substances, liquid-crystalline states were reported for a broad range of therapeutics, such as anti-inflammatory, antifungal, anticancer, or antibiotic agents [105]. Depending on the degrees positional and orientational order, liquid crystals are classified as: nematic (no positional order, long-range directional order), smectic (positionally ordered along one direction), and chiral nematic (molecules aligned in parallel to the director but twisted perpendicular to the director). Schematic alignments in different liquid-crystalline phases are given in Figure 11.36.

It is well known that the shape of a molecule has an important impact on intermolecular interactions and, therefore, on its molecular dynamics as well. An elongated, banana- or rod-like shape of a drug molecule might be the first sign that a given pharmaceutical compound can show liquid-crystalline phases. Letz *et al.* [106] proposed a very simple model that can be used predict whether the glass-transition dynamics is driven by the orientational degree of freedom, which is known for liquid-crystalline ordering. The sole input parameter of the model is the ratio (R) of size of the molecule along the major and the minor axes. If R is >2, then the model predicts that the liquid–glass transition should be governed mainly by an orientational degree of freedom. Furthermore, the probability of formation of

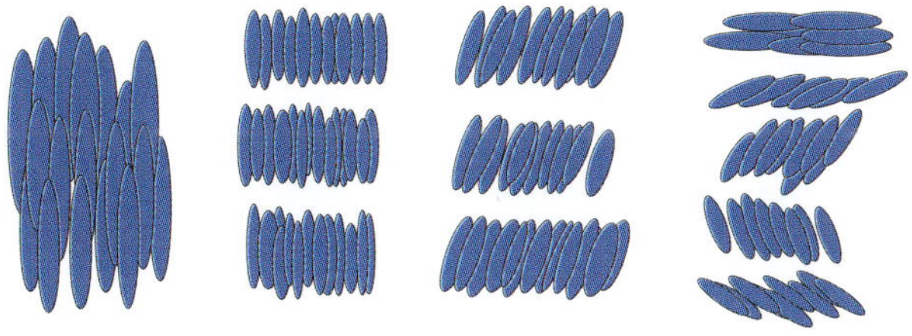

Figure 11.36 Schematic of alignment in (a) nematic, (b) smectic A and C, and (c) chiral nematic phases.

nanoscale domains with nematic order upon cooling is very high. Very recently, it has been successfully tested in the case of few supercooled pharmaceutical substances, such as vitamin A, itraconazole, and posaconazole [107–109].

The existence of liquid-crystalline mesophases has typically a significant impact on the relaxation dynamics probed by dielectric spectroscopy, which is different from that of a normal glass former. In the isotropic liquid, a single dielectric loss peak is observed as a result of cooperative reorientational motions of molecules. Then, when the liquid phase is cooled down, a nematic phase transition usually occurs and the main relaxation peak splits into two, which represent the complementary relaxations of a dipole moment between parallel and perpendicular arrangement with respect to the electric field. The slower process (δ-relaxation) is due to the rotational motion of the molecule around the short axis and is related to the parallel arrangement of the permittivity tensor. The faster process (tumbling mode) is related to the perpendicular arrangement of the permittivity tensor to the nematic director and reflects librational fluctuations of the mesogen around the long molecular axis. For some liquid crystals, the tumbling mode was called also α-relaxation, because it corresponds to the glassy dynamics [107, 110]. In some cases, in the nematic phase a further relaxation is observed. It origin is still not fully understood, but typically ascribed to reorientational motion of molecular aggregates [111]. This relaxation scenario typical for liquid-crystalline substances is demonstrated in Figure 11.37 for the example of the nematic supercooled phase of vitamin A.

The most prominent δ-relaxation process, of Debye-like character, is rather typical for supercooled nematic phases. In addition, for vitamin A two relaxation processes, α and δ, coalesce into one broadened relaxation as the glass transition temperature is reached. However, for some pharmaceutical liquid-crystalline materials, this situation might be slightly different. For example, Tarnacka *et al.* [108] have demonstrated that the δ-relaxation process in the supercooled liquid state of itraconazole has a much lower amplitude than the α-process, and a special DC cleaning procedure was even necessary to make it more outstanding (Figure 11.38a). Upon cooling, the α- and δ-processes tend to merge at a temperature around the calorimetric glass transition (Figure 11.38b).

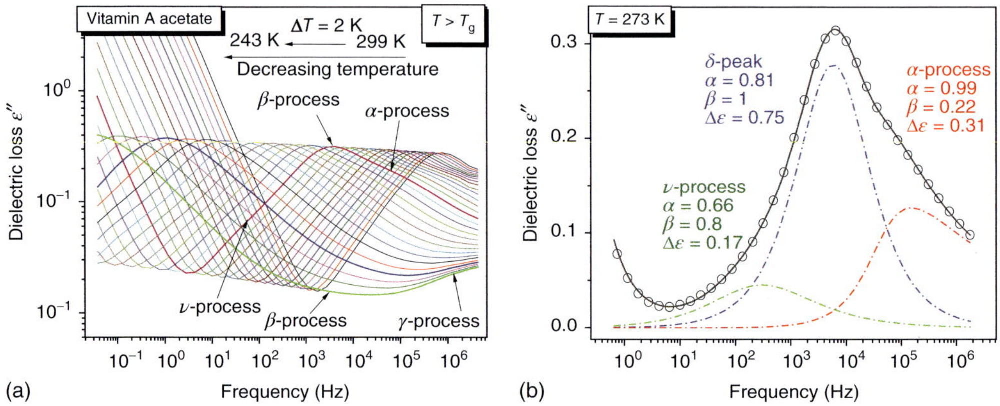

Figure 11.37 Dielectric loss versus frequency measured (a) at temperatures from 299 to 243 K and (b) at 273 K. The solid line is the fit of a sum of three HN functions to the experimental data. The other lines are shown to demonstrate the contribution of the individual process to the entire dielectric loss spectrum for vitamin A. (Adapted from Wojnarowska *et al.* [107]. Reproduced with permission of American Physical Society.

In addition to the nematic mesophase formation, some pharmaceutical systems might also reveal an additional liquid-crystalline moleculal ordering, as happens for itraconazole mentioned above. In that case, DSC thermograms reveal two phase transitions events occurring at 363 and 346 K, which on the basis of further and more detailed studies were assigned to isotropic liquid–nematic and nematic–SmA transitions, respectively. It is a very interesting finding that the δ-relaxation process associated with rotation around the short axis of the itraconazole molecule senses both transition events, whereas the α-relaxation time does not. This is clearly illustrated in Figure 11.38b. The other very interesting feature of above-mentioned liquid-crystalline pharmaceutical material is the evident lack of merging tendency for α- and δ-relaxations above the clearing temperature, which suggests that some kind of the nanoscale nematic-like domains persist even in the isotropic phase.

Finally, it is very important to mention that, in some cases, polarized light microscopy fails to detect the liquid-crystalline phases (especially when molecules are arranged perpendicular to the layer plane so the optic axis of the liquid crystal is also orientated parallel to the propagating light), and therefore one can benefit considerably from dielectric measurements being an appropriate technique used for detecting phase transitions in liquid-crystalline pharmaceuticals (but only polar ones). Additionally, it is worth noting that, initially, a crystalline substance might reveal liquid-crystalline mesophases not only when cooled down from the melt [112] but also upon processing (spray-drying, salt forms) [113, 114], in presence of moisture [115], or in aqueous solutions at a critical concentration [116]. In this context, appropriate methods for their characterization are of a key importance in the pharmaceutical technology area.

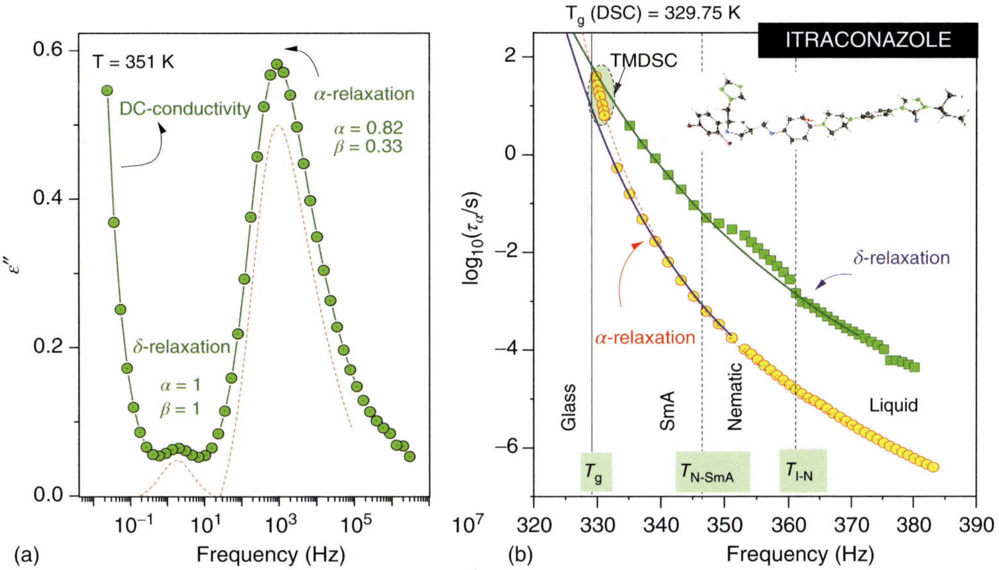

Figure 11.38 (a) Dielectric loss versus frequency at $T = 351$ K for itraconazole. The solid line represents the fit of two HN functions with the DC conductivity term to the experimental data. (b) Temperature dependence of α- and δ-relaxation times. The liquid-crystalline phase transformations are marked as vertical dotted lines Solid lines are the best VFT fits. (Adapted from Tarnacka *et al.* [108]. Reproduced with permission of Royal Society of Chemistry).

Acknowledgment

K.G. and M.P. are grateful for the financial support received within the Project No. 2015/16/W/NZ7/00404 (SYMFONIA 3) from the National Science Centre, Poland.

References

1. Kremer, F. and Schönhals, A. (2003) *Broadband Dielectric Spectroscopy*, Springer-Verlag.
2. Schaumburg, G. (1994) Overview: Modern measurement techniques in Broadband Dielectric Spectroscopy. Dielectric Newsletter, Issue March 1994, pp. 4–7.
3. Novocontrol Technologies www.novocontrol.de (accessed 26 November 2015).
4. Schaumburg, G. (1994) New broadband dielectric spectrometers. Dielectric Newsletter, Issue July 1994, pp. 8–12.
5. Roland, C.M., Hensel-Bielowka, S., Paluch, M., and Casalini, R. (2005) Supercooled dynamics of glass-forming liquids and polymers under hydrostatic pressure. *Rep. Prog. Phys.*, **68**, 1405–1478.
6. Floudas, G., Paluch, M., Grzybowski, A., and Ngai, K. (2011) *Molecular Dynamics of Glass-Forming Systems: Effects of Pressure*, Springer Series Advances in Dielectrics, series edited by F. Kremer, Springer-Verlag, Berlin.

7. Paluch, M., Patkowski, A., and Fischer, E.W. (2000) Temperature and pressure scaling of the α relaxation process in fragile glass formers: a dynamic light scattering study. *Phys. Rev. Lett.*, **85**, 2140.
8. Paluch, M., Sekula, M., Pawlus, S., Rzoska, S.J., Ziolo, J., and Roland, C.M. (2003) Test of the Einstein–Debye relation in supercooled dibutylphthalate at pressures up to 1.4 GPa. *Phys. Rev. Lett.*, **90**, 175702.
9. Paluch, M., Roland, C.M., Pawlus, S., Zioło, J., and Ngai, K.L. (2003) Does the arrhenius temperature dependence of the Johari-Goldstein relaxation persist above Tg? *Phys. Rev. Lett.*, **91**, 115701.
10. Hong, L., Begen, B., Kisliuk, A., Pawlus, S., Paluch, M., and Sokolov, A.P. (2009) Influence of pressure on quasielastic scattering in glasses: relationship to the boson peak. *Phys. Rev. Lett.*, **102**, 145502.
11. Wojnarowska, Z., Roland, C.M., Swiety-Pospiech, A., Grzybowska, K., and Paluch, M. (2012) Anomalous electrical conductivity behavior at elevated pressure in the protic ionic liquid procainamide hydrochloride. *Phys. Rev. Lett.*, **108**, 015701.
12. Riande, E. and Diaz-Calleja, R. (2004) *Electrical Properties of Polymers*, Marcel Dekker, New York.
13. Cole, K.S. and Cole, R.H. (1941) Dispersion and absorption in dielectrics I. Alternating current characteristics. *J. Chem. Phys.*, **9**, 341.
14. Davidson, D.W. and Cole, R.H. (1950) Dielectric relaxation in glycerine. *J. Chem. Phys.*, **18**, 1417.
15. Davidson, D.W. and Cole, R.H. (1951) Dielectric relaxation in glycerol, propylene glycol, and n-propanol. *J. Chem. Phys.*, **19**, 1484.
16. Lukenheimer, P. (1999) *Dielectric Spectroscopy of Glassy Dynamics, Praca Habilitacyjna*, Shaker Verlag, Aachen.
17. Havriliak, S. and Negami, S. (1966) A complex plane analysis of a-dispersions in some polymer systems. *Polym. Sci. Part C: Polym. Symp.*, **14**, 99.
18. Havriliak, S. and Negami, S. (1967) A complex plane representation of dielectric and mechanical relaxation processes in some polymers. *Polymer*, **8**, 161.
19. Williams, G. and Watts, D.C. (1970) Non-symmetrical dielectric relaxation behaviour arising from a simple empirical decay function. *Trans. Faraday Soc.*, **66**, 80–85.
20. Böhmer, R., Ngai, K.L., Angell, C.A., and Plazek, D.J. (1993) Non-exponential relaxations in strong and fragile glass-formers. *J. Chem. Phys.*, **99**, 4201–4209.
21. Donth, E. (2001) *The Glass Transition. Relaxation Dynamics in Liquids and Disordered Materials*, Springer-Verlag.
22. Stockmayer, W. (1967) Dielectric dispersion in solutions of flexible polymers. *Pure Appl. Chem.*, **15**, 539.
23. Johari, G.P. and Pathmanathan, K. (1986) An analysis for β-process in several molecular glasses. *J. Chem. Phys.*, **85**, 6811.
24. Casalini, R. and Roland, C.M. (2003) Pressure evolution of the excess wing in a type- B glass former. *Phys. Rev. Lett.*, **91**, 015702.
25. Gupta, P., Chawla, G., and Bansal, A.K. (2004) Physical stability and solubility of advantage from amorhous celecoxib: the role of thermodynamic quantities and molecular mobility. *Mol. Pharm.*, **1** (6), 406–413.
26. Bhugra, C. and Pikal, M.J. (2008) Role of thermodynamic, molecular, and kinetic factors in crystallization from the amorphous state. *J. Pharm. Sci.*, **97**, 1329–1349.
27. Shamblin, S.L., Tang, X., Chang, L., Hancock, B.C., and Pikal, M.J. (1999) Characterization of the time scales of molecular motion in pharmaceutically important glasses. *J. Phys. Chem. B*, **103**, 4113–4121.
28. Shamblin, S.L., Hancock, B.C., Dupuis, Y., and Pikal, M.J. (2000) Interpretation of relaxation time constant for amorphous pharmaceutical systems. *J. Pharm. Sci.*, **89**, 417–427.
29. Hancock, B.C. and Shamblin, S.L. (2001) Molecular mobility of amorphous pharmaceuticals determined using differential scanning calorimetry. *Thermochim. Acta*, **380**, 95–107.

30. Grzybowska, K., Paluch, M., Grzybowski, A., Wojnarowska, Z., Hawelek, L., Kolodziejczyk, K., and Ngai, K.L. (2010) Molecular dynamics and physical stability of amorphous anti-inflammatory drug: celecoxib. *J. Phys. Chem. B*, **114**, 12792–12801.
31. Vogel, H. (1921) Das Temperaturabhangigkeitgesetz der Viskosität von Flüssigkeiten. *Z. Phys.*, **22**, 645–646.
32. Fulcher, G.S. (1925) Analysis of recent measurements of the viscosity of glasses. *J. Am. Ceram. Soc.*, **8**, 339–355.
33. Tammann, G. and Hesse, W. (1926) Die Abhängigkeit der Viscosität von der Temperatur bie unterkühlten Flüssigkeiten. *Z. Anorg. Allg. Chem.*, **156**, 245–257.
34. Angell, C.A. (1997) Why c1 = 16-17 in the WLF equation is physical – and the fragility of polymers. *Polymer*, **38**, 6261–6266.
35. Adam, G. and Gibbs, J.H. (1965) On the temperature dependence of cooperative relaxation properties in glass-forming liquids. *J. Chem. Phys.*, **43**, 139–146.
36. Angell, C.A. (1991) Relaxation in liquids, polymers and plastic crystals – strong/fragile patterns and problems. *J. Non-Cryst. Solids*, **131-133**, 13–31.
37. Kaushal, A.M. and Bansal, A.K. (2008) Thermodynamic behavior of glassy state of structurally related compounds. *Eur. J. Pharm. Biopharm.*, **69**, 1067–1076.
38. Perera, D.N. (1999) Compilation of the fragility parameters for several glass-forming metallic alloys. *J. Phys.: Condens. Matter*, **11**, 3807–3812.
39. Lu, Z.P., Li, Y., and Liu, C.T. (2003) Glass-forming tendency of bulk La-Al-Ni-Cu-(Co) metallic glass-forming liquids. *J. Appl. Phys.*, **93**, 286–290.
40. Tanaka, H. (2005) Relationship among glass-forming ability, fragility, and short-range bond ordering of liquids. *J. Non-Cryst. Solids*, **351**, 678–690.
41. Shintani, H. and Tanaka, H. (2006) Frustration on the way to crystallization in glass. *Nat. Phys.*, **2**, 200–206.
42. Tombari, E., Ferrari, C., Johari, G.P., and Shanker, R.M. (2008) Calorimetric relaxation in pharmaceutical molecular glasses and its utility in understanding their stability against crystallization. *J. Phys. Chem. B*, **112**, 10806–10814.
43. Paluch, M., Ngai, K.L., and Hensel-Bielowka, S. (2001) Pressure and temperature dependences of relaxation dynamics of cresolphthaleindimethylether: evidence of contributions from thermodynamics and molecular interactions. *J. Chem. Phys.*, **114**, 10872–10883.
44. Koperwas, K., Grzybowski, A., Grzybowska, K., Wojnarowska, Z., Sokolov, A.P., and Paluch, M. (2013) Effect of temperature and density fluctuations on the spatially heterogeneous dynamics of glass-forming van der waals liquids under high pressure. *Phys. Rev. Lett.*, **111**, 125701.
45. Hodge, I.M. (1996) Strong and fragile liquids – a brief critique. *J. Non-Cryst. Solids*, **202**, 164–172.
46. Huang, D. and McKenna, G.B. (2001) New insights into the fragility dilemma in liquids. *J. Chem. Phys.*, **114**, 5621–5630.
47. Stevenson, J.D. and Wolynes, P.G. (2005) Thermodynamic-kinetic correlations in supercooled liquids: a critical survey of experimental data and predictions of random first-order transition theory of glasses. *J. Phys. Chem. B*, **109**, 15093–15097.
48. Wang, L.-M., Velikov, V., and Angell, C.A. (2002) Direct determination of kinetic fragility indicates of glassforming liquids by differential scanning calorimetry: kinetic versus thermodynamic fragilities. *J. Chem. Phys.*, **117**, 10184–10355.
49. Simon, S.L., Plazek, D.J., Sobieski, J.W., and McGregor, E.T. (1997) Physical aging of a polyetherimide: volume recovery and its comparison to creep and enthalpy measurements. *J. Polym. Sci., Part B: Polym. Phys.*, **35**, 929–936.
50. Wang, L.-M. and Angell, C.A. (2003) Response to "Comment on 'Direct determination of the fragility indexes of glassforming liquids by differential scanning calorimetry. Kinetic vs thermodynamic fragilities' ". *J. Chem. Phys.*, **118**, 10353–10355.

51. Lubchenko, V. and Wolynes, P.G. (2003) Barrier softening near the onset of nonactivated transport in supercooled liquids: implications for establishing detailed connection between thermodynamic and kinetic anomalies in supercooled liquids. *J. Chem. Phys.*, **119**, 9088–9105.
52. Cangialosi, D., Alegria, A., and Colmenero, J. (2006) A thermodynamic approach to the fragility of glass-forming polymers. *J. Chem. Phys.*, **124**, 024906.
53. Scopigno, T., Cangialosi, D., and Ruocco, G. (2010) Universal relation between viscous flow and fast dynamics in glass-forming materials. *Phys. Rev. B*, **81**, 100202 (R).
54. Grzybowska, K., Grzybowski, A., Ziolo, J., Rzoska, S.J., and Paluch, M. (2007) Anomalous behavior of secondary dielectric relaxation in polypropylene glycols. *J. Phys.: Condens. Matter*, **19**, 376105.
55. Johari, G.P. and Goldstein, M. (1970) Viscous liquids and the glass transition. II. Secondary relaxations in glasses of rigid molecules. *J. Chem. Phys.*, **53**, 2372.
56. Johari, G.P. (2002) Localized molecular motions of β-relaxation and its energy landscape. *J. Non-Cryst. Solids*, **307-310**, 317.
57. Williams, G. and Watts, D.C. (1971) Molecular motion in the glassy state. The effect of temperature and pressure on the dielectric? relaxation of polyvinyl chloride. *Trans. Faraday Soc.*, **67**, 1971.
58. Ngai, K.L. and Paluch, M. (2004) Classification of secondary relaxation in glass-formers based on dynamic properties. *J. Chem. Phys.*, **120**, 857–873.
59. Ngai, K.L. (2003) An extended coupling model description of the evolution of dynamics with time in supercooled liquids and ionic conductors. *J. Phys.: Condens. Matter*, **15**, S1107.
60. Ngai, K.L. and Tsang, K.Y. (1999) Similarity of relaxation in supercooled liquids and interacting arrays of oscillators. *Phys. Rev. E*, **60**, 4511.
61. Ngai, K.L. and Rendell, R.W. (1997) in *Supercooled Liquids: Advances and Novel Applications*, ACS Symposium Series, Chapter 4, vol. 676 (eds J.T. Fourkas, D. Kivelson, U. Mohanty, and K. Nelson), American Chemical Society, Washington, DC, p. 45.
62. Hikima, T., Hanaya, M., and Oguni, M. (1999) Microscopic observation of peculiar crystallization in the glass transition region and β-process as potentially controlling the growth rate in triphenylethylene. *J. Mol. Struct.*, **479**, 245–250.
63. Hikima, T., Hanaya, M., and Oguni, M. (1996) β-molecular rearrangements process, but not an α-process, as governing the homogenous crystal-nucleation rate in a supercooled liquid. *Bull. Chem. Soc. Jpn.*, **69**, 1863–1868.
64. Paladi, F. and Oguni, M. (2002) Anomalous generation and extinction of crystal nuclei in nonequilibrium supercooled liquid o-benzylphenol. *Phys. Rev. B*, **65**, 144202–144207.
65. Paladi, F. and Oguni, M. (2003) Generation and extinction of crystal nuclei in an extremely non-equilibrium glassy state of salol. *J. Phys.: Condens. Matter*, **15**, 3909–3917.
66. Adrjanowicz, K., Wojnarowska, Z., Wlodarczyk, P., Kaminski, K., Paluch, M., and Mazgalski, J. (2009) Molecular mobility in liquid and glassy states of Telmisartan (TEL) studied by broadband dielectric spectroscopy. *Eur. J. Pharm. Sci.*, **38**, 395–404.
67. Casalini, R. and Roland, C.M. (2009) Aging of the secondary relaxation to probe structural relaxation in the glassy state. *Phys. Rev. Lett.*, **102**, 035701.
68. Kaminski, K., Adrjanowicz, K., Kaminska, E., and Paluch, M. (2011) Probing of structural relaxation times in the glassy state of sucrose and trehalose based on dynamical properties of two secondary relaxation processes. *Phys. Rev. E*, **83**, 061502.
69. Ngai, K.L. (1998) Correlation between the secondary β-relaxation time at T g with the Kohlrausch exponent of the primary α relaxation or the fragility of glass-forming materials. *Phys. Rev. E*, **57**, 7346.
70. Ngai, K.L. and Capaccioli, S. (2004) Relation between the activation energy

of the Johari-Goldstein β relaxation and T_{g} of glass formers. *Phys. Rev. E*, **69**, 031501.

71. Kołodziejczyk, K., Paluch, M., Grzybowska, K., Grzybowski, A., Wojnarowska, Z., Hawelek, L., and Ziolo, J.D. (2013) Relaxation dynamics and crystallization study of sildenafil in the liquid and glassy states. *Mol. Pharm.*, **10**, 2270–2282.
72. Viciosa, M.T., Correia, N.T., Salmerón Sanchez, M., Carvalho, A.L., Romão, M.J., Gómez Ribelles, J.L., and Dionísio, M. (2009) Real-time monitoring of molecular dynamics of ethylene glycol dimethacrylate glass former. *J. Phys. Chem. B*, **113**, 14209–14217.
73. Avrami, M. (1939) Kinetics of phase change. I. General theory. *J. Chem. Phys.*, 7, 1103.
74. Avrami, M. (1940) Kinetics of phase change. II. Transformation-time relations for Random distribution of nuclei. *J. Chem. Phys.*, **8**, 212.
75. Avramov, I., Avramova, K., and Russel, C. (2005) New method to analyze data on overall crystallization kinetics. *J. Cryst. Growth*, **285**, 394–399.
76. Shi, X.M., Zhang, J., Jin, J., and Chen, S.J. (2008) Non-isothermal crystallization and melting of ethylene-vinyl acetate copolymers with different vinyl acetate contents. *eXPRESS Polym. Lett.*, **2**, 623–629.
77. Böhm, N. and Kulicke, W.-M. (1999) Rheological studies of barley (13)(14)-b-glucan in concentrated solution: mechanistic and kinetic investigation of the gel formation. *Carbohydr. Res.*, **315**, 302–311.
78. Caron, V., Bhugra, C., and Pikal, M.J. (2010) Prediction of onset of crystallization in amorphous pharmaceutical systems: phenobarbital, nifedipine/PVP, and phenobarbital/PVP. *J. Pharm. Sci.*, **99**, 3887–3900.
79. Korhonen, O., Bhugra, C., and Pikal, M.J. (2008) Correlation between molecular mobility and crystal growth of amorphous phenobarbital and phenobarbital with polyvinylpyrrolidone and L-proline. *J. Pharm. Sci.*, **97**, 3830–3841.
80. Bhugra, C., Shmeis, R., Krill, S.L., and Pikal, M.J. (2008) Prediction of onset of crystallization from experimental relaxation times. II. Comparison between predicted and experimental onset times. *J. Pharm. Sci.*, **97**, 455–472.
81. Napolitano, S. and Wübbenhorst, M. (2007) Effect of a reduced mobility layer on the interplay between molecular relaxations and diffusion limited crystallization rate in ultrathin polymer films. *J. Phys. Chem. B*, **111**, 5775–5780.
82. Dantuluri, A.K.R., Amin, A., Puri, V., and Bansal, A.K. (2011) Role of α – relaxation on crystallization od amorphous celecoxib above T_g probed by dielectric spectroscopy. *Mol. Pharm.*, **8**, 814–822.
83. Berthier, L., Biroli, G., Bouchaud, J.-P., Cipelletti, L., El Masri, D., L'Hôte, D., Ladieu, F., and Pierno, M. (2005) Accompanying the glass transition direct experimental evidence of a growing length scale. *Science*, **310**, 1797–1800.
84. Dalle-Ferrier, C., Thibierge, C., Alba-Simionesco, C., Berthier, L., Biroli, G., Bouchaud, J.-P., Ladieu, F., L'Hôte, D., and Tarjus, G. (2007) Spatial correlations in the dynamics of glassforming liquids: experimental determination of their temperature dependence. *Phys. Rev. E*, **76**, 041510-1–041510-15.
85. Gordon, M. and Taylor, J.S. (1952) Ideal copolymers and the second-order transitions of synthetic rubbers 1: non-crystalline copolymers. *J. Appl. Chem.*, **2**, 493–498.
86. Couchman, P.R. and Karasz, F.E. (1978) A classical thermodynamic discussion on the effect of composition on glass-transition temperatures. *Macromolecules*, **11**, 117–119.
87. Gupta, P., Thilagavathi, R., Chakraborti, A.K., and Bansal, A.K. (2005) Role of molecular interaction in stability of celecoxib – PVP amorphous systems. *Mol. Pharm.*, **2**, 384–391.
88. Gupta, P., Kakumanu, V.K., and Bansal, A.K. (2004) Stability and solubility of celecoxib–PVP amorphous dispersions: a molecular perspective. *Pharm. Res.*, **21**, 1762–1769.

89. Grzybowska, K., Paluch, M., Wlodarczyk, P., Grzybowski, A., Kaminski, K., Hawelek, L., Zakowiecki, D., Kasprzycka, A., and Jankowska-Sumara, I. (2012) Enhancement of amorphous celecoxib stability by mixing it with octaacetylmaltose: the molecular dynamics study. *Mol. Pharm.*, **9** (4), 894–904.

90. Fleischman, S., Morales, L., Moulton, B., Rodríguez-Hornedo, N., Bailey Walsh, R., and Zaworotko, M.J. (2002) Crystal engineering of the composition of pharmaceutical phases. *Chem. Commun.*, **3**, 186–187.

91. Fleischman, S.G., Kuduva, S.S., McMahon, J.A., Moulton, B., Walsh, B., Rosa, D., Rodríguez-Hornedo, N., and Zaworotko, M.J. (2003) Crystal engineering of the composition of pharmaceutical phases: multiple component crystalline solids involving carbamazepine. *Cryst. Growth Des.*, **3**, 909–919.

92. Haleblian, J. and McCrone, W. (1969) Pharmaceutical applications of polymorphism. *J. Pharm. Sci.*, **58**, 911–929.

93. Grant, D.J.W. (1999) Theory and origin of polymorphism, in *Polymorphism in Pharmaceutical Solids* (ed H.G. Brittain), Marcel Dekker Inc., New York.

94. Desiraju, G.R. (1995) Supramolecular synthons in crystal engineering—a new organic synthesis. *Angew. Chem. Int. Ed. Engl.*, **34** (21), 2311–2327.

95. Otsuka, M., Hasegawa, H., and Matsuda, Y. (1997) Effect of polymorphic transformation during the extrusion-granulation process on the pharmaceutical properties of carbamazepine granules. *Chem. Pharm. Bull.*, **45**, 894–898.

96. Phadnis NV, Suryanarayanan R. Polymorphism in anhydrous theophylline—implications on the dissolution rate of theophylline tablets. *J. Pharm. Sci.* 1997;**86**:1256–1263, doi: 10.1021/js9701418.

97. Parsonage, N.G. and Staveley, L.A.K. (1978) *Disorder in Crystals*, Chapters 9 and 10, Clarendon Press, Oxford.

98. Dionísio, M.S.C., Diogo, H.P., Farinha, J.P.S., and Moura-Ramos, J.J. (2005) Rotational mobility in a crystal studied by dielectric relaxation spectroscopy. an experiment for the physical chemistry laboratory. *J. Chem. Educ.*, **82** (9), 1355.

99. Kuntsche, J., Westesen, K., Drechsler, M., Koch, M.H.J., and Bunjes, H. (2004) Supercooled smectic nanoparticles—a potential novel carrier system for poorly water soluble drugs. *Pharm. Res.*, **21**, 1836–1845.

100. Nesseem, D.I. (2001) Formulation and evaluation of itraconazole via liquid crystal for topical delivery system. *J. Pharm. Biomed. Anal.*, **26**, 387–399.

101. Smyth, C.P. (1955) *Dielectric Behaviour and Structure*, McGraw-Hill, New York.

102. Pethrick, R.A. (1979) in *The Plastic Crystalline State*, Chapter 4 (ed J.N. Sherwood), John Wiley & Sons, Ltd., Chichester, pp. 123–146.

103. Descamps, M., Correia, N.T., Derollez, P., Danede, F., and Capet, F. (2005) Plastic and glassy state of caffeine. *J. Phys. Chem. B*, **109**, 16092–16098.

104. Lehto, V.P. and Laine, E. (1998) A kinetic study of polymorphic transition of anhydrous caffeine with microcalorimeter. *Thermochim. Acta*, **317**, 47.

105. Stevenson, C.L., Bennett, D.B., and Lechuga-Ballesteros, D. (2006) Pharmaceutical liquid crystals: the relevance of partially ordered systems. *J. Pharm. Sci.*, **94**, 1861–1880.

106. Letz, M., Schilling, R., and Latz, A. (2000) Ideal glass transitions for hard ellipsoids. *Phys. Rev. E*, **62**, 5173.

107. Wojnarowska, Z., Paluch, M., Wlodarczyk, P., Hawelek, L., Wrzalik, R., Ziolo, J., Wygledowska-Kania, M., Bergler-Czop, B., Brzezinska-Wcislo, L., and Bujak, P. (2011) Nanoscale domains with nematic order in supercooled vitamin-A acetate: molecular dynamics studies. *Phys. Rev. E*, **83**, 051502.

108. Tarnacka, M., Adrjanowicz, K., Kaminska, E., Kaminski, K., Grzybowska, K., Kolodziejczyk, K., Wlodarczyk, P., Hawelek, L., Garbacz, G., Kocot, A., and Paluch, M. (2013) Molecular dynamics of itraconazole at ambient and high pressure. *Phys. Chem. Chem. Phys.*, **15**, 20742–20752.

109. K. Adrjanowicz, K. Kaminski, P. Wlodarczyk, K. Grzybowska, M. Tarnacka, D. Zakowiecki, G. Garbacz, M. Paluch and S. Jurga (2013) Molecular dynamics of supercooled pharmaceutical agent posaconazole studied via differential scanning calorimetry, dielectric and mechanical spectroscopies, *Mol. Pharm.*, **10**, 3934–3945.

110. Brás, A.R., Dionisio, M., Huth, H., Schick, C., and Schönhals, A. (2007) Origin of glassy dynamics in a liquid crystal studied by broadband dielectric and specific heat spectroscopy. *Phys. Rev. E*, **75**, 061708.

111. Wacrenier, J.M., Druon, C., and Lippens, D. (1981) A study of experimental results obtained for dielectric relaxation of several mesomorphous members of the alkylcyanobiphenyl series. *Mol. Phys.*, **43**, 97.

112. Six, K., Verreck, G., Peeters, J., Augustijns, P., Kinget, R., and Van den Mooter, G. (2001) Investigation of thermal properties of glassy itraconazole: identification of a monotropic mesophase. *Thermochim. Acta*, **376**, 175–181.

113. Lechuga-Ballesteros, D., Abdul-Fattah, A., Stevenson, C.L., and Bennett, D.B. (2003) Properties and stability of a liquid crystal form of cyclosporine—the first reported naturally occurring peptide that exists as a thermotropic liquid crystal. *J. Pharm. Sci.*, **92**, 1821–1831.

114. Rades, T. and Muller-Goymann, C. (1994) Melting behavior and thermotropic mesomorphism of fenoprofen salts. *Eur. J. Pharm. Biopharm.*, **40**, 277–282.

115. Cox, J.S.G., Woodard, G.D., and McCrone, W.C. (1997) Solid state chemistry of cromolyn sodium (disodium cromoglycate). *J. Pharm. Sci.*, **60**, 1453–1465.

116. Ciuchi, F., Nicola, G.D., Franz, H., Gottarelli, G., Mariani, P., Bossi, M.G.P., and Spada, G.P. (1994) Self-recognition and self-assembly of folic acid salts: columnar liquid crystalline polymorphism and the column growth process. *J. Am. Chem. Soc.*, **116**, 7064–7707.

12
Raman Spectroscopy in Disordered Molecular Compounds: Application to Pharmaceuticals

Alain Hedoux

12.1
Introduction

Raman spectroscopy gives the unique opportunity to analyze the three categories of molecular motions existing in organic solids, in the framework of the rigid-body model: (i) internal motions of atoms within the molecule, mainly distinctive of the molecular conformation in the fingerprint region usually lying between 500 and 1800 cm^{-1}, and closely related to the local environment; (ii) semi-internal (or semi-external) motions corresponding to large-amplitude rotations of the molecule or a group of atoms within the molecule, giving a Raman signature in the very low frequency range; and (iii) external motions between a molecule and its first molecular neighbor. In ordered crystalline phases, external motions give rise to the phonon spectrum corresponding to the lattice vibrations. Disordering induces an inhomogeneous broadening of the phonon spectrum, and the low-frequency spectrum of the amorphous state is observed to be a very close representation of the vibrational density of states (VDOS), usually determined by inelastic neutron scattering. Molecular compounds are characterized by the pronounced contrast between the strong covalent intramolecular interactions and soft van der Waals intermolecular attractions or hydrogen bonding interactions. This contrast is responsible for the frequency gap between internal and external vibrations. Intermolecular interactions are responsible for specific physical properties of molecular compounds (polymorphism, low melting temperature, etc.), while covalent bonds maintain the cohesion within the molecule. Molecular flexibility favors disorder and reduces the gap between the external and internal vibrational modes. In this context, the investigation of the whole Raman spectrum provides structural description of disordered states (crystalline, microcrystalline/nanocrystalline, and amorphous states) and the understanding of the mechanism of phase transformation in relation to the molecular conformation.

The development of pharmaceutical solid formulations requires important information on the characteristics and the physical nature of solid phases, including crystalline and amorphous. Molecular compounds go through several stages

Disordered Pharmaceutical Materials, First Edition. Edited by Marc Descamps.

of processing (milling, freeze-drying, spray-drying, pressurizing) for packaging as the solid-dosage form. During their storage, drugs may be exposed to a wide range of temperature and humidity, which can lead to different kinds of phase transformations (crystallization from an amorphous state or crystalline–crystalline transitions). The stability conditions of the polymorphs and the amorphous state have to be determined, and polymorph selection and control is crucial.

The most common methods applied for solid-state analysis are X-ray powder diffraction, differential scanning calorimetry, and different spectroscopic techniques. Raman spectroscopy is a fast and nondestructive method and probes structural changes at different levels: molecular conformation [1, 2], short-range order (local molecular environment [3], molecular associations through hydrogen bonding [4–7]) and long-range order. Therefore, Raman spectroscopy plays a major role in the analysis and material characterization in the pharmaceutical industry. The increasing use of the Raman spectroscopy in the pharmaceutical field is partly related to the small contribution of the signal of water to the spectrum of the active pharmaceutical ingredients (APIs). Raman intensity is related to the polarizability of the molecule, which is much larger for π systems than σ-bonded structures [8]. Generally, APIs have π bonds and excipients are saturated molecules (σ bonded), and thus Raman spectroscopy is more sensitive to API molecules. It is widely used as a quality control/assurance tool for chemical identification or for analyzing quantitatively drug substances [9] in the context of their solid-state form, mainly from measurements performed in the 300–1800 cm^{-1} region, which is considered as molecular structure fingerprints of an analyte. In this chapter we show how Raman spectroscopy can be used for

- detecting different types of disorder
- distinguishing nanocrystals from amorphous materials
- describing disordered metastable states and mixed coexisting phases, and understanding the mechanisms of phase transformations

by measuring a wide frequency domain of the Raman spectrum, including the low-frequency range down to about 3–5 cm^{-1}.

12.2 Raman Spectroscopy

12.2.1 Basic Theory

The requirement for spontaneous Raman activity is a change in the electronic polarizability of the analyzed molecule. Monochromatic light from a laser is generally used to irradiate the sample. A photon interacts with the electron cloud and the bonds of the molecule. In classical terms, the interaction can be viewed as a perturbation of the molecule's electric field. When a molecule is exposed to an electric field, a dipole moment is induced, which is proportional to the electric field strength and to the molecular polarizability α. A molecular vibration described by the normal coordinate

$$q_i = q_i^0 \cos(2\pi \nu_i t) \tag{12.1}$$

can be observed in the Raman spectrum if the condition

$$\left(\frac{\partial \alpha}{\partial q}\right)_0 \neq 0 \tag{12.2}$$

is satisfied. In the quantum mechanical view, a photon excites the molecule from the ground state to a virtual energy state. When the molecule relaxes, it emits a photon and returns to an energy state different from the original state. The difference in energy between the original and final states leads to a frequency shift ($\Delta\nu$) between the emitted photon's frequency (ν_i) and that of the excitation wavelength (ν_0), the so-called Raman shift. Figure 12.1 shows two possible cases of energy transfer. If the final state is more energetic than the initial state, the Raman shift ($\Delta\nu = \nu_0 - \nu_i$) is designated as a Stokes shift. In the opposite case, the energy transfer corresponds to $\Delta\nu = \nu_0 + \nu_i$, called anti-Stokes shift. Most of the light is elastically scattered in different directions (Rayleigh scattering). The inelastically backscattered light corresponding to Stokes shift is usually analyzed, since it is characterized by an intensity higher than that corresponding to the anti-Stokes spectrum. This characteristic can be explained by considering that the probability

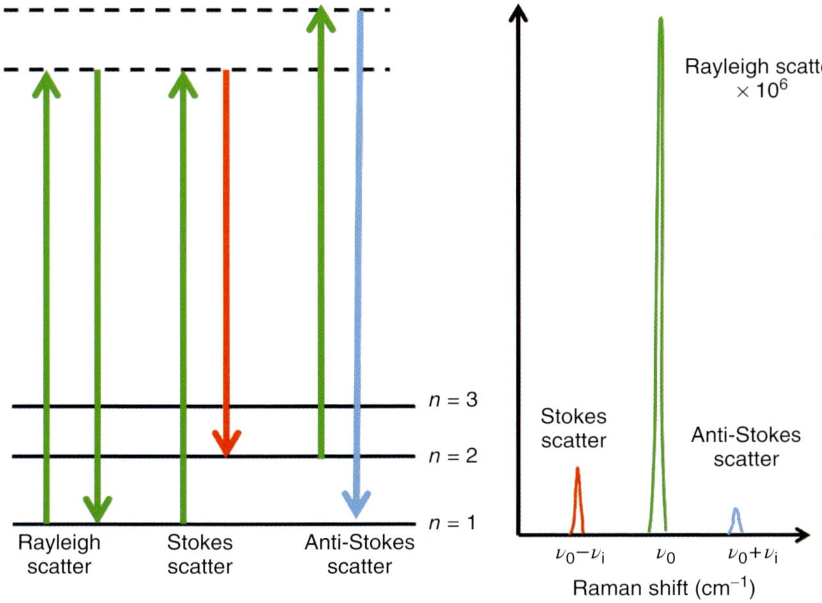

Figure 12.1 Description of Raman scattering.

of the Stokes transition is greater than that of the anti-Stokes transition. The ratio of Stokes to anti-Stokes intensities is given by [10]

$$\frac{(\nu_0 - \nu_i)^4}{(\nu_0 + \nu_i)^4} \exp(h\nu_i/kT) \tag{12.3}$$

This equation can be used for the determination of intrinsic sample temperature, by fitting simultaneously the Stokes and anti-Stokes intensities. The intensity of the Stokes Raman spectrum is mainly proportional to [10]

$$I_{\text{Raman}} \propto NI_0[n(\nu) + 1](\nu_0 - \nu_i)^4 \left[\vec{e}_f \left(\frac{\partial \overline{\overline{\alpha}}}{\partial q_i}\right) \vec{e}_j\right]^2 \tag{12.4}$$

where N is the number of molecules and

$$n(\nu) = [1 - \exp(h\nu/kT)]^{-1} \tag{12.5}$$

is the Bose–Einstein factor, $\overline{\overline{\alpha}}$ is the polarizability tensor corresponding to the (q_i, ν_i) vibration, and $\vec{e}_j$, $\vec{e}_f$ correspond to the unit vectors of the incident and scattered electric field. It is worth noting that $n(\nu)$ has a great influence only on the low-frequency Raman line shape ($\nu < 200\,\text{cm}^{-1}$).

12.2.2
Equipment and Procedures

Raman spectrometers are usually composed of four main components: (i) the laser, which is the source of monochromatic radiation, (ii) a sample illumination system and light collection optics, (iii) a device for the analysis of the scattered light (spectrometer), (iv) and a sensitive detector such as a charge-coupled device (CCD). Two kinds of spectrometers can be used for obtaining Raman spectra: the Fourier transform (FT)-Raman or the dispersive Raman spectrometer. The difference between the two types is based essentially on the device: a grating dispersing system for dispersive Raman spectrometers, or an interferometer for FT-Raman spectrometers. Most of FT-Raman spectrometers use a Nd:YAG laser (1.064 μm) with a near-infrared interferometer coupled to either a liquid-nitrogen-cooled germanium or indium gallium arsenide detector. The main advantages are (i) the reduction in the number of samples that exhibit laser-induced fluorescence, (ii) ease of operation, and (iii) high spectral resolution with good wavelength accuracy. However, FT-Raman spectrometers have the disadvantage that they are unsuitable for certain sample environments (temperature, pressure, humidity, etc.). Dispersive Raman spectrometers use multichannel (CCD) detectors, with extremely low intrinsic noise and high quantum efficiency. The benefits of dispersive Raman spectrometers are their far higher sensitivities and far lower detection limits than FT-Raman instruments. The shorter wavelengths of lasers associated with the dispersive Raman spectrometer increase the sensitivity of the instrument, because of the ν^4 scattering efficiency dependence. As a consequence, the data acquisition times are much shorter than for FT-Raman methods, which is important for the study of fast transformation kinetics. The performance of dispersive Raman spectrometers results from the tradeoff between good resolution and rejection of the laser line and sufficient intensity imposed by the dispersion system. Modern instrumentation almost universally employs notch or edge filters for laser rejection, and a spectrograph (monochromator). However, in this case the analysis of the low-frequency range ($< 150\,cm^{-1}$) is generally not possible. The use of multiple dispersion stages (double and triple monochromators) with long focal lengths allow the detection of Raman-active modes with frequencies as low as $2 - 5\,cm^{-1}$. New advances in ultra-narrow-band notch filter technology [11] with holographic gratings enable measurements of low-frequency spectra from a relatively compact, easy-to-use, and cost-effective system [12]. The development of this kind of dispersive system composed of a single-stage spectrograph opens newer possibilities in the use of Raman spectroscopy compared to IR spectroscopy in various industrial sectors.

Depending on the nature of the analysis, the important volumes or heterogeneities, or microscopic samples, we shall distinguish two types of installation: the conventional installation or the macroanalysis, and the micro-Raman installation suitable for micro analyses. Raman investigations under pressure are typically performed on microscopic samples located in diamond anvil cells (DACs), and then micro-Raman installation is systematically used for pressure analyses. For micro-Raman, the sample holder is a XYZ table equipped with a microscope composed of objectives of high numerical aperture. This type of equipment allows quantitative analyses of mixtures [13] or the mapping of solid dispersions [14].

12.2.3
Analysis of Raman Spectra of Molecular Compounds

In organic molecular compounds, the intramolecular bonding is very strong compared to the weak intermolecular bonding, inducing a spectral gap between the external and internal bands in the Raman spectrum, which is mainly dependent on the molecular flexibility. A free molecule composed of N atoms has $3N - 6$ internal vibrations and six degrees of freedom. The number of internal Raman-active modes results from considerations of the molecular and the crystalline symmetries, reflecting the local molecular arrangement. In a crystalline state characterized by a unit cell composed of Z molecules, $6Z$ external motions are expected, including three acoustic vibrations corresponding to collective translation motions of molecules, which are not detectable by Raman spectroscopy. The number of Raman-active external motions is usually lower than $6Z - 3$ and depends on the crystalline symmetry. Consequently, Raman spectroscopy can be considered as an indirect structural probe, providing information about the long-range order and the local molecular environment from the respective analyses of the low-frequency Raman spectrum (LFRS) and the spectrum of internal modes in the crystal. The fingerprint region provides information on the structure of the molecule, while at higher frequencies ($>2000\,cm^{-1}$) the

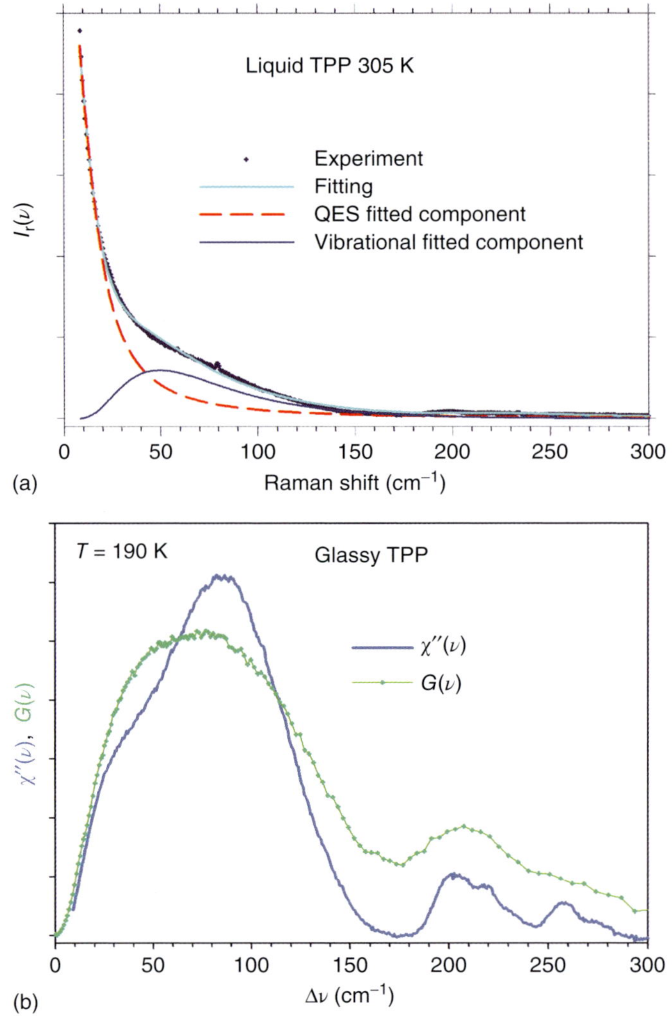

Figure 12.2 Representations of the low-frequency spectrum of TPP. (a) Reduced intensity and fitting procedure of experimental spectrum. (b) Raman susceptibility compared with the VDOS ($G(\nu)$) obtained by neutron scattering.

intramolecular C–H, N–H, and O–H stretching regions provide information on molecular associations via H-bonding.

12.2.3.1 Analysis of the Low-Frequency Raman Spectrum

The analysis of the LFRS of disordered molecular materials is very difficult since it is characterized by the overlapping contributions of the quasielastic intensity (I_{QES}) and the vibrational intensity (I_{VIB}), as shown in Figure 12.2. However, it can be considered as containing rich information (i) on the structural organization in disordered systems and (ii) on the mechanism of (order–disorder) phase transformation, confined to a narrow frequency range (3–200 cm^{-1}). The Raman intensity ($I_{Raman}(\nu, T)$) in the LFRS is related to the VDOS $G(\nu)$ according to [15, 16]

$$I_{Raman}(\nu, T) = \frac{[n(\nu, T) + 1]}{\nu} C(\nu)G(\nu) \tag{12.6}$$

and

$$\chi''(\nu) = \frac{C(\nu)}{\nu} G(\nu) \tag{12.7}$$

where $C(\nu)$ is the light–vibration coupling coefficient, and $\chi''(\nu)$ is the Raman susceptibility. To obtain the $\chi''(\nu)$ spectrum (plotted in Figure 12.2b), which contains the structural information, the Raman intensity is first transformed into reduced intensity $I_r(\nu)$ plotted in Figure 12.2a using

$$I_r(\nu) = \frac{I_{Raman}(\nu, T)}{[n(\nu, T) + 1]\nu} \tag{12.8}$$

In this representation, the QES intensity is the dominant contribution of the LFRS. This component is usually determined from a fitting procedure using a Lorentzian shape centered at $\nu = 0$. The vibrational component is usually well described with a lognormal function. After subtracting the QES component, $I_r(\nu)$ can be converted into Raman susceptibility by

$$\chi''(\nu) = \nu \cdot I_r(\nu) \tag{12.9}$$

The frequency dependence of $C(\nu)$ has been highly debated [16–20]. Several studies have converged to a linear dependence of $C(\nu)$, suggesting that $\chi''(\nu) \propto G(\nu)$ [21–23]. This relation was corroborated by the similarity between the $\chi''(\nu)$ and $G(\nu)$ spectra plotted in Figures 12.2b and 12.3 for two molecular glasses triphenyl phosphite [22] (TPP) and *ortho*-terphenyl [24] (OTP). Figure 12.3a,b shows the comparison between the VDOS of the glassy state and the phonon spectrum of the OTP crystal obtained by neutron experiments and Raman spectroscopy. These figures confirm that $\chi''(\nu)$ spectra can be considered as a close representation of the VDOS and demonstrate that the spectral resolution of the $\chi''(\nu)$ spectra is significantly higher than that of $G(\nu)$ spectra. Figure 12.3a shows the temperature dependence of $\chi''(\nu)$ spectra in TPP, plotted in the liquid, undercooled liquid, and glassy states. The temperature dependence in the temperature range of 100 K is relatively weak, reflecting the quasi-harmonic behavior of collective motions.

The temperature dependence of the QES component is analyzed using the fitting procedure described in Figure 12.2. This contribution to the LFRS corresponds to fast anharmonic motions, the so-called β-fast relaxational motions. The $I_r(\nu)$ spectra are normalized by the integrated intensity of the vibrational component, which has a weak temperature dependence, and plotted at various temperatures (in the liquid, undercooled liquid, and glassy states) in Figure 12.4a. This figure contrasts with Figure 12.4b, indicating a strong temperature dependence distinctive of anharmonic motions. $I_{QES}(T)$ is similar to the temperature dependence of the mean-square displacement [25] $\langle u^2 \rangle(T)$, which is usually

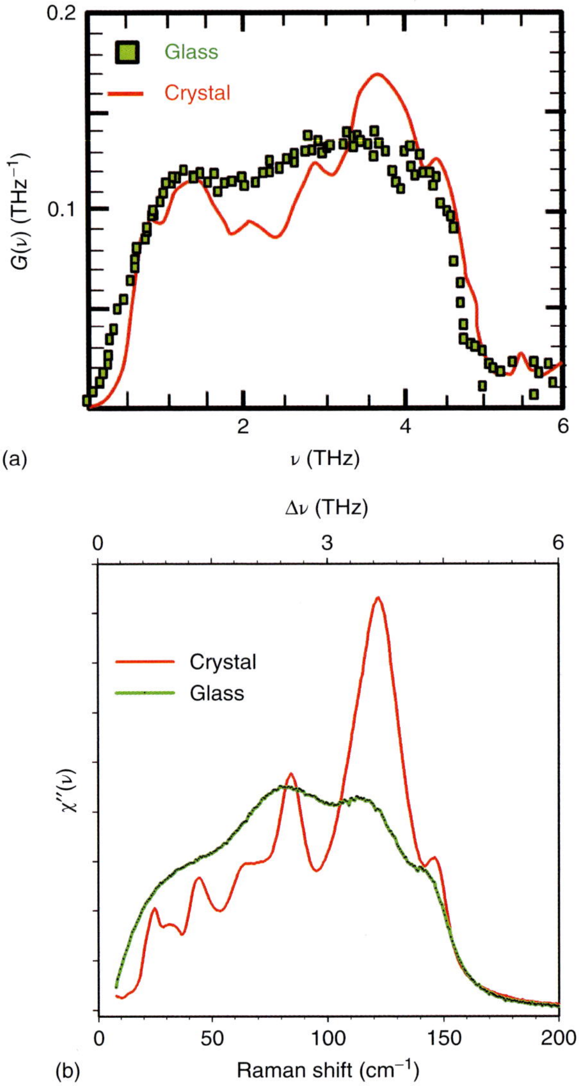

Figure 12.3 Representations of the low-frequency vibration spectra in the glassy and crystalline states of OTP obtained (a) by inelastic neutron scattering from [24] (b) by Raman spectroscopy.

determined by neutron scattering experiments. The temperature dependence of these kinds of anharmonic motions is considered as the driving force in the glass transition in fragile molecular liquids [25], and the relation

$$1/\langle u^2 \rangle \propto \log \eta \tag{12.10}$$

was experimentally determined in different glass-forming systems [26, 27]. The temperature dependence of $\log \eta$ is plotted in the inset of Figure 12.4a with the

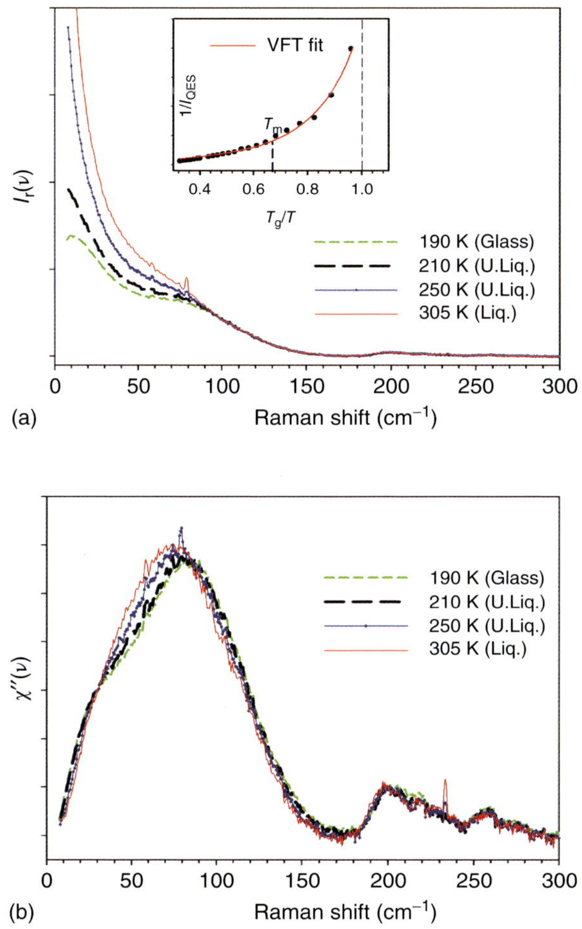

Figure 12.4 Temperature dependence of the LFRS of TPP (a) in the representation of reduced intensity. The inset shows the behavior of $<u^2>$ versus T_g/T. (b) Representation of Raman susceptibility.

fitting curve corresponding to the Vogel–Fulcher–Tammann (VFT) function given by

$$\exp(A \cdot T_0/(T - T_0)) \tag{12.11}$$

where A reflects the curvature of the plot distinctive of the fragility of the system.

12.2.3.2 **Analysis of the High-Frequency Spectrum**

Raman investigations are commonly performed at frequencies higher than 200 cm^{-1}, that is, in the domain that requires no specific high-dispersive equipment (triple monochromator) for the correct rejection of the elastically scattered light. For rigid molecules, internal modes are detected above 150–200 cm^{-1}. Two regions are usually distinguished in the spectrum of internal vibrations:

(i) the fingerprint region covering the 500–1800 cm^{-1} range, distinctive of the molecular conformation, and (ii) the intramolecular C–H and O–H stretching region lying in the range 2200–3800 cm^{-1}, recognized to be very sensitive to molecular associations via H-bonding [5]. In both regions, no intensity correction is required, since the Bose–Einstein factor becomes close to 1 above 100 cm^{-1}. However, to obtain information on the formation of H-bonds, related to the Raman intensity of X–H stretching bands [5] (with X=O, N, C), the spectrum must be normalized with respect to Raman bands that are temperature independent. The fingerprint region is used in multivariate analysis (MVA) for conformational state analysis of proteins or smaller pharmaceutical molecules for identifying polymorphs. Univariate spectral analyses are based on the peak height at a specified frequency of the untreated Raman data. The selected spectral response must be not biased by contributions from other sources (presence of water, buffer components, etc.). MVA consists of using newly defined variables that are linear combinations of the original ones, and are mutually independent. Principal component analysis (PCA) is a well-known statistical method used to determine the number of linear, independent components in a given system. PCA can be used to analyze a series of spectra collected in different samples, or a series of spectra collected in a mapping procedure to obtain a Raman image. Applications PCA in the analysis of pharmaceuticals will be described in Section 12.3.4. Direct classical least squares (DCLS) is another frequently used MVA method for quantitative analyses. DCLS finds the linear combination of spectra from the pure components that most closely matches the Raman spectrum of the sample. This method requires, as the starting step, collecting Raman spectra of pure components of the system. MVA methods are widely used for obtaining images from Raman mapping.

12.3 Analysis of Molecular Compounds by Raman Spectroscopy

12.3.1 The Use of the LFRS to Distinguish Micro/Nanocrystals from the Amorphous State

12.3.1.1 Analysis of Polyamorphic Situations

TPP, $P(OC_6H_5)_3$, is a molecular compound widely investigated by various experimental methods because it is considered as a notable example of polyamorphism [28–30]. Indeed, the undercooled liquid state isothermally transforms, just above $T_g = 202$ K, into an apparently second amorphous state, named the glacial state (GS). Low-frequency Raman experiments [31, 32] have given clear evidence that GS is not really an amorphous state, by comparing $\chi''(\nu)$ spectra of the undercooled liquid, the glacial state, and the crystalline states at the same temperature, as shown in Figure 12.5.The LFRS of GS has an amorphous-like band shape, with no apparent phonon peak. However, it rigorously corresponds to the envelope of the phonons of the crystal, and the fitted components of the GS are directly related

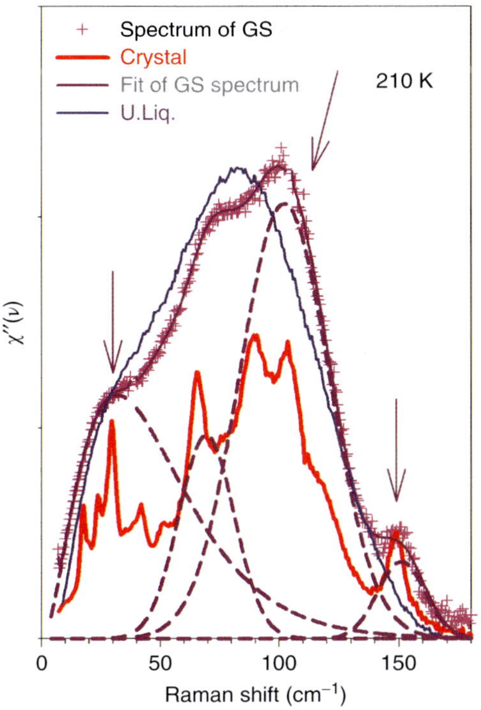

Figure 12.5 Comparison between Raman susceptibilities in the undercooled liquid (U.Liq.) glacial (GS) and crystalline states of TPP at 210 K. The dashed lines correspond to the fitted components of the spectrum in the GS. The arrows show the main spectral differences between U.Liq. and GS, which can be considered as crystalline signatures.

to phonons or phonon groups of the crystal. It was shown that the band shape of LFRS in GS, reflecting the size of crystallized domains, was highly dependent on the temperature of the isothermal transformation, with the domain size varying from 30 Å at 210 K up to 100 Å at 218 K [32]. A similar situation was observed in *n*-butanol [4, 5].

This study has shown the capability of the LFRS to detect and identify the earliest stage of crystallization, since the size of first detected crystallized domains in TPP was ~30 Å [32], that is, about the size of the unit cell [33].

12.3.1.2 Detection and Quantification of Crystallization in Ground Powder

It was observed that a ground amorphous powder of indomethacin (IMC) rapidly crystallizes at room temperature [34, 35], that is, ~20° below the glass transition temperature ($T_g \approx 42\,°C$). Crystallization of amorphous powder was analyzed by FT-Raman spectroscopy [35] in the 1650–1750 cm^{-1} frequency range, where the crystalline phase could be distinguished from the amorphous state, as shown in Figure 12.6a. However, the quantitative analysis of the crystallization requires the plot of a calibration curve (intensity of crystalline features versus the degree of crystallization) from the analysis of physical mixtures [35]. LFRS

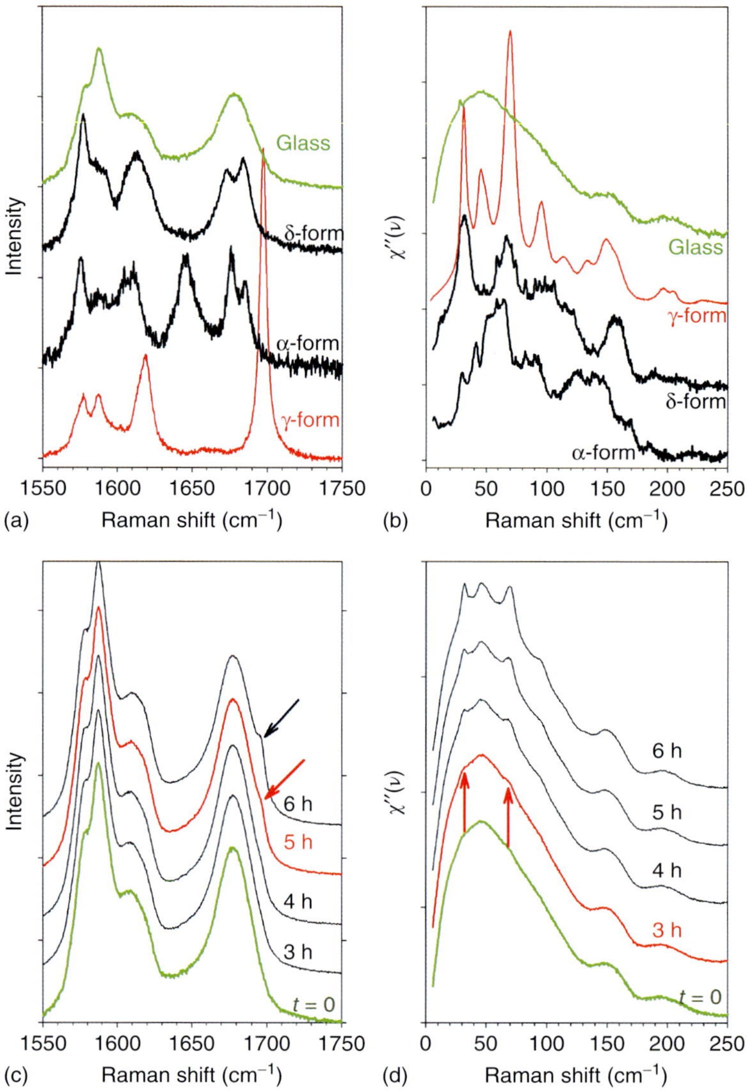

Figure 12.6 Raman spectra of indomethacin in the C=O stretching and low-frequency regions. (a, b) Crystalline forms and the glassy state of IMC. (c, d) Raman spectra of milled γ-IMC collected at various times of milling.

is of course distinctive of the different crystalline and amorphous states, as shown in Figure 12.6b. Additionally, studies on polyamorphic materials (see the previous section) have shown the sensitivity of LFRS to detect early stages of crystallization. The comparison of Figure 12.6c,d unequivocally reveals the high sensitivity of LFRS compared to the 1650–1750 cm^{-1} range to detect the first traces of crystallization. These traces correspond to the most intense phonon peaks of the γ-phase of IMC, which identifies the crystalline state. The intensity

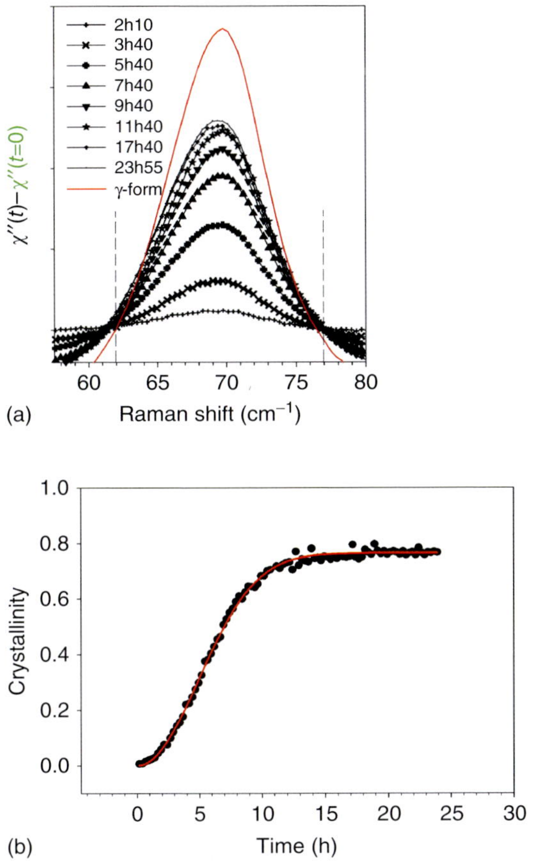

Figure 12.7 Analysis of the crystallization mechanism after amorphization of γ-IMC by cryo-milling. (a) $\chi''(\nu)$ spectral differences during the kinetics of crystallization at room temperature. (b) Time evolution of the volume fraction of γ-IMC (black circles) from the integration of $\chi''(\nu)$ spectral differences between the vertical dashed lines (plotted in a). The line represents the Avrami-like function fitted to experimental data.

of the VDOS spectrum in the amorphous state is equivalent to that of the phonon spectrum in the crystalline state. The time dependence of the intensity of the most intensely growing phonon peak of the γ-phase was determined by subtracting the spectrum $\chi''(t)$ from that of the amorphous state $\chi''(t=0)$, as plotted in Figure 12.7a. It is worth noticing that nonzero intensity can be detected at times earlier than 2 h and 10 min [36]. The integrated intensity of the growing peak, compared to that in the γ-phase, directly gives the degree of crystallization, which is plotted against time in Figure 12.7b. It can be observed that the degree of crystallinity is significantly less than 1, indicating partial crystallization. Data were fitted using an Avrami-like function

$$\rho(t) \propto \left(1 - \exp\left[-\left(\frac{t}{\tau}\right)^{n}\right]\right) \tag{12.12}$$

where τ is the time of half transformation, leading to the determination of the Avrami exponent close to 2 [36], instead of 4, for the classical nucleation and growth process (3 for 3D growth and 1 for nucleation). This deviation from the usual behavior was interpreted as corresponding to a two-dimensional crystallization from pre-existing nuclei [36].

12.3.2
The use of the LFRS to Reveal and Analyze Disordered States

12.3.2.1 Ibuprofen: Chirality and Disorder

Ibuprofen (*iso*-butyl-propanoic-phenolic acid, $C_{13}H_{18}O_2$, IBP) has a chiral center. Between the (*S*) and (*R*) enantiomers, only the (*S*) form is pharmacologically active and used as a nonsteroidal anti-inflammatory drug. The racemic mixture (*R*,*S*), composed of equal quantities of the (*S*) and (*R*) enantiomers, can be considered as a source of disorder with respect to the (*S*) and (*R*) forms. In this context, the Raman spectra of amorphous (glassy) and crystalline states of the (*S*) form and racemic mixture are compared in Figure 12.8 in three spectral windows: (i) the low-frequency range, (ii) a domain of the fingerprint region, and (iii) the intramolecular C–H and O–H stretching regions. The spectra of the amorphous state in (*S*)- and (*R*,*S*)-IBP are partially superimposed in the entire frequency range (0–3100 cm^{-1}). The similarity of Raman spectra in amorphous (*S*)- and (*R*,*S*)-IBP indicates the same local order in the glassy state in presence of only one type of enantiomer or both enantiomers. In contrast, spectra of stable crystalline states are clearly distinguishable in the three spectral windows. As can be expected, Figure 12.8a shows two different phonon spectra, corresponding to two different space groups, ($P2_1/c$) [37] for phase I of (*RS*)-IBP and ($P2_1$) [38] for crystalline (*S*)-IBP. In the fingerprint region, the most intense band around 1610 cm^{-1}, assigned to vibrations of aryl ring C=C bonds [39, 40], is not sensitive to the type of enantiomer, contrasting to the Raman bands at lower frequencies, ~1450 cm^{-1}. Figure 12.8c shows significant changes between crystalline (*S*)- and (*RS*)-IBP, in a region where C–H stretching bands overlap with stretching bands of O–H involved in intermolecular O–H···O associations. The distinction between spectra of (*S*)- and (*RS*)-IBP in the 2800–3100 $cm{-}^1$ mainly results from the existence of different molecular associations in the two crystalline states. A metastable crystalline state (phase II) was recently detected in the racemic (*RS*) mixture by calorimetric and X-ray investigations [41]. It was shown [41] that phase II can be obtained only by heating the glassy state obtained by a deep quench of the liquid below T_g, leading to the formation of cracks. LFRS of phase II has directly revealed evidence of a substantial disorder [42]. This spectrum, plotted in Figure 12.9, is typical of a highly disordered state with the characteristic absence of phonon peaks, contrasting with the observation of Bragg peaks in the X-ray powder diagram [43]. Additionally, Figure 12.9 shows that the spectrum of phase II corresponds to the envelope of the phonon peaks of the stable crystal (phase I), and the main distinctions between spectra of phases I and II correspond

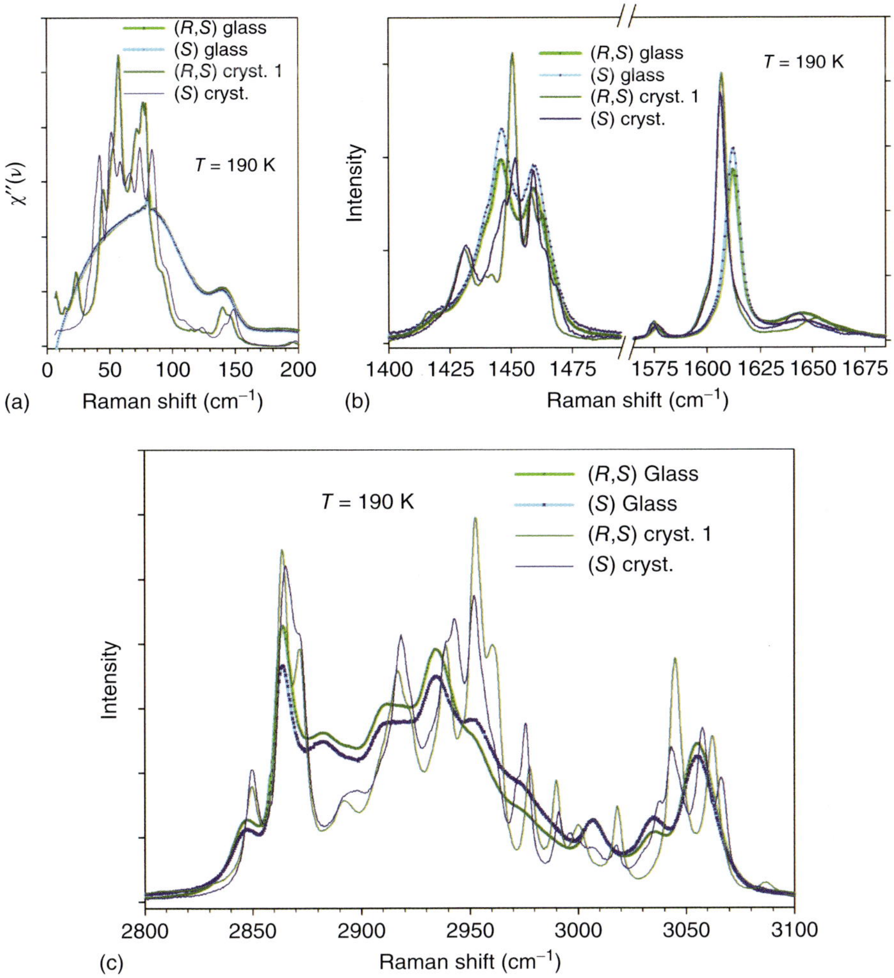

Figure 12.8 Raman spectra of amorphous and stable phases of (*S*)- and (*RS*)-IBP. (a) Raman susceptibility in the low-frequency region. (b) Selected parts of the fingerprint region. (c) Intramolecular C–H and O–H stretching regions.

to the crystalline features of phase I (see the arrow in Figure 12.9). LFRS of phase II was interpreted as resulting from an inhomogeneous broadening of phonon peaks in phase I, probably generated by a disordered distribution of the two different kinds of dimers existing in the stable crystalline states of (*S*) [38] and (*RS*) [37] IBP. It is worth noting that IBP is predominantly composed of hydrogen atoms, which have a poor electronic weight. Their position cannot be determined in a refinement procedure using X-ray data, and thus Raman spectroscopy can be considered as an alternative for structural description and for analyzing molecular associations via H-bonding.

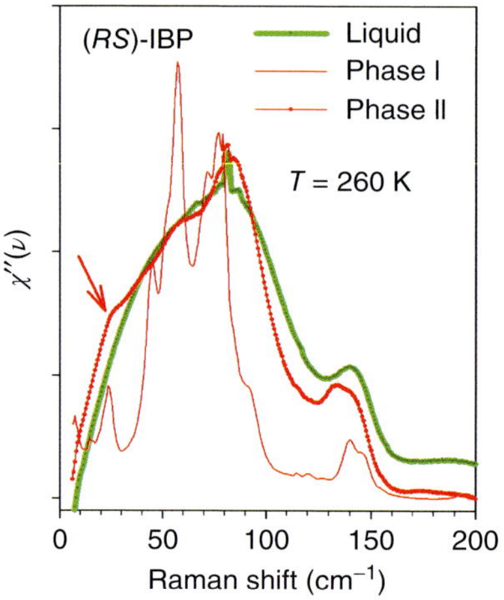

Figure 12.9 Raman susceptibility of (*RS*)-IBP in the liquid state, the stable crystalline phase (I), and the metastable phase (II). The arrow highlights the pronounced shoulder of the VDOS in phase II, which does not exist in phase I, and corresponding to the low-frequency phonon peak in phase I.

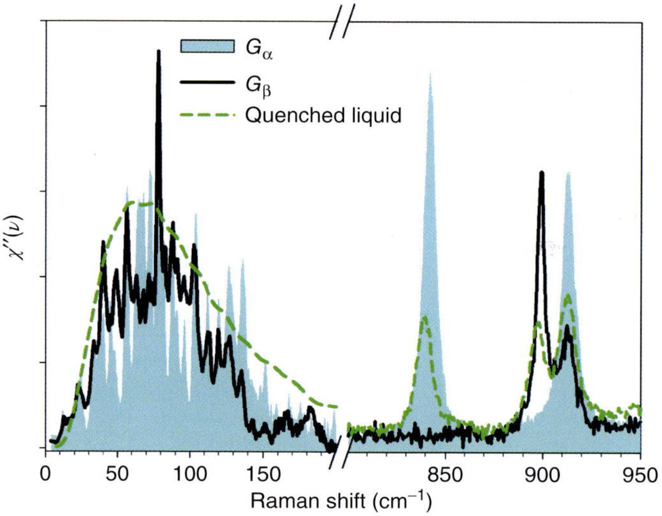

Figure 12.10 Raman spectra of G_α and G_β crystalline states, and the solid amorphous state obtained by quenching G_α or G_β melts, in the low-frequency range and the anomeric region.

12.3.2.2 **Evidence of Different Structural Organizations in Amorphous Glucose Obtained by Thermal and Non-Thermal Routes**

Both G_α [44] and G_β [45] crystalline states have two close structural descriptions, characterized by the same space group $P2_12_12_1$ and $Z = 4$. Consequently, similar phonon spectra are expected for crystalline G_α and G_β, with the same number of phonon peaks, as seen in Figure 12.10. However, G_α and G_β can be easily distinguished in the fingerprint region, especially between 800 and 950 cm^{-1}, plotted in Figure 12.10, corresponding to C–O and C–C group vibration modes [46]. It can be noticed that the spectrum of the glassy state obtained by quenching

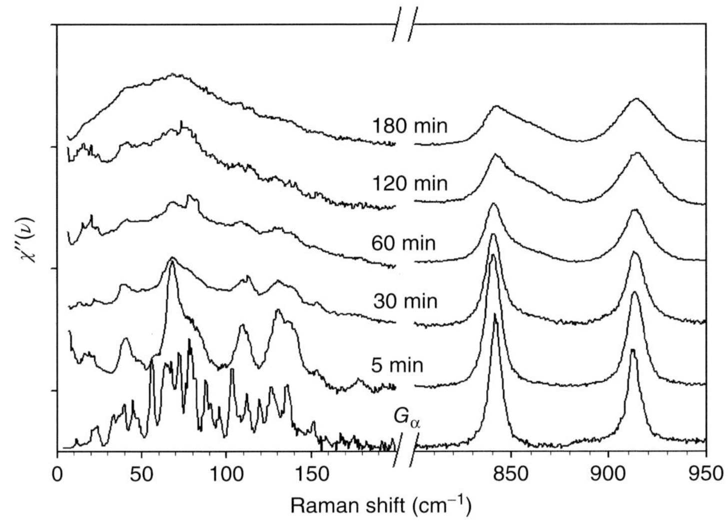

Figure 12.11 Raman spectra collected at various stages of cryo-milling from a G_α crystal.

(QC) the liquid corresponds to the envelope of external and internal vibration modes of the G_α and G_β crystalline states. It was shown that amorphization under milling can occur at temperatures significantly lower than T_g ($\sim T_g - 50\,°C$) [1]. Raman measurements require no special preparation of the sample, and the rapid acquisition times (6 min) give the unique opportunity to analyze, instantaneously, the behavior of the phonon spectrum under milling, by monitoring mutarotation. $\chi''(\nu)$ spectra are plotted in Figure 12.11 in the low-frequency range and in the anomeric region for different milling times. A flattening of the phonon peaks is observed, indicating the loss of the long-range order until obtaining the VDOS of the amorphous state. At higher frequencies, the continuous broadening of the Raman band distinctive of G_α (840 cm^{-1}) is observed but not the band around 900 cm^{-1}associated with the β anomer. These observations in both spectral regions give the indication that an anomerically pure G_α amorphous powder is obtained. The same kind of crystal to amorphous G_β transformation was observed. Figure 12.11 demonstrates the high sensitivity of the LFRS for detecting residual crystalline material. Indeed, LFRS collected after 120 and 180 min of milling are clearly distinguishable, but at higher frequencies both spectra are very similar. Raman spectra of milled G_α and G_β are compared to the glassy state reached by quenching the liquid state in Figure 12.12. The LFRS of the three samples are superimposed, indicating similar VDOS of the amorphous states. However, each spectrum is distinguishable from the others at higher frequencies. It is the demonstration of the existence of three different local organizations in the three amorphous states of glucose, evidenced only by Raman spectroscopy. Surprisingly, the internal bands are significantly broader in pure amorphous G_α and G_β than in the quenched liquid, where both conformations coexist. This indicates the sensitivity of this fingerprint region in probing the local organization of anomers, as well as

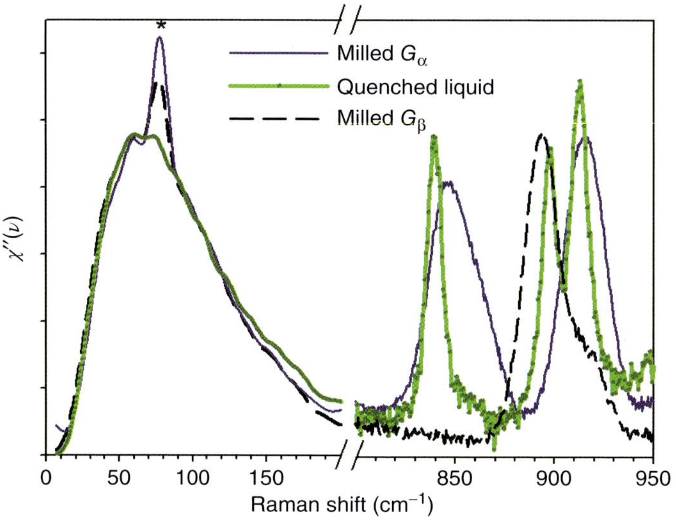

Figure 12.12 Raman spectra of amorphous states of glucose obtained by cryo-milling G_α and G_β crystals, and by quenching G_α and G_β melts. The star indicates a laser line.

the high degree of disorder of molecular orientations in pure anomerically amorphous states. The existence of different local organizations is not detected in the low-frequency domain, probably because the VDOS reflects the average distribution of molecular orientations involved in collective motions.

12.3.3
Use of LFRS for the Structural Description of Disordered Phases and the Analysis of Phase Transformations

12.3.3.1 Structural Description of Form II of Caffeine

Caffeine ($C_8H_{10}N_4O_2$) has both agrochemical and therapeutic applications. Forms I and II of anhydrous caffeine constitute an enantiotropic system [47]. The commercial form II is thermodynamically stable at room temperature and transforms upon heating into form I at 153 °C [48]. Form I is an orientationally [49, 50] and dynamically [50] disordered phase, called the rotator phase, where molecules slowly rotate around their molecular C_6 axis. Refinements of X-ray powder data collected in form I [49] were carried out to obtain a structural description of the molecular packing of caffeine molecules, which was very similar to that determined in the case of theophylline [51]. The main differences between the two compounds, which belong to the family of methylxanthines, is the absence of disorder and a tilt of molecules out of the molecular plane in theophylline. Form II is quite intriguing, since, despite more than 50 years of investigations [52], its structural description remains unclear. Indeed, the last refinements from X-ray powder diffraction data [53, 54] led to the determination of an unusually large unit cell (~4277 Å^3) composed of 20 molecules, suggesting

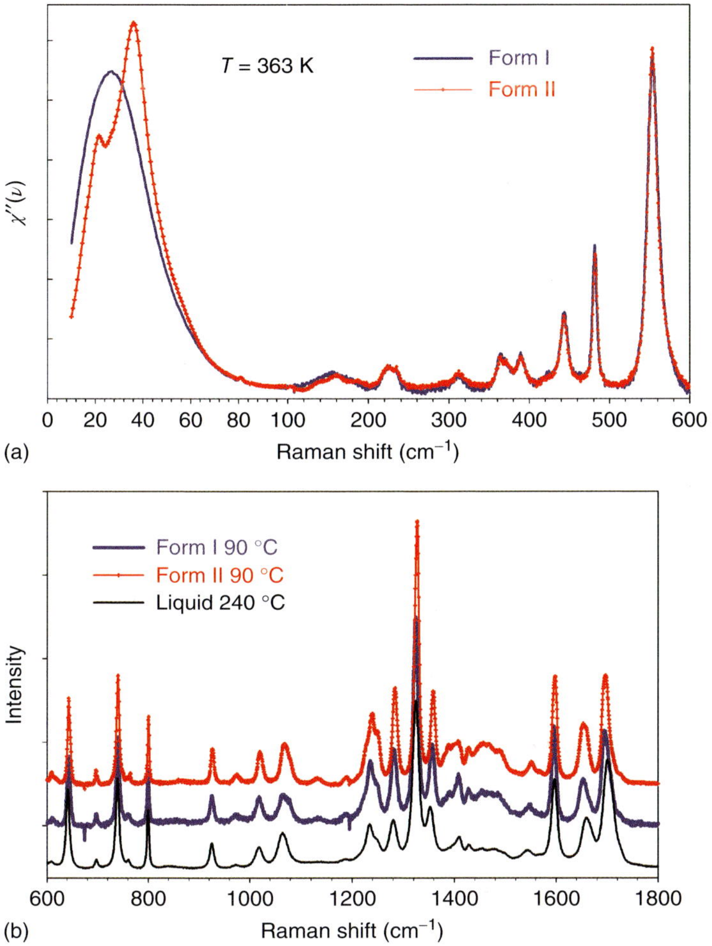

Figure 12.13 Raman spectra of forms I and II of caffeine. (a) In the low-frequency region: Raman susceptibility. (b) In the fingerprint region: spectra of crystalline forms are compared with the spectrum of the liquid.

that the existence of disorder similar to that existing in phase I [55] should be considered. After purification by cold sublimation, caffeine is in the metastable form I at room temperature. The I → II transformation is considerably hindered near room temperature, and faster transformations are observed at ~90 °C [56, 57]. Consequently, the Raman spectra of both caffeine forms are directly plotted in Figure 12.13 at 90 °C, after isothermal transformation of form I into form II. Figure 12.13a,b contains important information on the structural description of disordered phases:

1) The LFRS of form I is composed of a single and broad band and characterized by the absence of phonon peaks, contrasting with the observation of

sharp Bragg peaks in the X-ray powder diffraction diagram of form I [49, 54]. Bragg peaks result only from the periodic organization of the molecular mass centers, while lattice modes are related to intermolecular atom–atom potentials. Consequently, the orientational molecular disorder induces an inhomogeneous broadening of phonon peaks corresponding to a librational density of states. The $\chi''(\nu)$ spectrum of form II is characterized by a splitting of the broad low-frequency band into two components, which are too broad to be considered as phonon peaks, suggesting the existence of substantial disorder.

2) Both polymorphic forms of caffeine can be distinguished only below 100 cm^{-1}; the fingerprint regions of both polymorphic forms are very similar, and do not display highlights with respect to that of the liquid. These considerations indicate similar local order in both forms, and then similar description of the disorder in forms I and II. The close relationship between the fingerprint regions of polymorphic phases and the liquid state confirms the high degree of orientational disorder of caffeine molecules.

$\chi''(\nu)$ spectra of theophylline and form II of caffeine are plotted in Figure 12.14 at room temperature. This figure reveals that the Raman susceptibility of form II of caffeine corresponds to the rigorous envelope of the low-frequency phonon peaks in theophylline. The close similarity between the LFRS of caffeine and only the low-frequency phonon spectrum of theophylline can be related to the predominant contribution of librational motions in rotator phases. The relationship between both spectra indicates that in form II, caffeine molecules are located in positions similar to those occupied by theophylline molecules but rotate around their molecular axis, probably more slowly than in form I, in agreement with

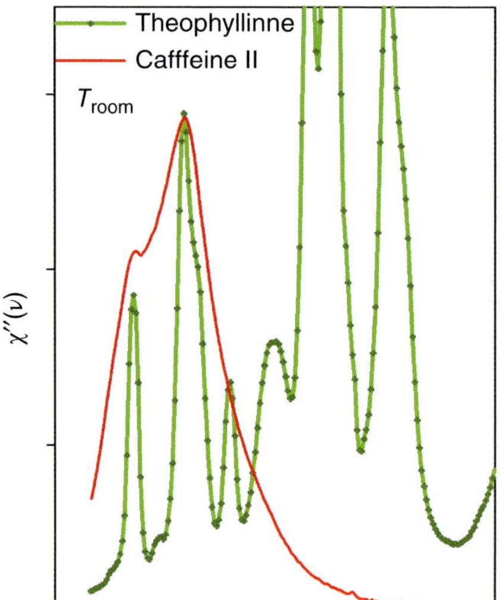

Figure 12.14 Comparison at room temperature of Raman susceptibilities of form II of caffeine and theophylline.

previous studies [55]. The I → II transformation can be interpreted as the tilt of molecules out of the hexagonal plane in the lattice of form I, giving molecular positions similar to those determined by last X-ray investigations [53, 54].

12.3.3.2 Analysis of Polymorphic Transformations in Caffeine

Form II can be distinguished from form I of caffeine only below 100 cm^{-1}, mainly in the 30–60 cm^{-1} spectral range. In this region, the area (light blue in Figure 12.15a) of spectral difference can be considered as representative of a complete polymorphic transformation. For partial transformation, the area (black area in Figure 12.15a) of the spectral difference of the partially transformed state and form I gives the degree of transformation after renormalization with the blue area.

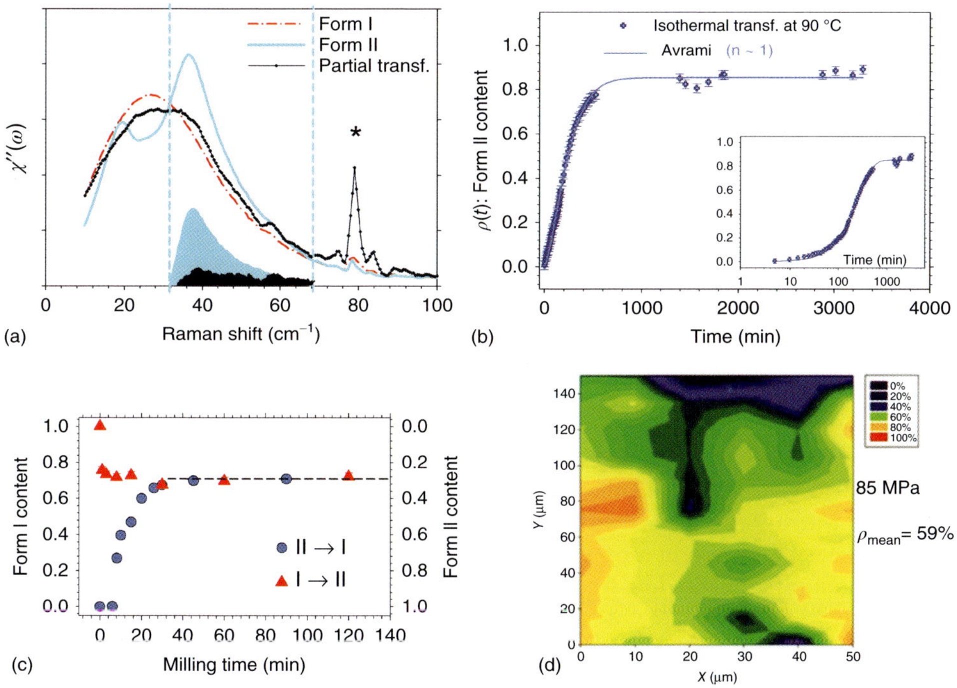

Figure 12.15 Determination of the degree of transformation between forms I and II (a) from the identification of the Raman signature distinctive of the partial transformation. The blue and black areas, respectively, correspond to the spectral difference between forms II and I, and between a partially transformed state and form I. The degree of transformation is determined from the ratio ρ = (blue area)/(black area). The arrow indicates a laser line. (b) During the isothermal I → II isothermal transformation at 90 °C using the procedure described in (a). The line represents the fitting curve using an Avrami-like function. (c) During milling at room temperature, using the procedure described in (a) applied to spectra collected instantaneously after different times of milling. (d) On tablet surface, after a compression of 85 MPa. The use of procedure (a) provides a distribution of the degree of transformation (I → II).

Analysis of the I → II Isothermal Transformation at 90 °C

The estimation of the ratio of both the areas (black/blue) leads to a plot of the degree of transformation ($\rho(t)$) during the isothermal transformation of form I to form II at 90 °C, as plotted in Figure 12.15b. This crystallization kinetics exhibits an unusually stretched exponential shape, characterized by an Avrami exponent, estimated to be $n = 1.2$ (±0.1) by fitting using the function defined in Eq. (12.12). Such a value close to 1 indicates a transformation controlled by nucleation without growth, suggesting a phase transformation within small crystallites. It is worth noting that the isothermal transformation appears incomplete.

Analysis of Phase Transformations Induced by Milling, Tableting, and Compression

Caffeine is well recognized as the prototype of pharmaceuticals that exhibit a I → II phase transformation under tableting [58, 59], compression [58, 60], and milling [58, 60, 61]. Raman spectroscopy is the only method that can directly determine the degree of transformation induced by such kinds of stress, since a quantitative X-ray analysis requires the rigorous knowledge of the structural description of both crystalline forms. A crucial advantage of Raman spectroscopy is that the spectra can be rapidly acquired on milled powder without any need for sample preparation, avoiding rapid relaxation of the metastable milled powder. It was thus possible to demonstrate the existence of a metastable state only stabilized under milling at room temperature, independently the initial state (form I or form II), by quasi-instantaneous analysis of the milled powder. Raman spectroscopy indicates that this metastable state is a partially transformed form I or form II. Using the method described in Figure 12.15a, the degree of transformation under milling was plotted against the time of milling (in Figure 12.15c), indicating that the metastable state corresponds to 30% of the complete I → II transformation, or 70% of the II → I transformation [60]. The stabilization of this metastable state under milling was related to the nature of the intriguing disorder–disorder transformation and the stability conditions of form I at room temperature [60].

Micro-Raman spectroscopy allows microscopic analysis of tablets obtained by the compression of form I of caffeine. However, micro-Raman mapping must be performed in the low-frequency range because no significant change between forms I and II can be observed above 100 cm^{-1}. Usually, micro-Raman mapping is performed at higher frequencies, and multivariate methods are used for analyzing the Raman data [62]. Analyzing the low-frequency domain of caffeine spectra requires the use of the method previously described in Figure 12.15a, to obtain a Raman image (Figure 12.15d) of the surface of the tablet corresponding to a distribution of the degree of transformation between phases I and II.

12.3.4 The Use of Multivariate Analysis

12.3.4.1 Analysis of Spectra Collected in a Series of Samples

MVA is very well suited to the analysis of a series of very similar spectra, characterized by several spectral changes in the error bar of the measurement (frequency, intensity, or width) or by a low signal-to-noise ratio in one variable. In MVA, it is not required to select the most appropriate spectral variables, and a noise reduction can be obtained because multiple measurements of the same phenomenon are used.

Analysis of the Amorphous State of Indomethacin (IMC)

The PCA method, schematically described in Figure 12.16, was used by Karmwar *et al.* [63] to analyze the spectral variation between the different amorphous states of IMC prepared via different transformation methods: transformation via melt quenching (QC), spray-drying (SD), ball-milling (BM), and cryo-milling (CM) the α and γ crystalline forms. The spectrum of each sample was recorded in the 1000–1720 and 2800–3100 cm^{-1} regions (Figure 12.17a). For each method, the samples were prepared and analyzed in triplicate. In this study, three principal components accounted for 99% of the variation in the standard normal variant transformed data, performed on the spectra to remove intensity differences unrelated to the sample composition, and the spectra were then mean-centered. Trends in the data can be visualized by the scores plot, constructed as shown in Figure 12.16. The scores describe the location of the data points in the principal component space, and the loadings describe the relationship between the original data space and the principal component space. Karmwar *et al.* [63] have shown the existence of three clusters in the score plot (Figure 12.17b), which mirror the differences observed in the diffractograms of the differently prepared samples plotted in Figure 12.18. The strong similarity between the spectra plotted in Figure 12.17a clearly demonstrates the capability of PCA to point out subtle structural differences in amorphous IMC samples prepared by different transformation methods. Information on the sample differences can obtained from the spectral loading plots (Figure 12.17c) and from the assignment of Raman bands [7] in the spectral regions where spectral differences are observed. Figure 12.17c shows that the systematic differences associated with the differently prepared amorphous forms cannot be attributed on the basis of one particular vibration but of all vibrations in regions distinctive of the molecular fingerprint and molecular associations. It is observed that Raman bands in the 1500–1700 and 2930–3100 cm^{-1} regions significantly contribute to PC1.

Similar analysis using the PCA method was applied to IMC glassy states prepared by cooling the liquid at different cooling rates [64]. This analysis revealed significant structural changes in the different amorphous samples, in the fingerprint region and the high-frequency spectrum, suggesting different local organizations depending on the cooling rate used to undercool the liquid until glass formation.

Online Raman Spectroscopy

In situ Raman spectroscopy is an invaluable tool to provide rapid and noninvasive monitoring of processing-induced amorphization or crystallization [65]. It is recognized that drugs have higher kinetic solubility and dissolution rate than the crystalline forms. Indeed, it was shown that the dissolution rate was higher in amorphous IMC compared to the α and γ crystalline forms. However, *in situ*

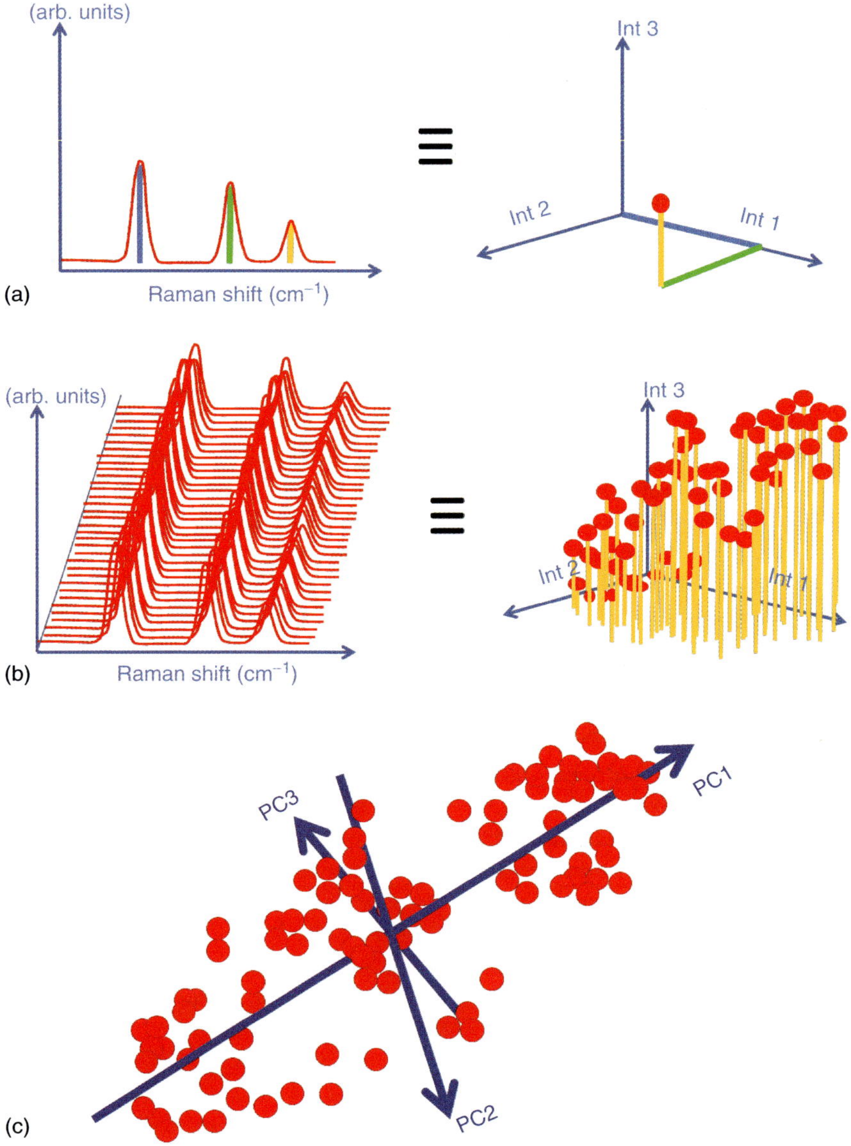

Figure 12.16 Schematic description of PCA. A spectrum of *n* peaks can be represented by single point in a *n*-dimensional space (a). A set of spectra is equivalent to a set of points (b). PCA transforms the original coordinate system into new coordinates called principal components (PCs) (c). The origin of the new coordinate system is located in the center of the data points, the first PC points in the direction of the highest variance, the second PC points in the direction of the second highest variance, and so on. The values that the spectra (red points) have in the PC coordinate system are called scores, and the plot (c) is then the scores plot. It can be estimated how much of the old coordinates (peaks) contribute to each of the new ones (PCs). These values are called loadings. The spectral loading plot provides information of how peaks contribute to the PCs.

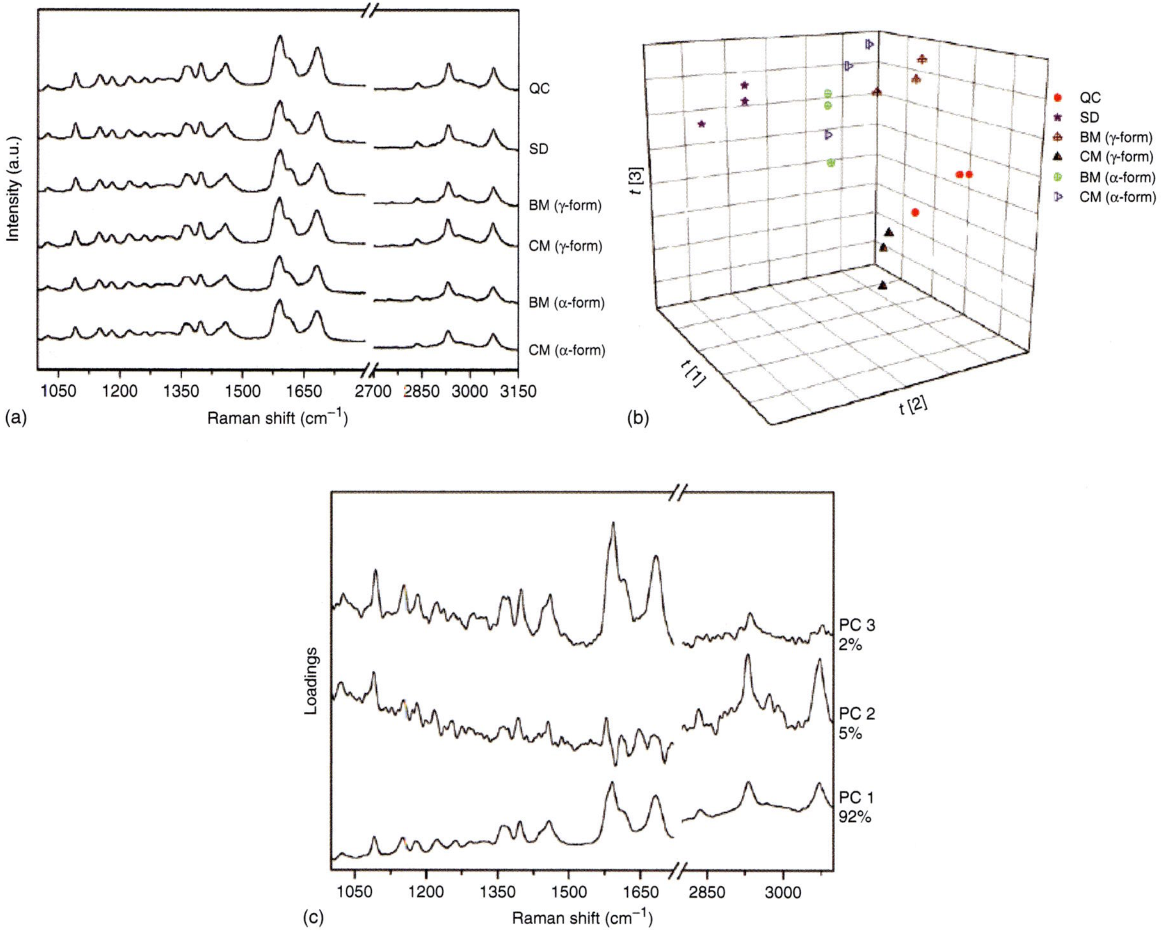

Figure 12.17 Raman spectra of amorphous indomethacin prepared by different transformation methods (a), scores plot (b), and their corresponding loadings (c). Adapted from Karmwar *et al.* 63. Reproduced with permission of Elsevier.

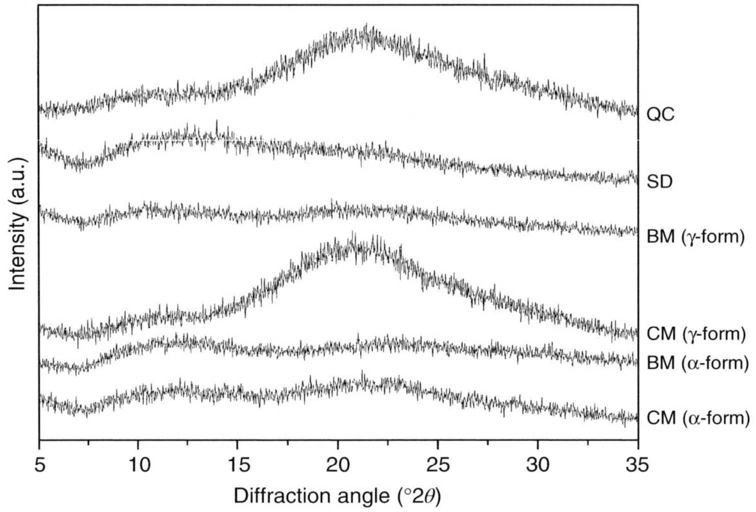

Figure 12.18 Diffractograms of freshly prepared amorphous form of indomethacin by different transformation methods. Adapted from Karmwar *et al.* 63. Reproduced with permission of Elsevier.

Raman investigations, monitoring the crystallization process, revealed that the dissolution rate started to slow down during dissolution toward the crystalline α-form [66]. The advantage of MVA analysis is that all the spectral information in the 1085–1480 and 1590–1690 cm^{-1} regions can be used for the analysis.

12.3.4.2 **Raman Mapping**

Raman mapping generally requires a data collection from several thousand spectra, and thus requires the use of MVA.

Raman microspectroscopy can be used as aid to understand the drug release mechanism from matrix dosage forms. Both the distribution of the matrix components and physical changes during dissolution can impact the drug release behavior. Haaser *et al.* [67] used Raman microscopy to investigate different extrudate formulations in terms of component distribution and structural changes during dissolution testing. Figure 12.19 shows Raman maps of the cross-section of the extrudate corresponding to a binary system, based on a tripalmitin matrix containing equal amount of the model drug theophylline anhydrate, after a dissolution period of 120 min. Figure 12.19a showing the Raman map of the single drug reveals the absence of the drug signal in the area delimited by semicircles, which is at a depth of about 45 μm from the lateral surface of the extrudate. It is observed that the drug boundary does not uniformly recede (Figure 12.19a) and the lipid matrix remains intact during dissolution (Figure 12.19b,c). Figure 12.19e, corresponding to the map recorded with a better spatial resolution over a smaller area, shows drug particles completely surrounded by lipid, making dissolution impossible.

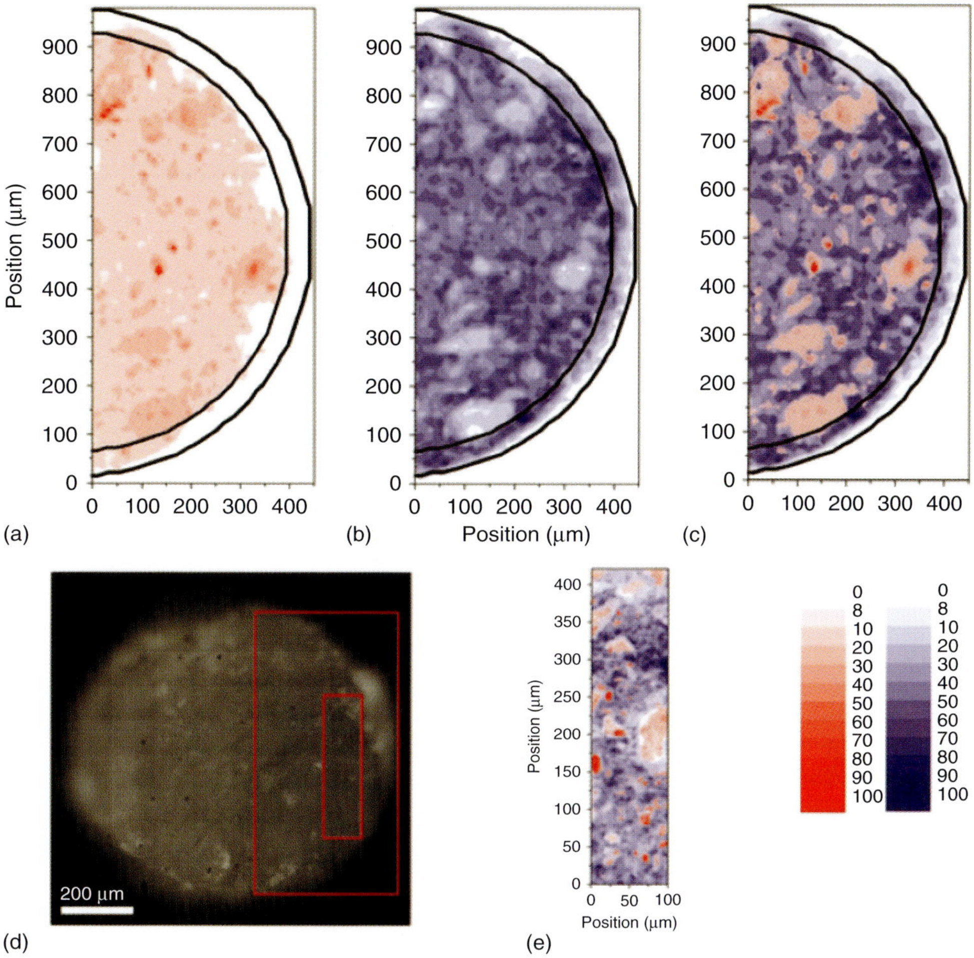

Figure 12.19 Cross-section of an extrudate containing tripalmitin (blue) and theophylline anhydrate (red) after 120 min of dissolution testing. Raman map of theophylline anhydrate (a), tripalmitin (b), and both components (c) recorded with a 20× objective, optical microscopy image (d), and small area recorded with a 50× objective (e); rectangles localize the areas mapped; semicircles indicate the edge and the calculated uniformly receding drug boundary. Adapted from Haaser *et al.* 67. Reproduced with permission of Wiley.

Raman spectroscopy is frequently used for inline monitoring manufacturing processes, such for inline polymer–drug quantification and for the solid-state characterization during a pharmaceutical holt-melt extrusion process [68]. It can be also used for inline process monitoring of a freeze-drying process [69, 70]. Raman spectroscopy can provide information on operating parameters such as the time of the sublimation kinetics by measuring the Raman bands of ice. The determination of the end of the first drying stage is crucial, since it is a time- and energy-consuming stage. Moreover, Raman spectroscopy provides information on the physical state of APIs and excipients during a freeze-drying cycle and, further, during the storage of the freeze-dried product.

Raman spectroscopy is also widely used for a rapid and noninvasive surface analysis of tablets, since tableting can induce transformation of drugs or/and excipients [71, 72], or merely to detect the existence of counterfeits [73].

12.4 Conclusion

This chapter has shown the potential use of Raman spectroscopy as an indirect structural probe into the disordered states of molecular compounds. In these materials, which are rich in hydrogen atoms, Raman spectroscopy is an essential complementary technique to X-ray diffraction, which is unsuitable for localizing H atoms. Raman spectroscopy is one of the most suited techniques to analyze molecular associations via H-bonding, which are closely related to the polymorphism of molecular compounds. By capturing a wide spectral window including the low-frequency, the fingerprint, and the intramolecular X – H (X=H, C, N) stretching regions, we have shown the capabilities of the Raman spectroscopy to reveal the existence of disorder, the presence of different kinds of amorphous solid states, and the structural description of the disordered states. It is worth noting that investigations in the low-frequency range ($\sim$5–200 cm^{-1}) contribute significantly to the understanding of phase transformation mechanisms and to the structural description of disordered states. However, analyses on caffeine, glucose, and ibuprofen have shown the complementarity and the advantages of performing simultaneously low- and high-frequency investigations. Both domains are sensitive structural probes. The low-frequency region is very useful to detect and identify disordering in an ordered state, or vice versa. At higher frequencies, Raman bands that are closely related to the molecular conformation are sensitive probes of the local order and can be used to discriminate the structural organization of different amorphous states. This low-frequency domain is easily accessible with a simple one-stage spectrograph. More generally, the significant advances in technology that surround the technique (miniaturization of lasers, spectrographs, detectors), in combination with the development of notch filters, make Raman spectroscopy an important analytical tool in the pharmaceutical industry [74].

References

1. Dujardin, N. *et al.* (2008) Solid state vitrification of alpha and beta D-glucose by mechanical milling. *Solid State Commun.*, **148**, 78–82.
2. Hedoux, A. *et al.* (2006) Evidence of a two-stage thermal denaturation process in lysozyme: a Raman scattering and differential scanning calorimetry investigation. *J. Chem. Phys.*, **124** (1), 14703.
3. Denicourt, T. *et al.* (2003) Raman scattering investigations of the stable and metastable phases of cyanoadamantane

glassy crystal. *J. Phys. Chem. B*, **107** (33), 8629–8636.

4. Wypych, A., Guinet, Y., and Hédoux, A. (2007) Isothermal transformation of supercooled liquid n-butanol near the glass transition: polyamorphic transitions in molecular liquids investigated using Raman scattering. *Phys. Rev.*, **B76** (14), 144202.
5. Hédoux, A. *et al.* (2013) Vibrational and structural properties of amorphous n-butanol: a complementary Raman spectroscopy and X-ray diffraction study. *J. Chem. Phys.*, **138**, 214506.
6. Strachan, C.J., Rades, T., and Gordon, K.C. (2007) A theoretical and spectroscopic study of gamma-crystalline and amorphous indometacin. *J. Pharm. Pharmacol.*, **59** (2), 261–269.
7. Taylor, L.S. and Zografi, G. (1997) Spectroscopic characterization of interactions between PVP and indomethacin in amorphous molecular dispersions. *Pharm. Res.*, **14** (12), 1691–1698.
8. Colles, M.J. and Griffiths, J.E. (1972) Relative and absolute cross sections in liquids. *J. Chem. Phys.*, **56**, 3384–3391.
9. Strachan, C.J. *et al.* (2007) Raman spectroscopy for quantitative analysis of pharmaceutical solids. *J. Pharm. Pharmacol.*, **59** (2), 179–192.
10. Long, D.A. (1977) *Raman Spectroscopy*, McGraw-Hill International Book Company.
11. Moser, C. and Havermeyer, F. (2009) Ultra-narrow band tunable laserline notch filter. *Appl. Phys. B: Lasers Opt.*, **95** (3), 597–601.
12. Heyler, R., *et al.* (2013) THz-Raman – Accessing molecular structure with Raman spectroscopy for enhanced chemical identification, analysis and monitoring. Conference on Next-Generation Spectroscopic Technologies VI, Proceedings of SPIE, Baltimore, MD, 2013.
13. Strachan, C.J. *et al.* (2004) Quantitative analysis of polymorphic mixtures of carbamazepine by Raman spectroscopy and principal components analysis. *J. Raman Spectrosc.*, **35** (5), 347–352.
14. Breitenbach, J., Schrof, W., and Neumann, J. (1999) Confocal Raman-spectroscopy: analytical approach to solid dispersions and mapping of drugs. *Pharm. Res.*, **16** (7), 1109–1113.
15. Galeener, F.L. and Sen, P.N. (1978) Theory for the first-order vibrational spectra of disordered solids. *Phys. Rev. B*, **17** (4), 1928–1933.
16. Shuker, R. and Gammon, R. (1970) Raman-scattering selection-rule breaking and the density of states in amorphous materials. *Phys. Rev. Lett.*, **25** (4), 222–225.
17. Martin, A.J. and Brenig, W. (1974) Model for brillouin scattering in amorphous solids. *Phys. Status Solidi B*, **64**, 163.
18. Sokolov, A.P. *et al.* (1993) Evaluation of density of vibrational states of glasses from low-frequency Raman spectra. *Phys. Rev. B*, **48**, 7692–7695.
19. Fontana, A. *et al.* (1990) Low-frequency dynamics in superionic borate glasses by coupled Raman and inelastic neutron scattering. *Phys. Rev. B*, **41** (6), 3778–3785.
20. Surovtsev, N.V. *et al.* (2004) Density of vibrational states and light-scattering coupling coefficient in the structural glass and glassy crystal of ethanol. *J. Phys.: Condens. Matter*, **16**, 223–230.
21. Achibat, T., Boukenter, A., and Duval, E. (1993) Correlation effects on Raman scattering from low-energy vibrational modes in glasses. II. Experimental results. *J. Chem. Phys.*, **99** (3), 2046–2051.
22. Hedoux, A. *et al.* (2001) Low-frequency vibrational excitations in the amorphous and crystalline states of triphenyl phosphite: a neutron and Raman scattering investigation – art. no. 144202. *Phys. Rev. B*, **6314** (14), 4202,NIL_169-NIL_175.
23. Novikov, V.N. and Duval, E. (1994) A model of low-frequency Raman scattering in glasses: comparison of Brillouin and Raman data. *J. Chem. Phys.*, **102** (11), 4691–4698.
24. Tolle, A. *et al.* (2000) Vibrational states of glassy and crystalline orthoterphenyl. *Eur. Phys. J. B*, **16**, 73–80.
25. Sokolov, A.P. *et al.* (1993) Dynamics of strong and fragile glass formers: differences and correlation with low-temperature properties. *Phys. Rev. Lett.*, **71**, 2062–2065.

26. Buchenau, U. and Zorn, R. (1992) A relation between fast and slow motions in glassy and liquid selenium. *Europhys. Lett.*, **18**, 523–528.
27. Ribeiro, M.C.C. (2010) Low-frequency Raman spectra and fragility of imidazolium ionic liquids. *J. Chem. Phys.*, **133**, 24503–24506.
28. Ha, A. *et al.* (1996) Supercooled liquids and polyamorphism. *J. Phys. Chem.*, **100**, 1–4.
29. Senker, J. and Rossler, E. (2001) Triphenyl phosphite: a candidate for liquid polyamorphism. *Chem. Geol.*, **174** (1–3), 143–156.
30. Tanaka, H., Kurita, R., and Mataki, H. (2004) Liquid–liquid transition in the molecular liquid triphenyl phosphite. *Phys. Rev. Lett.*, **92** (2), 025701/1–025701/4.
31. Hedoux, A., Guinet, Y., and Descamps, M. (1998) Raman signature of polyamorphism in triphenyl phosphite. *Phys. Rev. B: Condens. Matter*, **58** (1), 31–34.
32. Hedoux, A., Guinet, Y., and Descamps, M. (2001) Size dependence of the Raman spectra in an amorphous-nanocrystalline mixed phase: the glacial state of triphenyl phosphite. *J. Raman Spectrosc.*, **32** (8), 677–688.
33. Hernandez, O. *et al.* (2002) Ab initio structure determination of triphenyl phosphite by powder synchrotron X-ray diffraction. *J. Appl. Crystallogr.*, **35**, 212–219.
34. Schmidt, A.G., Wartewig, S., and Picker, K.M. (2004) Polyethylene oxides: protection potential against polymorphic transitions of drugs? *J. Raman Spectrosc.*, **35** (5), 360–367.
35. Taylor, L.S. and Zografi, G. (1998) The quantitative analysis of crystallinity using FT-Raman spectroscopy. *Pharm. Res.*, **15** (5), 755–761.
36. Hédoux, A. *et al.* (2009) Using the low-frequency Raman spectroscopy to analyze the crystallization of amorphous indomethacin. *Eur. J. Pharm. Sci.*, **38** (2), 156–164.
37. Shankland, N. *et al.* (1997) Refinement of ibuprofen at 100 K by single-crystal pulsed neutron diffraction. *Acta Crystallogr., Sect. C: Cryst. Struct. Commun.*, **53**, 951–954.
38. Freer, A.A. (1993) Structure of (S)-(+)-ibuprofen. *Acta Crystallogr., Sect. C: Cryst. Struct. Commun.*, **49**, 1378–1380.
39. Jubert, A. *et al.* (2006) Vibrational and theoretical studies of non-steroidal anti-inflammatory drugs ibuprofen; naproxen and tolmetin acids. *J. Mol. Struct.*, **783**, 34–51.
40. Rossi, B. *et al.* (2009) Ibuprofen-cyclodextrin inclusion complexes investigated by Raman scattering and numerical simulation. *J. Raman Spectrosc.*, **40**, 453.
41. Dudognon, E. *et al.* (2008) Evidence for a new crystalline phase of racemic ibuprofen. *Pharm. Res.*, **25** (12), 2853–2858.
42. Hédoux, A. *et al.* (2011) Raman spectroscopy of racemic ibuprofen: evidence of molecular disorder in phase II. *Int. J. Pharm.*, **421**, 45–52.
43. Derollez, P. *et al.* (2010) Ab initio structure determination of phase II of racemic ibuprofen by X-ray powder diffraction. *Acta Crystallogr., Sect. B: Struct. Sci.*, **66**, 76–80.
44. McDonald, T.R.R. and Beevers, C.A. (1950) The crystal structure of alpha-D-glucose. *Acta Crystallogr.*, **3**, 394–395.
45. Chu, S.S.C. and Jeffrey, G.A. (1968) The refinement of the crystal structures of beta-D-glucose and cellobiose. *Acta Crystallogr.*, **B24**, 830–838.
46. Araujo-Andrade, C. *et al.* (2005) Infrared and Raman spectra, conformational stability, ab initio calculations of structure, and vibrational assignment of alpha and béta glucose. *J. Mol. Struct.*, **714**, 143–146.
47. Bothe, H. and Cammenga, H.K. (1979) Phase transitions and thermodynamic properties of anhydrous caffeine. *J. Therm. Anal.*, **16** (2), 267–275.
48. Cesaro, A. and Starec, G. (1980) Thermodynamic properties of caffeine crystal forms. *J. Phys. Chem.*, **84**, 1345–1346.
49. Derollez, P. *et al.* (2005) Ab initio structure determination of the high-temperature phase of anhydrous caffeine by X-ray powder diffraction. *Acta Crystallogr.*, **B61**, 329–334.
50. Descamps, M. *et al.* (2005) Plastic and glassy crystal states of caffeine. *J. Phys. Chem. B*, **109**, 16092–16098.

51. Smith, E.D.L. *et al.* (2001) The determination of the crystal structure of anhydrous theophylline by X-ray powder diffraction with a systematic search algorithm, lattice energy calculations, and 13C and 15 N solid-state NMR: a question of polymorphism in a given unit cell. *J. Phys. Chem. B*, **105**, 5818–5826.
52. Sutor, D.J. (1958) The structures of pyrimidines and purines. VII. The crystal structure of caffeine. *Acta Crystallogr.*, **11**, 453–458.
53. Lehmann, C. and Stowasser, F. (2007) The crystal structure of anhydrous beta-caffeine as determined from X-ray powder-diffraction data. *Chem. Eur. J.*, **13**, 2908–2911.
54. Enright, G. *et al.* (2007) The structure of two anhydrous polymorphs of caffeine from single-crystal diffraction and ultrahigh-field solid-state C NMR spectroscopy. *Cryst. Growth Des.*, **7** (8), 1406–1410.
55. Moura Ramos, J. *et al.* (2006) Dielectric study of the slow motional processes in the polymorphic states of anhydrous caffeine. *J. Phys. Chem.*, **B110**, 8268–8273.
56. Lehto, V.P. and Laine, E. (1998) A kinetic study of polymorphic transition of anhydrous caffeine with micocalorimeter. *Thermochim. Acta*, **317**, 47–58.
57. Hédoux, A. *et al.* (2011) Low- and high-frequency investigations on caffeine: polymorphism, disorder and phase transformation. *J. Phys. Chem. B*, **115**, 5746–5753.
58. Pirttimaki, J. *et al.* (1993) Effects of grinding and compression on crystal structure of anhydrous caffeine. *Int. J. Pharm.*, **95**, 93–99.
59. Hubert, S. *et al.* (2011) Process induced transformations during tablet manufacturing: phase transition analysis of caffeine using DSC and low frequency micro-Raman spectroscopy. *Int. J. Pharm.*, **420**, 76–83.
60. Hédoux, A. *et al.* (2013) Polymorphic transformation of anhydrous caffeine upon grinding and hydrostatic pressurizing analyzed by low-frequency Raman spectroscopy. *J. Pharm. Sci.*, **102** (1), 162–170.
61. Mazel, V. *et al.* (2011) Polymorphic transformation of anhydrous caffeine under compression and grinding: a re-evaluation. *Drug Dev. Ind. Pharm.*, **37** (7), 832–840.
62. Zhang, L., Henson, M.J., and Sekulic, S.S. (2005) Multivariate analysis for Raman imaging of a model tablet. *Anal. Chim. Acta*, **545**, 262–278.
63. Karmwar, P. *et al.* (2011) Investigations of properties and recrystallisation behaviour of amorphous indomethacin samples prepared by different methods. *Int. J. Pharm.*, **417**, 94–100.
64. Karmwar, P. *et al.* (2011) Investigations on the effect of different cooling rates on the stability of amorphous indomethacin. *Eur. J. Pharm. Sci.*, **44**, 341–350.
65. Chieng, N., Rades, T., and Aaltonen, J. (2011) An overview of recent studies on the analysis of pharmaceutical polymorphs. *J. Pharm. Biomed. Anal.*, **55**, 618–644.
66. Savolainen, M. *et al.* (2009) Better understanding of dissolution behaviour of amorphous drugs by in situ solid-state analysis using Raman spectroscopy. *Eur. J. Pharm. Biopharm.*, **71**, 71–79.
67. Haaser, M. *et al.* (2011) Analysis of matrix dosage forms during dissolution testing using Raman microscopy. *J. Pharm. Sci.*, **100** (10), 4452–4459.
68. Saerens, L. *et al.* (2011) Raman spectroscopy for the in-line polymer-drug quantification and solid state characterization during a pharmaceutical holt-melt extrusion process. *Eur. J. Pharm. Biopharm.*, **77**, 158–163.
69. De Beer, T. *et al.* (2009) In-line and real-time process monitoring of a freeze drying process using Raman and NIR spectroscopy as complementary process analytical technology tools. *J. Pharm. Sci.*, **98**, 3430–3446.
70. Hédoux, A. *et al.* (2013) Mechanism of protein stabilization by trehalose during freeze-drying analyzed by in-situ micro-Raman spectroscopy. *J. Pharm. Sci.*, **102**, 2484–2494.
71. Johansson, J. *et al.* (2007) Quantitative transmission Raman spectroscopy of pharmaceutical tablets and capsules. *Appl. Spectrosc.*, **61**, 1211–1218.

72. Lee, S.H. *et al.* (2012) End point determination of blending process for trimebutine tablets using principle component analysis and partial least squares regression. *Arch. Pharmacal Res.*, **35**, 1599–1607.
73. Li, L. *et al.* (2014) Identification of anisodamine tablets by Raman and near spectroscopy with chemometrics. *Spectrochim. Acta, Part A*, **127**, 91–97.
74. McGoverin, C.M., Rades, T., and Gordon, K.C. (2008) Recent pharmaceutical applications of Raman and terahertz spectroscopies. *J. Pharm. Sci.*, **97**, 4598–4621.

13
Study of Disordered Materials by Terahertz Spectroscopy

Juraj Sibik and J. Axel Zeitler

13.1
Introduction

With the advent of terahertz time-domain spectroscopy (THz-TDS) in the 1970s and 1980s [1–5], a completely new experimental approach to explore molecular excitations at terahertz frequencies became available.[1] Given its energy, electromagnetic radiation at terahertz frequencies strongly interacts with systems that have characteristic transitions in the millielectronvolt range and lifetimes on picosecond time scales. This includes phonons in crystalline solids, transient molecular dipoles, relaxational dynamics in aqueous liquids, hydrated biological matter, and weakly bonded systems such as hydrogen-bonded networks or van der Waals forces, among others. A number of reviews have been published recently that introduce the general background of the technique in the context of spectroscopic applications [6, 7].

Terahertz spectroscopy is a time-domain technique typically built around a near-infrared laser that provides pulses of femtosecond duration. The laser pulse train is split into the generation (or pump) and detection (or probe) pulse. The pump pulse excites charge carriers in a semiconducting material with a suitable bandgap (e.g., GaAs), and terahertz radiation is generated from the transient current that forms upon accelerating these charge carriers in a DC electric field. After propagation through the sample, the terahertz pulse is combined with the probe pulse and its electric field is measured by means of a time-gated detection process. This can be achieved, for example, by the reverse of the generation process, where the transient current is measured and the acceleration of the charges is carried out by the incoming terahertz pulse. Given that the duration of the generated terahertz pulse is usually about 1 ps, the time-domain signal contains exactly one oscillation, as shown in Figure 13.1. Following the

1) In the context of this chapter we refer to the terahertz region as electromagnetic radiation with frequencies of 0.1 – 4 THz or wavenumbers of 3 – 130 cm^{-1}, which corresponding photon energies of 0.4 – 16 meV.

Disordered Pharmaceutical Materials, First Edition. Edited by Marc Descamps.

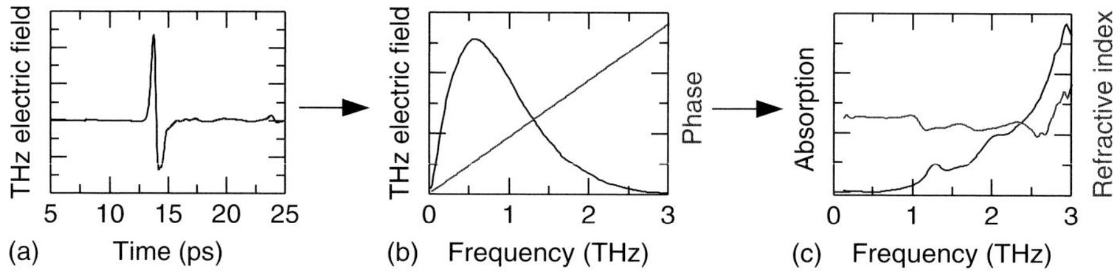

Figure 13.1 Schematic of data collection by terahertz time-domain spectroscopy. The measured time-domain signal (a) is converted into the frequency-domain amplitude and phase via Fourier transform (b), allowing extraction of both absorption coefficient and refractive index, or real and imaginary part of dielectric function (c), simultaneously.

measurement, the time-domain signal is converted to the frequency domain by Fourier transformation, resulting in both an amplitude and a phase of the signal. By comparing the frequency-domain data from sample and reference, it is possible to extract the absorption coefficient and the refractive index, or real and imaginary part of dielectric function, of the sample simultaneously.

A number of sampling geometries, such as transmission, reflection, and attenuated total reflection, can be implemented for terahertz spectroscopy, but for simplicity we will restrict our discussion to the transmission geometry (see [6] for details of other implementations). In the transmission setup, the terahertz pulse propagates through the sample at an angle corresponding to normal incidence. The (supercooled) liquid sample material can be loaded into a cuvette made of two windows separated by a thin plastic spacer [8]. For such a sample, a good reference structure would be to use two identical windows without any spacer, thus allowing the neglect of any phase shift or absorption of the window material when extracting the optical properties from the sample. The window material is chosen to ensure high terahertz transparency and negligible temperature dependence of its optical properties over the desired temperature range. A commonly used material is crystalline z-cut quartz. For temperature control, the sample and reference sandwich structures are attached to the cold finger of a cryostat. Further details on the practicalities of measuring high-quality spectra of organic amorphous materials can be found elsewhere [9].

13.2 Exploration of Terahertz Dynamics Prior to THz-TDS

13.2.1 Poley Absorption

The dynamics of disordered systems that occur at terahertz frequencies attracted scientific interest well before the development of modern THz-TDS techniques. Perhaps the very first prediction for the existence of terahertz absorption in noncrystalline materials was made in 1955 by Poley in the context of a spectroscopic investigation of low-viscosity polar liquids [10]. When carefully analyzing the experimental data, Poley noticed an inconsistency between the extrapolated dielectric constant ε_∞, which was measured in the microwave region, and the square of the refractive index n^2 when extrapolated to lower frequencies from measurements in the visible and infrared region. The extrapolations did not match, and based on this discrepancy, Poley deduced that an additional dipolar absorption process in liquids must exist at terahertz frequencies. This absorption is now often referred to as Poley absorption [11].

13.2.2 Far-Infrared Spectroscopy

Experimentally, the first absorption spectra that extended to the terahertz range were reported a couple of decades after Poley's paper, and they were acquired by means of Fourier transform far-infrared spectroscopy (FIR). The FIR spectra of a series of structurally and electronically different inorganic glasses exhibited a universal temperature-independent absorption that increased in intensity with frequency as ν^β ($\beta \lesssim 2$) [12]. This frequency-squared dependence of absorption was explained in terms of disorder-induced coupling of the far-infrared radiation to a density of low-frequency Debye modes.

Of particular importance in this context are the dielectric and FIR studies reported by Reid and Evans in the late 1970s and early 1980s [13]. By this time, the dielectric spectroscopy community was well aware of the existence of two universal dielectric relaxation mechanisms: a primary (α-) as well as a secondary (Johari–Goldstein β-) relaxation process [14]. Based on the experimental data from FIR studies of supercooled liquids and glasses of decalin solutions, a third universal absorption feature was resolved, which fell into the terahertz region, and which Reid and Evans referred to as a γ process following the terminology used in the dielectric spectroscopy community [13]. While the frequency of this process matches that of the coupling of radiation to a low-frequency Debye mode in inorganic glasses, there are two major differences between these two processes: First, the γ process exhibits a well-developed peak in the absorption spectra, as opposed to the Debye mode which typically only clearly resolves when plotted as frequency-squared absorption. Second, the γ process is not temperature-independent.

An overview of the frequency dependent dynamics is shown in Figure 13.2 expressed in units of dielectric loss. The dielectric loss $\varepsilon''(\omega)$ and the absorption coefficient $\alpha(\omega)$ are related by

$$\varepsilon''(\omega) = \frac{\alpha(\omega)\, n(\omega)\, c}{\omega} \tag{13.1}$$

where ω is the angular frequency, c is the speed of light, and $n(\omega)$ is the refractive index.

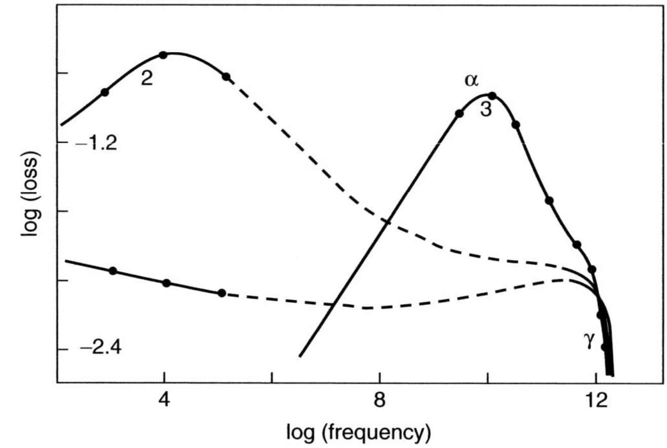

Figure 13.2 Dielectric loss spectrum for 10% bromobenzene/decalin. As the liquid is cooled from 293 K, the main relaxation (α) moves to lower frequencies while the peak at terahertz frequencies (γ process considered by Reid and Evans) remains in the terahertz region. (1) 110 K (glass). (2) 145 K (ultraviscous liquid). (3) 293 K. (Adapted from Reid and Evans [13]. Reproduced with permission of American Physical Society.)

The spectra clearly demonstrate the difference in temperature dependence between the relaxation processes: as the solutions are cooled, the dielectric relaxation moves from the gigahertz region to lower frequencies while the γ process shifts slightly up in frequency but remains in the terahertz region throughout the temperature change.

13.2.3
Raman and Neutron Scattering

Complementing the FIR measurements, a range of scattering techniques, such as Raman and neutron scattering, were used to probe the molecular dynamics at terahertz frequencies (e.g., [15–20]. Typically, the spectra reveal a single peak that develops at terahertz frequencies (or corresponding energy values of a few millielectronvolt) as the sample liquid becomes supercooled or a glass. Such a spectral feature is often referred to as the Boson peak.

Given the similar frequency and energy range of Poley absorption, the γ process, and the Boson peak, it seems intuitive that there is a link in the molecular origin for all these processes. Indeed, as Johari discusses, these features appear remarkably similar when ignoring any molecule-specific effects [11]. However, in this context it is important to keep in mind that it is not always straightforward to compare the results obtained by different experimental techniques in absolute terms, as the resonances excited in each method relate to fundamentally different physical processes. The molecular dynamics that are resolved by the respective measurements hence originate from different forces. For example, dielectric spectroscopy will excite dipolar moments, while Raman spectroscopy is sensitive to the polarizability of the molecules. Therefore, it is not too surprising that the shape, width, and center frequency of the spectral features observed at terahertz frequencies in disordered systems by various techniques fall into a similar energy range but do not match exactly [21, 22]. This is further complicated by the complex nature of the terahertz dynamics in (supercooled) liquids and glasses, where rotational, vibrational, and relaxational movements of molecules often overlap and are difficult to separate.

13.3 Response of Supercooled Liquids and Glasses at Terahertz Frequencies

Although terahertz spectra that are acquired by the current generation of commonly used THz-TDS setups span only about one decade of frequency (0.2–4 THz) many different absorption mechanisms contribute toward the observed losses in this range. The dielectric relaxation processes dominate the spectra at gigahertz frequencies and below. However, these processes result in very broad spectral features, and at sufficiently high temperatures when the supercooled liquid molecules still have significant mobility, these dielectric relaxations can penetrate into the frequency window accessible by THz-TDS. At the same time, low-energy vibrational motions are also excited (Figure 13.3).

13.3.1 Primary (α-) and Secondary (β-) Relaxations

Given the broad nature of the spectral features associated with the different relaxation processes, a wide frequency response must be investigated in order to differentiate the overlapping dynamics of supercooled liquids and glasses. One

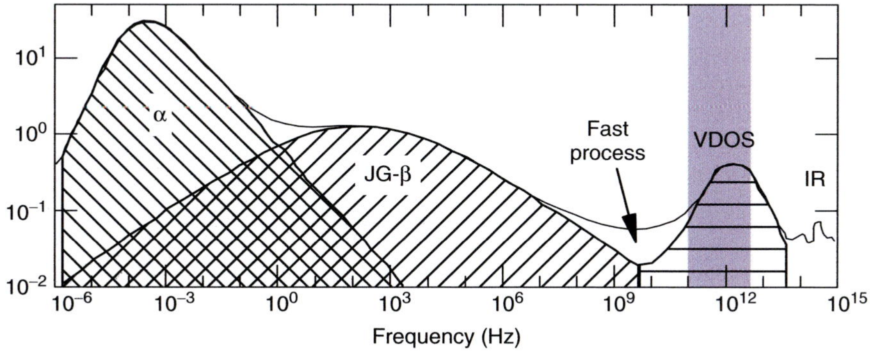

Figure 13.3 Overview of the dielectric response of disordered materials: (α) primary relaxation, (JG-β) JG secondary relaxation, (fast process) fast secondary relaxation, (VDOS) vibrational density of state peak, and (IR) intramolecular infrared modes. The blue shading highlights the frequency region that can be accessed by typical THz-TDS instruments. Adapted from Ref. [23].

technique that is very well suited in this context is dielectric spectroscopy. It can access frequencies of $10^{-6}-10^{9}$ Hz (and possibly up to 10^{12} Hz using modern interferometric techniques [24]). Using this approach, the complex dielectric constant is measured, and the real and imaginary parts of the dielectric constant are subsequently used to quantify the molecular dynamics by fitting the spectral response to Debye (in normal liquids), Havriliak–Negami [25], or Cole–Cole [26] (in supercooled liquids) functions, which yield the characteristic molecular relaxation times.

At terahertz frequencies, dielectric spectroscopy measurements become very difficult because of the lack in sensitivity of the detectors, the low power output of electronic sources, as well as the requirement for suitable waveguides. In this context, one of the main strengths of THz-TDS compared to other optical techniques, such as FIR, is that it directly measures both the real and the imaginary part of the dielectric constants without having to resort to Kramers–Kronig relations, and hence THz-TDS can be used as a high-frequency extension of dielectric spectroscopy.

As mentioned previously, the dielectric spectra of liquids and glasses generally reveal primary as well as secondary relaxation features, all of which span a very wide frequency range that extends into the frequency window accessible by THz-TDS. The primary relaxation is associated with molecular diffusion processes, and it vanishes as the viscosity of a supercooled liquid increases when approaching the glass transition temperature T_g [27]. Several types of secondary relaxation mechanisms have been proposed in the literature such as intramolecular flexibility [28], but of particular importance in the context of this chapter is the so-called Johari–Goldstein (JG) secondary relaxation. The JG β-relaxation is a genuine property of the disordered state and exists even in glasses formed by completely rigid molecules [14]. It should also be noted that the JG β-relaxation is not always well separated from the primary relaxation. Sometimes it manifests itself only as a high-frequency shoulder on the wing of the primary relaxation peak [29]. It is thought that the JG β-relaxation originates from a movement of the entire molecules and is strictly intermolecular in its nature [14, 28]. In contrast, the intramolecular secondary relaxations originate from movements involving only a subset of the entire molecule. Such processes are generally classified as non-JG β-processes. Because of its intermolecular nature, the JG β-process is understood to play a major role in the crystallization of glasses, and hence it plays a critical role in the stability of amorphous pharmaceuticals [28].

13.3.2
Fast Secondary Relaxation and Caged Dynamics

Several theoretical models have been proposed to explain the general dynamics of the amorphous state. A very nice review of the most commonly used theories can be found elsewhere [30]. Out of all the models described in the literature, two theoretical frameworks are particularly popular to describe the amorphous state: one is the mode-coupling theory developed by Götze [31], and the second is the coupling model of Ngai [32, 33]. It should be, however, noted that the validity of Götze's mode-coupling theory is somewhat restricted at temperatures below T_g and that it requires a nontrivial extension [31], making Ngai's coupling model a more popular approach at the present time. Both theories predict the existence of a fast molecular motion that is explained in terms of a molecule rattling in a cage formed by its neighboring molecules. This process is typically referred to as the caged dynamics.[2] More accurately, the caged dynamics span the frequency range between the primitive relaxations (which shows remarkable similarity to the JG secondary relaxation process [34]) and the vibrational density of states (VDOS) discussed in the next section.

The caged dynamics is still an active field of research with significant fundamental importance toward the understanding of the glassy state. Growing evidence supports the critical role that the caged dynamics plays in the stabilization of proteins in glassy matrices [18, 35], highlighting the practical importance of this process.

13.3.3 Vibrational Density of States

The major absorption mechanism of terahertz radiation (> 0.1THz) by amorphous solids is via coupling of the photons to the VDOS [36], which is sometimes also referred to as the microscopical peak. This microscopical peak is relatively broad and in spans most of the terahertz range [37] (e.g., ~0.5–7 THz in the case of glycerol, see [38]). This peak is, however, not necessarily a single peak, and it may show an underlying multi-peak nature [39]. As can be seen from the comparison presented in Figures 13.2 and 13.3, the VDOS peak refers to the same feature as the γ process in the work of Reid and Evans (see [13]). For clarity, we will not use the term "γ process" in the remainder of this chapter.

Debye theory predicts a frequency-squared dependence of the VDOS, which is in good agreement with FIR observations in inorganic glasses [12]. The excess density of states above the Debye level is generally termed the Boson peak [36, 40]. It is important to emphasize in this context that some authors refer to the whole VDOS peak as the "Boson peak" [37], and further care must be taken as the term "Boson peak" is also very commonly used in the context of Raman and neutron scattering spectra. In contrast to these scattering techniques, where the Boson peak is typically quite strong, it is not easily distinguishable in the dielectric/terahertz absorption spectra. However, a signature of the actual boson peak can be observed in the terahertz absorption coefficient divided by the square of frequency and refractive index, as outlined recently [9].

2) Sometimes this process is also called the fast relaxation, although this term is technically not correct. In reality, no relaxation time can be associated with the caged motion because it spans a broad range of frequencies and exhibits only little frequency dependence, resulting in the nearly constant loss region [32].

13.4
Terahertz Studies of Disordered Molecular Solids

13.4.1
Disordered Crystals

Terahertz spectroscopy has been used extensively for the past decade to probe low-frequency phonon modes in crystalline materials in general and hydrogen bonding molecular crystals in particular [41–44]. These low-frequency modes are very sensitive to even slight structural changes, allowing terahertz spectroscopy to identify structurally very similar crystals and often even exceed the sensitivity of other spectroscopic techniques commonly used for polymorph identification, such as Raman spectroscopy [45–48].

Apart from the identification of crystalline polymorphs, terahertz spectroscopy is an effective tool to probe subtle disorder within solid-state structures. An example is the tautomeric polymorphism of irbesartan [49]. Irbesartan is known to exist in at least two different polymorphs, forms A and B. Irbesartan form B exhibits conformational disorder within the *n*-butyl hydrocarbon chain. Such disorder of the crystal structure may alter some of the physical properties of the material, such as its solubility, physical stability, or compressibility, and thus can have significant impact on the quality of pharmaceutical products. Terahertz spectra not only revealed distinct features for each polymorph but also allowed the characterization of the conformational disorder. The subsequent computational simulations of the vibrational modes showed that the disorder in this crystal structure arises from a competition between internal conformational strain and external cohesive binding [49].

In another study, terahertz spectroscopy was shown to be highly sensitive to the location of the hydrogen atom position in the dimers formed in the solid state between the carboxylic acid groups of crystalline benzoic acid [50]. Within the context of their crystal structure environment, the carboxylic acids are not identical and hence the hydrogen position introduce structural disorder, the extent of which changes with temperature. The evolving disorder in the hydrogen atom positions influences the terahertz spectrum profoundly: discrete vibrational modes break and develop into a large number of weaker features as a result of the broken local symmetry in the structure. The main spectral features are strongly influenced by changes in the inter-dimer interaction, and by comparing the experimental results with computational simulations, it was possible to show the high sensitivity of terahertz spectroscopy to this type of disorder [50].

Succinonitrile is an orientationally disordered yet spatially ordered molecular plastic crystal at room temperature [51]. The orientational disorder arises from the rotation of its central carbon–carbon bond, allowing for gauche–trans isomerization dynamics. Upon cooling succinonitrile below temperatures of 238 K, the high-temperature rotational disorder of the plastic crystal phase is frozen out and succinonitrile forms a rigid crystal with a monoclinic unit cell that exhibits two distinct phonon modes at terahertz frequencies (Figure 13.4). Both modes

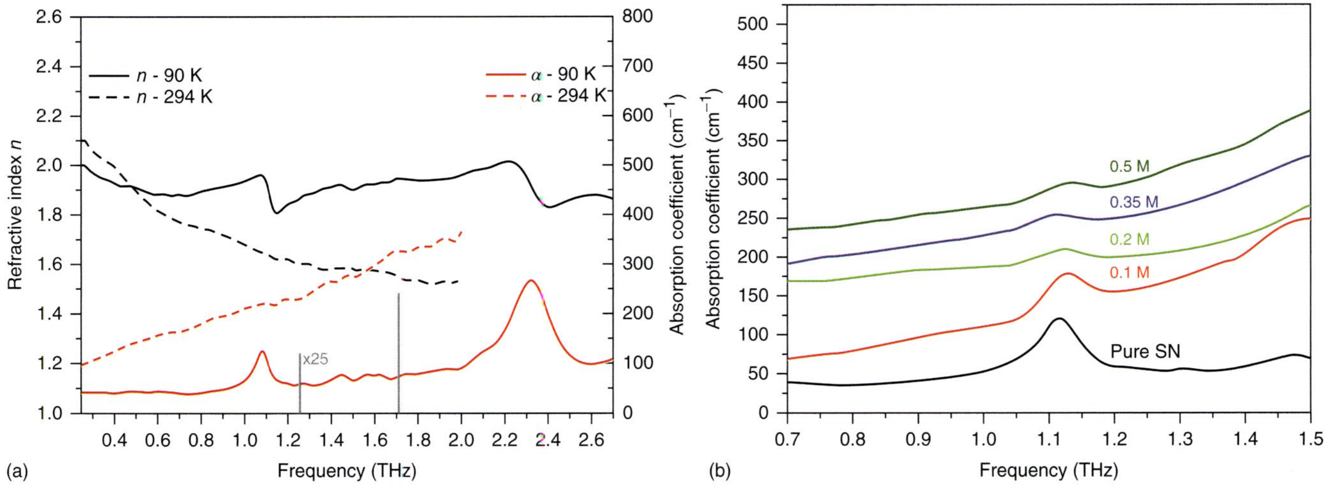

Figure 13.4 (a) Refractive index (black) and absorption coefficients (red) of succinonitrile (SN) at 294 K (disordered plastic crystal, dashed lines) and 90 K (ordered rigid crystal, solid lines). The gray vertical lines are predictions of vibrational modes by density functional theory simulations. (b) Absorption spectra of pure and LiTFSI-doped SN with varying concentrations at 90 K, offset for clarity. In mole percentages ($n/n\%$), the concentrations are 0.5 M = 3.9%, 0.35 M = 2.7%, 0.2 M = 1.6%, and 0.1 M = 0.8%. .(Nickel *et al.* [51]. Reproduced with permission of American Chemical Society.)

are suppressed upon adding a lithium salt (LiTFSI) to succinonitrile, which highlights that the presence of an ionic dopant strongly affects the long-range order of the rigid crystal structure and induces disorder [51].

Simvastatin in another material in which subtle changes in disorder of the crystal structure are of great importance [52]. The THz-TDS study revealed that the polymorphism in simvastatin is a direct result of the molecular disorder due to the rotational freedom of a single functional group in the molecule.

Another interesting group of solids comprises phase-change materials (PCMs). In PCMs, reversible phase transitions between the amorphous and the (metastable) semiconducting crystalline state can be triggered on nanosecond timescales. An example of such materials is germanium telluride or chalcogenide alloys. PCMs are attractive as data storage devices [53], where phase switching can be induced either thermally (e.g., as exploited in CD and DVD technology) or electrically. One of the open questions for such materials is the physical mechanism of conductivity. The conductivity of PCMs exhibits very prominent changes at terahertz frequencies, and THz-TDS was found to be a very convenient method to study the conductivity of PCMs [54]. The differences between the terahertz conductivity of amorphous and crystalline phases of PCMs seem to be linked to the distortions of the crystal unit cell from cubic symmetry.

13.4.2 Polymers

Wietzke *et al.* demonstrated the applicability of terahertz spectroscopy to study the glass transition in polymers [55–58]. In polymers, one interpretation of the glass transition is in terms of the change of its free volume. The free volume is defined as the temperature-dependent space in a polymer sample that is not occupied by the macromolecular chains due to imperfect packing inside their amorphous domains [55]. Upon cooling, the free volume decreases. At the glass transition temperature, T_g, the remaining space is no longer sufficient for segmental motions to occur along the polymer chain axis.

The authors argued that a changing free volume in the polymer sample is reflected in its refractive index n and that the temperature dependence of the refractive index can be used to measure T_g. An example is shown in Figure 13.5 for the case of semicrystalline poly(oxymethylene) (POM) [55]. For temperatures below T_g, the sensitivity of n to changes in temperature is significantly lower, resulting in the change of the thermal gradient $\partial n/\partial T$ at $T = T_g$. A similar approach has been applied to a range of semicrystalline, highly transparent polymers [58].

While technical polymers, such as polyethylene or polytetrafluoroethylene, are generally transparent to terahertz radiation, many polymers for pharmaceutical applications as well as biodegradable polymers exhibit relatively strong absorption in the terahertz range. The losses typically originate form strong hydrogen bonding that is characteristic in such polymers. In these cases, terahertz spectroscopy is able to help to understand how large molecules can form higher

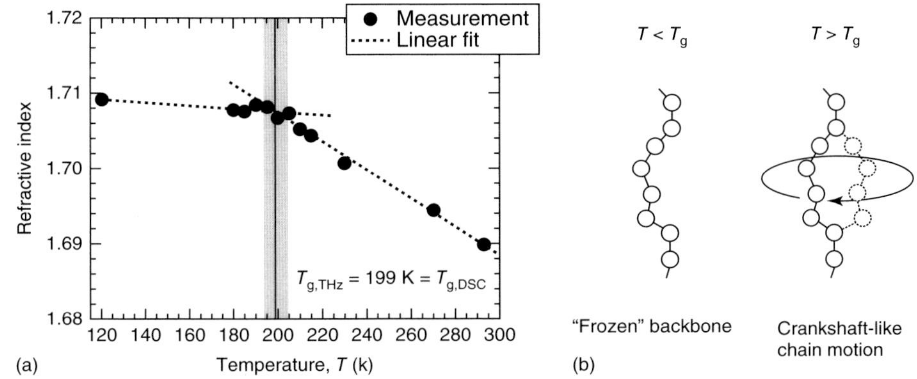

Figure 13.5 (a) The temperature dependence of the THz refractive index revealing a fractional glass transition of POM at $T_{g,THz}$ = 199K as the intersection of two linear fits extrapolating the low- and the high-temperature regime. For comparison, the glass transition temperature interval, obtained from differential scanning calorimetry measurements, is denoted by the gray area. The values match perfectly. (b) Phenomenological visualization of the glass transition. (Wietzke *et al.* [55]. Reproduced with permission of The Optical Society.)

order conformations, such as investigated in the example of the crystallization of poly(3-hydroxybutyrate) [59].

In the case of partially crystalline polymers, the glass transition is a feature of the amorphous domains only, and it was shown that in cases of highly crystalline polymers terahertz spectroscopy can surpass the sensitivity of standard characterization techniques such as differential scanning calorimetry (DSC): in high-density polyethylene, DSC does not observe a clear T_g, yet the glass transition region can still be identified from the shift of the crystalline lattice modes [56].

13.4.3
Inorganic Glasses

In the 1960s, Maradudin and Wallis [60] developed the theoretical formalism to describe the interaction of light with atomic vibrations in both ordered (i.e., crystalline) and disordered (i.e., amorphous) systems. It can be summarized as follows [61]: In a solid, an external electromagnetic field will interact with both the charged cores of atoms and the valence electrons. The two interaction processes, however, are very different in energy. Atomic vibrations fall into the low terahertz frequency range, while electronic interactions will typically be observed an order of magnitude higher in frequency. Therefore, the electron response can be approximated as instantaneous, leading to a time-independent screening effect on the ions. This model is also known as the rigid-ion model. There are two net contributions to the optical absorption in the FIR region: one is due to the vibrational eigenmodes, which can be characterized by a Debye-like VDOS (up to low terahertz frequencies), and the other is due to the fluctuating atomic charges within the sample [61]. In this context, the low-frequency absorption coefficient can be expressed as [36]

$$\alpha(\nu) = C(\nu)g(\nu) \tag{13.2}$$

Glasses generally exhibit terahertz absorption proportional to ν^{β} ($\beta \lesssim 2$) [12]. Here, $g(\nu)$ is the VDOS in the glass, $g(\nu) \propto \nu^2$, and the coefficient $C(\nu)$ quantifies the degree of coupling between photons and atomic vibrations. The atomic charge fluctuations can be split into two different contributions: an uncorrelated, random component due to structural disorder on intermediate and long-range length scales (A), and a correlated component of charge fluctuations caused by variations in the local structure that obey local charge neutrality within the structural units (B) [61]. The coupling coefficient C can then be expressed as $C(\nu) \simeq A + B\nu^2$, thus yielding [36]

$$\alpha(\nu) \simeq \nu^2(A + B\nu^2) \tag{13.3}$$

where A and B are material-dependent constants, which can be extracted by terahertz spectroscopy, such as demonstrated for the case of sodosilicate glasses [62].

While the generalized power-law absorption $\alpha(\nu) \propto \nu^2$ has been demonstrated by several terahertz spectroscopy studies [63, 64], Eq. 13.3 is valid only for frequencies below the Ioffe–Regel crossover, which in many materials is close to the Boson peak [36]. The breakdown of the power law at higher frequencies has been shown experimentally using an ultra-broadband plasma terahertz spectroscopy system that can access frequencies up to 18THz [65].

13.5 Organic Glass-Forming Liquids

Because of the high T_g of inorganic glasses, usually only the amorphous solid state is examined in experimental terahertz studies to date. In contrast, for organic glass formers the T_g is often close to or (far) below room temperature, and thus it is easier in such materials to examine the properties of the liquid, supercooled liquid, as well as the amorphous solid state.

13.5.1 Hydrogen-Bonded Liquids and Solutions

Given the energy range that can be probed by THz-TDS, it is well suited to excite weak intermolecular forces, such as van der Waals forces and hydrogen bonds [6]. In the liquid state, relaxations and vibrational motions are typically coupled and, as outlined before, multicomponent analysis of MHz–GHz dielectric and terahertz spectra that span a vast frequency range is required to describe the overall dielectric function. The spectra are dominated by the losses due to reorientational motions of polar molecules, and it is possible to extract interesting information on the intermolecular interactions from the terahertz spectra [66, 67]. For example, it is known that polar small organic molecules form hydrogen-bonded chain structures and such structures exhibit relaxation processes over a range of timescales: the cooperative rearrangement of the alcohol–alcohol chain structure ($\approx$ 100ps), the reorientation of an individual alcohol molecule situated at the end of the alcohol–alcohol chain ($\approx$ 10ps), and the relaxation of liquid molecules in the process of hydrogen-bond formation and decomposition ($\approx$ 1ps) [68–70].

In order to resolve and quantify such dielectric relaxation processes, Debye models have been used:

$$\hat{\varepsilon}(\omega) = \varepsilon_{\infty} + \frac{\Delta\varepsilon_1}{1 + i\omega\tau_1} + \frac{\Delta\varepsilon_2}{1 + i\omega\tau_2} \tag{13.4}$$

In this simplest case of a two-component relaxation model, it is possible to resolve two relaxation processes in water at room temperature. Spectral analysis shows that they correspond to relaxation times of $\tau_1 \approx 8$ps and $\tau_2 \approx 170$fs (here $\hat{\varepsilon}$ is the complex dielectric loss, ω is the angular frequency, and $\Delta\varepsilon_i$ is a measure of the strength of the respective relaxation process). Additional Debye and oscillator terms can be added to this equation to account for molecules undergoing hydrogen-bond formation and breaking as well as intermolecular stretching vibrations. A representative example is that of liquid water, which has been studied extensively [71, 72]. Here, both a fast (fs) and a slow (ps) Debye relaxation process are observed together with a vibrational mode attributed to intermolecular stretching as well as an intermolecular librational mode (see Figure 13.6), which can be described by Yada *et al.* [72]

$$\tilde{\varepsilon}(\omega) = \frac{\Delta\varepsilon_1}{1 + i\omega\tau_1} + \frac{\Delta\varepsilon_2}{1 + i\omega\tau_2} + \frac{A_S}{\omega_S^2 - \omega^2 + i\omega\gamma_S} + \frac{A_L}{\omega_L^2 - \omega^2 + i\omega\gamma_L} + \varepsilon_{\infty} \tag{13.5}$$

Here, $\Delta\varepsilon_1$ and $\Delta\varepsilon_2$ are the respective relaxation strengths of the two Debye relaxation modes with τ_1 (slow, ps) and τ_2 (fast, fs) being the relaxation times. The stretching vibration and libration modes have amplitudes of A_S and A_L, angular frequencies ω_S and ω_L, and damping constants γ_S and γ_L.

The two Debye relaxation processes are indicative of the heterogeneity of the water structure [72]. Both processes depend strongly on temperature and become significantly faster at elevated temperatures. The observation of the temperature behavior is very important in order to understand the physical mechanism behind the dielectric relaxation processes in water. The slower relaxation time τ_1 shows a linear change with viscosity η and absolute temperature T, and obeys the Einstein–Stokes–Debye relation $\tau_1 = 4\pi\eta R^3/k_B T$ [71]. The temperature dependence of the slow relaxation time observed in the study of protonated water and deuterated water [66] obeyed a power law and indicated that the slow relaxation process in water is the structural relaxation process, related to the liquid as a whole. In contrast, the fast relaxation process showed no power-law dependence on temperature [71, 72], and it is reasonable to ascribe the fast relaxation to the relaxation of individual water molecules [6].

The temperature dependence of the dielectric relaxation processes in liquid water and its mixtures with other liquids was the subject of a number of studies

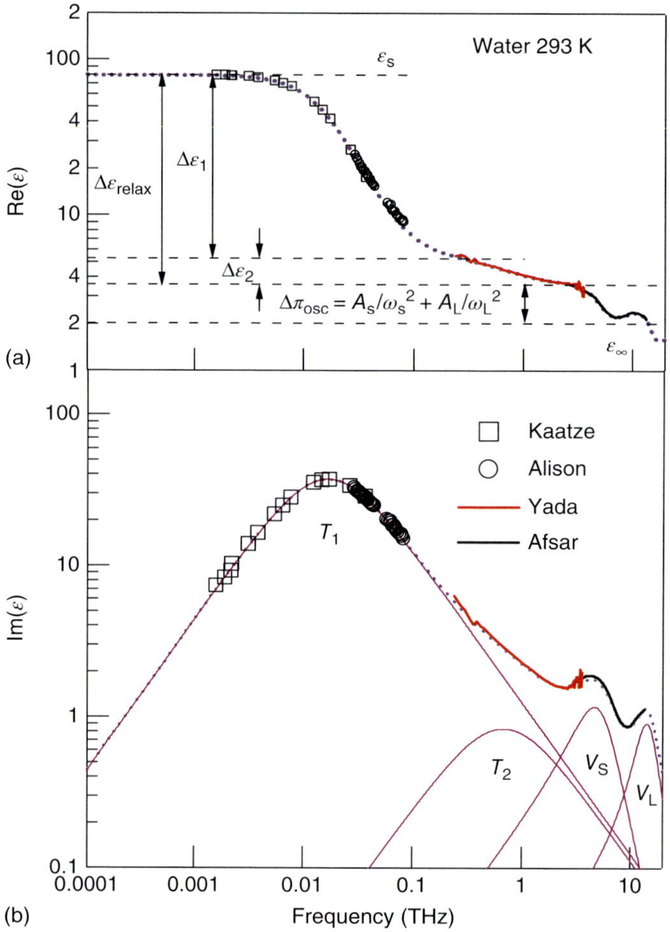

Figure 13.6 (a) Real and (b) imaginary parts of the complex dielectric constants of liquid water at 293 K. The plots combine results of microwave spectroscopy by Kaatze [73] and Alison (*et al*). [74], far-infrared spectroscopy by Afsar and Hasted [75], and terahertz spectroscopy by Yada *et al.* [72]. The fitting result by using Eq. 13.5 is denoted by a dashed blue line in both figures, and the results of the spectral decomposition are shown in (b) by green lines. (Adapted from Yada *et al.* [72]. Reproduced with permission of Elsevier.)

[71, 72, 76–78]. It was possible to determine the size of the hydration shells around solvated protons in water [76]. In combination with infrared and molecular dynamics (MD) simulations, THz-TDS was used to study the structure of methanol/water mixtures [77] and acetonitrile/water mixtures [79]. In addition to the studies of aqueous mixtures, a number of studies have focused on the interaction between nonpolar solvents with polar solvents [80–82]. The dielectric response of tetramethylurea was measured using a combination of terahertz and gigahertz dielectric spectroscopy, and the measurements were compared to femtosecond infrared pump–probe studies [83]. Yomogida *et al.* found direct evidence for a vibrational mode around 1.5 THz and an intermolecular stretching mode around 2.5 THz in measurements for a series of seven pentanol isomers [84, 85]. Together with the dielectric relaxation data at microwave frequencies, the studies showed the complex interplay between reorientation motion and vibration dynamics in hydrogen-bonded liquids. In a separate study, Yamaguchi *et al.* even found evidence for an intermolecular vibrational mode originating from benzoic acid, which was found to form dimers when dissolved in carbon tetrachloride [86].

More recently, terahertz spectroscopy measurements were complemented with pulsed field gradient NMR diffusometry [87] and neutron scattering measurements [88] to understand the dynamics of alcohol/water mixtures at the molecular level. It was found that the hydrogen-bonding structure and dynamics in binary mixtures of protic–protic, protic–aprotic, and aprotic–aprotic systems can all be investigated at terahertz frequencies [89].

Similar models, either with two or three Debye components, have been applied to alcohols and their aqueous mixtures [90, 91], as well as sugar/water mixtures [92] or ionic liquids [93, 94]. For example, it was found that a three-component Deybe model is required to describe the dielectric spectra of water/ethanol mixtures [91]. In this case, it is not the slow process but the intermediate process that is related to the structural relaxation of the mixture. Adding ethanol to water reduces the concentration of hydrogen bonds in the solution, and thus the total relaxational strength of the solution is lower than that of pure water. The relaxation times of ethanol/water mixtures are also slower than those in pure water [6, 91].

13.5.2 Supercooled Liquids and Glasses

Supercooling of liquids leads to a dramatic shift of the frequencies at which the dielectric relaxation processes are observed. Therefore, the losses at terahertz frequencies typically decrease upon cooling liquid samples. The effect of temperature on the vibrational processes is much less pronounced. This vitrification of the dielectric relaxations is characterized by several universalities, which have been investigated in detail for the case of the polyols glycerol, threitol, xylitol, and sorbitol.

As already outlined earlier (see Figure 13.2), a broad peak emerges as the dielectric relaxation losses vanish. This peak originates from the low-frequency librational/torsional modes in the amorphous material (i.e., its VDOS) and is sometimes referred to as the Boson peak [95]. It reaches its maximum typically at frequencies of a few THz (Figure 13.7).

Upon cooling further, the primary dielectric relaxation vitrifies at T_g, when the primary relaxation time reaches $\tau_\alpha \approx 10^2 - 10^3$s. This vitrification process can be observed easily also by terahertz spectroscopy when plotting the dielectric losses over the rescaled temperature T/T_g (Figure 13.8). The dielectric losses decrease linearly with temperature. When the samples are cooled below $T_g = T_{g\alpha}$, the

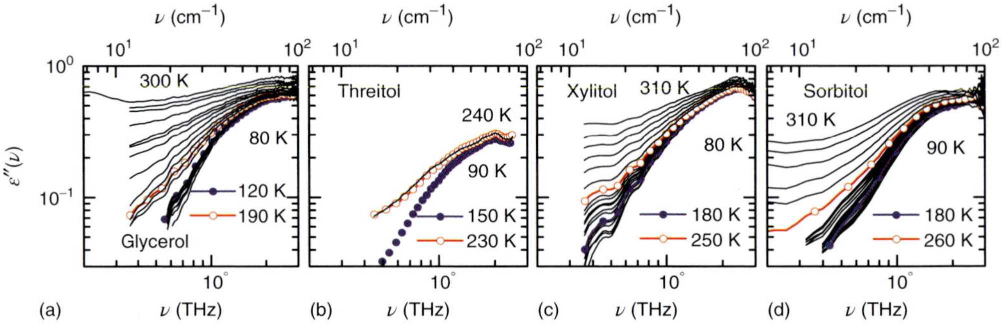

Figure 13.7 Dielectric losses $\varepsilon''(\nu)$ of (a) glycerol, (b) threitol, (c) xylitol, and (d) sorbitol at terahertz frequencies 0.2–3 THz in the temperature range 80–310 K with 10 K temperature increments. The blue and red circles highlight the losses in the proximity of 0.65 T_g and T_g, respectively. (Sibik *et al.* [96]. http://pubs.acs.org/doi/abs/10.1021/jz5007302. Used under CC-BY. https://creativecommons.org/licenses/by/2.0/.)

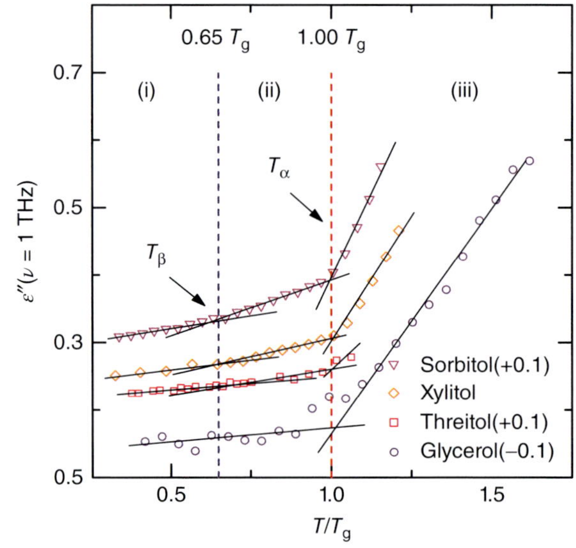

Figure 13.8 Dielectric losses $\varepsilon''(T/T_g)$ at frequency $\nu = 1$THz. The solid lines represent linear fits of regions (i), (ii), and (iii) as explained in the text. The dashed vertical blue and red lines highlight 0.65 T_g and 1.00 T_g, respectively. T_β and T_α represent the crossing point of the linear fits from area (i) to area (ii) and from area (ii) to area (iii), respectively. The dielectric loss data for sorbitol, threitol, and glycerol are offset in the positive (+0.1) and negative (−0.1) direction for clarity. .(Sibik *et al.* [96]. http://pubs.acs.org/doi/abs/10.1021/jz5007302. Used under CC-BY. https://creativecommons.org/licenses/by/2.0/.)

primary relaxation becomes too slow to contribute to the molecular mobility and hence a change of gradient at $T = T_g$ is observed.

The study of polyols revealed that at temperatures below the glass transition there are still temperature-variable changes in the spectra below T_g [8, 96]. This change in the losses $\varepsilon''(T/T_g)$ is again linear, and the slope of the fit was found to strongly correlate with the strength of the JG β-relaxation [96]. The linear trend stops at some temperature below $T_{g\beta} < T_{g\alpha} \equiv T_g$, where the secondary relaxation time reaches $\tau_{JG-\beta} \approx 10^2 - 10^3$s [96], or in other terms, when the JG β-relaxation vitrifies.

The sensitivity of terahertz spectroscopy to $T_{g\alpha}$ and $T_{g\beta}$ is, however, striking, because the α and JG-β relaxations are too slow to directly contribute to terahertz absorption at these temperatures. The true origin of $T_{g\alpha}$ and $T_{g\beta}$ at terahertz frequencies is the influence of the vitrification on the caged dynamics, and in particular on the cages themselves [97]. The ability to observe the vitrification of the relaxations at terahertz frequencies in amorphous materials is significant in that it indicates that there must be a change in the intermolecular bonding (such as hydrogen bonding) and/or intermolecular dynamics at these temperatures, which affects the properties of the glasses. These results triggered the re-examination of the data from other high-frequency spectroscopies including neutron scattering, light scattering, Brillouin scattering, and inelastic X-ray scattering [97–99].

At temperatures below 0.65 T_g, the terahertz spectra of hydrogen-bonded glass-forming liquids are dominated by the VDOS and only a very weak variation with temperature is observed [8, 85, 95, 96, 100]. Overall, the intensity of the peak in secondary alcohols (2-propanol, 2-butanol, and 2-pentanol) is higher than that in normal alcohols (1-propanol, 1-butanol, and 1-pentanol). The peak varied only weakly with the number of carbon atoms [85] while it was strongly affected by the position of the -OH group in a molecule. Thus it appears that the position of the -OH group has a strong effect on the local structures in monohydric alcohols [85].

As discussed before, the terahertz absorption (or more accurately, the real part of the conductivity) of glasses at temperatures far below T_g is well described by a power law ν^β ($\beta \lesssim 2$) [12]. A slightly more advanced model is obtained by modifying this power law as [8]

$$\sigma' = n(\nu)\alpha(\nu) = A + C(\nu - \nu_0)^q \tag{13.6}$$

Here, σ' stands for the real part of the conductivity $\hat{\sigma} = \sigma' + i\sigma''$, which can also be expressed as the real part of the refractive index n multiplied by the absorption coefficient α. The parameters ν_0 and A are introduced to account for the experimental low-frequency cutoff in the terahertz spectra at ν_0 and any absorption offset at this frequency. The parameter C describes the coupling of terahertz radiation to disordered matter. The exponent q is generally around 2 for amorphous solids, decreases continuously upon heating a glass through its glass transition, and tends toward 1 in the normal liquid state [8].

Equation 13.6 allows a good fit of the terahertz spectra of amorphous solids, supercooled liquids, as well as normal liquids. It is worth mentioning in this context that the scattering background in crystalline samples is also generally well described by a power law with exponent $q \approx 2$. Applying the power-law model is therefore a sensitive technique to study the crystallization of amorphous samples upon heating, as heating will tend to lower the exponent q toward 1, while crystallization will push q toward 2 [101].

13.6 Characterization of Disordered Biological and Pharmaceutical Systems

The unique fingerprints of crystalline materials at terahertz frequencies predetermine the sensitivity of terahertz spectroscopy to the presence of different crystalline polymorphs [102, 103], hydrates [104, 105], or co-crystals [48, 106] in a sample, and allow the observation of phase transitions between such solid-state modifications. This information is very useful for the characterization of active pharmaceutical ingredients (APIs). Terahertz spectroscopy has several advantages that make it attractive for API characterization: (i) its insensitivity to thermal interference allows measurements of materials and processes to be acquired at high temperatures; (ii) performing time-resolved studies on the sub-picosecond timescale, potentially allowing insight into dynamic systems such as amorphous systems; and (iii) the low energy used in terahertz spectroscopy minimizes the risk of sample degradation [102, 107].

13.6.1 Quantifying Crystallinity

Strachan *et al.* have demonstrated the ability of terahertz spectroscopy, in combination with partial least-squares analysis, to determine the levels of crystallinity in a mixture of amorphous and crystalline drugs as low as 1% [102]. With increasing crystallinity, the intensity of the crystalline vibrational modes increases, while the absorption baseline (featureless absorption) decreases, as illustrated in Figure 13.9.

The high sensitivity of terahertz spectroscopy to detect crystallinity has also proven useful for the characterization of amorphous drug systems prepared by different methods such as quench cooling or grinding [108], as well as spray-drying [109] and freeze-drying [110–112]. For example, amorphous samples prepared by grinding were found to contain nanoscopic crystalline particles, which could be resolved in terahertz spectra but were not apparent in the XRD analysis [108]. The crystallinity detection is not restricted to glasses only, but has been demonstrated to work well also in the case of aqueous suspensions, where the water absorption is strongly affected by the presence of the crystalline phase [113].

13.6.2 Crystallization of Amorphous Drugs Above T_g

Several terahertz studies explored the crystallization kinetics of amorphous drugs. In the example of carbamazepine, it was shown that terahertz spectroscopy is

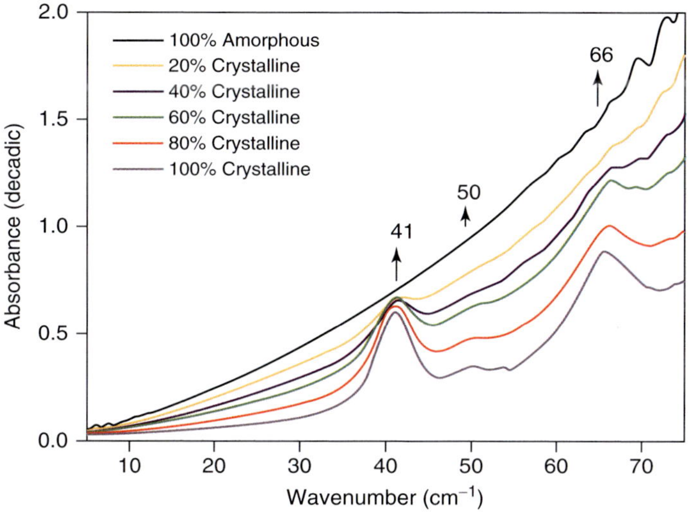

Figure 13.9 Terahertz pulsed spectra of binary mixtures of infomethacin amorphous and crystalline forms (20% intervals, 0–100% crystalline form). The arrows indicate spectral changes as the amorphous form concentration decreases and the crystalline form concentration increases. (Strachan *et al.* [102]. Reproduced with permission of Wiley.)

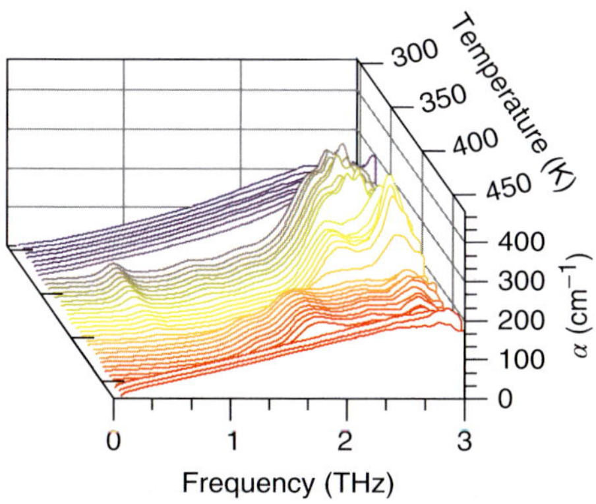

Figure 13.10 Absorption spectra of paracetamol at terahertz frequencies upon heating the sample from the supercooled liquid phase over the temperature range of 295–470 K in 5 K increments. (Sibik *et al.* [101]. http://pubs.acs.org/doi/abs/10.1021/jz5007302. Used under CC-BY. https://creativecommons.org/licenses/by/2.0/.)

very well suited to detect the crystallization of the amorphous drug together with any subsequent phase transitions between crystalline polymorphs [114]. A similar study was performed on amorphous paracetamol [101]. Paracetamol is known to exhibit a rich crystallization behavior, as shown in Figure 13.10. For a sample that is fully covered from the atmosphere by two windows, paracetamol crystallizes upon heating to temperatures above $T_g = 296$K from the amorphous phase into its crystalline form III (~ 335K), followed by a phase transition to form II (~ 375K), form I (~ 405K), and subsequent melting (~ 455K). In another study, McIntosh *et al.* monitored the crystallization of amorphous lactose with terahertz spectroscopy at elevated humidity at room temperature [115].

Several methods can be used to monitor the conversion of an amorphous sample into a crystalline solid. Perhaps the most straightforward approach is to monitor the change in height or intensity of a crystalline resonance peak as the crystallization occurs [114]. For a more detailed analysis, McIntosh *et al.* showed that it was possible to describe the change in peak intensity with time during crystallization by employing Avrami-type kinetics [115]. They also proposed to describe the terahertz spectra during the course of crystallization by a linear combination of a Lorentzian function centered at the frequency of a crystalline resonance peak in the spectrum ν_c of specified half width at half-maximum γ and a power law describing the background absorption, resulting in Ref. [115]

$$\alpha(\nu) = \frac{A}{1 + \left(\frac{\nu - \nu_c}{\gamma}\right)^2} + B\nu^a + C \tag{13.7}$$

This model assumes that the crystalline resonance peak is well separated from any other crystalline modes, the frequency window is restricted to only include the peak at ν_c, and the width of the peak does not change significantly with temperature. The changes in the fitting parameters can then be used to track the progress of the crystallization.

A very similar method was proposed by Sibik *et al.* to describe the crystallization of amorphous paracetamol [101]. The spectra were fitted with the power law only rather than including an additional oscillator term. The results showed that the change in the exponent is an excellent indicator for the progress of crystallization and can be fitted using a Avrami–Erofeev function after normalization of the exponent value.

A third method to track the progress of crystallization is to decompose the spectra directly into a linear combination of fully amorphous (α_a) and fully crystalline (α_c) spectra as [101]

$$\alpha(T, \nu) = r_a(T)\alpha_a(\nu) + r_c\alpha_c(\nu) \tag{13.8}$$

Here, T may represent either the temperature, in the case of a variable-temperature experiment, or time, in the case of an isothermal experiment. The coefficients r_c and r_a directly characterize the respective crystalline and amorphous content and can be described directly using an Avrami–Erofeev function. Again, the frequency window needs to be limited in order to satisfy the condition $r_a + r_c \approx 1$ with regard to the scattering background in the crystalline absorption data.

13.6.3 Crystallization of Amorphous Drugs Below T_g

So far in this chapter we have only discussed crystallization processes at temperatures above T_g. For the stability of amorphous products, it is, however, equally important to understand the mechanisms that govern the crystallization of glasses below T_g. Several dielectric spectroscopy studies have highlighted the importance of the secondary relaxation processes in this context, in particular that of the JG β-relaxation [28, 116, 117].

A recent terahertz spectroscopy study of naproxen confirms these results [118]. As outlined in Section 13.5.2, THz-TDS is sensitive to the vitrification of JG β-relaxation process in glasses, which has been found to occur in sorbitol, xylitol, and threitol at around $T \approx 0.6-0.7\ T_g$ (see Figure 13.8 and [96]). This vitrification appears to be a universal feature of hydrogen-bonded glasses and can also be used to characterize the secondary relaxation in amorphous drug molecules. Naproxen is very unstable in the amorphous phase and has a strong tendency to crystallize. When a sample of naproxen melt was quenched in liquid nitrogen, a sample was obtained that was mostly amorphous, but the spectra in Figure 13.11 show that a weak spectral feature is present at around 1.2THz at 100K, the starting temperature of the experiments. This subtle feature is caused by trace crystallites that are still present in the quench-cooled sample. In contrast to the temperature-dependent absorption observed for pure, amorphous phases, where an increase in absorption due to the emergence of the relaxation processes occurs with increasing temperature, the sample of amorphous naproxen that contains seed crystals steadily crystallizes even at temperatures well below T_g. Nucleation has already occurred, and crystal growth commences with increasing temperature. This continued crystallization leads to a steady decrease in the overall absorption losses as

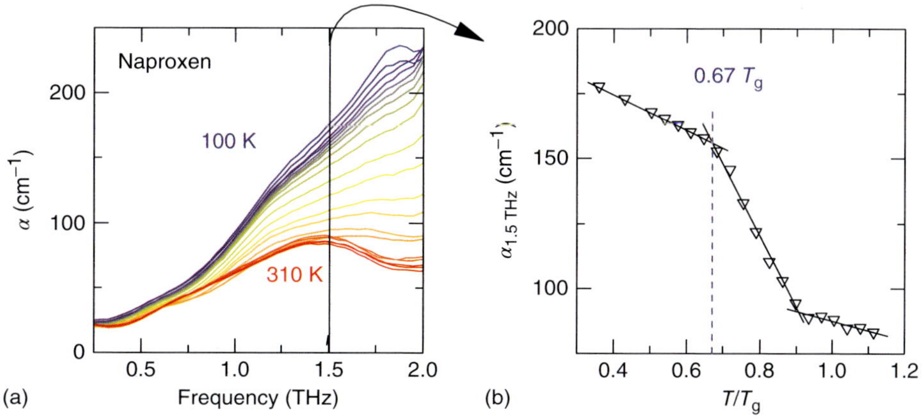

Figure 13.11 (a) Terahertz spectra of amorphous naproxen from 100 to 310K with 10K step increments. (b) Terahertz absorption of naproxen at 1.5 THz as a function of rescaled temperature T/T_g. Modified from Ref. [118].

the amorphous fraction in the sample decreases, and hence the VDOS vanishes (Figure 13.7). Over time, a peak at 1.4THz emerges from the weak spectral feature. What is extremely interesting in this context is that the results show that the crystallization of amorphous naproxen commences at temperatures well below T_g in the presence of seed crystals.

When the absorption is plotted again as a function of the rescaled temperature T/T_g, a striking fact is revealed. At temperatures above 0.67 T_g the decrease in absorption is about 3.5 times faster than below this temperature. This coincides exactly with the onset of the JG β-relaxation and its contribution to the intermolecular mobility which plays a crucial role in the crystallization of glasses below T_g. Thus, one must be very careful when evaluating the stability of a glass above 0.67 T_g, or more accurately above the vitrification temperature of the JG β-relaxation, as the molecules may have enough mobility for crystallization to occur (albeit the fact that the crystallization rate of naproxen is quite an extreme case of course).

13.6.4
Stability of Amorphous Drugs Below T_g

In pharmaceutics, the formulation of drug molecules into an amorphous solid is an attractive route to overcome the poor aqueous solubility of many drug candidate molecules. However, in order to use this strategy successfully, it is essential to understand the molecular factors that govern the crystallization of amorphous solids and to be able to quantify them. Because of the lack of molecular understanding, the long-term storage of the amorphous products poses a major risk at present. Recrystallization is a complex process that involves an intricate interplay between crystal nuclei formation and subsequent crystal growth. The recrystallisation kinetics, and thus the physical stability of amorphous systems, has been the subject of detailed analysis and was related to thermodynamic [119, 120, 121], kinetic [122, 123], and molecular driving forces [124–126]. What is clear is that the energy difference between the amorphous and crystalline state is the driving force of the crystallization process overall, while the molecular mobility plays an important role in facilitating the process by enabling nuclei formation and crystal growth [127].

In order to avoid crystallization, it is commonly acknowledged that in order to reduce molecular mobility, amorphous systems should be stored at temperatures far below their respective T_g as well as at low humidity [128, 129]. In the past, empirical rules such as that a glass is completely stable below the Kauzmann temperature T_K [130], which can be roughly estimated as T_g − 50K, were commonly used [120, 128, 131]. The rationale behind such thinking was that it was assumed that the primary molecular mobility, the characteristic for liquids, no longer exists on the relevant timescales of drug product storage. It is now acknowledged that sufficient molecular mobility, which originates from the JG β-relaxation [28, 117, 132, 133], still exists at such temperatures and hence crystallization can still take place at T_g− 50K. By far the most commonly used and convenient technique to observe these relaxation processes directly is dielectric spectroscopy. More recently, it has become clear that neutron and light scattering measurements are equally sensitive to changes in the molecular dynamics that occur on the nanosecond-to-picosecond timescale during the glass transition [134, 135] and that such measurements reflect on the overall stability of glassy systems [35].

Given the ability of terahertz spectroscopy to measure accurately the onset of the local mobility in a sample at temperatures of $T_{g\beta} < T_g$, it provides an excellent alternative technique to quantify the risk of crystallization between different samples and to determine molecular mobility. A recent study established the correlation between the thermal gradient of terahertz absorption and amorphous stability [118]: the higher the thermal gradient of terahertz absorption between the $T_{g\beta}$ and $T_{g\alpha}$, the larger the contribution of the local mobility and hence the poorer the stability of the amorphous system (see Figure 13.12). THz-TDS could be used as an analytical tool to rank and quantify the stability properties of amorphous drug molecules in a manner similar to neutron and light scattering. Given the comparatively wide availability, relatively low cost, and ease of use of THz-TDS compared to the other techniques, this could prove to be a very attractive technique for such measurements in future.

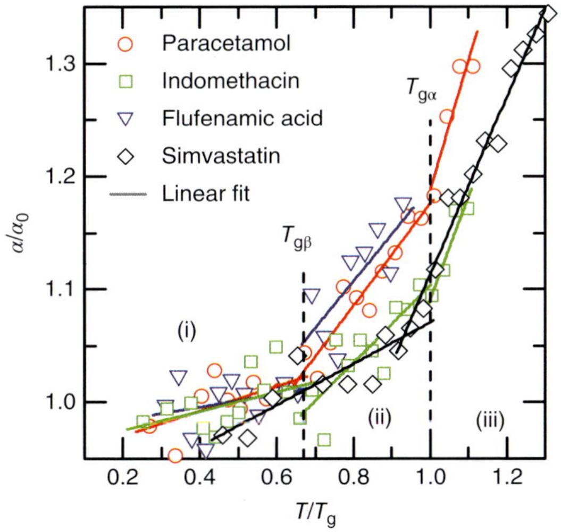

Figure 13.12 Terahertz absorption of amorphous paracetamol, flufenamic acid, indomethacin, and simvastatin measured at 1.0THz. The absorption coefficient is rescaled by its low-temperature average α_0 and the temperature is rescaled by $T_g = T_{g\alpha}$. The solid lines represent linear fits. Paracetamol and flufenamic acid both exhibit a large increase in absorption upon heating between $T_{g\beta}$ and $T_{g\alpha}$. They are highly unstable and recrystallize within a few minutes at ambient conditions. In contrast, simvastatin shows only a small increase in terahertz absorption upon heating between $T_{g\beta}$ and $T_{g\alpha}$ and remains amorphous for over 220 days under ambient conditions. Indomethacin represents an intermediate case, remaining stable for about 7 days before recrystallization. Modified from Ref. [118].

13.6.5
Multicomponent Disordered Pharmaceutical Systems

From a scientific point of view, understanding the behavior of single-component amorphous systems is the first step to systematically advance the field. However, from a practical perspective, the performance of multicomponent systems that are designed to keep the drug molecules in the amorphous state are of critical interest. Good examples of such systems are polymer/drug dispersions and protein/sugar mixtures.

A recent study investigated the effects of trehalose, inulin, and dextran on the stabilization of bovine serum albumin (BSA) in freeze-dried protein/sugar mixtures [136]. The THz-TDS data confirmed that the intermolecular bonding between the protein and sugar plays a key role in the stability of BSA in the freeze-dried matrix. The strength of molecular interaction between protein and sugar can be estimated on the basis of the temperature dependence of terahertz absorption: significant bonding between protein and sugar molecules, such as in the case of BSA and trehalose, leads to a deviation of the absorption coefficient of the mixture from the simple sum of its constituent components (see Figure 13.13 (a)). In contrast, for samples where the intermolecular interaction between the protein and sugar molecules is small, the mixture behaves like a phase-separated physical mixture and the measured absorption coefficient is close to a linear combination of the weighted component spectra (see Figure 13.13 (b)) [136].

It is likely that such effects also apply to polymer/drug systems, and that THz-TDS could be used to determine the maximum amount of drug loading, above which partial phase separation of the drug component might occur. This would be quite important in the context of amorphous stability, as the crystallization process for the phase-separated mixture would again depend on the molecular mobility of the drug molecule alone and these molecules are no longer stabilized by the presence of polymer molecules. However, such analysis remains a topic for future investigation.

13.7
Outlook

As discussed in this chapter, albeit spanning only one decade of frequency, terahertz spectra of disordered systems can be used to resolve a range of phenomena that originate mainly from intermolecular interactions. The field is rapidly developing at present, and in the future we anticipate to see advances in the experimental technology, a deeper understanding of terahertz kinetics, and new applications.

Although several commercial instruments are already available on the market, terahertz technology is still relatively recent and under active development. At present, in most standard THz-TDS systems terahertz pulses are generated from photoconductive antennae on semiconductor substrates, such as GaAs or InP, driven by femtosecond near-infrared lasers. The semiconductor material limits

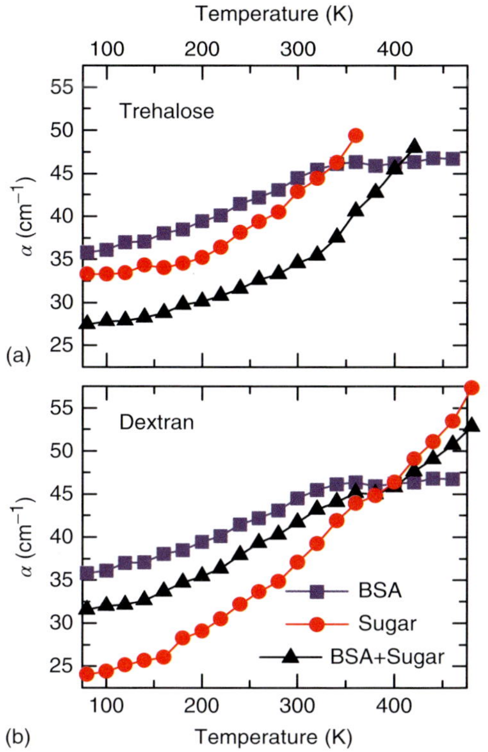

Figure 13.13 Terahertz absorption of (a) BSA/trehalose mixture and its components, and (b) BSA/dextran mixture and its components, at 1THz. The molecular weight of dextran is 70kDa. After [136].

the frequency bandwidth because of the phonon modes from the material itself (e.g., at 8.06THz in GaAs) and also their relatively low output of terahertz power. A novel approach that omits semiconductors is based on the generation and detection of terahertz radiation from plasmas [137]. Such systems can span frequencies from 0.2 to over 30THz, offering access to the complete VDOS as well as a significant part of the FIR region. Obtaining the full VDOS peak for disordered materials is crucial for any molecular simulations to be performed, and hence this technology is very promising in the context discussed in this chapter. However, at present these plasma-based systems are still very expensive and quite unstable in their output power compared the more commonly used techniques.

THz-TDS is, however, by far not the only technique allowing access to the terahertz dynamics. Recent development of low-frequency Raman spectroscopy, as well as new technologies that extend the range accessible by dielectric spectroscopy to higher frequencies, will result in more readily available and more reliable experimental data available at terahertz frequencies. Such techniques are more and more complemented by experiments based on the optical Kerr effect, which is an established measurement technique to probe the structural relaxation in both complex and simple liquids [138]. In this context, it is, however, important to emphasize again that it is not necessarily straightforward to compare the results obtained by different techniques in the same frequency range, as the physical principles that govern the interactions are fundamentally different between the techniques: for example, modes that can be measured in terahertz and Raman spectroscopy follow different selection rules, some techniques are absorption measurements while others are scattering techniques, and the measurements in the infrared can also be complemented by techniques such as visible light and neutron scattering.

References

1. Auston, D.H., Glass, A.M., and Ballman, A.A. (1972) Optical rectification by impurities in polar crystals. *Phys. Rev. Lett.*, **28** (14), 897–900.
2. Auston, D.H. (1983) Subpicosecond electro-optic shock waves. *Appl. Phys. Lett.*, **43** (8), 713–715.
3. Fattinger, C. and Grischkowsky, D.R. (1989) A Cherenkov source for freely-propagating terahertz beams. *IEEE J. Quantum Electron.*, **25** (12), 2608–2610.
4. Fattinger, C. and Grischkowsky, D. (1989) Terahertz beams. *Appl. Phys. Lett.*, **54** (6), 490–492.
5. Grischkowsky, D., Keiding, S., Exter, M.V., and Fattinger, C. (1990) Far-infrared time-domain spectroscopy with terahertz beams of dielectrics and semiconductors. *J. Opt. Soc. Am. B*, **7** (10), 2006–2015.
6. Jepsen, P., Cooke, D., and Koch, M. (2011) Terahertz spectroscopy and imaging – modern techniques and applications. *Laser Photon. Rev.*, **166** (1), 124–166.
7. Parrott, E.P.J. and Zeitler, J.A. (2015) Terahertz time-domain and low frequency Raman spectroscopy of organic materials. *Appl. Spectrosc.*, **69** (1), 1–24.
8. Sibik, J., Shalaev, E.Y., and Zeitler, J.A. (2013) Glassy dynamics of sorbitol solutions at terahertz frequencies. *Phys. Chem. Chem. Phys.*, **15** (28), 11931–11942.
9. Sibik, J. and Zeitler, J.A. (2016) Terahertz response of organic amorphous systems: experimental concerns and perspectives. *Philos. Mag.*, doi: 10.1080/14786435.2015.1111528.
10. Poley, J.P. (1955) Microwave dispersion of some polar liquids. *J. Appl. Sci. B*, **4**, 337–387.
11. Johari, G.P. (2002) Molecular inertial effects in liquids: poley absorption, collision-induced absorption, low-frequency Raman spectrum and Boson peaks. *J. Non-Cryst. Solids*, **307-310**, 114–127.
12. Strom, U., Hendrickson, J.R., Wagner, R.J., and Taylor, P.C. (1974) Disorder-induced far infrared absorption in amorphous materials. *Solid State Commun.*, **15**, 1871–1875.
13. Reid, C.J. and Evans, M.W. (1982) Dielectric and far infrared study of solutions in the glassy state from 100 Hz to 10 THz: discovery and characterization of the universal γ process. *J. Chem. Phys.*, **76** (5), 2576–2584.
14. Johari, G.P. (1970) Viscous liquids and the glass transition. II. Secondary relaxations in glasses of rigid molecules. *J. Chem. Phys.*, **53** (6), 2372–2388.
15. Wang, C.H. and Wright, R.B. (1971) Raman scattering studies of liquids and glasses. I. Liquid and supercooled liquid glycerol. *J. Chem. Phys.*, **55** (4), 1617–1624.
16. Wang, C.H. and Wright, R.B. (1971) Raman scattering studies in liquids and glasses. II. Liquid and supercooled liquid glycerol. *J. Chem. Phys.*, **55** (7), 3300–3307.
17. Wuttke, J., Kiebel, M., Bartsch, E., Fujara, F., Petry, W., and Sillescu, H. (1993) Relaxation and phonons in

viscous and glassy orthoterphenyl by neutron scattering. *Z. Phys. B*, **91**, 357–365.

18. Cicerone, M.T. and Soles, C.L. (2004) Fast dynamics and stabilization of proteins: binary glasses of trehalose and glycerol. *Biophys. J.*, **86** (6), 3836–3845.
19. Kalampounias, A., Yannopoulos, S.N., and Papatheodorou, G.N. (2006) A low-frequency Raman study of glassy, supercooled and molten silica and the preservation of the Boson peak in the equilibrium liquid state. *J. Non-Cryst. Solids*, **352** (42-49), 4619–4624.
20. Yannopoulos, S., Andrikopoulos, K., and Ruocco, G. (2006) On the analysis of the vibrational Boson peak and low-energy excitations in glasses. *J. Non-Cryst. Solids*, **352** (42-49), 4541–4551.
21. Perova, T.S., Christensen, D.H., Rasmussen, U., Vij, J.K., and Nielsen, O.F. (1998) Far-infrared spectra of highly viscous liquids: glycerol and triacetin (glycerol triacetate). *Vib. Spectrosc.*, **18**, 149–156.
22. Perova, T., Vij, J., Christensen, D., and Nielsen, O. (1999) A comparison of the far-infrared and low-frequency Raman spectra of glass-forming liquids. *J. Mol. Struct.*, **479** (2-3), 111–122.
23. http://www.physik.uni-augsburg.de/lehrstuehle/exp5/Topics/dismatter/glassfl/ (Accessed 17 May 2014).
24. Kastner, S., Köhler, M., Goncharov, Y., Lunkenheimer, P., and Loidl, A. (2011) High-frequency dynamics of type B glass formers investigated by broadband dielectric spectroscopy. *J. Non-Cryst. Solids*, **357** (2), 510–514.
25. Havriliak, S. and Negami, S. (1967) A complex plane representation of dielectric and mechanical relaxation processes in some polymers. *Polymer*, **8**, 161–210.
26. Cole, K.S. and Cole, R.H. (1941) Dispersion and absorption in dielectrics I. Alternating current characteristics. *J. Chem. Phys.*, **9** (4), 341–351.
27. Angell, C.A., Ngai, K.L., McKenna, G.B., McMillan, P.F., and Martin, S.W. (2000) Relaxation in glassforming liquids and amorphous solids. *J. Appl. Phys.*, **88** (6), 3113–3157.
28. Grzybowska, K., Paluch, M., Grzybowski, A., Wojnarowska, Z., Hawelek, L., Kolodziejczyk, K., and Ngai, K.L. (2010) Molecular dynamics and physical stability of amorphous anti-inflammatory drug: celecoxib. *J. Phys. Chem. B*, **114** (40), 12792–12801.
29. Döß, A., Paluch, M., Sillescu, H., and Hinze, G. (2002) From strong to fragile glass formers: secondary relaxation in polyalcohols. *Phys. Rev. Lett.*, **88** (9), 095701.
30. Cavagna, A. (2009) Supercooled liquids for pedestrians. *Phys. Rep.*, **476** (4-6), 51–124.
31. Götze, W. (2009) *Complex Dynamics of Glass-Forming Liquids*, Oxford University Press, New York.
32. Ngai, K.L. (2003) An extended coupling model description of the evolution of dynamics with time in supercooled liquids. *J. Phys. Condens. Matter*, **15**, S1107–S1125.
33. Ngai, K.L. and Capaccioli, S. (2007) On the relevance of the coupling model to experiments. *J. Phys. Condens. Matter*, **19**, 205114.
34. Ngai, K.L. and Paluch, M. (2004) Classification of secondary relaxation in glass-formers based on dynamic properties. *J. Chem. Phys.*, **120**, 857–873.
35. Cicerone, M.T. and Douglas, J.F. (2012) β-relaxation governs protein stability in sugar-glass matrices. *Soft Matter*, **8** (10), 2983.
36. Taraskin, S., Simdyankin, S., Elliott, S., Neilson, J., and Lo, T. (2006) Universal features of terahertz absorption in disordered materials. *Phys. Rev. Lett.*, **97**, 1–4.
37. Lunkenheimer, P. and Loidl, A. (2003) Response of disordered matter to electromagnetic fields. *Phys. Rev. Lett.*, **91**, 20–23.
38. Schneider, U., Lunkenheimer, P., Brand, R., and Loidl, A. (1998) Dielectric and far-infrared spectroscopy of glycerol. *J. Non-Cryst. Solids*, **235-237** (3), 173–179.
39. Angell, C.A., Yue, Y., Wang, L.-M., Copley, J.R.D., Borick, S., and Mossa, S. (2003) Potential energy, relaxation, vibrational dynamics and the boson peak, of hyperquenched glasses.

J. Phys. Condens. Matter, **15** (11), S1051–S1068.
40. Lunkenheimer, P. and Loidl, A. (2006) High-frequency excitations in glassy crystals. *J. Non-Cryst. Solids*, **352**, 4556–4561.
41. Walther, M., Fischer, B., Schall, M., Helm, H., and Jepsen, P. (2000) Far-infrared vibrational spectra of all-trans, 9-cis and 13-cis retinal measured by THz time-domain spectroscopy. *Chem. Phys. Lett.*, **332** (3-4), 389–395.
42. Walther, M., Plochocka, P., Fischer, B., Helm, H., and Jepsen, P. (2002) Collective vibrational modes in biological molecules investigated by terahertz time-domain spectroscopy. *Biopolymers*, **67** (4-5), 310–313.
43. Taday, P.F., Bradley, I., and Arnone, D.D. (2003) Terahertz pulse spectroscopy of biological materials: L-glutamic acid. *J. Biol. Phys.*, **29** (2-3), 109–115.
44. Upadhya, P., Shen, Y., Davies, A., and Linfield, E. (2003) Terahertz time-domain spectroscopy of glucose and uric acid. *J. Biol. Phys.*, **29** (2-3), 117–121.
45. Taday, P.F., Bradley, I., Arnone, D.D., and Pepper, M. (2003) Using terahertz pulse spectroscopy to study the crystalline structure of a drug: a case study of the polymorphs of ranitidine hydrochloride. *J. Pharm. Sci.*, **92** (4), 831–838.
46. Strachan, C.J., Rades, T., Newnham, D., Gordon, K.C., Pepper, M., and Taday, P.F. (2004) Using terahertz pulsed spectroscopy to study crystallinity of pharmaceutical materials. *Chem. Phys. Lett.*, **390** (1-3), 20–24.
47. Yamaguchi, M., Miyamaru, F., Yamamoto, K., Tani, M., and Hangyo, M. (2005) Terahertz absorption spectra of L-, D-, and DL-alanine and their application to determination of enantiometric composition. *Appl. Phys. Lett.*, **86** (5), 053903.
48. Parrott, E.P.J., Zeitler, J.A., Fris?c?ic?, T., Pepper, M., Jones, W., Day, G.M., and Gladden, L.F. (2009) Testing the sensitivity of terahertz spectroscopy to changes in molecular and supramolecular structure: a study of structurally similar cocrystals. *Cryst. Growth Des.*, **9** (3), 1452–1460.
49. Delaney, S.P., Pan, D., Galella, M., Yin, S.X., and Korter, T.M. (2012) Understanding the origins of conformational disorder in the crystalline polymorphs of irbesartan. *Cryst. Growth Des.*, **12** (10), 5017–5024.
50. Li, R., Zeitler, J.A., Tomerini, D., Parrott, E.P.J., Gladden, L.F., and Day, G.M. (2010) A study into the effect of subtle structural details and disorder on the terahertz spectrum of crystalline benzoic acid. *Phys. Chem. Chem. Phys.*, **12** (20), 5329–5340.
51. Nickel, D.V., Delaney, S.P., Bian, H., Zheng, J., Korter, T.M., and Mittleman, D.M. (2014) Terahertz vibrational modes of the rigid crystal phase of succinonitrile. *J. Phys. Chem. A*, **118** (13), 2442–2446.
52. Tan, N.Y. and Zeitler, J.A. (2015) Probing phase transitions in simvastatin with terahertz time-domain spectroscopy. *Mol. Pharm.*, **12**, 810–815.
53. Siegrist, T., Jost, P., Volker, H., Woda, M., Merkelbach, P., Schlockermann, C., and Wuttig, M. (2011) Disorder-induced localization in crystalline phase-change materials. *Nat. Mater.*, **10** (3), 202–208.
54. Kadlec, F., Kadlec, C., and Kužel, P. (2012) Contrast in terahertz conductivity of phase-change materials. *Solid State Commun.*, **152** (10), 852–855.
55. Wietzke, S., Jansen, C., Jung, T., Reuter, M., Baudrit, B., Bastian, M., Chatterjee, S., and Koch, M. (2009) Terahertz time-domain spectroscopy as a tool to monitor the glass transition in polymers. *Opt. Express*, **17** (21), 19006–19014.
56. Wietzke, S., Jansen, C., Reuter, M., Jung, T., Hehl, J., Kraft, D., Chatterjee, S., Greiner, A., and Koch, M. (2010) Thermomorphological study of the terahertz lattice modes in polyvinylidene fluoride and high-density polyethylene. *Appl. Phys. Lett.*, **97** (2), 022901.
57. Wietzke, S., Reuter, M., Nestle, N., Klimov, E., Zadok, U., Fischer, B.M., and Koch, M. (2011) Analyzing morphology and thermal history of polybutylene terephthalate by THz

time-domain spectroscopy. *J. Infrared Millimeter Terahertz Waves*, **32** (7), 952–959.
58. Wietzke, S., Jansen, C., Reuter, M., Jung, T., Kraft, D., Chatterjee, S., Fischer, B., and Koch, M. (2011) Terahertz spectroscopy on polymers: a review of morphological studies. *J. Mol. Struct.*, **1006** (1-3), 41–51.
59. Hoshina, H., Ishii, S., Morisawa, Y., Sato, H., Noda, I., Ozaki, Y., and Otani, C. (2012) Isothermal crystallization of poly(3-hydroxybutyrate) studied by terahertz two-dimensional correlation spectroscopy. *Appl. Phys. Lett.*, **100** (1), 011907.
60. Maradudin, A.A. and Wallis, R.F. (1961) Lattice anharmonicicty and optical absorption in polar crystals. II. Classical treatment in the linear approximation. *Phys. Rev.*, **123** (3), 777–789.
61. Parrott, E.P.J., Zeitler, J.A., and Gladden, L.F. (2012) Terahertz spectroscopy of inorganic glasses and carbon nanotubes. *Spectrosc. Prop. Inorg. Organomet. Compd.*, **42**, 157–183.
62. Parrott, E., Zeitler, J., Simon, G., Hehlen, B., Gladden, L., Taraskin, S., and Elliott, S. (2010) Atomic charge distribution in sodosilicate glasses from terahertz time-domain spectroscopy. *Phys. Rev. B*, **82** (14), 3–6.
63. Naftaly, M. and Miles, R. (2005) Terahertz time-domain spectroscopy: a new tool for the study of glasses in the far infrared. *J. Non-Cryst. Solids*, **351** (40-42), 3341–3346.
64. Naftaly, M. and Miles, R.E. (2007) Terahertz time-domain spectroscopy of silicate glasses and the relationship to material properties. *J. Appl. Phys.*, **102** (4), 043517.
65. Zalkovskij, M., Bisgaard, C., Novitsky, A., Malureanu, R., Savastru, D., Popescu, A., Jepsen, P., and Lavrinenko, A.V. (2012) Ultrabroadband terahertz spectroscopy of chalcogenide glasses. *Appl. Phys. Lett.*, **100** (3), 031901.
66. Ro?nne, C., Astrand, P.-O., and Soren, R.K. (1999) THz spectroscopy of liquid H2O and D2O. *Phys. Rev. Lett.*, **82** (14), 2888–2891.
67. Turton, D.A., Hunger, J., Stoppa, A., Thoman, A., Candelaresi, M., Hefter, G., Walther, M., Buchner, R., and Wynne, K. (2011) Rattling the cage: micro- to mesoscopic structure in liquids as simple as argon and as complicated as water. *J. Mol. Liq.*, **159**, 2–8.
68. Barthel, J., Bachhuber, K., Buchner, R., and Hetzenauer, H. (1990) Dielectric spectra of some common solvents in the microwave region. Water and lower alcohols. *Chem. Phys. Lett.*, **165**, 369–373.
69. Barthel, J.M.G. and Buchner, R. (1991) High frequency permittivity and its use in the investigation of solution properties. *Pure Appl. Chem.*, **63**, 1473–1482.
70. Keutsch, F.N. and Saykally, R.J. (2001) Water clusters: untangling the mysteries of the liquid, one molecule at a time. *Proc. Natl. Acad. Sci. U.S.A.*, **98**, 10533–10540.
71. Ronne, C., Thrane, L., Astrand, P.-O., Wallqvist, A., Mikkelsen, K.V., and Keiding, S.R. (1997) Investigation of the temperature dependence of dielectric relaxation in liquid water by THz reflection spectroscopy and molecular dynamics simulation. *J. Chem. Phys.*, **107** (14), 5319–5331.
72. Yada, H., Nagai, M., and Tanaka, K. (2008) Origin of the fast relaxation component of water and heavy water revealed by terahertz time-domain attenuated total reflection spectroscopy. *Chem. Phys. Lett.*, **464** (4-6), 166–170.
73. Kaatze, U. (1989) Complex permittivity of water as a function of frequency and temperature. *J. Chem. Eng. Data*, **34** (4), 371–374.
74. Alison, J.M. and Sheppard, R.J. (1991) A precision waveguide system for the measurement of complex permittivity of lossy liquids and solid tissues in the frequency range 29 GHz to 90 GHz-III. The liquid system for 57 to 82 GHz: an investigation into water and formamide. *Meas. Sci. Technol.*, **2** (10), 975–979.
75. Afsar, M.N. and Hasted, J.B. (1977) Measurements of the optical constants of liquid H20 and D20 between 6 and

450 cm-1. *J. Opt. Soc. Am.*, **67** (7), 902–904.

76. Tielrooij, K., Timmer, R.L.A., Bakker, H.J., and Bonn, M. (2009) Structure dynamics of the proton in liquid water probed with terahertz time-domain spectroscopy. *Phys. Rev. Lett.*, **102**, 198303–198304.
77. Venables, D. and Schmuttenmaer, C. (2000) Spectroscopy and dynamics of mixtures of water with acetone, acetonitrile, and methanol. *J. Chem. Phys.*, **113**, 11222–11236.
78. Schmidt, D.A., Birer, O., Funkner, S., Born, B.P., Gnanasekaran, R., Schwaab, G.W., Leitner, D.M., and Havenith, M. (2009) Rattling in the cage: ions as probes of sub-picosecond water network dynamics. *J. Am. Chem. Soc.*, **131**, 18512–18517.
79. Venables, D. and Schmuttenmaer, C. (1998) Far-infrared spectra and associated dynamics in acetonitrile-water mixtures measured with femtosecond THz pulse spectroscopy. *J. Chem. Phys.*, **108**, 4935–4944.
80. Dutta, P. and Tominaga, K. (2009) Obtaining low-frequency spectra of acetone dissolved in cyclohexane by terahertz time-domain spectroscopy. *J. Phys. Chem. A*, **113** (29), 8235–8242.
81. Dutta, P. and Tominaga, K. (2009) Terahertz time-domain spectroscopic study of the low-frequency spectra of nitrobenzene in alkanes. *J. Mol. Liq.*, **147** (1-2), 45–51.
82. Dutta, P. and Tominaga, K. (2009) Dependence of low frequency spectra on solute and solvent in solutions studied by terahertz time-domain spectroscopy. *Mol. Phys.*, **107** (18), 1845–1854.
83. Tielrooij, K.-J., Hunger, J., Buchner, R., Bonn, M., and Bakker, H.J. (2010) Influence of concentration and temperature on the dynamics of water in the hydrophobic hydration shell of tetramethylurea. *J. Am. Chem. Soc.*, **132**, 15671–15678.
84. Yomogida, Y., Sato, Y., Yamakawa, K., Nozaki, R., Mishina, T., and Nakahara, J. (2010) Comparative dielectric study of pentanol isomers with terahertz time-domain spectroscopy. *J. Mol. Struct.*, **970**, 171–176.
85. Yomogida, Y., Sato, Y., Nozaki, R., Mishina, T., and Nakahara, J. (2010) Comparative study of boson peak in normal and secondary alcohols with terahertz time-domain spectroscopy. *Physica B*, **405** (9), 2208–2212.
86. Yamaguchi, S., Tominaga, K., and Saito, S. (2011) Intermolecular vibrational mode of the benzoic acid dimer in solution observed by terahertz time-domain spectroscopy. *Phys. Chem. Chem. Phys.*, **13** (32), 14742–14749.
87. Li, R., Agostino, C.D., McGregor, J., Mantle, M.D., Zeitler, J.A., and Gladden, L.F. (2014) Mesoscopic structuring and dynamics of alcohol/water solutions probed by terahertz time-domain spectroscopy and pulsed field gradient nuclear magnetic resonance. *J. Phys. Chem. B*, **118** (34), 10156–10166.
88. McGregor, J., Li, R., Zeitler, J.A., D'Agostino, C., Collins, J.H.P., Mantle, M.D., Manyar, H., Holbrey, J.D., Falkowska, M., Youngs, T.G.A., Hardacre, C., Stitt, E.H., and Gladden, L.F. (2015) Structure and dynamics of aqueous 2-propanol: a THz-TDS, NMR and neutron diffraction study. *Phys. Chem. Chem. Phys.*, **17** (45), 30481–30491.
89. Tan, N.Y., Li, R., Bräuer, P., D'Agostino, C., Gladden, L.F., and Zeitler, J.A. (2015) Probing hydrogen-bonding in binary liquid mixtures with terahertz time-domain spectroscopy: a comparison of Debye and absorption analysis. *Phys. Chem. Chem. Phys.*, **17**, 5999–6008.
90. Kindt, J.T. and Schmuttenmaer, C.A. (1996) Far-infrared dielectric properties of polar liquids probed by femtosecond terahertz pulse spectroscopy. *J. Phys. Chem.*, **100** (24), 10373–10379.
91. Möller, U., Cooke, D.G., Tanaka, K., and Jepsen, P.U. (2009) Terahertz reflection spectroscopy of Debye relaxation in polar liquids [Invited]. *J. Opt. Soc. Am. B*, **26** (9), A113.
92. Jepsen, P.U., Møller, U., and Merbold, H. (2007) Investigation of aqueous alcohol and sugar solutions with reflection

terahertz time-domain spectroscopy. *Opt. Express*, **15** (22), 14717–14737.
93. Asaki, M.L.T., Redondo, A., Zawodzinski, T.A., and Taylor, A.J. (2002) Dielectric relaxation of electrolyte solutions using terahertz transmission spectroscopy. *J. Chem. Phys.*, **116** (19), 8469–8482.
94. Koeberg, M., Wu, C.-C., Kim, D., and Bonn, M. (2007) THz dielectric relaxation of ionic liquid:water mixtures. *Chem. Phys. Lett.*, **439** (1-3), 60–64.
95. Yomogida, Y., Sato, Y., Yamakawa, K., Nozaki, R., Mishina, T., and Nakahara, J. (2010) Comparative dielectric study of pentanol isomers with terahertz time-domain spectroscopy. *J. Mol. Struct.*, **970** (1-3), 171–176.
96. Sibik, J., Elliott, S.R., and Zeitler, J.A. (2014) Thermal decoupling of molecular-relaxation processes from the vibrational density of states at terahertz frequencies in supercooled hydrogen-bonded liquids. *J. Phys. Chem. Lett.*, **5** (11), 1968–1972.
97. Capaccioli, S., Ngai, K.L., Thayyil, M.S., and Prevosto, D. (2015) Coupling of caged molecule dynamics to JG β-relaxation: I. *J. Phys. Chem. B*, **119**, 8800–8808.
98. Ngai, K.L., Capaccioli, S., Prevosto, D., and Wang, L.-M. (2015) Coupling of caged molecule dynamics to JG β-relaxation II: polymers. *J. Phys. Chem. B*, **119** (38), 12502–12518.
99. Ngai, K.L., Capaccioli, S., Prevosto, D., and Wang, L.-M. (2015) Coupling of caged molecule dynamics to JG β-relaxation III:van der Waals glasses. *J. Phys. Chem. B*, **119**, 12519–12525.
100. Yomogida, Y., Sato, Y., Nozaki, R., Mishina, T., and Nakahara, J. (2010) Comparative dielectric study of monohydric alcohols with terahertz time-domain spectroscopy. *J. Mol. Struct.*, **981** (1-3), 173–178.
101. Sibik, J., Sargent, M.J., Franklin, M., and Zeitler, J.A. (2014) Crystallization and phase changes in paracetamol from the amorphous solid to the liquid phase. *Mol. Pharm.*, **11** (4), 1326–1334.
102. Strachan, C.J., Taday, P.F., Newnham, Da., Gordon, K.C., Zeitler, J.A., Pepper, M., and Rades, T. (2005) Using terahertz pulsed spectroscopy to quantify pharmaceutical polymorphism and crystallinity. *J. Pharm. Sci.*, **94** (4), 837–846.
103. Zeitler, J.A., Newnham, D.A., Taday, P.F., Threlfall, T.L., Lancaster, R.W., Berg, R.W., Strachan, C.J., Pepper, M., Gordon, K.C., and Rades, T. (2006) Characterization of temperature-induced phase transitions in five polymorphic forms of sulfathiazole by terahertz pulsed spectroscopy and differential scanning calorimetry. *J. Pharm. Sci.*, **95**, 2486–2498.
104. Zeitler, J.A., Kogermann, K., Rantanen, J., Rades, T., Taday, P.F., Pepper, M., Aaltonen, J., and Strachan, C.J. (2007) Drug hydrate systems and dehydration processes studied by terahertz pulsed spectroscopy. *Int. J. Pharm.*, **334** (1-2), 78–84.
105. Liu, H.-B., Chen, Y., and Zhang, X.-C. (2007) Characterization of anhydrous and hydrated pharmaceutical materials with THz time-domain spectroscopy. *J. Pharm. Sci.*, **96** (4), 927–934.
106. Nguyen, K.L., Friscić, T., Day, G.M., Gladden, L.F., and Jones, W. (2007) Terahertz time-domain spectroscopy and the quantitative monitoring of mechanochemical cocrystal formation. *Nat. Mater.*, **6** (3), 206–209.
107. Shah, B., Kakumanu, V.K., and Bansal, A.K. (2006) Analytical techniques for quantification of amorphous/crystalline phases in pharmaceutical solids. *J. Pharm. Sci.*, **95** (8), 1641–1665.
108. Otsuka, M., Nishizawa, J.-I., Fukura, N., and Sasaki, T. (2012) Characterization of poly-amorphous indomethacin by terahertz spectroscopy. *J. Infrared Millimeter Terahertz Waves*, **33** (9), 953–962.
109. Takeuchi, I., Tomoda, K., Nakajima, T., Terada, H., Kuroda, H., and Makino, K. (2012) Estimation of crystallinity of trehalose dihydrate microspheres by usage of terahertz time-domain spectroscopy. *J. Pharm. Sci.*, **101** (9), 3465–3472.
110. Darkwah, J., Smith, G., Ermolina, I., and Mueller-Holtz, M. (2013) A THz spectroscopy method for quantifying

the degree of crystallinity in freeze-dried gelatin/amino acid mixtures: an application for the development of rapidly disintegrating tablets. *Int. J. Pharm.*, **455** (1-2), 357–364.

111. Ermolina, I., Darkwah, J., and Smith, G. (2014) Characterisation of crystalline-amorphous blends of sucrose with terahertz-pulsed spectroscopy: the development of a prediction technique for estimating the degree of crystallinity with partial least squares regression. *AAPS PharmSciTech*, **15** (2), 253–260.
112. Vieira, F.S. and Pasquini, C. (2014) Determination of cellulose crystallinity by terahertz-time domain spectroscopy. *Anal. Chem.*, **86** (8), 3780–3786.
113. Takebe, G., Kawada, Y., Akiyama, K., Takahashi, H., Takamoto, H., and Hiramatsu, M. (2013) Evaluation of drug crystallinity in aqueous suspension using terahertz time-domain attenuated total reflection spectroscopy. *J. Pharm. Sci.*, **102** (11), 4065–4071.
114. Zeitler, J.A., Taday, P.F., Pepper, M., and Rades, T. (2007) Relaxation and crystallization of amorphous carbamazepine studied by terahertz pulsed spectroscopy. *J. Pharm. Sci.*, **96** (10), 2703–2709.
115. McIntosh, A.I., Yang, B., Goldup, S.M., Watkinson, M., and Donnan, R.S. (2013) Crystallization of amorphous lactose at high humidity studied by terahertz time domain spectroscopy. *Chem. Phys. Lett.*, **558**, 104–108.
116. Chang, L.L., Shepherd, D., Sun, J., Ouellette, D., Grant, K.L., Tang, X.C., and Pikal, M.J. (2005) Mechanism of protein stabilization by sugars during freeze-drying and storage: native structure preservation, specific interaction, and/or immobilization in a glassy matrix? *J. Pharm. Sci.*, **94** (7), 1427–1444.
117. Bhattacharya, S. and Suryanarayanan, R. (2009) Local mobility in amorphous pharmaceuticals–characterization and implications on stability. *J. Pharm. Sci.*, **98** (9), 2935–2953.
118. Sibik, J., Löbmann, K., Rades, T., and Zeitler, J.A. (2015) Predicting crystallization of amorphous drugs with terahertz spectroscopy. *Mol. Pharm.*, **12**, 3062–3068.
119. Marsac, P.J., Konno, H., and Taylor, L.S. (2006) A comparison of the physical stability of amorphous felodipine and nifedipine systems. *Pharm. Res.*, **23**, 2306–2316.
120. Zhou, D., Zhang, G.G.Z., Law, D., Grant, D.J.W., and Schmitt, E.A. (2002) Physical stability of amorphous pharmaceuticals: importance of configurational thermodynamic quantities and molecular mobility. *J. Pharm. Sci.*, **91**, 1863–1872.
121. Graeser, K.A., Patterson, J.E., Zeitler, J.A., Gordon, K.C., and Rades, T. (2009) Correlating thermodynamic and kinetic parameters with amorphous stability. *Eur. J. Pharm. Sci.*, **37** (3–4), 492–498.
122. Andronis, V. and Zografi, G. (1997) Molecular mobility of supercooled amorphous indomethacin, determined by dynamic mechanical analysis. *Pharm. Res.*, **14** (4), 410–414.
123. Di Martino, P., PAlmieri, G.F., and Martelli, S. (2000) Molecular mobility of the paracetamol amorphous form. *Chem. Pharm. Bull.*, **48** (8), 1105–1108.
124. Kaushal, A.M. and Bansal, A.K. (2008) Thermodynamic behavior of glassy state of structurally related compounds. *Eur. J. Pharm. Biopharm.*, **69**, 1067–1076.
125. Ambike, A.A., Mahadik, K.R., and Paradkar, A. (2005) Physico-chemical characterization and stability study of glassy simvastatin. *Drug Dev. Ind. Pharm.*, **31**, 895–899.
126. Fukuoka, E., Makita, M., and Nakamura, Y. (1989) Glassy state of pharmaceuticals. IV. Studies on glassy pharmaceuticals by thermomechanical analysis. *Chem. Pharm. Bull.*, **37** (10), 2782–2785.
127. Descamps, M. and Dudognon, E. (2014) Crystallization from the amorphous state: nucleation-growth decoupling, polymorphism interplay, and the role of interfaces. *J. Pharm. Sci.*, **103**, 2615–2628.
128. Hancock, B.C. and Zografi, G. (1997) Characteristics and significance of the amorphous state in pharmaceutical systems. *J. Pharm. Sci.*, **86** (1), 1–12.

129. Yu, L. (2001) Amorphous pharmaceutical solids: preparation, characterization and stabilization. *Adv. Drug Delivery Rev.*, **48**, 27–42.
130. Kauzmann, W. (1948) The nature of the glassy state and the behavior of liquids at low temperatures. *Chem. Rev.*, **43**, 219–256.
131. Hancock, B.C., Shamblin, S.L., and Zografi, G. (1995) Molecular mobility of amorphous pharmaceutical solids below their glass transition temperatures. *Pharm. Res.*, **12** (6), 799–806.
132. Okamoto, N. and Oguni, M. (1996) Discovery of crystal nucleation proceeding much below the glass transition temperature in a supercooled liquid. *Solid State Commun.*, **99** (1), 53–56.
133. Hikima, T., Hanaya, M., and Oguni, M. (1999) Microscopic observation of a peculiar crystallization in the glass transition region and β-process as potentially controlling the growth rate in triphenylethylene. *J. Mol. Struct.*, **479**, 245–250.
134. Sokolov, A., Kisliuk, A., Novikov, V., and Ngai, K. (2001) Observation of constant loss in fast relaxation spectra of polymers. *Phys. Rev. B*, **63**, 172204.
135. Ngai, K.L. (2004) Why the fast relaxation in the picosecond to nanosecond time range can sense the glass transition. *Philos. Mag.*, **84**, 1341–1353.
136. Mensink, M.A., Sibik, J., van der Voort Maarschalk, F.H.W.K., Hinrichs, K.L.J., and Zeitler, J.A. (2015) Terahertz time domain spectroscopy as a tool for measuring intermolecular protein-sugar interactions in the solid state, in preparation.
137. Dai, J., Liu, J., and Zhang, X.-C. (2011) Terahertz wave air photonics: terahertz wave generation and detection with laser-induced gas plasma. *IEEE J. Sel. Top. Quantum Electron.*, **17** (1), 183–190.
138. Turton, D.A. and Wynne, K. (2009) Universal nonexponential relaxation: complex dynamics in simple liquids. *J. Chem. Phys.*, **131** (20), 201101.

14
Study of Disorder by Solid-State NMR Spectroscopy

Marco Geppi, Silvia Borsacchi, and Elisa Carignani

14.1
Introduction

Since the discovery of the nuclear magnetic resonance (NMR) phenomenon in 1945, three main methodologies have been developed: liquid-state NMR spectroscopy, magnetic resonance imaging (MRI), and solid-state NMR spectroscopy (SSNMR). The first is nowadays one of the most used characterization techniques in chemistry and pharmaceutics, while MRI is a very important imaging technique in medical diagnostics and research. SSNMR is at present one of the most powerful techniques to investigate solid systems, including small organic and inorganic molecules, polymers, biomacromolecules, advanced materials, and so on.

SSNMR is widely used in pharmaceutics, and it is particularly suitable for the investigation of disorder in drugs. In the development of this chapter we will try to explain why, but some key concepts can be anticipated.

First of all, SSNMR is a nondestructive technique equally suitable to investigate crystalline and amorphous phases, hard and soft matter and samples in different forms (powders, films, fibers, etc.). Moreover, it can give very detailed information about structure and dynamics on very broad spatial and timescales, respectively.

On one side, this allows "static" (associated with structure) and "dynamic" (due to molecular motions) disorder to be clearly distinguished and individually investigated. On the other side, both static and dynamic disorder can be studied with different spatial or time "magnifying glasses."

Different types of structural order/disorder can be dealt with on spatial scales ranging from 1Å to about 100 nm: molecular (conformational, positional), supramolecular (in molecular complexes or adducts), intraphase (either amorphous or crystalline), interphase (interfaces), and supraphase (distribution of domains of different phases). Some consequent advantages of SSNMR are, for instance, the possibility of detecting nanocrystalline domains ("transparent" to traditional calorimetric techniques when the domain dimensions are smaller than about 40 nm) and the capability to ascertain the homogeneity/inhomogeneity of a heterophasic material on the nanometer scale (particularly useful when electron microscopic techniques cannot be applied).

Disordered Pharmaceutical Materials, First Edition. Edited by Marc Descamps.

Furthermore, molecular motions (in a broad sense, of the whole molecule, of molecular fragments, or of collections of molecules) can be studied, often individually and with a great detail, over 12–13 orders of magnitude broad timescale, ranging from picoseconds to seconds. The combination of this accessible broad range of characteristic times with the possibility of detecting and quantitatively describing each specific motion makes SSNMR unique in the characterization of dynamic processes. In fact, although, for instance, dielectric spectroscopy can access an even broader time range, only motions changing the orientation of the molecular electric dipole moment can be detected, and the identification and characterization of individual motions of fragments is usually not possible. On the other hand, the characteristic motional times accessible by X-ray diffraction are much shorter compared to SSNMR, and therefore this technique provides a sort of instantaneous "picture" of the system, often preventing the possibility of distinguishing static from dynamic disorder, that is, static distributions of frozen conformations from the presence of interconformational motions.

The organization of this chapter is as follows: First, an elementary introduction to the basics of SSNMR is given for nonspecialist. The two subsequent sections will describe how static and dynamic disorder in pharmaceuticals can be studied by means of SSNMR. Finally, a case study of Na-ibuprofen, where most of the concepts previously presented have been experimentally exploited, will be illustrated in detail.

14.2 Basics of Solid-State NMR

With the aim of facilitating the comprehension of this chapter, some basic concepts of solid-state NMR are briefly described here, but for a complete description of the theory and experiments the reader should refer to fundamental literature [1–4].

SSNMR, even though arising from the same fundamental phenomenon, presents substantial differences with respect to its more popular counterpart in the liquid state. These differences, which have a theoretical origin, also determine important practical differences in terms of experimental apparatuses, design of the experiments, and the information obtainable.

The basic NMR phenomenon arises from the Zeeman interaction between the magnetic moment of nuclei characterized by a non-null spin ($\mathbf{I}$) and an external static magnetic field ($\mathbf{B}_0$), which is able to remove the natural degeneracy of the nuclear spin states. Transitions between adjacent spin states can be induced by the interaction with the oscillating magnetic field associated with a suitable radiation, the frequency of which falls in the radio frequency (rf) region of the electromagnetic spectrum and is commonly referred to as the Larmor frequency. At a given static magnetic field $\mathbf{B}_0$, the Larmor frequency uniquely depends on the gyromagnetic ratio (γ_I), which is a property specific to each nuclear species. Using a classical description, for an ensemble of like nuclei, the equilibrium state consists in the alignment of the nuclear magnetization (vectorial sum of the individual nuclear magnetic moments) with the direction of the external magnetic field (conventionally indicated as the z-direction), that is, in the maximum longitudinal magnetization (parallel to $\mathbf{B}_0$), and zero transverse magnetization (perpendicular to $\mathbf{B}_0$). The basic NMR experiment consists of the so-called 90° rf pulse, that is an rf pulse at the Larmor frequency that flips the nuclear magnetization by 90° from the direction of the external magnetic field to the xy plane (direct excitation or DE experiment). When the rf pulse is turned off, the evolution of the transverse magnetization generates a time-dependent electric current in a conducting coil surrounding the sample, called free induction decay (FID), representing the NMR signal. After Fourier transformation, the time-dependent FID becomes the NMR spectrum in the frequency domain. Different nuclear species, characterized by different γ_I values, have different Larmor frequencies, and therefore separate spectra are always collected.

The enormous power of NMR spectroscopy for structural studies arises from the extremely high sensitivity of the NMR frequency to the very local environment of a nucleus. This is due to the existence of the so-called internal interactions between the nuclear magnetic moment and local fields, which are much weaker than the Zeeman interaction. The internal interactions allow nuclei of the same species but experiencing different environments in terms of chemical, structural, and dynamic properties to be clearly distinguished and characterized. The main internal interactions in diamagnetic samples are shielding, dipolar, and indirect (or scalar, or J) couplings and, for nuclei with spin $I > 1/2$ (quadrupolar nuclei), quadrupolar coupling. In this chapter we will not deal with the case of quadrupolar nuclei, which are extensively studied by SSNMR but need a completely separate treatment.

Briefly, the shielding interaction is the indirect magnetic coupling between the external magnetic field and the nuclear spin through the involvement of the surrounding electrons, which partially shield $\mathbf{B}_0$, modifying the magnetic field really felt by the nucleus. This interaction generates detectable variations in the spin resonance frequency, commonly referred to as the chemical shift. Chemical shift is extremely sensitive to the chemical and structural properties of the very local environment of the nucleus.

Dipolar coupling is the direct, through-space magnetic interaction between nuclear spins. This interaction can be either homo- or heteronuclear, intra- or intermolecular. It is inversely proportional to the third power of the distance between the coupled nuclei, and therefore it can provide important structural information. On the other hand, the indirect coupling is a magnetic interaction among nuclear spins occurring through bond electrons, and thus it is an important probe of the molecular skeleton.

All the internal interactions contain an anisotropic and an isotropic part, differing because the former depends on the orientation of the molecular fragment involved in the interaction with respect to the direction of the external magnetic field, while the latter is independent of it. The isotropic part is the only one contributing to the NMR spectrum in the liquid state: indeed, the fast and isotropic molecular motions average out the anisotropic parts of the interactions. Since the isotropic term is nonzero only for the shielding and the indirect coupling, NMR spectra in liquids are only determined by these two interactions and are characterized by narrow signals with various multiplicities.

On the contrary, in solids, fast and isotropic molecular motions are not present, and therefore all the interactions contribute to the spectrum with both their isotropic and anisotropic parts. Considering that in common solid samples all possible molecular orientations with respect to the external static magnetic field are statistically present, the NMR signal of each nuclear site results in a broad distribution of resonance frequencies (powder pattern). Since several interactions are experienced by each nucleus, and nuclei in many different molecular environments are usually present, the NMR spectrum of a solid sample is a very complex superposition of powder patterns, leading to very poor spectral resolution, which has been the limiting factor in the early development of SSNMR.

Very important parameters affecting SSNMR experiments are, as in the case of liquids, nuclear relaxation times, which are discussed in detail in Section 14.4. Three main relaxation times must be considered: the transverse or spin–spin relaxation time T_2, the longitudinal or spin–lattice relaxation time T_1, and the spin–lattice relaxation time in the rotating frame $T_{1\rho}$. They all describe the recovery of an equilibrium state after a perturbation, determined, for instance, by rf irradiation. T_2 describes the decay to zero of the transverse magnetization generated by suitable rf pulses, while T_1 describes the recovery of the equilibrium value of the longitudinal magnetization. $T_{1\rho}$ is the characteristic time of decay of the magnetization kept aligned to a given direction by an rf "spin-lock" pulse. In all cases, $T_1 \geq T_2$, but while in liquids the two relaxation times are of the same order of magnitude, in liquids, $T_1 \gg T_2$ in solids; for instance, for ^{1}H nuclei T_2 typically ranges from tens of microseconds to tens of milliseconds whereas T_1 is typically larger than 100 ms.

Before introducing the basic experimental SSNMR approaches, it is useful to make a distinction between abundant and rare nuclei. While dealing with SSNMR, it is convenient to consider as abundant nuclei all those nuclei not only characterized by a high isotopic abundance but also present in the sample in a high concentration: the typical example of this category in pharmaceuticals is given by ^{1}H nuclei. On the contrary, nuclei with a low isotopic abundance or present in low concentration in a sample can be considered as rare. From an SSNMR standpoint, for rare nuclei the homonuclear interactions can be mostly neglected. In solid pharmaceuticals, the most commonly investigated rare nuclei are ^{13}C and, to a minor extent, ^{31}P, ^{19}F, and ^{15}N. In SSNMR, rare and abundant nuclei (from now on exemplified by ^{13}C and ^{1}H, respectively) require quite different theoretical descriptions and experimental approaches.

The most important internal interactions for rare nuclei are the shielding and the couplings (dipolar and indirect) with abundant nuclei. As already noted, because of the anisotropy of the internal interactions, the ^{13}C SSNMR spectrum of an organic powder will completely lack resolution. In the mid-1970s, two fundamental techniques were combined, which made possible the recording of high-resolution SSNMR spectra of rare nuclei: magic angle spinning (MAS) and high-power decoupling (HPD). MAS involves a fast macroscopic rotation of the sample about an axis oriented at the so-called magic angle (54.74°) with respect to the external magnetic field direction. It requires special experimental apparatus, nowadays routinely present in NMR spectrometers equipped for solid-state studies. MAS is a brilliant method that reproduces, on solid samples, the effect of the fast Brownian motions in the liquids. Without entering into theoretical details, MAS is in principle able to remove (when spinning at an infinite frequency) the anisotropy of all the internal interactions. In practice, the achievable MAS frequencies are limited (commonly ranging from 3 to 30 kHz, even though the value of 110 kHz has been reached), thus often causing a substantial or complete reduction of the anisotropy only for the shielding interaction: if the sample is spun at a frequency much larger than the shielding anisotropy expressed as static line width, only a narrow signal resonating at the isotropic chemical shift value is observed. If the MAS frequency is smaller, on both sides of the isotropic signal "copies" of it, with different intensities, appear at integer multiples of the MAS frequency. These signals (called spinning sidebands) can be easily recognized since, differently from the isotropic signal, they change position by changing the MAS frequency.

The other important interaction that prevents ^{13}C SSNMR spectra to be highly resolved is the dipolar coupling with ^{1}H nuclei. This can be completely removed by HPD, which, in its simplest form, consists of continuous rf irradiation at the ^{1}H Larmor frequency during the acquisition of ^{13}C FID. Nowadays, many different improved HPD pulse sequences are available on common SSNMR spectrometers (TPPM, SPINAL, etc.).

By combining MAS and HPD, high-resolution ^{13}C spectra of solids can be quite easily obtained. Even though the interactions anisotropy is removed, the high-resolution SSNMR spectrum preserves important differences with respect to the corresponding solution NMR spectrum: the spectrum of the sample is recorded directly in the solid state, and thus it is representative of the crystalline or amorphous state. The isotropic chemical shift is also extremely sensitive to conformations as well as intra and intermolecular interactions. Moreover, the anisotropy of the interactions can be suitably exploited even in high-resolution experiments for obtaining detailed structural or dynamic information. All these aspects will be exemplified in the next sections.

A last aspect that is worth considering in the case of rare nuclei is the intrinsic low sensitivity of NMR (due to both low concentration of the nuclei and unfavorable Boltzmann population distribution), which requires the same experiment ("transient") to be repeated many times and the collected FIDs to be summed up to increase the signal-to-noise ratio. For example, in a ^{13}C DE experiment, the time between two subsequent transients is determined by the time necessary to recover the equilibrium conditions and, therefore, by ^{13}C T_1, which is typically very long for solids (up to minutes). This often implies very long experimental times to record a ^{13}C DE spectrum with a good signal-to-noise ratio. A brilliant solution

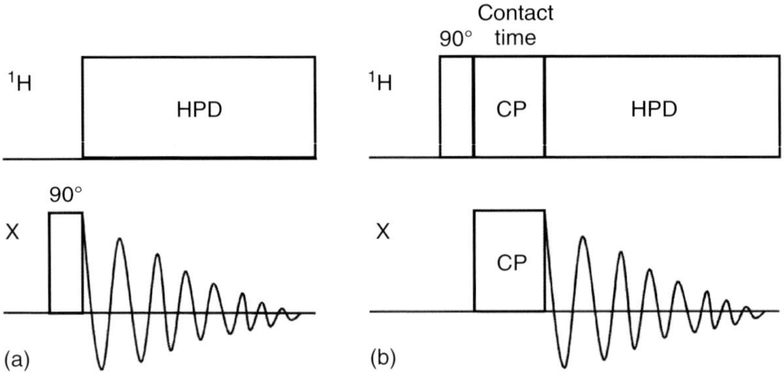

Figure 14.1 Schemes of (a) DE and (b) CP pulse sequences for recording a spectrum of a rare nucleus X, with HPD from ^{1}H nuclei. Rectangles indicate rf pulses, while the oscillating curve represents the FID.

to this problem is provided by a technique called cross polarization (CP), which is commonly used in SSNMR of rare nuclei. CP is a pulse sequence that, if suitably set, allows transverse magnetization of ^{13}C nuclei to be built by transferring the magnetization of ^{1}H nuclei coupled to them through dipolar interaction, previously excited by a 90° rf pulse, rather than by direct ^{13}C irradiation. The magnetization transfer occurs by suitable simultaneous rf irradiation (Hartmann–Hahn conditions) of ^{13}C and ^{1}H nuclei during the so-called contact time. In Figure 14.1, both DE and CP pulse sequence schemes are illustrated.

The use of CP provides a double sensitivity gain. For each transient, there is a gain equal to the ratio between the gyromagnetic ratios of the abundant and the rare nuclei (this effect is related to the Boltzmann population distribution). Moreover, in CP the time between two subsequent transients is determined by ^{1}H T_1, which is usually much shorter than ^{13}C T_1, thus allowing a larger number of transients to be recorded in the same time with respect to the ^{13}C DE experiment. The drawback of this technique is that CP spectra are not quantitative; that is, the areas underlying the signals are not proportional to the number of nuclei giving rise to the signals. This is due to the fact that the intensity of ^{13}C signals, created by magnetization transfer from ^{1}H nuclei, depends on the strength of the ^{1}H–^{13}C dipolar couplings (which in turn depend on the chemical and geometrical structure and dynamics). Quantitation issues will be treated in the following section.

On the other hand, abundant nuclei behave quite differently: homonuclear dipolar couplings are the predominant interactions and they have a very heavy impact on the spectra. Just to give a semiquantitative idea, an ^{1}H SSNMR spectrum is typically composed of unresolved signals giving a single 50-kHz broad line, while a complete ^{1}H NMR spectrum of a liquid sample, composed of tens of resolved signals, usually spans over 10–12 ppm (4 kHz at a typical Larmor frequency of 400 MHz). Obtaining high-resolution solid state spectra of abundant nuclei is more complex than for rare nuclei, as the homonuclear dipolar couplings are intrinsically more difficult to be removed either by MAS or rf decoupling. The use of both very fast MAS and complex homonuclear decoupling schemes (such as LG, DUMBO, etc.) has allowed big steps forward: in many cases, high-resolution ^{1}H spectra are obtained, even though some drawbacks can be the requirement of very expensive apparatuses and suitable chemical shift scaling, strong heating of the sample, and reduced sensitivity.

It must be mentioned that the study of ^{1}H nuclei in the solid state, especially the measurement of relaxation times, can also be carried out by sacrificing high resolution and working directly in the time domain (without performing Fourier transform). This will be described in more detail in Section 14.4.

An additional tool routinely available is constituted by two-dimensional (2D) experiments. The exploitation of a second dimension can significantly enlarge the information obtainable. 2D NMR spectra are, in general, obtained by applying pulse sequences in which the signal is made to depend on two time variables, and then subjected to double Fourier transformation. 2D experiments can be roughly divided into two categories: separation and correlation experiments. In separation experiments, the isotropic spectrum is recorded in one dimension, while the other is exploited for reintroducing an anisotropic interaction. In particular, in the second dimension, the 1D spectrum relative to a specific anisotropic interaction for each individual chemically distinguishable site is separately observed. Thus it is possible, for example, to characterize the chemical shift anisotropy of different carbon nuclei as well as the homonuclear dipolar interaction of ^{1}H nuclei coupled with specific rare nuclei. In correlation experiments, instead, isotropic spectra are recorded in both dimensions and cross-peaks are observed between nuclei related by some specific interactions (hetero- or homonuclear dipolar couplings, J couplings, etc.).

14.3 Static Disorder

Among the nuclear properties exploitable by SSNMR, the isotropic chemical shift constitutes the fundamental and most powerful tool for recognizing the static disorder occurring in solid systems, in particular in pharmaceuticals. Isotropic chemical shift is extremely sensitive to the very local environment of the observed nucleus, being determined by the "average" electronic distribution around it. The electronic distribution is, in turn, strongly dependent not only on the chemistry but also on the molecular 3D structure, as well as on the intermolecular interactions. The strong sensitivity of the isotropic chemical shift makes SSNMR one of the most powerful techniques for the fast discrimination and identification of crystalline polymorphs and pseudopolymorphs of active pharmaceutical ingredients (APIs). It is well known that many small organic molecules, such as the majority of APIs, are commonly able to crystallize in several different forms, which

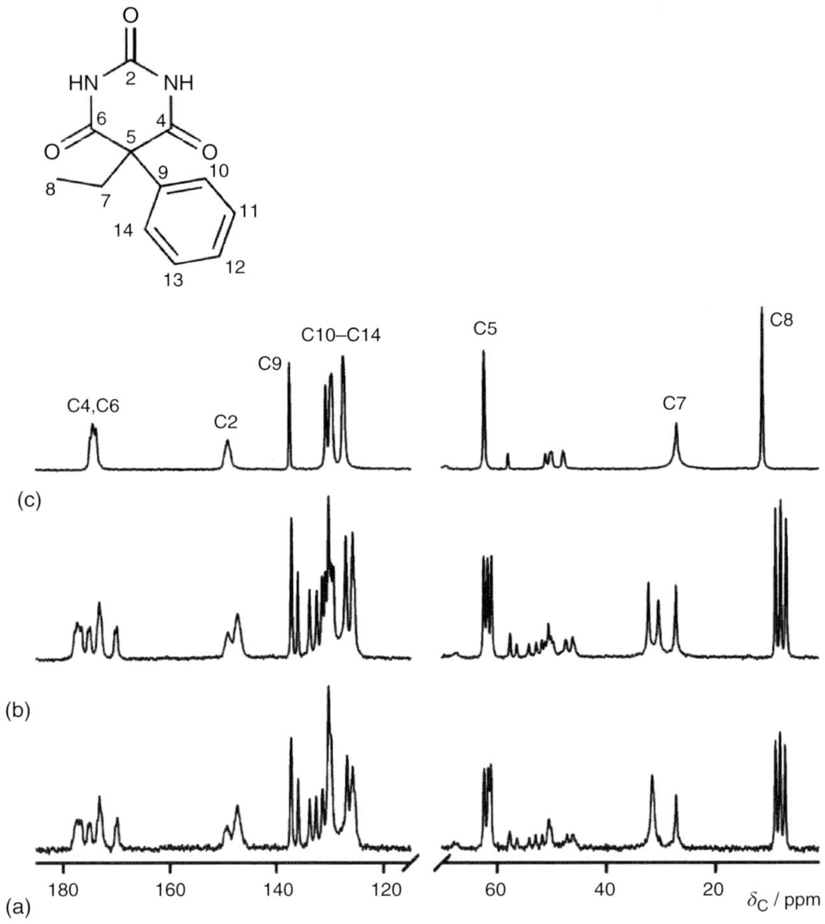

Figure 14.2 ^{13}C CP-MAS spectra of phenobarbital forms (a) I, (b) II, and (c) III. Adapted from Zencirci *et al.* [5]. Reproduced with permission of American Chemical Society.

can be different in the 3D structure exhibited by the single molecules (conformational polymorphism) or in crystal packing features (packing polymorphism). Different polymorphs give rise to significantly different ^{13}C high-resolution spectra (^{13}C is just the most exploited nucleus, but in principle many other nuclei can be used), since most of the nuclei resonate at a chemical shift particular to a certain polymorph. For example, in Figure 14.2 the ^{13}C CP-MAS spectra of three different polymorphs of the API phenobarbital [5] are reported. In the case of crystalline phases, the use of ^{13}C MAS spectra as a polymorph fingerprint is facilitated by the very small line width characterizing the signals, provided that a suitable high-power decoupling from ^{1}H nuclei is applied. This is due to the very high degree of order typical of crystalline phases: each nucleus experiences a well-defined and, in principle, perfectly repeated environment, which is associated with a precise value of the isotropic chemical shift.

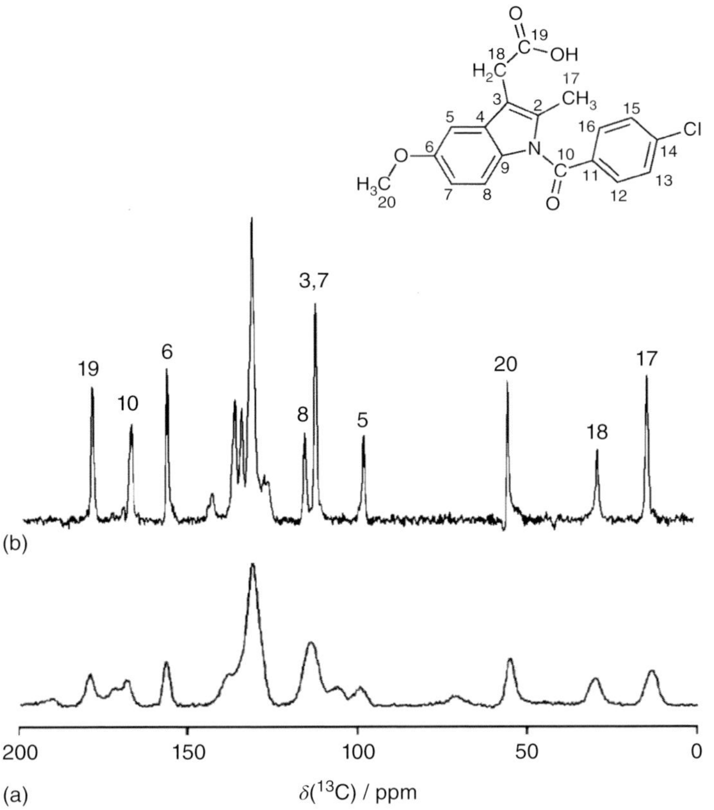

Figure 14.3 ^{13}C CP-MAS spectra of indomethacin. (a) Amorphous quench-cooled melt. (b) Microcrystalline (γ form) powders. Adapted from Apperley *et al.* [6]. Reproduced with permission of American Chemical Society.

If a certain degree of static disorder, that is, a distribution of different static situations for the same nucleus, is introduced, a corresponding distribution of isotropic chemical shift values arises.

The extreme case is that of completely amorphous phases. In an amorphous phase, any long-range order is lost and, in most cases, molecules experience a large static distribution of conformations, environments, and interactions. This determines a correspondingly broad distribution of isotropic chemical shift values for each chemically inequivalent nuclear site of a molecule. The consequence is a large broadening of the signals, which gives rise to the typical poorly resolved aspect of the spectrum of an amorphous phase. The comparison between a spectrum of a crystalline and an amorphous phase is shown in Figure 14.3, which reports the ^{13}C CP-MAS spectra of the API indomethacine as microcrystalline (γ form) and quench-cooled melt amorphous powders [6].

The above statement applies not only to small organic molecules, such as those often used as APIs, but also to excipients, many of which are organic natural or

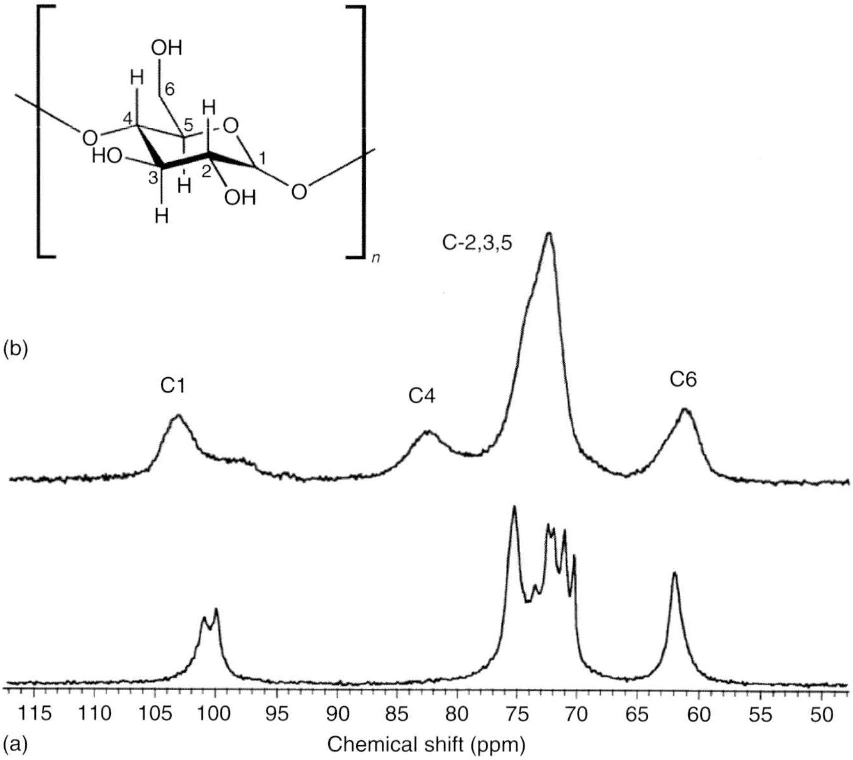

Figure 14.4 ^{13}C CP-MAS spectra of (a) hydrated B-type crystalline and (b) amorphous amylose. Adapted from Paris *et al.* [7]. Reproduced with permission of Elsevier.

synthetic polymers. For instance, polysaccharides are often used in solid formulations and, depending on their origin and on the treatments experienced, they can exhibit crystalline or amorphous phases as well as their mixtures. For example, in Figure 14.4 ^{13}C CP-MAS spectra of crystalline and amorphous forms of amylose are shown [7].

In the following, we will treat the different aspects of static disorder, dividing the cases into the two main categories of crystalline and amorphous systems.

14.3.1 Crystalline Systems

SSNMR is an excellent tool for the study of crystalline solids, as proven by the increasing number of applications of the so-called *NMR crystallography*, a term recently coined that refers to the use of SSNMR to obtain information about crystal structures (complementing the well-established diffraction methods) [8]. In particular, the combination of advanced experimental methods with quantum mechanics (QM) computation of NMR observables has been crucial, and nowadays NMR crystallography can be used to validate and complement X-ray diffraction results.

A peculiarity of SSNMR compared to other techniques is the ability to precisely identify and characterize different kinds of static disorder in crystals, such as *molecular site disorder*, involving only specific molecular moieties, and *one-phase disorder*, which includes crystalline lattice defects and continuous conformational disorder.

With *molecular site disorder*, here we refer to the disorder limited to one or more molecular moieties but not considering the whole molecule. In many cases, this coincides with *conformational site disorder*, since the disorder concerns the presence of different conformations in different molecules. SSNMR is particularly powerful for the investigation of molecular site disorder: in this case, in the high-resolution spectra of rare nuclei with spin $1/2$ (typically ^{13}C, ^{31}P, ^{15}N, etc.), selective splitting or broadening of some signals is observed. The signals affected by such splitting/broadening are those of the nuclei belonging to the disordered fragment. In particular, if a few well-defined conformations are present, for a specific nucleus a signal from each conformer is expected. On the other hand, in the presence of a distribution of conformations, a broad signal is observed, which arises from the sum of all the resonances corresponding to each conformer.

For example, conformational site disorder was observed in a derivative of natural product pleromutilin-*trans*-3-aminocyclobutyl (**A** in Figure 14.5) [9]. Two of the ^{13}C CP-MAS spectra shown in Figure 14.5 are those of (a) 1 : 1 **A**/tartrate and (b) 1.0 : 2.8 **A**/propionic acid (PA). The presence of sharp peaks in the spectra indicates that both these derivatives are crystalline, but in the case of the PA derivative a group of broader signals is observed at about 180 ppm, indicating the occurrence of site disorder selectively affecting the carboxylic carbons of PA (the other signals arising from CH and CH_2 PA carbons cannot be distinguished because of strong superposition with other signals).

Conformational disorder can be studied in more detail by applying more advanced SSNMR methods, such as 2D correlation and separation experiments. In particular, NMR crystallographic methods can be applied to refine the structure of crystals containing molecular site disorder, for which often the data obtained by X-ray diffraction are of poor quality. Moreover, it is possible to characterize the different conformations and the distributions possibly present in terms of dihedral angles and conformational energy. For these studies, the use of QM calculations has become essential and, in some cases, selective isotopic enrichment is very useful. Even though, in principle, applicable to all kinds of pharmaceuticals, both advanced SSNMR techniques and QM calculations have been mainly used for the investigation of relatively small organic molecules. Indeed, in much larger and complicated systems, the application of these methods could be very difficult or it would require extensive and selective isotopic labeling.

A very recent example of the application of NMR crystallography to disordered crystals is the study of two naturally occurring peptides named enkephalins [10]. NMR crystallography was employed to refine the solid-state molecular structures

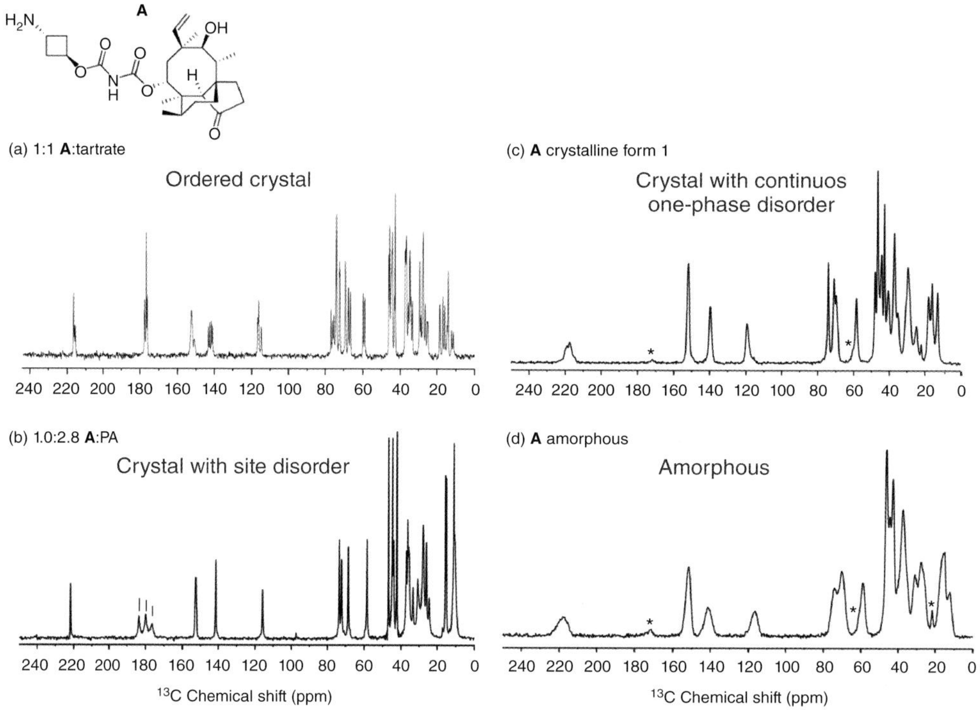

Figure 14.5 ^{13}C CP-MAS spectra of different solid forms of **A**: (a) 1:1 **A**:tartrate, (b) 1.0:2.8 **A**:PA (propionic acid), (c) **A** free base crystalline form 1, and (d) **A** free base amorphous. The kind of disorder is indicated for each form, and in (b) relevant signals are highlighted in red. ^{13}C signals of EtOAc trapped in the samples are indicated by asterisks. Adapted from Clawson *et al.* [9]. Reproduced with permission of Wiley.

of Leu- and Met-enkephalins. An approach based on many different slow and very fast MAS SSNMR experiments and QM calculations was used. Isotropic chemical shifts, chemical shift anisotropy parameters, and data obtained from 2D correlation experiments have been used as structural restraints in refining the peptide structures. In principle, this methodology appears very promising in the study of pharmaceutical compounds.

SSNMR is a useful tool also for the study of *one-phase disorder*. Contrary to those situations in which domains of crystalline and amorphous forms are mixed, the one-phase model describes a continuous, distributed loss of order within a single phase. This kind of disorder can be produced by crystal lattice defects: as the defects increase in number, domains of defect sites are created and, eventually, an amorphous solid phase is formed. One-phase disorder can be induced by grinding, milling, and other processing methods of the solid material. Desolvation of a solvate can also generate one-phase disorder. In these systems, the molecules progressively shift and distort, occupying the free volume left by the loss of the solvent molecules.

Starting from the maximum degree of order, we should consider the presence of *crystal lattice defects*, which are imperfections in the crystal structure. Lattice defects affect a few localized molecules, leaving the rest of the structure unchanged. Crystal lattice defects are almost always present and are formed in the crystallization process, but they can also be induced by grinding, milling, and other processing techniques of the solid material. Generally, crystal lattice defects are present in very small quantities compared to the crystal bulk, and this is the reason why they are not very widely studied by SSNMR, especially in molecular crystals. Indeed, SSNMR is not a very sensitive technique, and sensitivity problems prevent the easy study of very small portions of the sample. Nevertheless, some studies on crystal defects have been performed by SSNMR in inorganic systems, and a few studies have also been performed on molecular crystals, where, however, isotopic enrichment is required. For example, SSNMR methods have been exploited to study chiral impurities in crystals. One enantiomer can be selectively observed by ^{13}C labeling, and SSNMR can be used to investigate its environment. An interesting example is the study of D- and L-proline reported in [11]. ^{13}C CP-MAS spectra of a series of samples at increasing amount of ^{13}C-L-proline in D-proline are shown in Figure 14.6. In particular, peculiar peaks of the racemic co-crystal monohydrate (DL-MH), racemic co-crystal form I (DL-I), racemic co-crystal form II (DL-II) and L-crystal defects (L-CD) could be identified by comparison with the spectra of the corresponding pure forms, and the distribution of L-proline in the different forms could be characterized by the intensity of each peak.

With increasing levels of disorder, a largely distributed loss of order within a single phase can be identified by ^{13}C CP-MAS spectra by looking at the line width of the signals. As already seen in the case of molecular site disorder, a broadening of the ^{13}C CP-MAS lines can be ascribed to the presence of disorder, but contrary to the previous case, in the case of one-phase disorder the line broadening generally affects all the signals. An interesting comparison of different kinds of disorder detected by ^{13}C CP-MAS is reported in Figure 14.5. As previously observed, the spectra of crystalline tartrate (Figure 14.5a) and PA derivatives (Figure 14.5b) show narrow peaks (with the exception of site specific disorder in the latter case). On the contrary, in the spectrum of **A** free base in the amorphous form (Figure 14.5d), all the signals are broad, as expected. The interesting case is that of the spectrum of **A** crystalline form 1 (Figure 14.5c), in which the line widths are somewhat intermediate between those observed for the derivatives and the amorphous form. This is a case of a crystal system with continuous conformational disorder, as also confirmed by X-ray powder diffraction (XRPD) and single-crystal X-ray diffraction (SCXRD) [9].

In order to investigate the origin of the disorder, the broadening of the signals of rare nuclei can be measured and interpreted. In particular, in favorable cases, all the slightly different molecular conformations giving rise to the continuous disorder can be identified. This has been done, for example, in the case of a crystalline form of *N*,*N*-bis-(diphenylphosphino)-*N*-((*S*)-*R*-methylbenzyl)amine (compound **B** in Figure 14.7), with the support of QM calculations [12, 13]. On

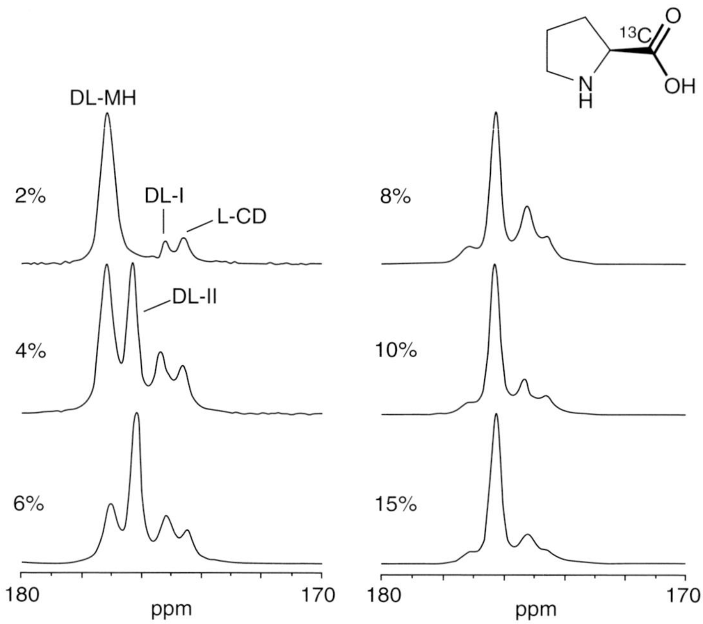

Figure 14.6 ^{13}C CP-MAS spectra of L-proline in samples of D-proline with different amounts of ^{13}C labeled L-proline. Peculiar peaks of racemic co-crystal monohydrate (DL-MH), racemic co-crystal form I (DL-I), racemic co-crystal form II (DL-II), and L-crystal defects (L-CD) are marked with the corresponding label. Adapted from Berendt and Munson [11]. Reproduced with permission of Wiley.

the basis of the crystal structure, eight distinct signals, arising from the presence of two inequivalent ^{31}P nuclei in each molecule and four molecules in the asymmetric crystallographic unit, should be observed in the ^{31}P MAS spectrum. The presence of a slight structural disorder causes a broadening of the lines and prevents the eight signals to be all distinguished. From the ^{31}P 2D 1Q–2Q (single quantum–double quantum) correlation spectrum (Figure 14.7), it is possible to observe a strong correlation between the changes in chemical shifts induced by disorder for coupled ^{31}P nuclei (see, e.g., the inset in Figure 14.7). A step forward in the analysis was obtained by using QM calculations and statistical analysis of the 2D line shapes. In particular, low-energy trial sets of local distortions of the crystalline structure have been used to describe candidate static distributions of local geometries, and by comparison between calculated and experimental line shapes the distributions of structures compatible with the experimental data have been individuated.

All the procedures that require a combination of advanced SSNMR experiments and QM calculations are quite complex and cannot be routinely applied, but they can provide very detailed information often not accessible by other techniques.

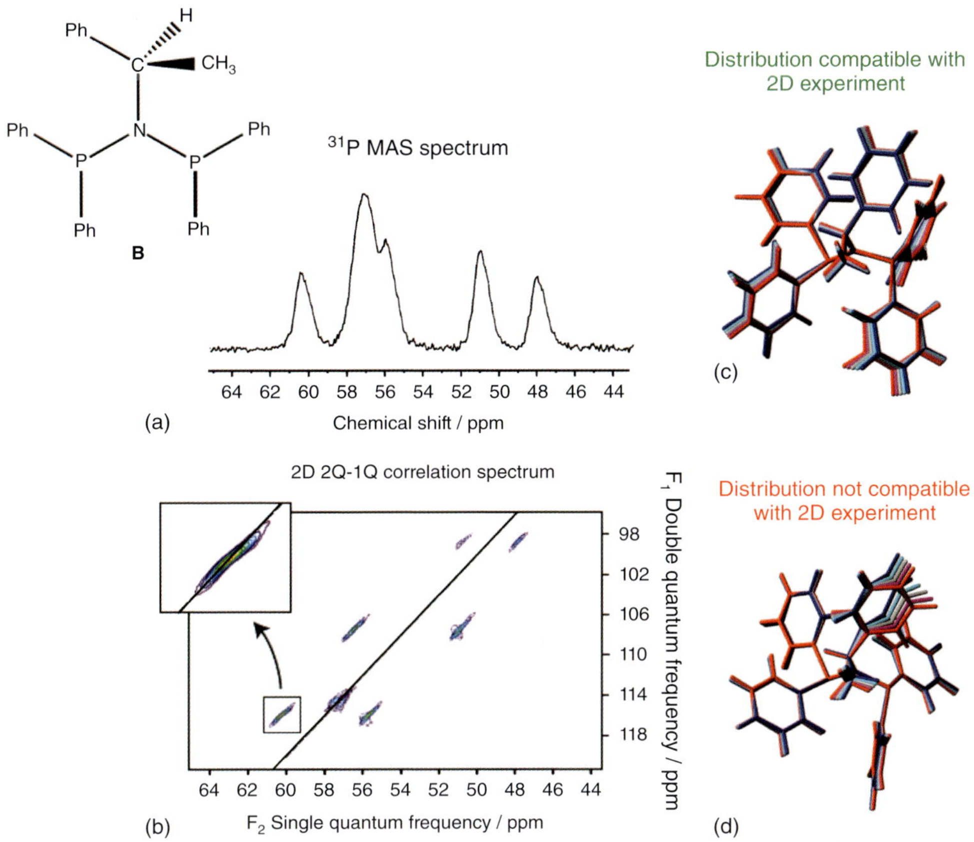

Figure 14.7 (a) Chemical structure and ^{31}P MAS spectrum of **B**. (b) 2D 1Q–2Q experiment correlating the chemical shifts of coupled ^{31}P nuclei. Two distributions of molecular structures taken into account in the analysis, resulted to be compatible (c) and not compatible (d) with the experimental data. (a,b) Adapted from Sakellariou *et al.* [12]. Reproduced with permission of American Chemical Society. (c,d) Adapted from Cadars *et al.* [13]. Reproduced with permission of American Chemical Society.

14.3.2
Amorphous Systems

Amorphous systems are characterized by the maximum extent of static disorder, with the complete absence of any long-range order. In amorphous systems, the static distribution of molecular structures not only affects all the molecular sites but it is also larger than in "disordered" crystalline systems. This causes a broad distribution of isotropic chemical shift values for each chemically inequivalent nuclear site, and therefore a large broadening of all the signals in the SSNMR spectrum, as already shown in Figures 14.3 and 14.4.

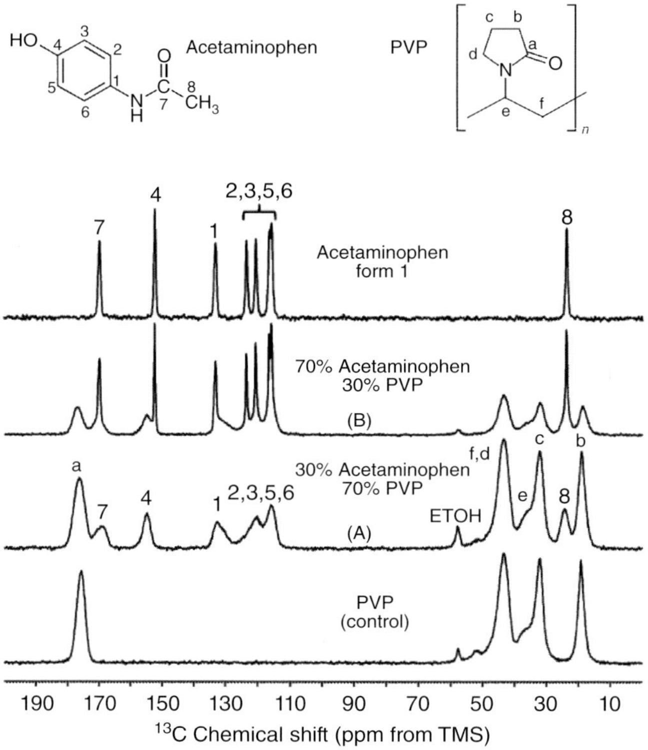

Figure 14.8 ^{13}C CP-TOSS spectra of **A** (30:70 w/w) and **B** (70:30 w/w) dispersions of acetaminophen in PVP, crystalline acetaminophen form 1, and PVP. Adapted from *Pham et al.* [14]. Reproduced with permission of American Chemical Society.

Often crystalline APIs experience an amorphization not only, as already shown, as a consequence of several treatments (in which the pure API is subjected to grinding, melt-quenching, etc.) but also during the preparation of a formulation with an excipient. The ^{13}C high-resolution SSNMR spectrum of a formulation, and in particular the comparison with that of the pristine API, provides a straightforward way of controlling possible phase modifications occurring during processing or storage. As an example, Figure 14.8 shows the spectra of two solid dispersions of acetaminophen in polyvinylpyrrolidone (PVP) (**A** 30:70 w/w, **B** 70:30 w/w), prepared by rapid solvent evaporation from ethanol solutions under reduced pressure [14]. From the comparison between the spectra of **A** and of neat acetaminophen and PVP, it is clear that acetaminophen signals are much broader, and in many cases shifted, in the spectrum of the dispersion, indicating that in its preparation complete amorphization of the API occurred. On the other hand, in the spectrum of **B**, acetaminophen signals appear to be a superimposition of narrow peaks due to the crystalline form and broader peaks ascribable to the amorphous form, thus indicating the coexistence in this dispersion of amorphous and crystalline domains of the API.

So far we have shown comparative cases in which a considerable broadening of the SSNMR signals of a compound with respect to those of its crystalline phase can be safely considered as an evidence of amorphization. On the other hand, when a spectrum of a crystalline phase of the compound under study is not available, the sole ^{13}C SSNMR spectrum does not always constitute a rigorous and self-consistent evidence of the amorphous state of a sample. However, especially for small molecules, line width at half of the maximum height of the peaks of, as a rule of thumb, a few parts per million is usually indicative of an amorphous phase, while typical signals of crystalline phases can easily exhibit a line width of less than 1 ppm. This criterion holds true provided that the spectra are recorded using a sufficiently strong HPD and in the absence of interferences between molecular dynamic processes and the decoupling field or MAS, cases that will be discussed in the next section.

Advanced studies can bring a detailed interpretation of the distribution of isotropic chemical shifts in amorphous phases in terms of distributions of molecular conformations. To this purpose, it is necessary the support of numerical simulations, which are more complex than in crystalline systems, due to the absence of XRD structures. This has been done, for example, in a work by Lefort *et al.* [15, 16], where the conformational energy landscape of glassy disaccharides was studied by ^{13}C CP-MAS NMR and numerical simulations. Under the hypothesis that the chemical shift surface for disaccharides is primarily dependent on the dihedral angles ϕ and ψ (Figure 14.9), simulated chemical shift surfaces were compared with the experimental data in order to obtain information on the populated conformers.

In many cases, crystalline and amorphous phases of an API coexist in the same sample. Even though, as previously stated, the signals of the two forms have markedly different line widths, the detection of signals arising from an amorphous component, if present in small amount, can be difficult. This can be due to the likely overlap with crystalline signals and to the weak intensity of the amorphous signals, deriving from their large line width. Therefore, when the presence of an amorphous phase is in doubt, it is convenient to perform experiments in which signals from amorphous phases are usually favored. One common strategy consists in decreasing the recycle delay between two consecutive transients in either ^{13}C DE-MAS or CP-MAS experiments. Indeed, amorphous phases are often characterized by a dynamics faster than that of crystalline phases, resulting in shorter spin–lattice relaxation times. Decreasing the recycle delay should therefore result in a partial suppression of the crystalline signals and a consequent emphasis of the amorphous ones.

14.3.2.1 **Quantitation of Amorphous Phases**

Once an amorphous phase has been detected, an important issue concerns its quantification. Quantification of the amorphous content in pharmaceutical samples is a very important and nontrivial problem. SSNMR can significantly contribute in this area, and indeed several papers dealing with specific

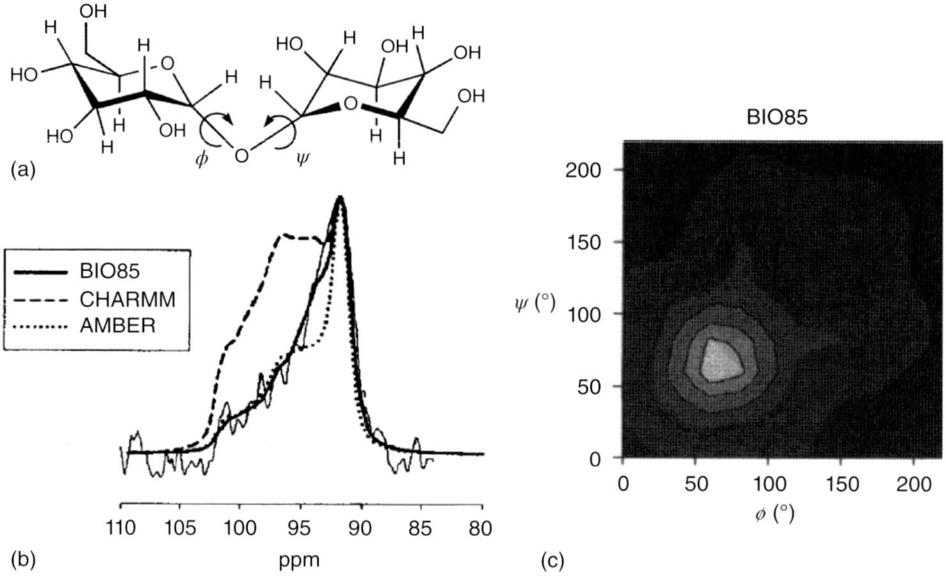

Figure 14.9 (a) Chemical structure of trehalose, where the conformational angles ϕ and ψ are indicated. (b) Best fit simulations of the glycosidic region of the ^{13}C CP-MAS spectrum of trehalose compared with the experiment (thin solid line), performed by calculating $E(\phi,\psi)$ with different force fields (BIO85, CHARMM, AMBER). (c) Populations, $\chi(\phi,\psi)$, of the different conformations calculated as $\chi(\phi,\psi) = \exp(E(\phi,\psi)/RT)$, where $E(\phi,\psi)$ has been calculated with the force field BIO85. Adapted from Lefort *et al.* [15]. Reproduced with permission of American Institute of Physics.

pharmaceutical cases have been published [17–19]. As already mentioned in Section 14.2, in principle SSNMR spectroscopy can straightforwardly provide quantitative information since the area underlying a signal is proportional to the number of nuclei giving rise to that signal. In practice, the fulfillment of this condition requires that SSNMR spectra are suitably recorded. There are basically two experimental approaches to obtaining quantitative information from ^{13}C spectra. The first consists in recording DE-MAS spectra using recycle delays between two consecutive transients at least five times longer than the longest ^{13}C T_1 of the sample (thus implying preliminary ^{13}C T_1 measurements). This approach can be extremely time consuming (one spectrum can require days of acquisition) due to the usually long values of ^{13}C T_1 in solids. The other approach is based on CP-MAS spectra: as already pointed out, these spectra are intrinsically not quantitative since ^{13}C signals are built up by magnetization transfer from dipolar-coupled ^1H nuclei. In particular, the intensity of a signal strongly depends on the contact time according to the CP dynamics curve (trend of signal intensity vs. contact time), which is usually specific to each chemically inequivalent carbon nucleus and can also vary from one phase to another. In Figure 14.10, the experimental CP dynamics curves of two crystalline and the amorphous forms of the artificial sweetener neotame are shown, obtained by

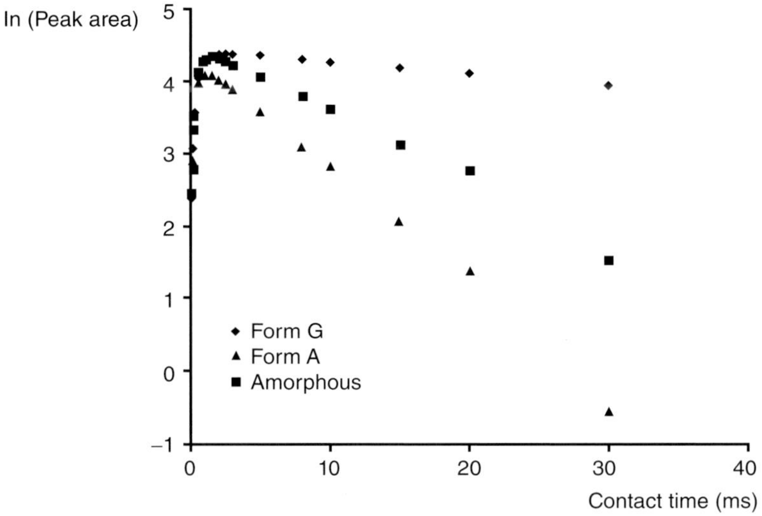

Figure 14.10 Experimental ^{13}C CP dynamics curves of various forms of neotame. Adapted from Offerdahl *et al.* [17]. Reproduced with permission of Wiley.

recording ^{13}C CP-MAS spectra at different values of contact time and plotting the values of the areas of peaks characteristic of the different phases.

CP dynamics curves can be, in general, described by suitable equations, differing in some approximations. One of them is

$$M(t) = M_0 \frac{\left[\exp\left(-\frac{t}{T_{1\rho\mathrm{H}}}\right) - \exp\left(-\frac{t}{T_{\mathrm{CH}}}\right)\right]}{1 - \left(\frac{T_{\mathrm{CH}}}{T_{1\rho\mathrm{H}}}\right)} \tag{14.1}$$

which can be applied when $T_{1\rho\mathrm{C}} \gg T_{\mathrm{CH}}$. $M(t)$ is the magnetization (proportional to the peak area) and t is the contact time. $T_{1\rho\mathrm{C}}$ and $T_{1\rho\mathrm{H}}$ are the spin–lattice relaxation times in the rotating frame of ^{13}C and ^{1}H nuclei, respectively, and T_{CH} is called cross-polarization time constant, which depends on the strength of the dipolar coupling between ^{13}C and ^{1}H nuclei. The increasing part of the curve is substantially described by T_{CH}, while the decreasing part mainly depends on the decay of the magnetization in the rotating frame regulated by $T_{1\rho\mathrm{H}}$. By fitting a CP dynamics experimental curve to Eq. (14.1), it is possible to determine T_{CH}, $T_{1\rho\mathrm{H}}$, and M_0. The latter parameter is the thermal equilibrium magnetization of the considered nucleus, and it is the quantitative parameter of interest. Therefore, in the presence of a mixture of amorphous and crystalline forms, by recording and analyzing CP dynamics curves for signals diagnostic of the different forms, it is possible to quantify them. Actually, in several cases reported in the literature, spectra recorded at a single value of contact time have been directly used for quantification of amorphous and crystalline phases. This is possible only if the CP dynamics curves are known for the single forms.

A problem often encountered is the previously mentioned overlap between signals of corresponding carbon nuclei in amorphous and crystalline phases, which usually do not show large differences in chemical shift. Therefore, for correctly measuring their underlying areas it is usually necessary to apply spectral fitting procedures. The availability of the spectra of pure amorphous and crystalline phases, recorded under the same experimental conditions, is helpful and sometimes necessary. Finally, a quantification procedure based on SSNMR is usually validated by applying it to purposely prepared mixtures of known amounts of amorphous and crystalline forms. In the literature, amorphous contents have been determined through SSNMR down to a few percent. The SSNMR results are often in very good agreement with results obtained by XRD and more accurate than those obtained by DSC.

So far we have considered the discrimination and quantitation of amorphous forms with respect to crystalline forms. A more difficult task is the discrimination, and relative quantification, of possibly different amorphous forms of the same compound. The intrinsic structural disorder of the amorphous phases and the corresponding relatively poor spectral resolution make the highlighting of significant differences very complex. Nonetheless, for instance, different amorphous excipients obtained by different processes and/or suppliers can show sufficient differences in ^{13}C high-resolution spectra, whose visual inspection can therefore represent a valuable tool for qualitative discrimination and control [19, 20]. A recent work [21] has reported a statistical approach, based on factor analysis of ^{19}F MAS spectra of atorvastatin, which allowed different amorphous forms of the drug, giving rise to only subtle spectral differences, to be recognized.

14.3.2.2 **Miscibility and Interactions in Heterophasic Systems and Formulations**

Moving on a larger spatial scale, we can include in the concept of static disorder also the mixing degree between different components of a pharmaceutical formulation, an issue of noticeable importance in the pharmaceutical field, especially because of its influence on solubility and release properties. It is remarkable that the preparation of amorphous solid dispersions, in which the amorphous form of a poorly water-soluble API is dispersed in a hydrophilic polymeric matrix, is at present one of the most successful strategies for stabilizing the amorphous form of an API and improving its solubility and dissolution rate. The achievement of these results is maximized when the API and polymer are mixed at a molecular level, preferably also with the establishment of API–polymer interactions.

SSNMR provides very powerful tools for investigating the mixing degree and also specific interactions among different components of a mixture [22]. The investigation of the mixing degree is primarily based on the spin-diffusion phenomenon, which is a spatial diffusion of nuclear magnetization, occurring without matter diffusion, based on spin polarization exchange among dipolar-coupled nuclei of the same species. This phenomenon, peculiar to the abundant nuclei in the solid state, has the important effect of tending to average to single values different intrinsic spin–lattice relaxation times in the laboratory (T_1) and in the rotating frame ($T_{1\rho}$). For three-dimensional diffusion, the mean-square distance traversed via spin diffusion ($\langle r^2 \rangle$) is related to the spin diffusion coefficient (D_s) and the diffusion time (t) by the equation

$$\langle r^2 \rangle = 6D_s t \tag{14.2}$$

D_s depends on the homonuclear dipolar couplings occurring in the system, and it is usually taken from the literature, even though it can also be calculated from ^{1}H T_2 values. When the average of the different intrinsic relaxation times is complete and a single T_1 (or $T_{1\rho}$) is measured, its value can be substituted in Eq. (14.2) to give the distance over which the system can be considered homogeneous. If single T_1 and $T_{1\rho}$ of the order of 1 s and 10 ms are measured, respectively, considering a typical D_s value of the order of 5×10^{-16} m^{2}s^{-1}, the system can be considered homogeneous on distances of 50 and 5 nm, respectively. On the contrary, if multiple T_1 (or $T_{1\rho}$) values are measured, it can be inferred that the sample is constituted by heterogeneous domains with size larger than 50 nm (or 5 nm). Considering the difficulty in obtaining ^{1}H high-resolution spectra of solids, it is usually convenient to measure ^{1}H T_1 and $T_{1\rho}$ by exploiting the spectral resolution easily achievable for ^{13}C nuclei, by applying common pulse sequences that combine the proton relaxation times measurement with CP-MAS. Nice recent examples of this kind of measurements can be found in [14, 23]. For instance, Policianova and coauthors [23] reported a thorough study of PVP/acetylsalicylic acid (ASA) amorphous dispersions. They observed that, starting from a proton T_1 of neat ASA of 59.0 s and PVP of 2.9 s, both ASA and PVP in their dispersion exhibited a T_1 of 4.2 s; the same equalization was observed for $T_{1\rho}$. These data indicated that the ASA/PVP dispersion was homogeneous on the nanometer scale. The presence of interactions between ASA and PVP at a molecular level was observed by looking at the isotropic ^{13}C chemical shift dispersion of the carbonyl carbon signal of PVP (Figure 14.11). The asymmetric broadening of this signal to higher chemical shifts on passing from neat PVP to the amorphous dispersion indicates the establishment of hydrogen bonds with ASA.

The very intimate mixing of ASA and PVP was further confirmed by the observation of strong correlations between peaks characteristic of the two components in a ^{1}H–^{1}H SQ/SQ DUMBO experiment (Figure 14.12), in which cross-peaks were observed between signals of strongly dipolar-coupled ^{1}H nuclei, which have a reciprocal spatial distance of a few angstroms.

This advanced experiment is one of the several existing 2D correlation experiments, mainly exploiting homo- or heteronuclear dipolar couplings, which can be used for investigating spatial proximities between chemically distinguished nuclei. In these experiments, the use of a second dimension and of homonuclear decoupling schemes allows obtaining a much improved resolution with respect to standard 1D experiments. Therefore, in some cases they provide important, detailed structural information even on fully amorphous systems, and indeed they have been successfully exploited for revealing molecular interactions and/or mixing between APIs and excipients in several amorphous dispersions [14, 23].

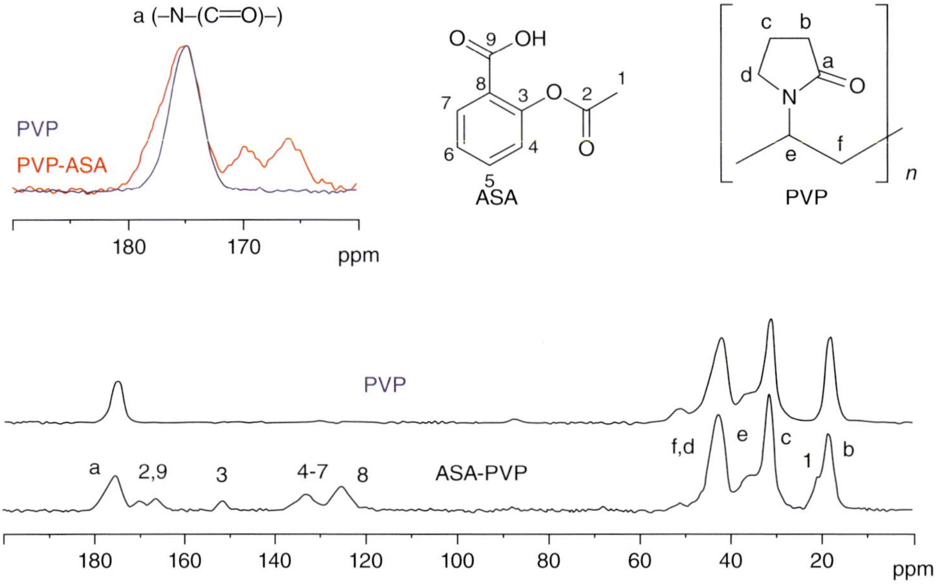

Figure 14.11 ^{13}C CP-MAS spectra of PVP and ASA/PVP (30:70 w/w) dispersion. In the inset an expansion of the PVP C=O signal is shown. Adapted from Policianova *et al.* [23]. Reproduced with permission of American Chemical Society.

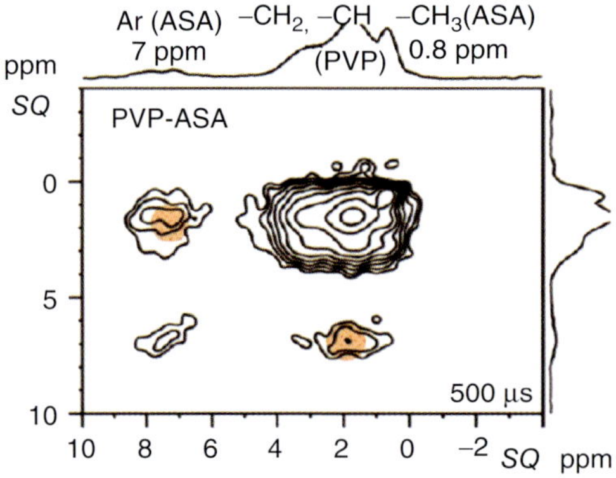

Figure 14.12 $^{1}H-^{1}H$ SQ/SQ DUMBO spectrum of ASA/PVP (30:70 w/w) dispersion. Adapted from Policianova *et al.* [23]. Reproduced with permission of American Chemical Society.

14.4
Dynamic Disorder

Molecules, even in solid phases, are characterized by the presence of several motions, which are responsible of the "dynamic disorder" affecting many solid pharmaceuticals. Vibrations are very active in the solid state, and their effects on nuclear parameters are important, but their timescale is too short with respect to the NMR timescale: the nuclear parameters measured by SSNMR techniques are therefore "vibrationally averaged," that is, only their average values over the vibrations can be measured. The translational motions in solids, contrary to liquids, are frozen or very slow. There are a few exceptions especially concerning the migration of small atoms or ions: these translational motions can be effectively studied by NMR techniques, especially using magnetic field gradients, but this subject will not be discussed here in detail. Rotational motions are generally limited to molecular fragments, and do not occur for the molecule as a whole, with a few exceptions. These motions essentially consist in jumps among different conformers, corresponding to potential energy minima. Interconformational motions can be, indeed, effectively investigated by SSNMR techniques, since on one side they induce a time dependence of the different nuclear interactions, and on the other they occur on the same timescales of some NMR parameters/techniques.

The main advantages of using SSNMR to characterize interconformational motions in solids are basically three:

a) The timescale accessible to SSNMR, thanks to a combination of different techniques and/or measurable nuclear properties, is very broad, ranging from picoseconds to seconds.
b) Each motion occurring in the systems can be, in principle, selectively investigated, thanks to the possibility of obtaining site-specific information, either exploiting different NMR-active nuclei (e.g., ^{13}C, ^{15}N, ^{31}P, etc.) or chemically inequivalent nuclei of the same species located in different molecular positions (e.g., ^{13}C nuclei in different chemical environments and therefore characterized by different NMR signals).
c) The information that can be obtained for each motion is often very detailed, including its geometry, correlation time, activation energy, and other parameters describing its nature (distribution of the correlation times, correlation among different motional processes, etc.). Of course, the level of detail that can be effectively obtained strongly depends on the complexity of the system investigated.

Each nuclear property is associated with a sort of "reference" frequency (ν_{ref}), which is typically an instrumental quantity, and it is sensibly affected by motional processes occurring with frequencies of the same order of the reference frequency. So, for instance, ^{1}H spin–lattice relaxation times in the laboratory frame have the proton Larmor frequency $\nu_0(^{1}\text{H})$ as reference frequency, and therefore they are sensitive to fast motions with characteristic frequencies of the order of 10 MHz – 10 GHz. On the contrary, the line shape due to ^{13}C chemical shielding anisotropy (CSA) is affected by motions with characteristic frequencies of the order of the carbon static line width due to this interaction (1 – 100 kHz). A list of measurable nuclear properties exploitable for obtaining dynamic information, and of the corresponding reference frequencies and range of motional frequencies that can be studied, is reported in Table 14.1.

Table 14.1 Measurable nuclear properties and corresponding reference frequencies and accessible motional ranges.

Measured nuclear property	Reference frequency	Value of reference frequency (Hz)	Approximate range of motional frequencies investigated (Hz)
^{13}C isotropic chemical shift (exchange)	Chemical shift difference between exchanging signals	10^2-10^3 [a]	10^1-10^4
^{13}C MAS-HPD line width	MAS or ^1H HPD frequencies	10^3-10^5 (MAS) $3\times10^4-2\times10^5$ (HPD)	10^2-10^5
^{13}C chemical shielding anisotropy (CSA)	CSA static line width	$1\times10^3-5\times10^4$ [a]	10^3-10^5
^1H $T_{1\rho}$	^1H spin-lock frequency	$2\times10^4-2\times10^5$	10^4-10^6
^1H T_1	^1H Larmor frequency	$1\times10^8-1\times10^9$ (standard) [a]	10^6-10^{10}
		10^3-10^7 (field cycling)	10^3-10^7
^{13}C T_1	^{13}C Larmor frequency	$1\times10^8-2.5\times10^8$ [a]	10^6-10^{10}
^1H T_2	Residual homonuclear dipolar coupling	$1\times10^3-5\times10^4$	10^2-10^5

a) These values also depend on the intensity of the external magnetic field: here fields ranging from 9.4 to 23.5 T have been taken into account for ^{13}C and from 2.3 to 23.5 T for ^1H.

The experimental nuclear properties that can be exploited to obtain dynamic information can be divided into two categories: (i) spectral parameters and (ii) relaxation times.

1) Spectral parameters are quantities that can be directly measured from the spectra, mostly as characteristic line shapes, and they are often directly ascribable to a specific interaction (isotropic or anisotropic chemical shielding, dipolar or quadrupolar interaction, etc.).
2) Relaxation times are time constants related to the way the thermal equilibrium of the spin magnetization is restored after it has been somehow perturbed (see Section 14.2), and they are measured by means of suitable pulse sequences, which monitor the signal intensity as a function of a given time, inserted as a variable within the pulse sequence.

In the following, the studies of dynamics in pharmaceuticals are divided in two sections, dealing with interconformational motions in crystalline systems and global motions in amorphous systems.

14.4.1 Interconformational Motions in Crystalline Systems

This subsection mainly concerns pure APIs in their crystalline form, which experience only one or a few interconformational motions due to reorientation of small molecular fragments. Such motions are very important in pharmaceuticals, as they can critically affect the properties of APIs, for instance, being strongly related to the occurrence and stability of different polymorphs. Moreover, such kind of dynamic disorder can be induced by the molecular interactions between the API and excipients or porous matrices or host molecules. It must be stressed again that the timescale explored by SSNMR methods is much longer than that felt by X-ray diffraction techniques: when the latter reveal a generic disorder, SSNMR can classify this as static or dynamic disorder and, in the latter case, furnish a "quantitative" description of the motion involved.

Individual motions in systems with restricted mobility can be often revealed and characterized by means of SSNMR "spectral" techniques. In most cases reported in the literature, the nucleus ^{13}C is exploited because of the ubiquitous character of the element carbon, combined with the good resolution of ^{13}C MAS spectra and the possibility of recording such spectra in "natural abundance" (i.e., without resorting to isotopic enrichment) within a reasonable time. The two phenomena that can be directly detected from simple 1D ^{13}C high-resolution spectra and that can give precious information about individual interconformational motions are (i) exchange, and (ii) interference with an instrumental (MAS or decoupling field) frequency. The observation and reproduction, through suitable spectral simulation procedures, of (i) and (ii) can allow specific motions to be recognized and fully characterized on timescales of the order of milliseconds to seconds and microseconds to milliseconds, respectively (Figure 14.13a). A third spectral parameter is CSA, which, however, typically requires the use of 2D techniques, such as 2D-PASS [24], to be selectively determined for each nonequivalent ^{13}C nucleus in the molecule. Once the CSA profile for a given carbon nucleus is measured, it can be simulated to give precious information on the correlation time τ and the geometry of a motion occurring in the range $10\ \mu s < \tau < 1$ ms (Figure 14.13b). Figure 14.13 shows how the three spectral parameters described above are affected by a motion in the proper timescale for the specific case of π-flip (or continuous rotation) of a para-di-substituted phenyl ring. Through suitable simulations of the spectral features, it is possible to obtain the correlation time of the motion and, if variable-temperature spectra are available, also its activation energy.

Additionally, specific motions of molecular fragments in crystalline phases can be studied resorting to spin–lattice relaxation times. Different timescales are accessible looking at spin–lattice relaxation times in the laboratory (T_1) or the rotating frame ($T_{1\rho}$), as already shown in Table 14.1. However, on one side, the use of ^{13}C $T_{1\rho}$ is often made difficult by the presence of nondynamic ("spin–spin") contributions, and, on the other side, ^{1}H T_1 and $T_{1\rho}$ are both strongly affected by spin diffusion, which "disguises" the dynamic information. ^{13}C T_1 is the only relaxation time that can give quantitative site-specific dynamic information through the use of suitable models of the motions. ^{1}H T_1 and $T_{1\rho}$ can still be used, in spite of being affected by spin diffusion, by calculating their "population-weighted rate average" (PWRA), which still contains quantitative, but not site-specific, dynamic information, as it will be explained later on.

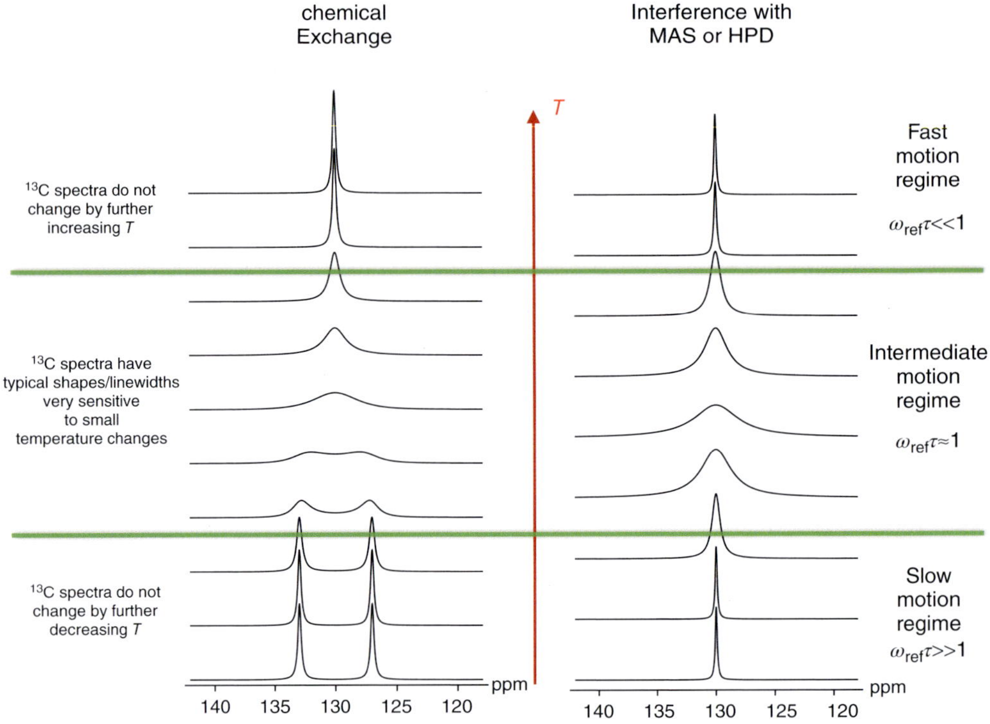

Figure 14.13 (a) Simulation of the effects of π-flip motion of a para disubstituted phenyl ring about its para axis on ^{13}C MAS spectra. On the left the effects of the motion on the signals of two inequivalent exchanging CH carbons are shown. On the right the effect on a CH signal of the interference of the same motion with ^{1}H decoupling field (or MAS) is reported. In the example shown $\nu_{\text{ref}} = \omega_{\text{ref}}/2\pi$ is 6 ppm (600 Hz at a magnetic field of 9.4 T, corresponding to the distance between the two exchanging signals in the slow motion regime) for the exchange, while it is 70 kHz (corresponding to the decoupling frequency) for the interference. τ is the correlation time of the motion. (b) Simulation of the effects of the motions of a para disubstituted phenyl ring on ^{13}C static line shape of a CH aromatic carbon. The effects of π-flip and continuous rotation (simulated by 120° flip) have been reported on the left and on the right, respectively. In the examples shown, the line width in the slow motion regime due to chemical shielding anisotropy ($\nu_{\text{CSA}} = \omega_{\text{CSA}}/2\pi$) is 210 ppm (21 kHz at a magnetic field of 9.4 T) and the isotropic chemical shift is 130 ppm. Changes of line shape discontinuities are highlighted by white dotted lines and arrows. It can be noticed how the different geometry of the phenyl motion brings about dramatic differences in the line shape in the intermediate and fast motion regimes.

A typical T_1 versus temperature curve is shown in Figure 14.14, where the dependence on the correlation time of the motion is also highlighted. The determination of parameters describing the motional processes affecting the relaxation from these curves is not straightforward and requires quite a lot of theory, whose detailed description is outside the scope of this book. Briefly, the theory of relaxation establishes the relationships between the relaxation times

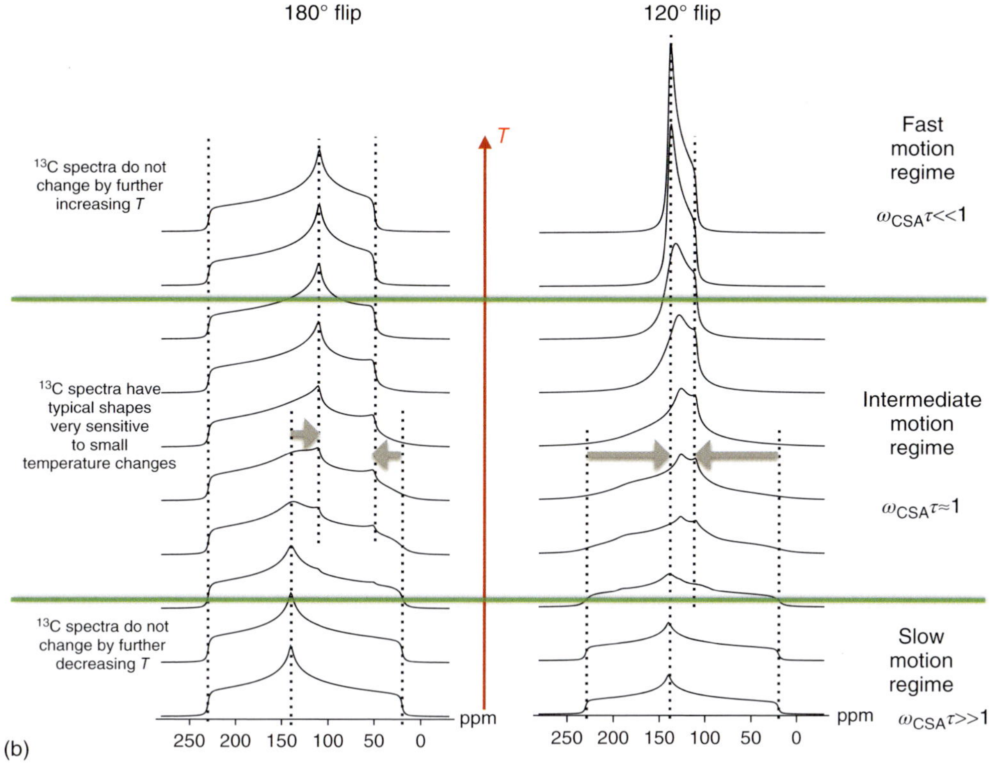

Figure 14.13 (*continued*)

and quantities called *spectral densities* [25]. These quantities can be written, using theoretical or semiempirical models (such as Bloembergen–Purcell–Pound, Davidson–Cole, Cole–Cole, Havriliak–Negami, etc.), in terms of the correlation times of the motion [26]. In turn, the dependence of such correlation times on temperature can be expressed through Arrhenius-like equations containing parameters such as the activation energy (E_a) and the correlation time at infinite temperature (τ_∞). Merging all these theories and models, it is possible to find analytical expressions linking the relaxation times to E_a and τ_∞, and therefore to the correlation time of the motion at each temperature. Such analytical expressions are fitted to the experimental curves, thereby obtaining optimized values for the motional parameters. An example of application of relaxation times to derive quantitative dynamic information will be described in Section 14.5, where the study of the interconformational motions of Na-Ibuprofen will be shown.

In the literature, the methods described in this paragraph have been applied to a variety of interconformational motions of molecular fragments, and in particular to phenyl ring π-flips [27–30]; reorientations of methyl, NH_3^+, or trifluoromethyl groups about their ternary symmetry axes [27–29, 31, 32]; reorientations of

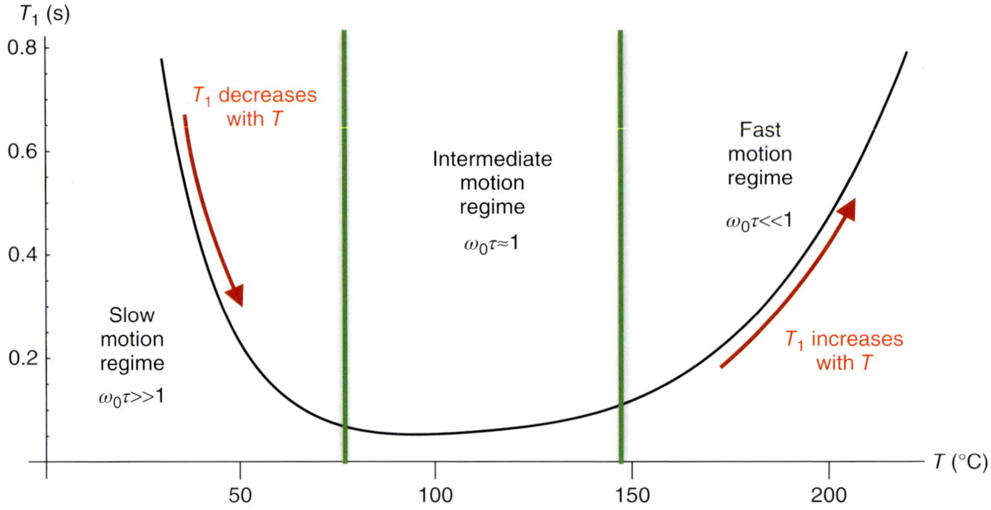

Figure 14.14 Simulation of the effects of a motion on ^{13}C spin–lattice relaxation time in the laboratory frame (T_1). $\omega_0 = 2\pi\ \nu_0$ where ν_0 is the ^{13}C Larmor frequency (about 100 rad MHz at a magnetic field of 9.4 T) and τ is the correlation time of the motion. Under the assumption that τ follows an Arrhenius dependence with temperature, the same curve describes the trend of T_1 versus ($-\ln\ \tau$) (red arrow on the top).

aliphatic chains [28, 32]; amino group jumps [33]; and interconversions between conformations of eight-member rings [34]. Moreover, a few studies have also been reported concerning the rotational motions of water molecules in hydrated crystalline pharmaceutical compounds (also exploiting D_2O and ^2H NMR) [35, 36] and the proton transfer in intermolecular hydrogen bonds [33].

14.4.2
Global Motions in Amorphous Systems

The study of molecular dynamics in amorphous pharmaceuticals is particularly interesting for several aspects, the most studied in the literature being (i) the investigation of glass transition, (ii) the presence of dynamic heterogeneities, and (iii) the possible links between molecular dynamics in an amorphous phase and its tendency to crystallize.

In the previous paragraph we have seen how ^{13}C spectral parameters and T_1 are the most employed to obtain site-specific and quantitative characterization of specific motions in crystalline pharmaceuticals. The same considerations partially hold true also for amorphous pharmaceuticals, and in particular for glassy phases, where the overall molecular system is rigid but some individual fragments can experience a fast local dynamics. However, for amorphous, especially non-glassy phases, often a "global" (non-site-specific) or "qualitative" characterization of dynamics can be sufficient. Therefore, in addition to ^{13}C T_1

and spectral parameters, ^{1}H T_2, T_1, and $T_{1\rho}$ are also largely employed in the study of amorphous phases.

Before dealing with relaxation properties in more detail, we must discuss the effects of fast motions on the spectral properties. Indeed, in the previous sections we have repeatedly stressed that we can easily recognize the presence of structural static disorder from a dramatic line broadening of the relevant peaks in ^{13}C MAS spectra. However, when an amorphous system experiences a fast and almost isotropic mobility ("liquid-like" dynamics), the line width is reduced to an extent similar to that observed in well-ordered phases. Indeed, each individual molecule experiences a variety of orientations with respect to the external magnetic field and, for a given nucleus, the observed chemical shift is an average of those associated with single orientations. If the motion is fast and isotropic, the same nucleus in different molecules shows the same average, so the line-broadening effect of the static disorder is somehow cancelled out by the dynamic disorder. All this considered, rigid, well-ordered, and very mobile disordered systems, albeit giving very similar line widths in ^{13}C MAS spectra, can be easily distinguished by a variety of experiments, showing remarkable differences in many nuclear parameters (anisotropic residual interactions and several relaxation times). For instance, rigid well-ordered systems, contrary to mobile systems, are typically characterized by long T_1, short ^{1}H T_2, and large residual chemical shielding and dipolar anisotropies. A simple ^{13}C static spectrum is usually sufficient to distinguish the two systems, since very mobile phases typically give some spectral resolution even without MAS.

Different proton relaxation times can be exploited. ^{1}H T_2 (which we will refer to in this chapter simply as the time decay of the FID) is not affected by spin diffusion and is mainly determined by the residual homonuclear dipolar interaction, resulting from the average performed by molecular motions. When the system is very rigid (the so-called *rigid-lattice regime*), that is, when the motions have frequencies smaller than the static line width (tens of kilohertz), T_2 (which is inversely proportional to the line width itself) is not affected by the motions and its value is determined by the extent of homonuclear dipolar couplings in the systems, being typically 10–20 μs (Figure 14.15). When the motions become faster (e.g., by increasing temperature), they partially reduce the line width, that is, they increase the T_2 value, to an extent that is dependent on the geometry of the motion, and that is larger the more isotropic-like the motion is. T_2 therefore monotonically increases with increasing temperature, but jumps will be observed in correspondence with activation of new motions, such as at solid–solid phase transitions. The most important example in amorphous systems is represented by the glass transition, which can be indeed very well investigated by ^{1}H T_2 (Figure 14.15). When comparing the "T_2" transition temperature with that determined by other techniques, such as DSC (where, on the other hand, the transition temperature is dependent on the temperature rate), it must be kept in mind that the T_2 transition is observed when the motions activated at the transition are able to average the static line width, that is, for instance, when they

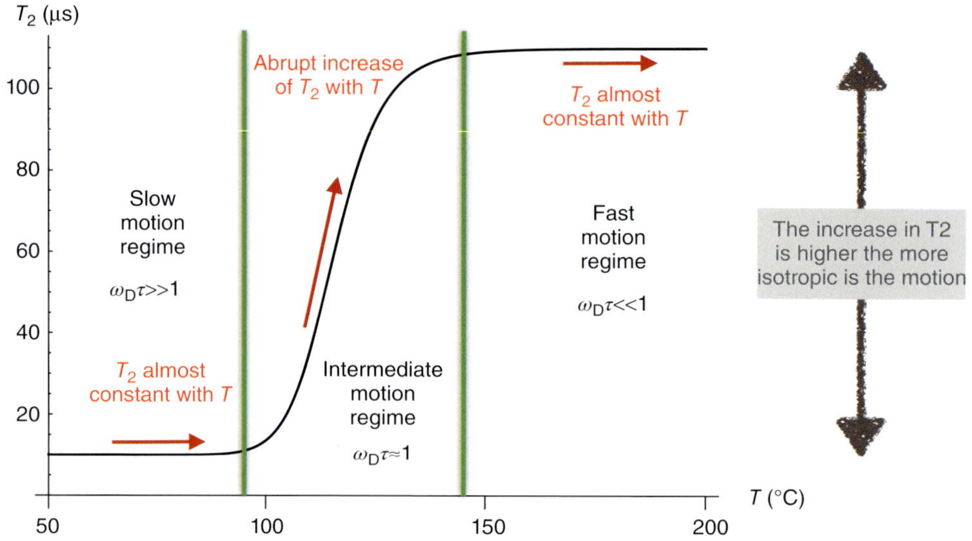

Figure 14.15 Simulation of the effects of a motion on ^{1}H spin–spin relaxation time (T_2). The static linewidth ($\nu_D = \omega_D/2\pi$) of the ^{1}H spectrum in the slow motion regime, mainly due to ^{1}H–^{1}H dipolar couplings, is of the order of 50 kHz for a rigid system. In the example shown, the system was considered in the "rigid lattice regime" below ≈ 90 °C, and the motion responsible for the T_2 jump had a strong anisotropic character.

reach a characteristic frequency of several tens of kilohertz. This condition can occur, depending on the activation energy of the motion, a few tens of degrees above the DSC-detected transition temperature. It must be noticed that, when a system experiences very fast and isotropic liquid-like motions, the homonuclear dipolar interaction is completely averaged out, the trend of T_2 with temperature becomes indefinitely and regularly increasing, and T_2 tends to become equal to T_1.

Given the poor resolution of ^{1}H spectra, ^{1}H T_2 can be experimentally determined using low-field low-resolution techniques, directly working on the FID, in the time domain, by setting up the instrumental conditions so that the FID is acquired on-resonance, that is, without any sinusoidal trend due to chemical shift. Measurements are typically carried out at low magnetic fields (Larmor frequencies of 20–60 MHz), on a single-channel spectrometer under static conditions. For acquiring FIDs with a very short T_2, the use of the solid-echo pulse sequence is required. Once recorded, the on-resonance FID is subjected to a mathematical analysis, which can be carried out using either continuous or discrete approaches. In the first case, use of the inverse Laplace transform is made, while in the second case, which we will take into account in the following, the FID is reproduced by fitting to a linear combination of suitable decaying functions, the most popular being the exponential and the Gaussian ones. When more than one function is needed to reproduce the FID, this indicates the presence of dynamic heterogeneities in the system investigated. Each function is characterized by a T_2 and a weight, so, in principle, the FID analysis gives the following information: (i) the number of dynamically distinguishable domains, given by the number of functions needed to reproduce the FID, (ii) the relative amount of protons in each domain, corresponding to the weight of each function, and (iii) the degree of mobility of each domain, from the corresponding T_2 value. Although the FID analysis does not provide any "spectral resolution," often very detailed dynamic information can be obtained. In order to link the different dynamic domains to the chemical structure, more sophisticated experiments can be applied, such as the high-resolution 2D-WISE [37].

^1H $T_{1\rho}$ and T_1 are, in principle, determined by motions in the kilohertz and megahertz ranges, respectively. However, in solids, spin diffusion tends to average the different relaxation times present in a sample to a single value. We have already seen in the previous section how the degree of average is exploited to obtain information on the miscibility of different domains on a nanometer scale. It is clear that most of the information arising from the dynamics is lost because of spin diffusion, but these relaxation times can still be used to obtain some "global" information about motions. Indeed, spin diffusion affects the single values of the relaxation times (either $T_{1\rho}$ or T_1), but not their PWRA (Population Weighted Rate Average), defined as

$$\mathrm{PWRA} = \sum_i \frac{w_i}{T_{1(\rho)i}} \tag{14.3}$$

where i runs over the number of T_1 or $T_{1\rho}$ components, and w_i is the fractional weight of the ith component. Of course, when a single relaxation time is measured as a result of a complete averaging effect of spin diffusion, $\mathrm{PWRA} \equiv 1/T_{1(\rho)}$.

PWRA does not change with spin diffusion and, therefore, independently of the completeness of its averaging effect, PWRA of the measured relaxation times is equal to that of the intrinsic relaxation times, which would be determined by dynamics only [38]. PWRA can therefore be seen as a quantity related to the global dynamics occurring in a system, and 1/PWRA trends with temperature can be analyzed with dynamic models similar to what done for ^{13}C T_1. Since ^1H spin–lattice relaxation times can be measured quite easily and quickly, their use is strongly advised whenever one is interested to study the global kilohertz and/or megahertz dynamics of a system, disregarding site-specific information. As an alternative to the measurement of proton T_1 using traditional spectrometers, "relaxometers" have been devised to quickly measure ^1H T_1 as a function of the Larmor frequency (dispersion curves) in the approximately 1 kHz–10 MHz range. This technique, called fast field cycling (FFC) [39, 40], allows the range of motional frequencies that can be investigated by proton T_1 to be extended down to the whole kilohertz range.

The proton relaxation times have been often used in the literature to study amorphous pharmaceuticals, in particular in attempts to correlate the dynamics of an amorphous phase to its tendency to crystallize. Good examples are shown in the papers by Aso *et al.* [41] and Apperley *et al.* [6, 42].

14.5
A Case Study

An interesting case of a pharmaceutical compound that shows different kinds of disorder, both static and dynamic, in different solid forms is that of sodium ibuprofen.

The polymorphism of sodium ibuprofen includes the chiral behavior of the crystalline forms: it is known that racemic sodium ibuprofen (*RS*) in the solid state forms both racemic compounds and racemic conglomerates. At ambient conditions, the dihydrated racemic compound (*RS*-H) is the stable form, while the anhydrous compound is known to be highly hygroscopic. By heating the dihydrated racemic compound at a temperature higher than 60 °C and ambient pressure, quite unexpectedly, the anhydrous racemic conglomerate (*RS*-A) is obtained, as recently demonstrated by calorimetric, XRD, and SSNMR techniques [43].

It is interesting to compare the ^{13}C CP-MAS spectra of *RS*-H and *RS*-A and of the corresponding forms of the pure *S* enantiomer: dihydrated (*S*-H) and anhydrous (*S*-A). The loss of the water molecules changes the crystal form of both *RS* and *S* and the corresponding ^{13}C spectra (Figure 14.16). It is worth noting that

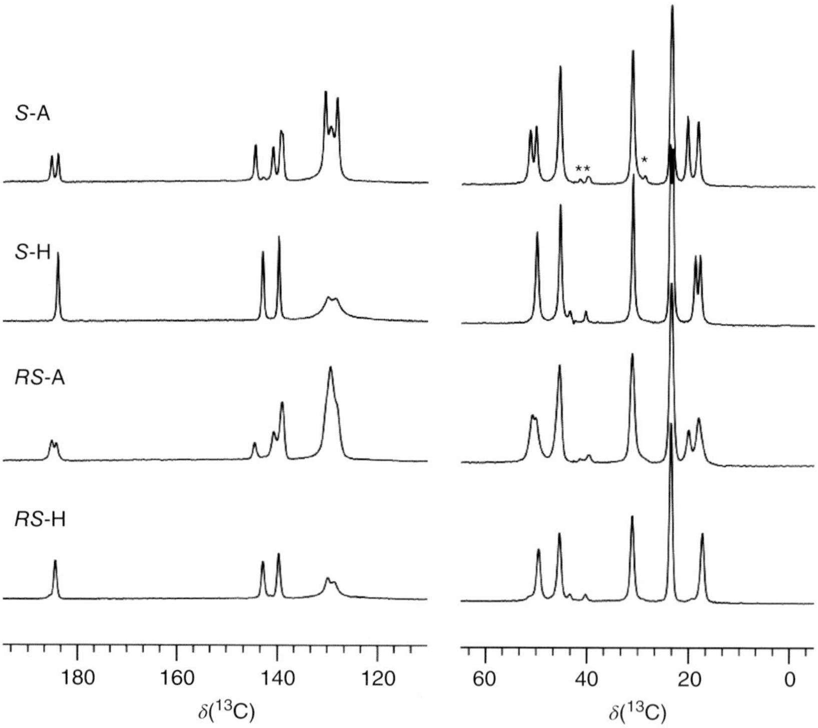

Figure 14.16 ^{13}C CP-MAS spectra of *RS*-H, *RS*-A, *S*-H and *S*-A. Asterisks denote spinning side-bands. Adapted from Rossi *et al.* [43]. Reproduced with permission of American Chemical Society.

all the signals of *RS*-A appear broader (on average by about 30%) than those of *RS*-H, *S*-H, and *S*-A. This suggests that, even though *RS*-A is crystalline, it exhibits a larger degree of structural disorder compared to the other forms. By comparison with the pure *S* enantiomer's behavior, this disorder must be mainly ascribed to the molecular rearrangement taking place on passing from the racemic compound to conglomerate, rather than to dehydration. This kind of disorder affects all the signals, so it can be interpreted as *one-phase disorder*.

On the other hand, site-specific conformational disorder of both static and dynamic nature has been observed and thoroughly characterized in *RS*-H. From the ^{13}C spectra of *RS*-H above room temperature (Figure 14.17), we can see that dynamic disorder is present in this form. In particular, the presence of a single signal for the two methyl carbons of the isobutyl group (C11, C12) and of only two partially superimposed signals for the tertiary aromatic carbons (C4, C7) clearly indicates that above room temperature the interconformational motions of the isobutyl group and the π-flip of the phenyl ring are fast with respect to the reference frequencies of the exchange processes (hundreds of hertz).

We can follow the slowing of the motions from the ^{13}C MAS spectra recorded at decreasing temperatures. However, performing MAS measurements at low temperature is not a trivial task. Commercial instruments are equipped with a system for reaching temperatures as low as 220–210 K, and some advanced instruments can reach about 100 K, but at the time of writing only few machines in the world can perform MAS experiments below 80 K. The results shown in the following have been obtained by using the cryoMAS equipment developed at the University of Southampton, which allows a minimum sample temperature of 9.6 K and a maximum MAS frequency of 15 kHz to be achieved.

In Figure 14.17, the ^{13}C CP-MAS spectra of *RS*-H recorded in the temperature range 20–358 K are shown. First, looking at the ^{13}C signals of the aromatic sites C4–C7 (Figure 14.17) we can observe a dramatic broadening as the temperature is reduced. This behavior is attributed to the slowing down of the π-flip of the phenyl ring, causing first an interference with the ^{1}H decoupling field, then with MAS frequency, and, finally, the coalescence due to the exchange between C4 and C5, and between C6 and C7. The intermediate regimes due to these three phenomena are not distinguishable, and the narrowing of the lines occurring below 185 K indicates that the slow motional regimes for both exchange and interferences have been reached. Below 185 K, the phenyl ring seems to be frozen in a single conformation, as indicated by the presence, for carbons C4–C7, of one peak at 128.4 ppm and three overlapping signals centered at 130.6, 129.8, and 128.9 ppm, identified by spectral fitting. The further broadening observed at a temperature lower than 139 K is partially due to the superimposition with a spinning sideband of the carboxylic carbon signal. This is an example of how SSNMR can identify the dynamic disorder for a specific molecular moiety, and highlights how it progressively evolves, as the temperature is lowered, into a situation of static order.

Looking at the isobutyl peaks (C8, C10, C11, C12), a dramatic broadening is also observed, which starts at lower temperatures (below 159 K) and reaches its

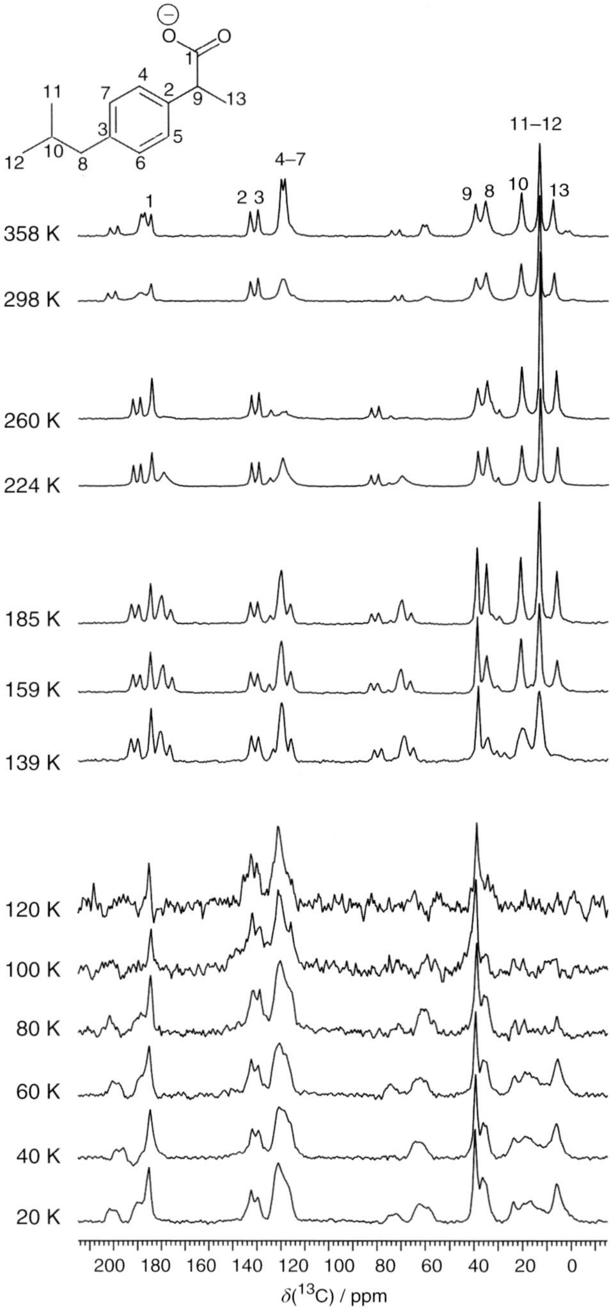

Figure 14.17 ^{13}C CP-MAS spectra of *RS*-H acquired in the temperature range from 358 down to 20 K. Signal assignment is shown on the spectrum at 358 K. Adapted from Concistrè *et al.* [44]. Reproduced with permission of American Chemical Society.

maximum at about 100 K. As in the case of the phenyl ring motion, this can be explained with the occurrence of interference and exchange phenomena due to the isobutyl reorientation. In addition, to explain the broadening of the ^{13}C peaks of the methyl sites C11 and C12, the slowing down of the rotation of the methyl groups about their ternary symmetry axes must be also taken into account. When the sample is cooled further, the peaks do not narrow; instead, the relevant spectral regions display a complex and broad inhomogeneous line shape (see the spectrum at 20 K in Figure 14.17). This indicates that this hydrophobic fragment of the molecule freezes in a wide distribution of conformations. Contrary to what was previously observed for the phenyl ring motion, in this case the analysis of ^{13}C MAS spectra shows how by lowering the temperature the local dynamic disorder evolves into a situation of static disorder.

As far as the hydrophilic part of the molecule is concerned, a very different behavior is observed. The peaks associated with the ^{13}C sites C1 and C9 remain relatively sharp in the whole range of temperatures investigated, indicating the presence of a single conformation, or at most of a narrow distribution of local conformations. The peak of the ^{13}C methyl site C13 clearly broadens below 180 K, becoming almost invisible at 120 K, due to the slowing down of the methyl rotation about its ternary symmetry axis. Different from the C11 and C12 methyl groups, the signal of C13 sharpens considerably as the temperature is reduced below 80 K, in agreement with the high degree of order of the isopropionic fragment, already inferred from the behavior of C1 and C9 signals.

The dynamic disorder present in *RS*-H has also been quantitatively characterized in the temperature range 220–350 K by means of relaxation measurements. In particular, by the analysis of variable-temperature ^{1}H T_1 and $T_{1\rho}$, and ^{13}C T_1, correlation times and activation energies of all the molecular motions occurring in this form have been determined.

Since the analysis of ^{13}C spectra indicated that the only motion having a characteristic frequency of the order of tens of kilohertz in this temperature range is the phenyl ring π-flip, ^{1}H $T_{1\rho}$ and FFC measurements could be used for selectively characterizing it. ^{1}H $T_{1\rho}$ versus temperature curves measured at three different spin-lock frequencies as well as ^{1}H T_1 versus Larmor frequency dispersion curves were simultaneously fitted to suitable equations containing correlation times and activation energy as fitting parameters. A good correspondence between calculated and experimental curves was obtained, as shown in Figure 14.18.

A quantitative analysis of the rotations of the three methyl groups about their ternary symmetry axes and the reorientation of the isobutyl fragment was also performed by fitting proton and carbon T_1 versus temperature curves. In order to obtain more reliable results, a global fitting procedure was applied to the ^{13}C T_1 curves of C8, C10, C11/C12, and C13 carbons as well as the ^{1}H T_1 curves obtained at two different Larmor frequencies. The activation energies of the C11/C12 and C13 methyl rotations were found to be very similar (about 12 kJ mol^{-1}), while that of the isobutyl group was higher (about 16 kJ mol^{-1}) (Figure 14.19).

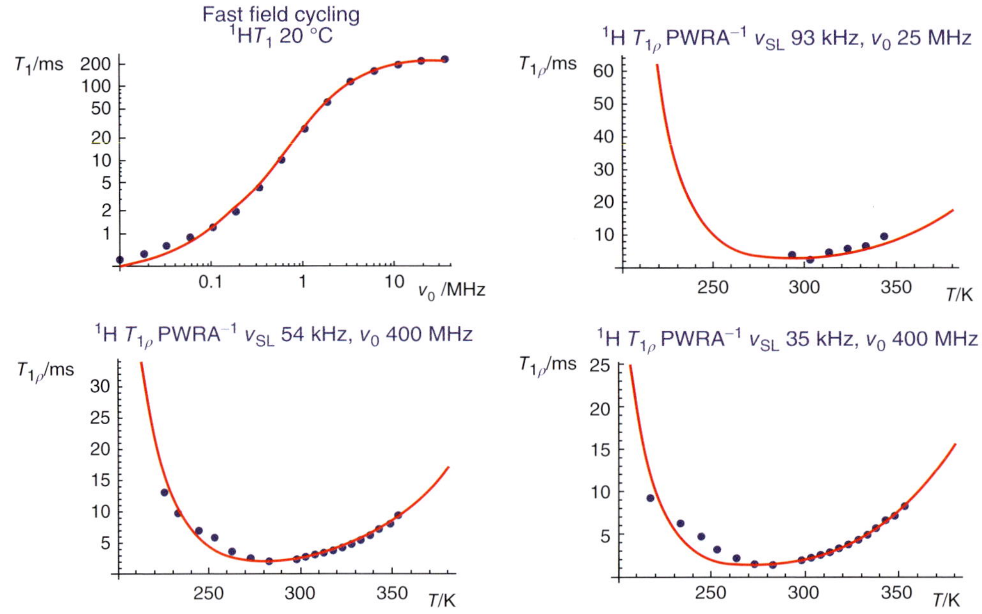

Figure 14.18 Best fitting curves of proton T_1 versus Larmor frequency and proton $T_{1\rho}$ versus temperature of *RS*-H. ν_0 and ν_{SL} indicate the Larmor and spin-lock frequencies, respectively. An activation energy of 32.8 kJ mol^{-1} and a correlation time at 298 K of 0.9 μs for the π-flip of the phenyl ring were determined. The experimental $T_{1\rho}$ data were published in [28].

14.6
Final Remarks and Future Perspectives

In this chapter we have shown how SSNMR spectroscopy and relaxometry can be exploited to investigate various aspects of disorder in pharmaceuticals. In particular, we showed how static disorder can be tackled mainly by the analysis of ^{13}C isotropic chemical shifts and line widths, and how it is possible to distinguish and characterize the different forms of static disorder: conformational site, one-phase, and amorphous. Moreover, relevant aspects such as the quantitation of phases with different degrees of disorder as well as miscibility and interactions in heterophasic systems have been treated.

As far as dynamic disorder is concerned, we have shown how it is possible to easily distinguish it from the static disorder and how very detailed information can be obtained on the interconformational motions affecting the molecules in crystalline phases and on the overall motions characterizing amorphous phases, resorting to a variety of nuclear parameters and experiments. Finally, the case study of ibuprofen has been treated in more detail to demonstrate the practical application of several concepts previously introduced.

Other NMR-related aspects, relevant for the study of pharmaceuticals, were not treated in this chapter, but they at least deserve to be mentioned here. For instance,

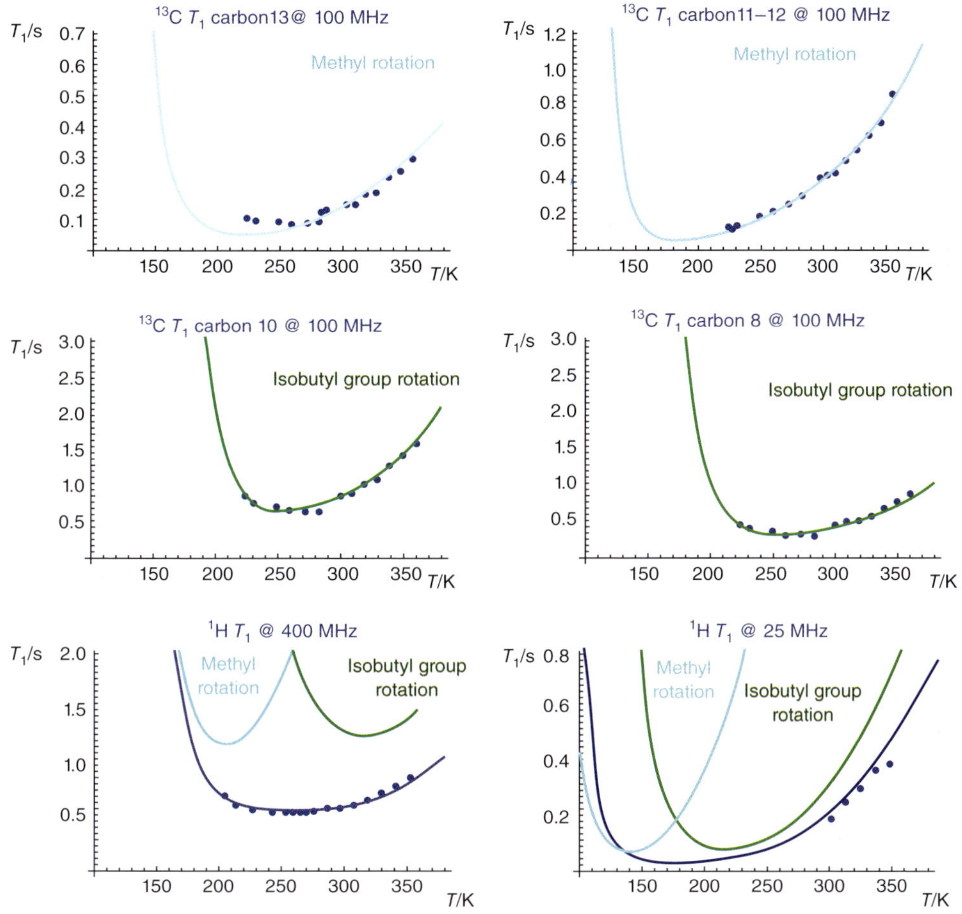

Figure 14.19 Best fitting curves of proton and carbon T_1 versus temperature of *RS*-H. The Larmor frequencies and the ^{13}C nuclei for which T_1 has been measured are indicated in the blue titles. The contributions arising from different motions are reported for each curve. Adapted with permission from [28]. Copyright (2011) American Chemical Society.

the use of magnetic field gradients enables the exploitation of NMR principles for the acquisition of images (MRI) and the study of translational diffusion. In addition to *in vivo* applications, MRI is largely used to study tablets, in order to monitor diffusion rates of dissolution medium or of active substances, as well as solvent penetration and erosion of the tablet. This is of clear importance in both the control of tablets during development and manufacturing, and to study the release of the active ingredient.

Moreover, some recent developments of the SSNMR techniques deserve to be cited, as they will probably largely affect future applications. In addition to the progressive improvements in magnetic field strengths and MAS frequencies, these include, but are not limited to, dynamic nuclear polarization (DNP), the use of microcoils and multiple receivers, and the complementary advances in density functional theory (DFT) calculations of nuclear properties. It is possible to see how most of the efforts are focused on sensitivity increases. Indeed, big sensitivity gains, in addition to saving experimental time, will enable the realization of difficult experiments, such as natural abundance ^{17}O and ^{15}N spectra, and complex multidimensional experiments highlighting correlations or specific anisotropic interactions. In particular, DNP is based on the saturation of the electronic spin transition of radicals, which brings about polarization enhancement of nuclei that are close to them. It has been recently shown how the sensitivity gain in the case of drugs can be huge. The typical drawback is the need for dispersing the drug in a glassy matrix of a solvent containing the polarization agent: this sample preparation, however, strongly affects the solid state properties of the drug. To this aim, solvent-free approaches to sample preparation have been proposed [45], which would represent a real breakthrough in the application of DNP in the pharmaceutical field.

Other technological advances allowing the increase of NMR sensitivity are represented by the use of microcoils, which are of fundamental importance for the study of samples available in small amounts, and multiple receivers, allowing several sets of compatible experiments to be simultaneously run on different nuclei.

Finally, the extraction of structural information in pharmaceuticals from SSNMR experiments will hugely benefit from increased accuracy and reliability of DFT calculations of nuclear parameters in the future, which will be hopefully extended to larger molecules and heavier nuclei.

References

1. Schmidt-Rohr, K. and Spiess, H.W. (1994) *Multidimensional Solid-State NMR and Polymers*, Academic Press, San Diego, CA.
2. Duer, M.J. (2001) *Solid-State NMR Spectroscopy Principles and Applications*, Blackwell Science Ltd., Oxford.
3. Levitt, M.H. (2008) *Spin Dynamics*, John Wiley & Sons, Ltd., Chichester.
4. Apperley, D., Harris, R., and Hodgkinson, P. (2012) *Solid-State NMR: Basic Principles & Practice*, Momentum Press, New York.
5. Zencirci, N., Gelbrich, T., Apperley, D.C., Harris, R.K., Kahlenberg, V., and Griesser, U.J. (2010) *Cryst. Growth Des.*, **10**, 302–313.
6. Apperley, D.C., Forster, A.H., Fournier, R., Harris, R.K., Hodgkinson, P., Lancaster, R.W., and Rades, T. (2005) *Magn. Reson. Chem.*, **43**, 881–892.
7. Paris, M., Bizot, H., Emery, J., Buzaré, J., and Buléon, A. (1999) *Carbohydr. Polym.*, **39**, 327–339.
8. Harris, R.K., Wasylishen, R.E., and Duer, M.J. (eds) (2009) *NMR Crystallography*, John Wiley & Sons, Ltd., Chichester.
9. Clawson, J.S., Kennedy-Gabb, S., Lee, A.Y., and Copley, R.C.B. (2011) *J. Pharm. Sci.*, **100**, 4302–4316.
10. Pawlak, T. and Potrzebowski, M.J. (2014) *J. Phys. Chem. B*, **118**, 3298–3309.
11. Berendt, R.T. and Munson, E.J. (2011) *J. Pharm. Sci.*, **100**, 1879–1891.
12. Sakellariou, D., Brown, S.P., Lesage, A., Hediger, S., Bardet, M., Meriles, C.A., Pines, A., and Emsley, L. (2003) *J. Am. Chem. Soc.*, **125**, 4376–4380.
13. Cadars, S., Lesage, A., Pickard, C.J., Sautet, P., and Emsley, L. (2009) *J. Phys. Chem. A*, **113**, 902–911.

14. Pham, T.N., Watson, S.A., Edwards, A.J., Chavda, M., Clawson, J.S., Strohmeier, M., and Vogt, F.G. (2010) *Mol. Pharm.*, 7, 1667–1691.
15. Lefort, R., Bordat, P., Cesaro, A., and Descamps, M. (2007) *J. Chem. Phys.*, **126**, 014510.
16. Lefort, R., Bordat, P., Cesaro, A., and Descamps, M. (2007) *J. Chem. Phys.*, **126**, 014511.
17. Offerdahl, T.J., Salsbury, J.S., Dong, Z., Grant, D.J.W., Schroeder, S.A., Prakash, I., Gorman, E.M., Barich, D.H., and Munson, E.J. (2005) *J. Pharm. Sci.*, **94**, 2591–2605.
18. Lefort, R., De Gusseme, A., Willart, J.-F., Danède, F., and Descamps, M. (2004) *Int. J. Pharm.*, **280**, 209–219.
19. Tan, I., Flanagan, B.M., Halley, P.J., Whittaker, A.K., and Gidley, M.J. (2007) *Biomacromolecules*, **8**, 885–891.
20. Sperger, D.M. and Munson, E.J. (2011) *AAPS PharmSciTech*, **12**, 821–833.
21. Brus, J., Urbanova, M., Sedenkova, I., and Brusova, H. (2011) *Int. J. Pharm.*, **409**, 62–74.
22. Paudel, A., Geppi, M., and Van Den Mooter, G. (2014) *J. Pharm. Sci.*, **103**, 2635–2662.
23. Policianova, O., Brus, J., Hruby, M., Urbanova, M., Zhigunov, A., Kredatusova, J., and Kobera, L. (2014) *Mol. Pharm.*, **11**, 516–530.
24. Antzutkin, O.N., Shekar, S.C., and Levitt, M.H. (1995) *J. Magn. Reson., Ser. A*, **115**, 7–19.
25. Abragam, A. (1961) *Principles of Nuclear Magnetism*, Oxford University Press, London.
26. Beckmann, P.A. (1988) *Phys. Rep.*, **171**, 85–128.
27. Carignani, E., Borsacchi, S., and Geppi, M. (2011) *ChemPhysChem*, **12**, 974–981.
28. Carignani, E., Borsacchi, S., and Geppi, M. (2011) *J. Phys. Chem. A*, **115**, 8783–8790.
29. Apperley, D.C., Markwell, A.F., Frantsuzov, I., Ilott, A.J., Harris, R.K., and Hodgkinson, P. (2013) *Phys. Chem. Chem. Phys.*, **15**, 6422–6430.
30. Rodríguez-Molina, B., Pozos, A., Cruz, R., Romero, M., Flores, B., Farfán, N., Santillan, R., and Garcia-Garibay, M.A. (2010) *Org. Biomol. Chem.*, **8**, 2993–3000.
31. Chandrappa, R.K., Ochsenbein, P., Martineau, C., Bonin, M., Althoff, G., Engelke, F., Malandrini, H., Castro, B., El Hajji, M., and Taulelle, F. (2013) *Cryst. Growth Des.*, **13**, 4678–4687.
32. Ren, P., Reichert, D., He, Q., Zhang, L., and Tang, H. (2011) *J. Phys. Chem. B*, **115**, 2814–2823.
33. Latosińska, J.N., Latosińska, M., and Medycki, W. (2009) *J. Mol. Struct.*, **931**, 94–99.
34. Glaser, R., Novoselsky, A., and Shiftan, D. (2000) *J. Org. Chem.*, **65**, 6345–6353.
35. Vogt, F.G., Brum, J., Katrincic, L.M., Flach, A., Socha, J.M., Goodman, R.M., and Haltiwanger, R.C. (2006) *Cryst. Growth Des.*, **6**, 2333–2354.
36. Vogt, F.G., Dell'Orco, P.C., Diederich, A.M., Su, Q., Wood, J.L., Zuber, G.E., Katrincic, L.M., Mueller, R.L., Busby, D.J., and Debrosse, C.W. (2006) *J. Pharm. Biomed. Anal.*, **40**, 1080–1088.
37. Schmidt-Rohr, K., Klauss, J., and Spiess, H.W. (1992) *Macromolecules*, **25**, 3273–3277.
38. Kenwright, A.M. and Say, B.J. (1996) *Solid State Nucl. Magn. Reson.*, **7**, 85–93.
39. Kimmich, R. and Anoardo, E. (2004) *Prog. Nucl. Magn. Reson. Spectrosc.*, **44**, 257–320.
40. Noack, F. (1986) *Prog. Nucl. Magn. Reson. Spectrosc.*, **18**, 171–276.
41. Aso, Y., Yoshioka, S., and Kojima, S. (2000) *J. Pharm. Sci.*, **89**, 408–416.
42. Forster, A., Apperley, D., Hempenstall, J., Lancaster, R., and Rades, T. (2003) *Pharmazie*, **58**, 761–762.
43. Rossi, P., Macedi, E., Paoli, P., Bernazzani, L., Carignani, E., Borsacchi, S., and Geppi, M. (2014) *Cryst. Growth Des.*, **14**, 2441–2452.
44. Concistrè, M., Carignani, E., Borsacchi, S., Johannessen, O.G., Mennucci, B., Yang, Y., Geppi, M., and Levitt, M.H. (2014) *J. Phys. Chem. Lett.*, **5**, 512–516.
45. Ong, T.-C., Mak-Jurkauskas, M.L., Walish, J.J., Michaelis, V.K., Corzilius, B., Smith, A.A., Clausen, A.M., Cheetham, J.C., Swager, T.M., and Griffin, R.G. (2013) *J. Phys. Chem. B*, **117**, 3040–3046.

15
Processing-Induced Disorder in Pharmaceutical Materials

Sheng Qi

15.1
Introduction

Disorder in pharmaceutical materials can be introduced through processing. A good example of unintentionally introduced disorder into the formulation is the generation of a small number of defects at the surface of the tablets during the compression of tableting process [1, 2]. The degree of the disorder can range widely and may be as high as that found in amorphous materials, or it may be relatively low such as is seen in localized defects. The defects can be in the form of point defects, line defects, or plane defects [3]. These defects often can exhibit different levels of effect in terms of altering the physical and mechanical properties of the finished product. If the amount of accumulated defects in the system reaches a critical level, the material can be converted into an amorphous state, and this conversion can impact on the clinical performance of the formulations. The nature and quantity of the disorder are dependent on the intrinsic properties of the material and the particular process used to manipulate the material. A large amount of data has indicated that many of the different processes used for pharmaceuticals may cause disorder of the material. As an example, many recent works have reported the different physical stabilities of amorphous formulations produced by different methods [4–6]. This is likely associated with the different levels of disorder created by different processing methods.

The amorphous state represents the highest level of disorder, which can be recognized through the lack of organized long-range order. Often, during the production of amorphous materials, other disordered states such as different degrees of defects (either different quantity or different type of defects) can be created. The production of amorphous materials can be classified into thermodynamic- and kinetic-based methods, as indicated in Figure 15.1 [3]. Thermodynamic methods are those that follow the equilibrium phase transformation from liquid to the solid state. In these methods, the processed materials are transformed to their liquid state via either melting or dissolving in a solvent. The thermodynamic methods include melt-based methods such as melt-quenching, melt-granulation, hot-melt extrusion (HME), spray-congealing, dry powder coating, thermal

Disordered Pharmaceutical Materials, First Edition. Edited by Marc Descamps.

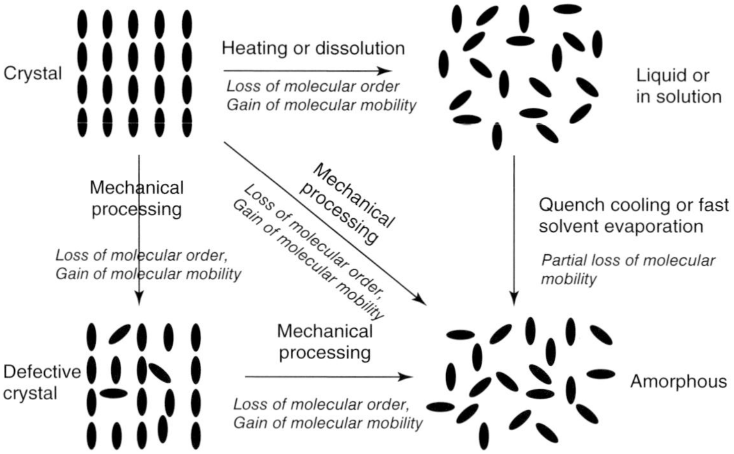

Figure 15.1 Process-induced disorder. (Adopted from Feng *et al.* [3]. Reproduced with permission of Elsevier)

sintering/curing, and solvent-evaporation-based methods such as spray-drying, freeze-drying and electrospinning. Kinetic-based methods rely on the gradual disruption of the molecular arrangement of a crystalline material by progressively creating defects. A typical kinetic-based method is milling. Once the accumulation of the defects reaches a critical level, the material is likely to be converted into an amorphous state. This chapter focuses on the major types of pharmaceutical processing methods that have been widely reported to produce disordered material either intentionally or unintentionally.

15.2 Pharmaceutical Processing

15.2.1 Milling

Milling is common and one of the most frequently used unit operations used by the pharmaceutical industry for reducing the particle size of solids. When milling is applied to crystalline solids, as a result of the mechanical stresses inherent to the milling process, disorder in the crystalline structure of the active pharmaceutical ingredient (API) can be introduced either unintentionally or intentionally. These disorders can take the form of defects in the crystal lattice, polymorphic transformations, complete loss of the crystalline lattice, and conversion to the amorphous state [7–11]. These structural changes may be located only at the surface of the solids, or they may penetrate through the bulk of the material. Depending on the proportion of the material affected, the presence of the disorder can often lead to changes in the physical and chemical properties of the milled solids [11].

The conversion of a crystalline solid to its amorphous state via milling has been explained by three possible mechanisms: accumulation of crystal defects with a higher free energy level than a certain critical level under kinetically favorable conditions [7, 12]; local overheating during milling, which leads to melting and rapid quenching of the API [7, 8, 11]; and obtaining the amorphous form of the API from the intermediate state formed during polymorphic conversion of the crystalline solid. For instance, the milling of form IV of fananserine could lead to its intermediate amorphous state, which precedes the transformation of the metastable form I polymorphs [11].

The possibility of intentionally generating an amorphous pharmaceutical product through milling or cryo-milling (milling at extremely low temperature) has been reported [7, 11–23]. The amount of disorder induced by milling is governed by many factors including the milling intensity, temperature, additives, and the intrinsic properties of the solid such as glass transition temperature and crystal structure [7, 11–23]. When milling at temperatures below the glass transition of the amorphous API, the increase in Gibbs free energy drives the conversion of the crystalline form to the amorphous form, as illustrated in Figure 15.2 [11]. In these cases (milling at temperatures below the T_g of the API), the milling intensity is crucial in determining the nature of the transformation [11]. Amorphous conversion occurred when the γ-form of crystalline indomethacin was milled with high intensity at room temperature, which is below the T_g of amorphous

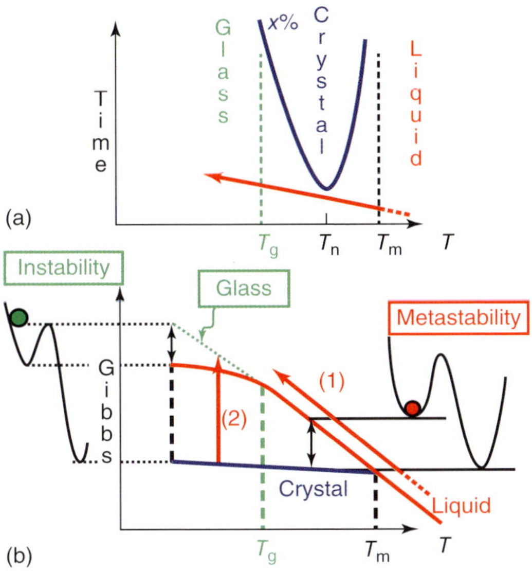

Figure 15.2 (a) Schematic TTT diagram for crystallization in an undercooled melt and conventional vitrification path. (b) Gibbs free enthalpy curve for crystal, liquid (stable and metastable), and glass. Red colour-coded paths 1 and 2 correspond to conventional and solid-state vitrification paths, respectively. (Adopted from Willart and Descamps [11]. Reproduced with permission of ACS)

indomethacin, whereas only polymorphic transformation occurred when milled with lower intensity below T_g [7]. In contrast, regardless of intensity, milling at a temperature above T_g only led to crystalline–crystalline polymorphic transformation [7].

Recently, mechanical milling has also been used to produce solid dispersions and co-amorphous systems [11, 24–29]. However, studies have shown that milling produces less physically stable solid dispersions compared to those prepared by other methods such as spray-drying and melt-quenching [30–32]. Some researchers have attributed this low physical stability to the heterogeneous relaxation of the milled solid dispersions, as indicated by the lower value of the relaxation distribution in comparison to the same systems prepared through spray-drying or melt-quenching [31]. It is also possible that a high proportion of nuclei are present in solid dispersions prepared by milling, which are available to undergo recrystallization on storage. Recently, the formation of co-amorphous materials by milling has been reported. These are complexes of poorly water-soluble drugs in a solid state with other small molecular entities, mainly via intermolecular interaction such as salt formation and hydrogen bonding [29] or interparticle hydrogen bonding [33]. The complexed solids are amorphous in nature. However, the co-amorphous materials produced by milling have been reported to have enhanced physical stability and dissolution rate in comparison to the amorphous APIs alone [26–29]. This has been attributed to the intermolecular interactions between the two drug species at certain stoichiometric ratios, which constrain the molecular mobility of amorphous APIs [26–29].

15.2.2
Thermal Processing Techniques

The basic principles of pharmaceutical thermal processing techniques are partially or completely based on liquefying either the API alone or the API with excipients via heating. Thermal processing techniques are mainly used for controlling or improving the release and the subsequent bioavailability of an API. In many formulations processed by thermal techniques, the disorder states were introduced into the API and the polymeric excipients during the cooling process. The conversion of the crystalline state to a more disordered state, whether completely amorphous or semicrystalline, can facilitate better control of the drug release by altering the properties of the excipients in the formulations. A wide range of thermal processing techniques have been developed or adopted from other industries to pharmaceutical use over the past few decades. Techniques such as melt-mixing, spray-congealing, sintering, melt-granulation, and HME have developed and evolved rapidly for large-scale pharmaceutical production [34]. New emerging thermal processing techniques such as dry powder coating, injection molding, melt-electrospinning, KinetiSol® dispersing, and melt-based 3D printing have been explored for their potential use for scaled-up pharmaceutical manufacturing [34, 35].

15.2.2.1 Simple Melt-Fusion Method

Melt-fusion is the simplest form of thermal processing. It directly liquefies the API and/or excipient, followed by the cooling of mixture into a solid form. If a carrier excipient is involved, mixing during the molten stage is often facilitated by stirring. The common carrier materials are usually lipids and low molecular weight (MW) biopolymers such as fatty acids, triglycerides and self-emulsifying lipids, and low-MW synthetic polymers, such as polyethylene glycol (PEG), Glycerol mono-oleate (GMO), and poloxamers. These materials are mostly crystalline or semicrystalline in nature and have relatively low melting points. The choice of the carrier material plays an important role in controlling or enhancing the dissolution of the API. The molten lipids or polymers can act as the solvent to dissolve the API at temperatures below the melting point of the pure crystalline API. The advantage of using such carrier materials is that they often can facilitate the amorphization of the drug via dissolving the drug in the carrier material during the molten stage.

After cooling, the crystalline state of the excipient material may also be disrupted by the incorporation of the API. In the case of triglycerides, the lipids often solidify as their metastable polymorphs and slowly transform into the stable polymorph on aging [37–39]. This often leads to physical stability concerns regarding the formulation [37–39]. For enhancing the dissolution of poorly soluble APIs, surface-active and self-emulsifying polymers such as Gelucire and poloxamer are often used as carrier materials. Although in some cases the poorly water-soluble APIs are not completely converted into the amorphous state, improvements in the dissolution of the APIs are often still evident [39, 40]. This has been attributed to a number of possible factors including the disorder introduced during the melt cooling (such as the formation of microfine crystals after recrystallization of the drug in the carrier material and partial amorphous conversion) as well as improved solubilization and wetting by the presence of surface-active polymers and lipids [39, 40].

The cooling rate of the melt is one of the most important processing parameters that can affect the degree of disorder. In the case of forming solid dispersions of API in a meltable carrier material, a faster cooling rate has been associated with a lower level of drug recrystallization on aging, rapid reaching of supersaturation state in the dissolution media after dissolution, and enhanced bioavailability of the formulated drug [41, 42]. Melt-quenching is also one of the commonly used methods for converting a crystalline drug to its amorphous state. The melting method can be applied only to compounds that are thermally stable upon melting [43, 44]. For example, piroxicam starts to degrade upon melting, and thus the use of melting methods may not be suitable for this compound [44]. In some cases, it was reported that even the time and temperature allowed for complete melting could affect the physical properties of the generated amorphous system [32, 45]. For example, Van den Brande *et al.* reported a significant degradation of loviride processed using slow cooling [45], but obtained a stable amorphous form by using quench-cooling.

15.2.2.2 Spray-Chilling/Congealing

Spray-chilling is a thermal-based atomization method for producing spherical microparticles containing lipid/polymer excipients and API. The basic principle of spray-chilling is illustrated in Figure 15.3. Droplets of the molten excipients containing either molten API or suspended undissolved API are formed by the atomizers and solidify in the collector [46]. The additional advantage of spray-chilled formulations in comparison to formulations prepared by the melt-fusion method is their small particle size, which leads to a high surface area to volume ratio. This large surface area, at least partially, contributed to the observed improved dissolution of spray-chilled Gelucire–piroxicam microspheres despite that the API was still largely in its crystalline state [40]. During the spray-chilling process, the lipid/polymer excipients are melted and cooled. The cooling process leads to the solidification and recrystallization of the processed materials. This often causes increased disorder of the carrier lipids (recrystallizing to a metastable polymorph) and the API [37–39]. Similar to the melt-fusion method, this can be attributed to the melting and recrystallization process of the formulations.

15.2.2.3 Melt-Granulation

Melt-granulation is a process that involves agglomeration of the API with meltable binders via direct heating or frictional heating during high-shear mixing. The binder material can be premelted and sprayed/dripped into the granulator to form granules, or premixed with the API and then melted during the granulation process [47]. The most popular industrial-scale melt-granulation can be achieved by using fluid-bed melt granulators and high-shear mixers [47]. The granules can be formed through either an immersion-dominated or a coalescence-dominated

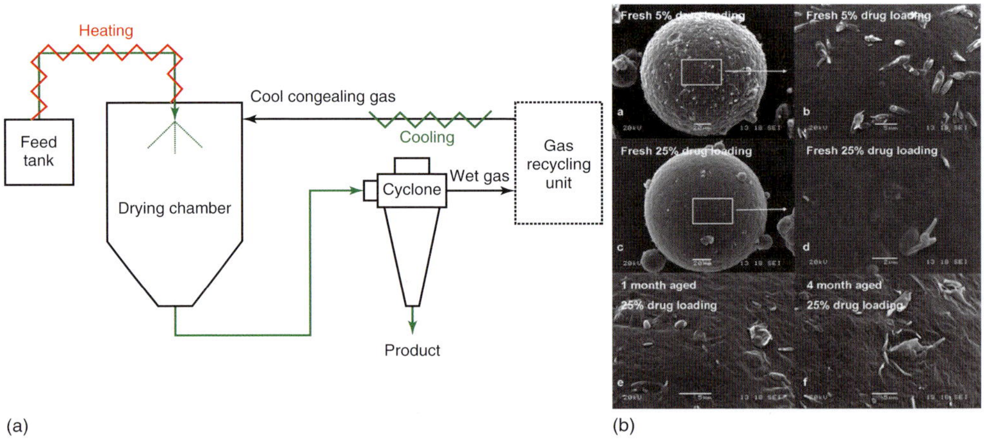

Figure 15.3 (a) Basic operational principle of spray-chilling/congealing (Adapted from *Qi et al.* [40]. Reproduced with permission of Wiley.) and (b) evident drug and lipid excipient instability over aging. (Adapted from Cordeiro *et al.* [46]. Reproduced with permission of Chimica Oggi-Chemistry Today.)

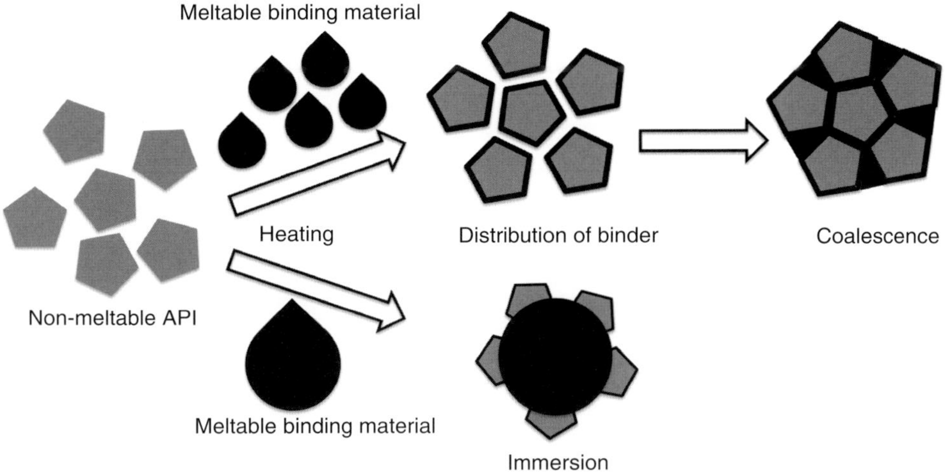

Figure 15.4 Possible mechanisms of granule formation via melt-granulation.

mechanism, as illustrated in Figure 15.4. The mechanism of the formation of granules is highly dependent on the amount of molten binder added and the granulation method used [47].

During the melt-granulation process, the processing temperature is often below the melting point of the API. Thus disorder is mainly introduced into the carrier/binder materials, which experience melting and cooling. The commonly used binder materials for melt-granulation are similar to those used for spray-chilling. These are meltable materials with relatively low melting temperatures (between 50 and 80 °C), such as Compritol 888 ATO, Gelucire, and low-MW PEG. A number of studies have reported the combined use of HME with melt-granulation in order to incorporate polymeric binders with high glass transition temperatures, such as hydroxypropyl methylcellulose (HPMC) and ethylcellulose (EC) into the granules. These polymeric additives have shown good capabilities for modulating the release and disintegration rates of tablets formed by the granules produced by melt-granulation [48, 49].

15.2.2.4 **Thermal Sintering/Curing**

Thermal sintering/curing is a less dynamic method in comparison to other thermal processing techniques. It is an additive curing process for preformed pellets, granules, and compressed tablets under elevated temperature [50, 51]. The process can be applied to matrices containing meltable excipients (such as lipids and semicrystalline polymers) or polymers that can be softened during curing [50, 51]. Sintering/curing is often used to modify, in most cases prolong, the drug release behavior of the formulation. It has been reported that the sintering process can cause redistribution of excipients that are thermally responsive [41, 50]. This redistribution of material is believed to be achieved by melting (for meltable excipients) or through increasing the molecular mobility of the excipients in the formulations.

The added advantages of this melting/fusion of excipients in the formulation are reduced porosity (as seen in Figure 15.5) and the subsequent increase in surface hydrophobicity of the formulation attributed to the coating [52]. This decreased porosity and increased surface hydrophobicity can reduce the wettability of the formulation and further contribute to the decrease in the drug release rate [52]. For meltable excipients, the thermal sintering/curing process leads to the recrystallization of the excipient after the curing process is completed. Although scarcely reported in the literature, it is reasonable to predict that lipid-based excipients will go through metastable to stable polymorphic transformation on aging, and the degree of crystallinity of a semicrystallized polymer may be reduced after the sintering process. For polymeric excipients with higher glass transition temperatures, the thermal curing process may introduce an increased degree of cross-linking and structural relaxation is expected after the curing process [50, 52].

15.2.2.5 Dry Powder Coating

Dry powder coating is a process using thermal sintering/curing that converts a dry powder deposited on a surface to a film coating on the surface. This is a solvent-free coating method requiring sufficient adherence of the dry powder to the surface at the first stage of the process. There are a number of dry powder coating techniques based on the adsorption mechanism of the powder layer including liquid-assisted powder deposition, thermal adhesion deposition, and electrostatic deposition [53, 54]. In order to achieve the desired product attributes, novel excipients are often required for these specific manufacturing processes. The success of a dry powder coating process strongly depends on the physical properties of the coating and substrate materials and the interaction between the interfaces formed during the process [53, 54]. Hydrophilic semicrystalline polymers with a low melting point (e.g., polyethylene glycol 3350) are the most frequently used excipients for dry powder coatings. However, other amphiphilic and hydrophobic excipients, such as Pluronic 127 and cetylstearyl alcohol, have also been used. When the processing temperature is above the glass transition temperature of a coating material, the formed film is more "liquid-like" and more susceptible to plastic deformation. As the relationship between the melt viscosity and temperature follows the Arrhenius equation, reduced viscosity can be achieved by using high processing and curing temperatures, which leads to the formation of a higher quality film. A sufficient reduction of the viscosity of the coating material can result in the formation of capillary forces, which facilitate in the adherence of the coating to the solid surface. Under high temperature conditions, surface energy differentials can act as the driving force for the spreading of the semi-molten polymer and enhance the coating efficiency [53, 54]. This process is demonstrated in Figure 15.6.

15.2.2.6 Hot-Melt Extrusion (HME) and Injection Molding

The first use of HME was reported in 1930s for plastic manufacturing as well as in the food industry. It was invented at the end of the eighteenth century by Joseph Brama for the production of lead pipes. This technique was then extensively used in the plastics industry in the middle of the nineteenth century for providing

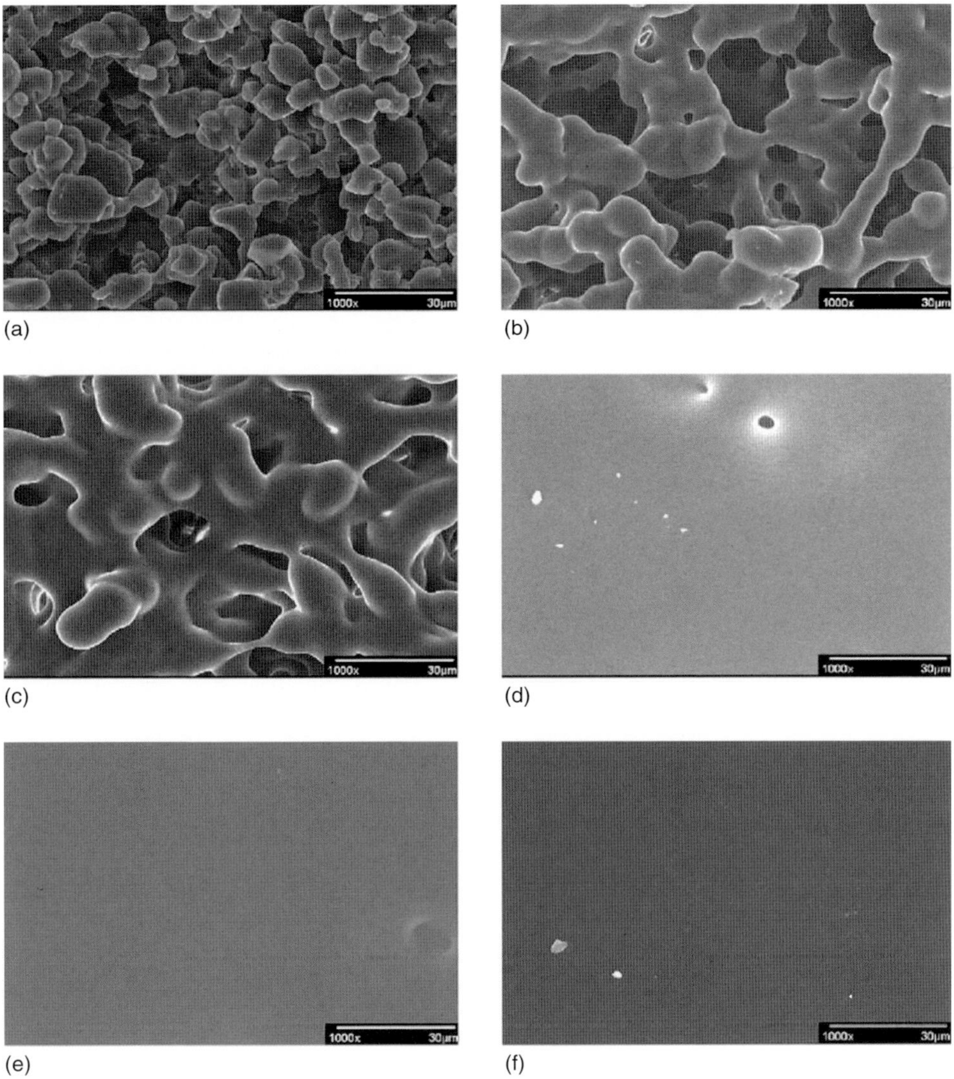

Figure 15.5 SEM images of Eudragit E PO powder films after thermally sintered/cured at 80 °C for (a) 1 h, (b) 2 h, (c) 4 h, (d) 8 h, (e) 12 h, and (f) 24 h. (Adapted from Cerea *et al.* [52]. Reproduced with permission of Elsevier.)

polymeric insulation of wires [55]. More recently, HME has been adopted for producing solid-dispersion-based systems [55, 56]. HME is a single continuous process that involves mixing, melting, homogenizing, and shaping [57]. The ability to process the materials in this way is an obvious requirement associated with HME, and the number of suitable excipients is currently limited. However, the method has the flexibility of being able to use multiple additives such as plasticizers as well as stabilization and solubility/absorption enhancers (such as

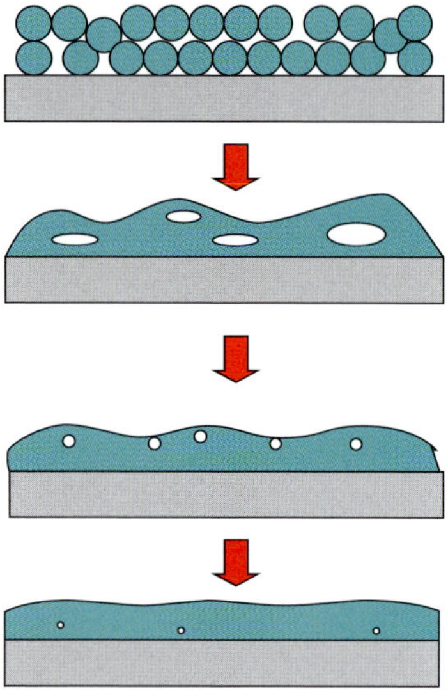

Figure 15.6 Illustration of the essential steps of dry powder coating (all steps occur under elevated temperature conditions). (Adapted from Sauer *et al.* [53]. Reproduced with permission of Elsevier.)

surfactants) to optimize the processing behavior and physical performance of the finished dispersions. Amorphous solid dispersions prepared by HME have shown to be able to enhance the dissolution and absorption of poorly soluble drugs, such as indomethacin, itraconazole, felodipine, celecoxib, and ketoprofen [55–57]. Other potential applications include targeted and controlled-release dosage forms, multiple unit dosage forms, floating dosage forms, implants, transdermal and transmucosal delivery systems, as well as the formation of nanocomposites for prolonged release [55–57].

The HME process can be divided into three typical zones: the feeding zone, the melting or compression zone, and the metering zone. Each of the zones has a different geometrical screw design that dictates the advance of the process for melting and mixing materials [56]. Because of the elevated operating temperature (at least 30–60 °C above the T_g of the material in order to sufficiently soften the processed materials) and the high level of shear generated by the rotation of the mixing barrels, high levels of disorder can be created during HME. Mixtures of polymers, API, and additives are passed through the feeding system and processed within a closed barrel to allow solubilizing of API in the polymeric matrix. At the end of barrel, high pressure applied within the metering zone forces the molten mass to be extruded through an orifice to produce a product of high density and uniformity [55–57]. Subsequently, the molten material is transported to the downstream equipment for final dosage form manipulation such as melt-pelletization, milling, and tableting.

As with other solid dispersion preparation methods, despite the high number of reported successful case studies in the academic literature, the number of hot-melt-extruded amorphous solid dispersion products is still very limited. This can be partially attributed to the poor understanding of the material behavior during HME processing and the post processes. Studies on the development of inline analysis of the dispersions during extrusion processing have been tried, showing promising potential for adoption in industrial production [58, 59]. These analytical tools can not only detect in-process degradation and uneven mixing of the drug and excipients but also be used to study the interaction between the drug and polymers in the molten state and under the shearing of the extrusion process. Advanced local and bulk characterization techniques, such as solid-state NMR (SSNMR), atomic force microscopy (AFM), photothermal-microspectroscopy (PT-MS), and thermal transition mapping (TTM), have been used to study hot-melt-extruded solid dispersions on a submicrometer scale in order to gain a better understanding of the disorder in extruded solid dispersions as a result of processing, such as higher level of surface drug recrystallization in comparison to the core of HME extrudates [60–62].

Injection molding is a manufacturing process of shaping thermoplastic or thermoset materials in a closed mold via the application of high temperature and pressure [63]. It is often used in conjunction with HME. It has the advantage over conventional HME in allowing a single-step shape formation of formulations that require precise and/or complex geometries, such as transdermal microneedles or intravaginal rings [64, 65]. Recently, there has been growing interest in using injection molding as a single-step manufacturing process to produce tablet- and capsule-shaped formulations for oral controlled release [66]. The injection molding process introduces additional stress to the hot-melt-extruded samples to allow the material to remelt, which is then forced into the mold under high pressure. However, in most cases, without additional quenching, the cooling process of injection molding is slower in comparison to HME alone. This is because the mold is a large piece of solid metal that has to be preheated to an elevated temperature. Therefore, once the material is injected into the mold, the cooling of the mold to ambient temperate takes much longer than the cooling of the strands of exudates. It is well known that for thermal-melting-based methods, the disorder created during the process is highly dependent on the cooling rate. Therefore injection-molded solid dispersions are expected to be different from HME dispersions in terms of the degree of disorder. However, this has not been extensively studied and requires more work for a better understanding and control of the injection molding process.

15.2.2.7 **Other Emerging Thermal Processing Techniques**

A range of new thermal processing techniques have been recently developed for manufacturing pharmaceutical formulations. Melt-electrospinning and KinetiSol dispersing (KSD) are two good examples that have shown promising potential in the preparation of solid-dispersion-based formulations. Melt-electrospinning is a modification of the conventional electrospinning technique [35]. This processing technique can produce micro-fibrous structures from polymer melts. The distinct difference between melt-electrospinning and solution-electrospinning (which will be discussed in a later section in this chapter) is that the electrified molten polymer jet lacks the random dynamic chaotic motion observed in solution-electrospinning. This means that the collection of melt-electrospun fibers can be highly controlled and aligned into predesigned patterns, as shown in Figure 15.7 [35, 37]. This has been attributed to the higher viscosity and lack of electrical conductivity of the molten polymer. However, the disadvantage of melt-electrospinning is that the diameters of the produced fibers are within the 5–40 μm range instead of in the nanometer range [35, 37]. Melt-electrospinning has been used to produce amorphous solid dispersions [67], which provided faster dissolution of a poorly soluble model drug in comparison to hot-melt-extruded dispersions. This was attributed to the high surface area of the fibrous formulation. However, the long-term physical stability of the amorphous drug in this form has not been widely studied. As a result of the high surface area, surface instability (such as drug crystallisation at the surfaces of the fibers) maybe expected for this type of formulations on aging.

Another good example of an emerging new thermal-based pharmaceutical process is the patented KSD technology. This process uses the frictional heat generated by the shear of sample powder against the paddles rotating at high speed within a cylindrical chamber to melt the API and the excipients to form solid dispersions [68]. In contrast to other thermal processing techniques for producing solid dispersions, the KSD method of elevating the temperature, which leads to the melting of the material, occurs within seconds. This gives this technique the unique advantage of being able to process many thermos-labile APIs that cannot be processed using other thermal methods. The ultrafast temperature elevation accompanied by high shear has also been reported to be able to achieve a higher level of mixing between the drug and polymer in comparison to other thermal processes such as HME [69]. For example, hot-melt-extruded itraconazole and HPMC show phase separation of drug-rich and polymer-rich domains, whereas only a single miscible phase can be detected by differential scanning calorimetry (DSC) in the KSD solid dispersions [69]. Therefore, KSD technology seems to be a good method for intentionally incorporating a high level of disorder into the processed formulations.

15.2.3
Solvent-Evaporation-Based Processing Techniques

Solvent-evaporation-based methods are important processing techniques for both raw materials, such as crystallization of the raw drug, and formulation manufacturing in the pharmaceutical industry. As crystallization is not within the scope of this chapter, we will solely focus on the processing that can potentially induce the formation of the disordered state during the manufacture of formulations. The widely used solvent-evaporation-based processing techniques in pharmaceutical formulation production include spray-drying, freeze-drying, film casting, and film coating. There are new, innovative techniques still at bench-scale development, such as spray-freeze-drying, solution electrospinning,

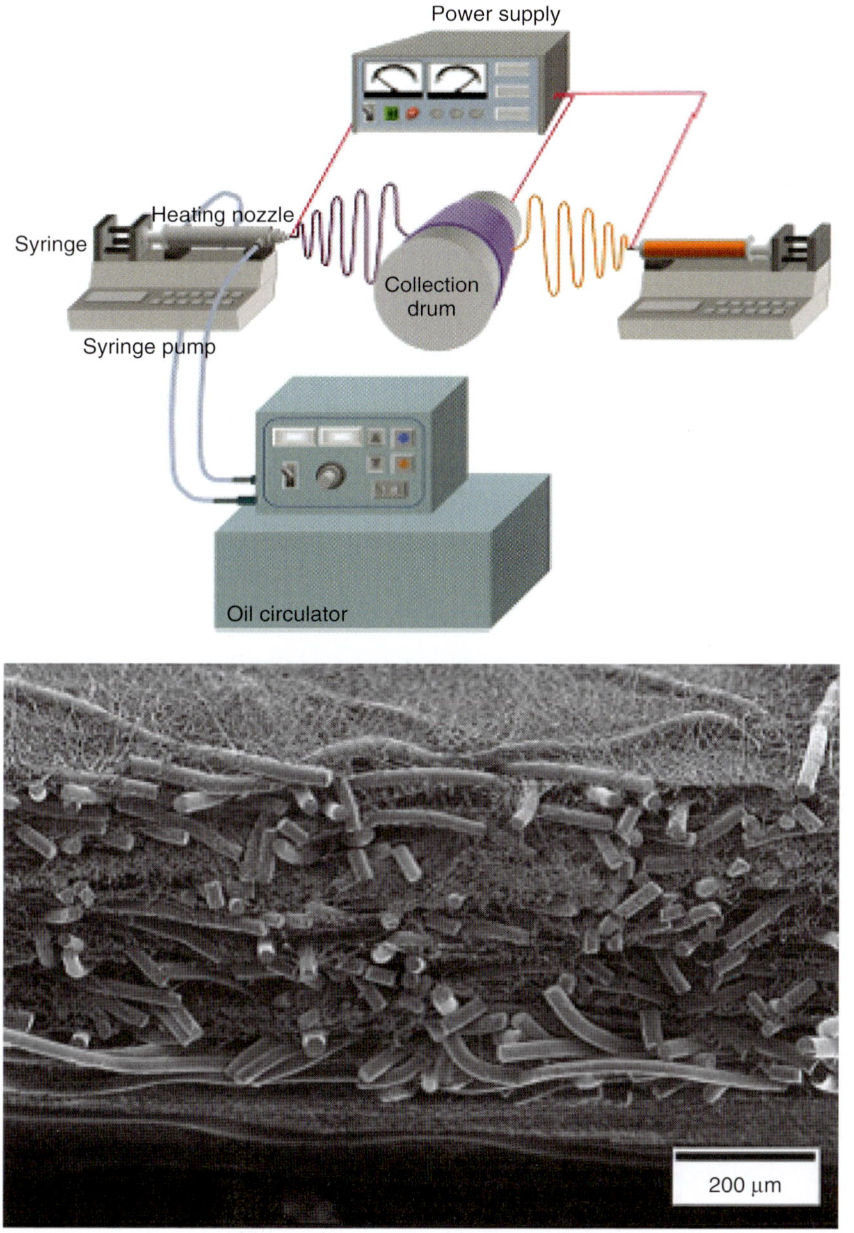

Figure 15.7 Melt-electrospinning setup (top panel) and the micro and nanofibers produced by melt electrospinning (bottom panel). (Adapted from Kim *et al.* [36]. Reproduced with permission of Elsevier.)

and spin coating, which have shown great potential for the preparation of highly sophisticated formulations.

15.2.3.1 Spray-Drying

Spray-drying is a unit operation for transforming liquid solutions or suspensions into solid products by rapidly atomizing the solutions/suspensions and drying them with a hot gas. Modern spray-drying technology can process thermo-liable materials, such as proteins and peptides through careful manipulation of the processing parameters and selection of excipients [70]. It is a well-established technology for the formation of solid microparticles, with applications in the delivery of poorly soluble drugs, proteins, vaccines, and inhalation therapies [70, 71]. The nature of solidifying the formulation via rapid, complete solvent removal leads to an inherent high potential for forming disordered states in the product either intentionally or unintentionally.

A spray-drying process can be divided into the atomization and drying steps. The atomization step produces droplets of the starting solution. The performance of the atomizer directly impacts on the size of the dried particles. There are four main types of industrially used atomizers: rotary atomizers, pressure nozzles, two-fluid nozzles, and ultrasonic atomizers [72]. The droplet size can also be influenced by the feed concentration and solution viscosity of the starting solution. This stage is less likely to contribute to the generation of disorder to the system in comparison to the drying step.

The drying stage is a complex transformation process from droplet to dry particles involving both heat and mass transport in a timescale of milliseconds [73]. During this step, there are rapid and significant changes at the surfaces of the droplets in terms of distribution and movement of the components [71, 73]. For solutions containing polymers and macromolecules, during drying significant surface adsorption of the molecules leads to surface enrichment and saturation followed by shell formation and complete solidification of the hollow particles [71]. However, if the diffusion of the solutes is relatively fast, they remain evenly distributed in the entire droplet. This leads to low surface enrichment and the formation of solid particles instead [71].

Regardless of the geometry of the particles, the formation of disorder during the drying stage is governed by the intrinsic properties of the solutes. For small molecules, whether the solidification process leads to crystallization or the formation of the amorphous form is highly dependent on the glass transition temperature in relation to the drying outlet temperature and the crystallization kinetics of the material. For large molecules with slower diffusion coefficients in solution, the disorder can be concentrated and immobilized at the shell of the particles after drying. Therefore, spray-drying has been used to prepare solid dispersions of small molecular weight drugs with polymeric carriers to intentionally form amorphous state of the drug in order to achieve faster dissolution [70, 71]. The intimate mixing following dissolution in a common solvent with the polymer and rapid drying can have the equivalent effect of fast quenching and "freeze" the drug molecules in a molecular dispersion with the polymeric carrier [70, 73]. For example, the second-generation non-nucleoside reverse transcriptase inhibitor, etravirine, for treating resistant HIV, was unable to provide efficient bioavailability if formulated into nanocrystals or solid dispersions via HME or film casting, but has proven to be successful when formulated as a spray-dried amorphous dispersion with HPMC [6].

15.2.3.2 Freeze-Drying

Freeze-drying, also known as *lyophilization*, is a well-established process that is traditionally used for used for stabilizing labile substances or processing pharmaceuticals of biological origin such as proteins, serum, vaccines, peptide drugs, and liposomes, often administered parenterally [74]. Recently, the application of freeze-drying has been significantly expanded to include manufacturing orally disintegrating tablets (ODTs) and fast dissolving tablets and films (FDTs), wafers for buccal delivery and wound healing, inhalable powders, and 3D scaffolds [75]. Many of the novel applications of freeze-drying are related to its ability to produce low-density and highly porous material.

The freeze-drying process can be divided into three stages: freezing, primary drying, and secondary drying. In the freezing step, the solvent (typically water) is separated from the solute to form ice crystals. This leads to the solute molecules concentrating as separate micro-islands, and this process is termed *freeze concentration* [74]. This formation of phase-separated, concentrated domains and ice crystals is the first step, which may initiate the formation of disordered phases. Therefore, the widespread opinion is that the primary drying stage should be conducted at temperatures below T_g to ensure the maintenance of the stability of macromolecules in particular [76]. However, some studies have demonstrated that freeze-drying at temperatures well above T_g without the need for vitrification can still preserve the activity of proteins as a result of the slow unfolding kinetics of proteins in relation to the timescale of the freeze-drying process [76].

The primary drying process involves the sublimation of ice under high vacuum with elevated shelf-temperature in order to supply the heat required for ice sublimation. The sublimated ice is then condensed on a cooled plate/coils ($< -50\,°C$) in the condenser [74]. The secondary drying phase removes any residual moisture content at a further increased temperature (i.e., 25–60 °C). Applying a low pressure at elevated temperature further dries any "unfrozen" water in the "freeze concentrate." For labile materials such as proteins, the drying temperatures and their duration are critical for retaining protein activity and post-process stability. These two drying steps are time consuming and can give the system time to further develop any disordered phase formed during the freezing step. However, the literature does not show any discussion on the effect of drying conditions on the stability of the freeze-dried, low-MW compounds.

In order to expand the applications of the freeze-drying process to produce microparticles, in recent years, freeze-drying has been used in conjunction with spray-drying. This combined technology is called spray-freeze-drying. Spray-freeze-drying has been used to prepare solid dispersions for the delivery of poorly soluble APIs [77]. A few studies have reported the ability of using freeze-drying to stabilize amorphous drugs in a solid dispersion and the superior physical stability and dissolution performance of freeze-dried solid dispersions compared to other solvent-evaporation-based methods [77, 78]. The underlying mechanism of this enhancement is still poorly understood. This may be attributed to the low operating temperatures in comparison to other solvent-based methods. However, the powder obtained from this method often have poor flowability as a result of its low density, which can potentially affect downstream processing for tableting or capsule-filling.

15.2.3.3 Film Coating and Casting

Film coating is one of the most commonly used pharmaceutical processes for coating solid dosage forms such as tablets and pellets for the purpose of taste-masking as well as controlled and targeted drug release [79, 80]. There are two main types of film coating: aqueous film coating and organic solvent film coating. The main components in a coating solution are the coating polymer and plasticizers. The coating liquid is sprayed on to the solid tablets/pellets in a coating pan with continuous (hot) air flow for drying. The process parameters, such as spray rate, atomizing air pressure, inlet air temperature, and the rotation speed of the pan, can significantly affect the quality of the film formed [79, 80]. If the heterogeneity of the coating film is considered to be a degree of disorder, alteration of process parameters and optimization of solution properties such as viscosity, solid content, and interfacial tension at the solid tablet/pellet surface will facilitate the minimization of the disorder created during the process.

Conventional film casting is a more static method compared to film coating, and is considerably different from spray-drying with regard to both film formation and the rate of solvent removal. The drying process of the film casting method can be accomplished with or without elevated temperatures and vacuum. However, the rate of solvent removal in the film casting method is much slower than in the spray-drying and film coating processes. In the cases of using film casting to prepare solid dispersions, the physical structure of the finished solid dispersion can be significantly influenced by the solvent removal rate. The physicochemical properties of the solid dispersions of itraconazole and etravirine prepared by film casting and spray-drying methods showed clear differences [4, 6]. The experimental solubility of the model drug in the casted films was much lower than that measured for the spray-dried dispersions. Although these results were ascribed to the effect of the manufacturing methodology used on the physical structure of the resulting solid dispersion, the mechanism underlying the observed deviation in solubility is still not well understood [4, 6]. It is possible that, during the casting process, the local drug concentration increases with continuous removal of solvents from the film. This may create domains with high drug concentration, which exceeds the saturation solubility of drug in the polymer. These domains may then act as nuclei for drug recrystallization. If the crystallization kinetics of the drug is faster than the solvent removal, a crystalline drug would be expected in the finished film.

15.2.3.4 Emerging Solvent-Evaporation-Based Processing Technologies

Electrospinning and spin coating are two emerging solvent-based processing technologies that have been recently adopted for pharmaceutical formulation preparation [81–83]. Electrospinning and electrospraying are two electrohydrodynamic techniques that have wide biomedical applications and have just recently been introduced to the pharmaceutical community for the preparation of amorphous solid dispersions [81, 82]. Electrospinning is a process in which a strong electrical potential is applied to a liquid (a solution or molten mixture of the drug and a polymer). At a critical voltage, a charged liquid jet of the sample solution is ejected from the tip of the liquid droplet and travels toward a collector connected to a grounded electrode [83–85]. On the journey to the collector, the solvent evaporates and nano- or microfibers can be formed and collected on the collector, which can be of different geometries. By altering the viscosity of the liquid solution and the processing parameters, electrospinning can be transformed into electrospraying, in which well-separated micro- and nanoparticles can be obtained [81–85]. In comparison to the amorphous solid dispersions formed using other solvent-evaporation-based methods, electrospun fibers can offer even better improvements in the dissolution enhancement of amorphous solid dispersion containing poorly soluble drugs as a result of their unique process-related features such as high surface area to volume ratio and highly porous interconnected network structure with tunable pore size (the pores refer to the spaces created between the fibers) [81, 82]. However, these characteristics may compromise the physical stability of the formulation. The high surface area exposed to the external environment can accelerate phase separation and crystallization of drug from the fibers. However, the physical instability and crystallization behavior in these nano- and microfibers are still poorly understood.

Spin coating is a rapid solvent-evaporation-based process that prepares thin and ultrathin films with controllable thickness (normally micrometer to nanometer thickness) on a substrate, as illustrated in Figure 15.8. It has been widely used in microfabrication in the semiconductor industry [86]. Unlike the traditional pharmaceutical film casting and film coating, the rate of solvent evaporation in spin coating is much higher because of the high spinning speed (normally 1000–10 000 rpm). This rapid solvent evaporation and solidification process can reduce the risk of clustering and phase separation of the drug in the films by shortening the time the system is not in the solid state (before the solvent has completely evaporated). With the advantages of high reproducibility, fast preparation, and precise control of the manufacturing process, spin coating has been used to prepare amorphous solid dispersions for experiments designed to provide a fundamental understanding of the physical stability of amorphous solid dispersions [83, 87–90]. For example, Ng *et al.* used spin coating to prepare solid dispersions containing model drugs including celecoxib, felodipine, fenofibrate, and carbamazepine and polymers with different hygroscopicities [88]. It has been reported that the key factors that affected the physical stability

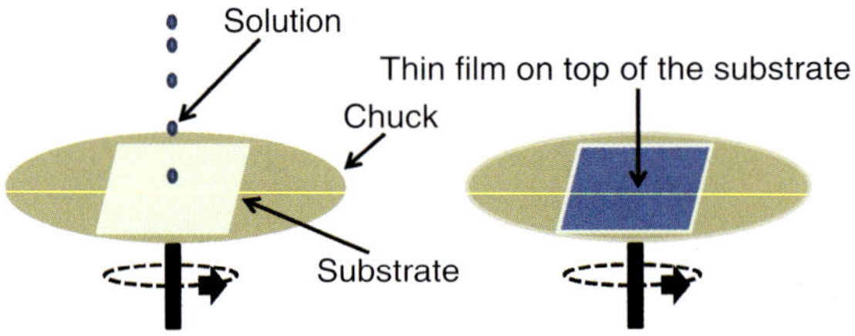

Figure 15.8 Procedure of using spin coating to prepare a thin film on a substrate.

of the solid dispersions under stressed humidity were the intrinsic physical stability of drugs alone and the hygroscopicity of polymers. In another study from the same group, drug migration from the bulk toward the surface of the spin-coated amorphous solid dispersion thin films of felodipine–PVP K29/32 was discovered during aging under stressed humidity [83, 88]. It has been reported that the film thickness can have significant effects on the crystal growth and hydrophobicities of the spin-coated crystalline polymeric films [83]. Therefore, it will be interesting to further investigate the relationship between the film thickness and the degrees of instability/crystallization on aging.

15.3 Conclusion

In a broad sense, disorder can be classified as any change of the molecular arrangement from its original form. If focusing on crystalline APIs, the degree of disorder increases with the change from crystal to crystal defects and eventually to the amorphous state. If focusing on semicrystalline polymers, the degree of disorder can be a decrease in the amount of crystalline domains in the polymer. In this chapter, commonly used and emerging pharmaceutical processing technologies were discussed with respect to their ability to create disordered states in the finished products either intentionally or unintentionally. Milling-based methods gradually build up the degree of disorder into the system, and significant accumulation of the disorder can eventually convert a crystalline material to an amorphous one. This is very different from the melting and solvent-based methods. Thermal melting and solvent-based methods start by introducing complete disorder into the systems either via melting the API with the excipients or dissolving them in a common solvent. The degree of disorder recovers during the second stage of the process, which is either cooling or drying. The conversion of disorder back to the low-energy crystalline state is often associated with physical instability of the formulations. A better understanding of the generation of disorders and their impact on the clinical outcome of the formulations requires a firm knowledge of the process and the characteristics of the material during and after processing.

References

1. Koivisto, M., Heinänen, P., Tanninen, V.P., and Lehto, V.P. (2006) Depth profiling of compression-induced disorders and polymorphic transition on tablet surfaces with grazing incidence X-ray diffraction. *Pharm. Res.*, **23** (4), 813–820.
2. Riippi, M., Tanninen, V., and Yliruusi, J. (2000) Effect of compression force on the crystal properties of erythromycin acistrate tablets. *Eur. J. Pharm. Biopharm.*, **50** (3), 365–371.
3. Feng, T., Pinal, R., and Carvajal, M.T. (2008) Process induced disorder in crystalline materials: differentiating defective crystals from the amorphous form of griseofulvin. *J. Pharm. Sci.*, **97** (8), 3207–3221.
4. Janssens, S., De Zeure, A., Paudel, A., Van Humbeeck, J., Rombaut, P., and Van den Mooter, G. (2010) Influence of preparation methods on solid state supersaturation of amorphous solid dispersions: a case study with itraconazole and eudragit E100. *Pharm. Res.*, **27** (5), 775–785.
5. Karmwar, P., Graeser, K., Gordon, K.C., Strachan, C.J., and Rades, T. (2011) Investigation of properties and recrystallisation behaviour of amorphous indomethacin samples prepared by different methods. *Int. J. Pharm.*, **417**, 94–100.
6. Weuts, I., Van Dycke, F., Voorspoels, J., De Cort, S., Stokbroekx, S., Leemans, R., Brewster, M.E., Xu, D., Segmuller, B., Turner, Y.T.A., Roberts, C.J., Davies, M.C., Qi, S., Craig, D.Q.M., and Reading, M. (2011) Physicochemical properties of the amorphous drug, cast films, and spray dried powders to predict formulation probability of success for solid dispersions: etravirine. *J. Pharm. Sci.*, **100**, 260–274.
7. Descamps, M., Willart, J.F., Dudognon, E., and Caron, V. (2007) Transformation of pharmaceutical compounds upon milling and comilling: the role of T_g. *J. Pharm. Sci.*, **96**, 1398–1407.
8. Willart, J.F., Carpentier, L., Danède, F., and Descamps, M. (2012) Solid-state vitrification of crystalline griseofulvin by mechanical milling. *J. Pharm. Sci.*, **101** (4), 1570–1577.
9. Trasi, N.S., Boerrigter, S.X., and Byrn, S.R. (2010) Investigation of the milling-induced thermal behavior of crystalline and amorphous griseofulvin. *Pharm. Res.*, **27** (7), 1377–1389.
10. Otte, A. and Carvajal, M.T. (2011) Assessment of milling-induced disorder of two pharmaceutical compounds. *J. Pharm. Sci.*, **100** (5), 1793–1804.
11. Willart, J.F. and Descamps, M. (2008) Solid state amorphization of pharmaceuticals. *Mol. Pharm.*, **5**, 905–920.
12. Elamin, A.A., Ahlneck, C., and Alderborn, G.C.N. (1994) Increased metastable solubility of milled griseofulvin, depending on the formation of a disordered surface structure. *Int. J. Pharm.*, **111**, 159–170.
13. Luisi, B.S., Medek, A., Liu, Z., Mudunuri, P., and Moulton, B. (2012) Milling-induced disorder of pharmaceuticals: one-phase or two-phase system? *J. Pharm. Sci.*, **101** (4), 1475–1485.
14. Qi, S., Weuts, I., De Cort, S., Stokbroekx, S., Leemans, R., Reading, M., Belton, P., and Craig, D.Q. (2010) An investigation into the crystallisation behaviour of an amorphous cryomilled pharmaceutical material above and below the glass transition temperature. *J. Pharm. Sci.*, **99** (1), 196–208.
15. Balani, P.N., Ng, W.K., Chan, S.Y., and Tan, R.B.H. (2010) Influence of excipients in co-milling on mitigating milling-induced amorphization or structural disorder of crystalline pharmaceutical actives. *J. Pharm. Sci.*, **99**, 2462–2474.
16. Gupta, M.K., Vanwert, A., and Bogner, R.H. (2003) Formation of physically stable amorphous drugs by milling with Neusilin. *J. Pharm. Sci.*, **92** (3), 536–551.
17. Bahl, D. and Bogner, R.H. (2006) Amorphization of indomethacin by co-grinding with Neusilin US2: amorphization kinetics, physical stability and mechanism. *Pharm. Res.*, **23** (10), 2317–2325.
18. Trasi, N.S. and Byrn, S.R. (2012) Mechanically induced amorphization

of drugs: a study of the thermal behavior of cryomilled compounds. *AAPS PharmSciTech*, **13** (3), 772–784.

19. Hu, Y., Macfhionnghaile, P., Caron, V., Tajber, L., Healy, A.M., Erxleben, A., and McArdle, P. (2013) Formation, physical stability, and quantification of process-induced disorder in cryomilled samples of a model polymorphic drug. *J. Pharm. Sci.*, **102** (1), 93–103.
20. Chattoraj, S., Bhugra, C., Telang, C., Zhong, L., Wang, Z., and Sun, C.C. (2012) Origin of two modes of non-isothermal crystallization of glasses produced by milling. *Pharm. Res.*, **29** (4), 1020–1032.
21. Willart, J.F., Dujardin, N., Dudognon, E., Danède, F., and Descamps, M. (2010) Amorphization of sugar hydrates upon milling. *Carbohydr. Res.*, **345** (11), 1613–1616.
22. Chieng, N., Zujovic, Z., Bowmaker, G., Rades, T., and Saville, D. (2006) Effect of milling conditions on the solid-state conversion of ranitidine hydrochloride form 1. *Int. J. Pharm.*, **327** (1–2), 36–44.
23. Otte, A., Zhang, Y., Carvajal, T., and Pinal, R. (2012) Milling induces disorder in crystalline griseofulvin and order in its amorphous counterpart. *CrystEngComm*, **14**, 2560–2570.
24. Kang, N., Lee, J., Choi, J.N., Mao, C., and Lee, E.H. (2014) Cryomilling-induced solid dispersion of poor glass forming/poorly water-soluble mefenamic acid with polyvinylpyrrolidone K12. *Drug Dev. Ind. Pharm.*, **22**, 1–11.
25. Al-Obaidi, H., Lawrence, M.J., Al-Saden, N., and Ke, P. (2013) Investigation of griseofulvin and hydroxypropylmethyl cellulose acetate succinate miscibility in ball milled solid dispersions. *Int. J. Pharm.*, **443** (1-2), 95–102.
26. Löbmann, K., Laitinen, R., Grohganz, H., Gordon, K.C., Strachan, C., and Rades, T. (2011) Co-amorphous drug systems: enhanced physical stability and dissolution rate of indomethacin and naproxen. *Mol. Pharm.*, **8** (5), 1919–1928.
27. Laitinen, R., Löbmann, K., Grohganz, H., Strachan, C., and Rades, T. (2014) Amino acids as co-amorphous excipients for simvastatin and glibenclamide: physical properties and stability. *Mol. Pharm.*, **11** (7), 2381–2389.
28. Löbmann, K., Strachan, C., Grohganz, H., Rades, T., Korhonen, O., and Laitinen, R. (2012) Co-amorphous simvastatin and glipizide combinations show improved physical stability without evidence of intermolecular interactions. *Eur. J. Pharm. Biopharm.*, **81** (1), 159–169.
29. Laitinen, R., Löbmann, K., Strachan, C.J., Grohganz, H., and Rades, T. (2013) Emerging trends in the stabilization of amorphous drugs. *Int. J. Pharm.*, **453** (1), 65–79.
30. Graeser, K.A., Strachan, C.J., Patterson, J.E., Gordon, K.C., and Rades, T. (2008) Physicochemical properties and stability of two differently prepared amorphous forms of simvastatin. *Cryst. Growth Des.*, **8**, 128–135.
31. Ke, P., Hasegawa, S., Al-Obaidi, H., and Buckton, G. (2012) Investigation of preparation methods on surface/bulk structural relaxation and glass fragility of amorphous solid dispersions. *Int. J. Pharm.*, **422** (1-2), 170–178.
32. Wojnarowska, Z., Grzybowska, K., Adrjanowicz, K., Kaminski, K., Paluch, M., Hawelek, L., Wrzalik, R., Dulski, M., Sawicki, W., Mazgalski, J., Tukalska, A., and Bieg, T. (2010) Study of the amorphous glibenclamide drug: analysis of the molecular dynamics of quenched and cryomilled material. *Mol. Pharm.*, **7** (5), 1692–1707.
33. Hu, Y., Gniado, K., Erxleben, A., and McArdle, P. (2014) Mechanochemical reaction of sulfathiazole with carboxylic acids: formation of a cocrystal, a salt, and coamorphous solids. *Cryst. Growth Des.* doi: 10.1021/cg401673z
34. Keen, J.M., McGinity, J.W., and Williams, R.O. III, (2013) Enhancing bioavailability through thermal processing. *Int. J. Pharm.*, **450** (1-2), 185–196.
35. Brown, T.D., Dalton, P.D., and Hutmacher, D.W. (2011) Direct writing by way of melt electrospinning. *Adv. Mater.*, **23** (47), 5651–5657.
36. Kim, S.J., Jang, D.H., Park, W.H., and Min, B.M. (2010) Fabrication and characterization of 3-dimensional PLGA

nanofiber/microfiber composite scaffolds. *Polymer*, **51** (6), 1320–1327.

37. Khan, N. and Craig, D.Q. (2004) Role of blooming in determining the storage stability of lipid-based dosage forms. *J. Pharm. Sci.*, **93** (12), 2962–2971.
38. Windbergs, M., Strachan, C.J., and Kleinebudde, P. (2009) Investigating the principles of recrystallization from glyceride melts. *AAPS PharmSciTech*, **10** (4), 1224–1233.
39. Windbergs, M., Strachan, C.J., and Kleinebudde, P. (2009) Understanding the solid-state behaviour of triglyceride solid lipid extrudates and its influence on dissolution. *Eur. J. Pharm. Biopharm.*, **71** (1), 80–87.
40. Qi, S., Marchaud, D., and Craig, D.Q. (2010) An investigation into the mechanism of dissolution rate enhancement of poorly water-soluble drugs from spray chilled gelucire 50/13 microspheres. *J. Pharm. Sci.*, **99** (1), 262–274.
41. Alonzo, D.E., Zhang, G.G., Zhou, D., Gao, Y., and Taylor, L.S. (2010) Understanding the behavior of amorphous pharmaceutical systems during dissolution. *Pharm. Res.*, **27**, 608–618.
42. Brough, C. and Williams, R.O. (2013) Amorphous solid dispersions and nano-crystal technologies for poorly water-soluble drug delivery. *Int. J. Pharm.*, **453**, 157–166.
43. Patterson, J., James, M., Forster, A., Lancaster, R., Butler, J., and Rades, T. (2004) The influence of thermal and mechanical preparative techniques on the amorphous state of four poorly soluble compounds. *J. Pharm. Sci.*, **94**, 1998–2012.
44. Forster, A., Hempenstall, J., Tucker, I., and Rades, T. (2001) The potential of small-scale fusion experiments and the Gordon-Taylor equation to predict the suitability of drug/polymer blends for melt extrusion. *Drug Dev. Ind. Pharm.*, **27** (6), 549–560.
45. Van den Brande, J., Weuts, I., Verreck, G., Peeters, J., Brewster, M., and Van den Mooter, G. (2004) DSC analysis of the anti-HIV agent loviride as a preformulation tool in the development of hot-melt extrudates. *J. Therm. Anal. Calorim.*, **77** (2), 523–530.
46. Cordeiro, P., Temtem, M., and Winters, C. (2013) Spray congealing: applications in the Pharmaceutical Industry. *Chim. Oggi-Chem. Today*, **31** (5), 69–73.
47. Passerini, N., Calogerà, G., Albertini, B., and Rodriguez, L. (2010) Melt granulation of pharmaceutical powders: a comparison of high-shear mixer and fluidised bed processes. *Int. J. Pharm.*, **391** (1-2), 177–186.
48. Weatherley, S., Mu, B., Thompson, M.R., Sheskey, P.J., and O'Donnell, K.P. (2013) Hot-melt granulation in a twin screw extruder: effects of processing on formulations with caffeine and Ibuprofen. *J. Pharm. Sci.*, **102** (12), 4330–4336.
49. Van Melkebeke, B., Vermeulen, B., Vervaet, C., and Remon, J.P. (2006) Melt granulation using a twin-screw extruder: a case study. *Int. J. Pharm.*, **326** (1-2), 89–93.
50. Alkhatib, H.S., Hamed, S., Mohammad, M.K., Bustanji, Y., Alkhalidi, B., Aiedeh, K.M., and Najjar, S. (2010) Effects of thermal curing conditions on drug release from polyvinyl acetate-polyvinyl pyrrolidone matrices. *AAPS PharmSciTech*, **11** (1), 253–266.
51. Sauer, D. and McGinity, J.W. (2009) Influence of additives on melt viscosity, surface tension, and film formation of dry powder coatings. *Drug Dev. Ind. Pharm.*, **35** (6), 646–654.
52. Cerea, M., Zheng, W., Young, C.R., and McGinity, J.W. (2004) A novel powder coating process for attaining taste masking and moisture protective films applied to tablets. *Int. J. Pharm.*, **279** (1-2), 127–139.
53. Sauer, D., Cerea, M., DiNunzio, J., and McGinity, J. (2013) Dry powder coating of pharmaceuticals: a review. *Int. J. Pharm.*, **457** (2), 488–502.
54. Luo, Y., Zhu, J., Ma, Y., and Zhang, H. (2008) Dry coating, a novel coating technology for solid pharmaceutical dosage forms. *Int. J. Pharm.*, **358** (1-2), 16–22.
55. Repka, M.A., Majumdar, S., Kumar Battu, S., Srirangam, R., and Upadhye, S.B. (2008) Applications of hot-melt extrusion for drug delivery. *Expert Opin. Drug Delivery*, **5** (12), 1357–1376.

56. Lang, B., McGinity, J.W., and Williams, R.O. III, (2014) Hot-melt extrusion - basic principles and pharmaceutical applications. *Drug Dev. Ind. Pharm.*, in press.
57. Breitenbach, J. (2002) Melt extrusion: from process to drug delivery technology. *Eur. J. Pharm. Biopharm.*, **54** (2), 107–117.
58. Saerens, L., Vervaet, C., Remon, J.P., and De Beer, T. (2014) Process monitoring and visualization solutions for hot-melt extrusion: a review. *J. Pharm. Pharmacol.*, **66** (2), 180–203.
59. De Beer, T., Burggraeve, A., Fonteyne, M., Saerens, L., Remon, J.P., and Vervaet, C. (2011) Near infrared and Raman spectroscopy for the in-process monitoring of pharmaceutical production processes. *Int. J. Pharm.*, **417** (1–2), 32–47.
60. Qi, S., Belton, P., Nollenberger, K., Gryczke, A., and Craig, D. (2011) Compositional analysis of low quantities of phase separation in hot melt extruded solid dispersions: a combined atomic force microscopy, photothermal Fourier-transform infrared microspectroscopy and localised thermal analysis approach. *Pharm. Res.*, **28** (9), 2311–2316.
61. Yang, Z., Nollenberger, K., Albers, J., Craig, D., and Qi, S. (2013) Microstructure of an immiscible polymer blend and its stabilization effect on amorphous solid dispersions. *Mol. Pharm.*, **10** (7), 2767–2780.
62. Lauer, M.E., Siam, M., Tardio, J., Page, S., Kindt, J.H., and Grassmann, O. (2013) Rapid assessment of homogeneity and stability of amorphous solid dispersions by atomic force microscopy—from bench to batch. *Pharm. Res.*, **30** (8), 2010–2022.
63. Zema, L., Loreti, G., Melocchi, A., Maroni, A., and Gazzaniga, A. (2012) Injection molding and its application to drug delivery. *J. Controlled Release*, **159** (3), 324–331.
64. Sammoura, F., Kang, J.J., Heo, Y.M., and Lin, L. (2007) Polymeric microneedle fabrication using a micro injection molding technique. *Microsyst. Technol. J.*, **13**, 517–522.
65. Clark, M.R., Johnson, T.J., McCabe, R.T., Clark, J.T., Tuitupou, A., Elgendy, H., Friend, D.R., and Kiser, P.F. (2012) A hot-melt extruded intravaginal ring for the sustained delivery of the antiretroviral microbicide UC781. *J. Pharm. Sci.*, **101** (2), 576–587.
66. Gazzaniga, A., Cerea, M., Cozzi, A., Foppoli, A., Maroni, A., and Zema, L. (2011) A novel injection-molded capsular device for oral pulsatile delivery based on swellable/erodible polymers. *AAPS PharmSciTech*, **12** (1), 295–303.
67. Nagy, Z.K., Balogh, A., Drávavölgyi, G., Ferguson, J., Pataki, H., Vajna, B., and Marosi, G. (2013) Solvent-free melt electrospinning for preparation of fast dissolving drug delivery system and comparison with solvent-based electrospun and melt extruded systems. *J. Pharm. Sci.*, **102** (2), 508–517.
68. DiNunzio, J.C., Brough, C., Miller, D.A., Williams, R.O. III, and McGinity, J.W. (2010) Fusion processing of itraconazole solid dispersions by kinetisol dispersing: a comparative study to hot melt extrusion. *J. Pharm. Sci.*, **99** (3), 1239–1253.
69. Dinunzio, J.C., Brough, C., Hughey, J.R., Miller, D.A., Williams, R.O. III, and McGinity, J.W. (2010) Fusion production of solid dispersions containing a heat-sensitive active ingredient by hot melt extrusion and Kinetisol dispersing. *Eur. J. Pharm. Biopharm.*, **74** (2), 340–351.
70. Sollohub, K. and Cal, K. (2010) Spray drying technique: II. Current applications in pharmaceutical technology. *J. Pharm. Sci.*, **99** (2), 587–597.
71. Vehring, R. (2008) Pharmaceutical particle engineering *via* spray drying. *Pharm. Res.*, **25** (5), 999–1022.
72. Cal, K. and Sollohub, K. (2010) Spray drying technique. I: hardware and process parameters. *J. Pharm. Sci.*, **99** (2), 575–586.
73. Guns, S., Dereymaker, A., Kayaert, P., Mathot, V., Martens, J.A., and Van den Mooter, G. (2011) Comparison between hot-melt extrusion and spray-drying for manufacturing solid dispersions of the graft copolymer of ethylene glycol and vinylalcohol. *Pharm. Res.*, **28** (3), 673–682.

74. Tang, X. and Pikal, M.J. (2004) Design of freeze–drying processes for pharmaceuticals: practical advice. *Pharm. Res.*, **21**, 191–200.

75. Lai, F., Pini, E., Corrias, F., Perricci, J., Manconi, M., Fadda, A.M., and Sinico, C. (2014) Formulation strategy and evaluation of nanocrystal piroxicam orally disintegrating tablets manufacturing by freeze-drying. *Int. J. Pharm.*, **467** (1-2), 27–33.

76. Santivarangkna, C., Aschenbrenner, M., Kulozik, U., and Foerst, P. (2011) Role of glassy state on stabilities of freeze-dried probiotics. *J. Food Sci.*, **76** (8), R152–R156.

77. He, X., Pei, L., Tong, H.H., and Zheng, Y. (2011) Comparison of spray freeze drying and the solvent evaporation method for preparing solid dispersions of baicalein with Pluronic F68 to improve dissolution and oral bioavailability. *AAPS PharmSciTech*, **12** (1), 104–113.

78. El-Badry, M. and Fathy, M. (2006) Enhancement of the dissolution and permeation rates of meloxicam by formation of its freeze-dried solid dispersions in polvvinylpyrrilidone k-30. *Drug Dev. Ind. Pharm.*, **32**, 141–150.

79. Felton, L.A. and Porter, S.C. (2013) An update on pharmaceutical film coating for drug delivery. *Expert Opin. Drug Delivery*, **10** (4), 421–435.

80. Porter, S.C. and Felton, L.A. (2010) Techniques to assess film coatings and evaluate film-coated products. *Drug Dev. Ind. Pharm.*, **36** (2), 128–142.

81. Sill, T.J. and von Recum, H.A. (2008) Electrospinning: applications in drug delivery and tissue engineering. *Biomaterials*, **29** (13), 1989–2006.

82. Zamani, M., Prabhakaran, M.P., and Ramakrishna, S. (2013) Advances in drug delivery via electrospun and electrosprayed nanomaterials. *Int. J. Nanomed.*, **8**, 2997–3017.

83. Qi, S., Moffat, J.G., and Yang, Z. (2013) Early stage phase separation in pharmaceutical solid dispersion thin films under high humidity: improved spatial understanding using probe-based thermal and spectroscopic nanocharacterization methods. *Mol. Pharm.*, **10**, 918–930.

84. Williams, G.R., Chatterton, N.P., Nazir, T., Yu, D.-G., Zhu, L.-M., and Branford-White, C.J. (2012) Electrospun nanofibers in drug delivery: recent developments and perspectives. *Ther. Delivery*, **3** (4), 515–533.

85. Chakraborty, S., Liao, I.C., Adler, A., and Leong, K.W. (2009) Electrohydrodynamics: a facile technique to fabricate drug delivery systems. *Adv. Drug Delivery Rev.*, **61** (12), 1043–1054.

86. Norrman, K., Ghanbari-Siahkali, A., and Larsen, N.B. (2005) 6 studies of spin-coated polymer films. *Annu. Rep. Prog. Chem. Sect. C: Phys. Chem.*, **101**, 174–201.

87. Van Eerdenbrugh, B., Baird, J.A., and Taylor, L.S. (2010) Crystallization tendency of active pharmaceutical ingredients following rapid solvent evaporation—classification and comparison with crystallization tendency from undercooled melts. *J. Pharm. Sci.*, **99**, 3826–3838.

88. Ng, Y.C., Yang, Z., McAuley, W.J., and Qi, S. (2013) Stabilisation of amorphous drugs under high humidity using pharmaceutical thin films. *Eur. J. Pharm. Biopharm.*, **84**, 555–565.

89. Van Eerdenbrughand, B. and Taylor, L.S. (2010) Small scale screening to determine the ability of different polymers to inhibit drug crystallization upon rapid solvent evaporation. *Mol. Pharm.*, **7**, 1328–1337.

90. Cui, W., Zhou, Y., and Chang, J. (2010) Electrospun nanofibrous materials for tissue engineering and drug delivery. *Sci. Technol. Adv. Mater.*, **11** (1), 014108(11)pp.

16
Patenting of Inventions Relating to Solid Forms, with Special Considerations on Disordered Forms

Bertrand Gellie

The quest for new solid forms of active pharmaceutical ingredients (APIs) has become very intense in the last 10 years. Encouraged by the success of faster dissolving polymorphs, researchers have investigated more sophisticated forms of API, particularly their disordered forms. Examples are amorphous forms, glassy forms, co-crystals, and dislocated crystals. Patenting of inventions relating to polymorphic forms has already been investigated by us (Solid State Characterization of Pharmaceuticals, Chapter 13, April 2011, Wiley-Blackwell) and others.

Inventors of a new solid form of an API should, as a general rule, always be well aware that their invention does not have the weight of an invention of a new molecule. They have only – however great the merits of their efforts are and the level of the difficulties they had to overcome – invented a new form of a substance that is already known. One has always to expect difficulties when, for instance, during litigation, it is necessary to explain the difference between the new form and the already known ones. Appreciating this difference requires good arguments that can be understood by nonexperts in this very special technical field. It is, unfortunately, a rather widespread belief among certain patent legislators/judges that solid forms are merely a standard way to achieve "evergreening" of the patent protection of the primary invention of a new API. This is, of course, largely an oversimplification, since patent laws provide for protection also of improvements. However, inventors will be well advised to bear this fact in mind. In certain countries, typically India, it has become almost impossible to patent a new API solid form, and this trend seems to spread to other countries also. In the following, we will accordingly outline – based on our experience – what should be respected by drafters of patent applications in order to optimize their chances to speed up the acquisition of patent protection for this kind of invention.

Patent practitioners are well able to cope with uncertainties that might still exist in the scientific world with respect to the exact definition and determination of certain forms exhibited by an API. The reason is fundamentally that, in case of litigation, the judge will focus his attention on similarities between what might be a counterfeit and the subject matter of a patent claim. Accordingly, a definition of

Disordered Pharmaceutical Materials, First Edition. Edited by Marc Descamps.

a new solid form might be acceptable in an efficient patent claim provided that its definition in that claim, first, can distinguish said form from those of the prior art – in order to be patentable – and, second, this definition contains essential features expressed in a way that allows the skilled person to ascertain without undue effort whether another, later appeared but similar product falls under the scope of the claim or not. The claim covering the solid form will then be regarded as clear by "the skilled person in the art," a fictitious person well known in all patent laws.

Clarity of the claims relating to the different types of solid forms in this book will, therefore, be dealt with in this chapter. Clarity, however, is only a necessary condition, not a sufficient one. A clear subject matter, if it is not sufficiently disclosed so that it can be reproduced by a skilled person, will fail to be patentable. Accordingly, it will be emphasized in this chapter how important the disclosure in a patent application relating to a solid form is of the means that can effectively lead to this form. This is particularly true for those forms that are difficult to define with absolute accuracy by physicochemical methods and parameters.

Often, patent applicants, be it by lack of awareness of this important criterion or by their unwillingness to disclose too much of their manufacturing secrets, fail to respect this patentability requirement, thereby incurring the risk of not obtaining their patent or losing it later in a court of law. A few rules aiming at avoiding the most frequent objections under insufficiency of disclosure will therefore be also given in the following.

Patent applicants, in addition to the above obligations that are under their sole control (in that they are not influenced by previous disclosures by others or themselves), should, however, not ignore what others have done before them. The prior art for a patent is anything that has been published prior to the filing of the patent application. With respect to this prior art, the solid form invention must fulfill two conditions: it must be different from the prior form and it must not obviously follow from the latter.

The first condition, novelty, has a direct impact on the way a patent claim should be preferably drafted. Thus the parameters used for defining the prior art form might encourage a patent applicant to use some of or the same parameters in order to easily demonstrate the novelty of his or her invention. Failing to do so is, however, not necessarily an obstacle to patenting but will definitely render the examination of their patent application more difficult and hence more costly and the final result of this examination uncertain.

The second condition, the inventive step, is regarded by practitioners as the most delicate since it involves the judgment of a skilled person, which inevitably implies some degree of subjectivity. This degree may vary from one patent system to another. Subjectivity is probably the lowest when applying the "problem–solution" approach used by the European Patent Office for more than 30 years, and is relatively high in patent systems requiring from forms that they impart an "enhanced efficacy" to the pharmaceutical compositions using them or constitute "a significant technical advance." In the following, considerations on the nonobvious character of certain forms will be made according to the problem–solution approach, that is, based on the consideration of the technical effect that is provided by the difference characterizing the invention by comparison with the closest prior art and considering the technical problem, derivable from the preceding technical effect, which had objectively to be solved.

As it provides the most natural transition to our preceding chapter on patenting polymorphic forms, we shall first deal with *disordered crystals*, and then with *co-crystals, amorphous forms, glass materials*, and, finally, *disordered nanocrystalline materials.* For each of these five types of form, each of the four preceding patentability criteria will be considered in order to take into consideration their specific peculiarities. For the reader of the five sections in a row, the impression of a certain repetition will unavoidably result, but the reader of one section alone will in this way be fully informed at once.

16.1 Patentability of Disordered Crystals

Patent applications relating to disordered crystals are, at the time writing this chapter, very rare. However, in view of their recognized definition that can be found in the literature, it is relatively easy to give reliable guidance as to what should be respected in order to ensure good patentability perspectives.

It must first be realized by inventors of disordered crystals that the differences distinguishing their creations from non-disordered crystals are tenuous, so that it requires, for them to be nevertheless patentable, a rigorous distinctive definition as well as the existence of a significant technical effect involved by them in order to justify that their finding implies an inventive step.

16.1.1 Clarity Problems in Claims for Disordered Crystals

Defects in crystals can be point, linear, dimensional, or multidirectional. Defining such crystals will be in most cases a difficult task.

Point defects involve only a limited number of atoms and do not extend in space or even any dimension. Unless these defects can be and are repeated throughout the crystals, so that the extent of their presence becomes measurable and imparts a truly different behavior to the crystals, it is difficult to find a distinctive physicochemical parameter characterizing disordered crystals from non-disordered ones. In addition, the (approximate) number of disorders per unit volume that a crystal exhibits should be quantified in order to ensure a solid definition.

Line defects concern a linear region around which some atoms are misaligned.

Plane defects involve changes in the crystallographic orientation in the lattice. Again, the defects must obviously be repeated in the crystals in order to be detectable and provide a noticeable effect. The same remarks as above apply.

Even in the case of *bulk defects*, which are essentially voids accumulating in order to provide vacancies in the crystal lattice, the disorders in the crystals are relatively localized and involve a limited number of molecules.

One common feature to all these defects is an increase of the free energy of the crystals. This increase is, however, not easily quantifiable in an absolute manner but can better be described in a relative manner, in comparison with defect-free crystals. This feature is therefore not sufficient to characterize the form since the free energy level can result from different types of defects.

Complementary spectrometric characterization appears to be indispensable, as well as any other characterization method that allows a proper differentiation of the disordered form from ordered forms (or from other disordered forms). In this respect, patent examiners prefer parameters that are widely used in a field because they are more familiar with them and because, in general, they allow an easier and more reliable comparison with the prior art. They are, however, not opposed to other characterization methods provided their use is accompanied by a reasonable and convincing justification. It is advisable in the case of a less usual or unusual characterization method to mention and, if necessary, to explain the method of measuring a characterizing parameter in the patent description either directly or by reference to a piece of literature that has an unambiguous content in this respect.

Certainly, the above-mentioned defects can have an impact on the mechanical properties, dissolution rates, moisture sorption, and storage stability. However, these properties may not be sufficient to characterize a crystalline substance if they can be achieved by other modifications of this substance. Consequently, the patent examiner might well consider the claims incorporating only such properties as unclear. Results to be achieved can be accepted in the definition of a substance but only if rather strict conditions are fulfilled.

Crystal imperfections affect only the superficial layers of a crystal and have been reported to trigger solid phase transitions that can be measured, and accordingly how this measurement is performed should be mentioned in a claim pertaining to crystals comprising such imperfections.

16.1.2
Sufficient Disclosure of Disordered Crystals

It is essential that the information given in the patent description allows a skilled person to reproduce the invention, thereby allowing the public, when the patent lapses, to benefit from the technique protected by the said patent.

At least one process for obtaining the disordered crystals must therefore be disclosed. It is not advisable, in this respect, to leave uncertainty zones in the description of the process. It is also important that the inventors determine the steps that are indispensable for obtaining the claimed disordered crystals, rather than other crystals, and disclose these steps. There is no obligation in most countries to underline the determinant character of a feature, but the feature itself must at least be disclosed.

If a seed crystal is needed to initiate the formation of a certain crystal form, it is indispensable to specify an accessible source for the seed. Failure to do so may result in the nullification or revocation of a patent. See T 1066/03 (decision of 11/07/2006, "Polymorphic Atorvastatin/WARNER-LAMBERT") which *mutatis mutandis* could be applied to any crystal form.

Whenever the use of an apparatus is necessary for obtaining a disordered crystal, it is always advisable to specify the type of the apparatus used as well as the working conditions, especially if varying the latter might result in a failure or in obtaining another crystalline form not covered by the claim. The use and reference to home-built apparatuses or scarcely available equipment should be avoided, as they can create considerable dispute.

16.1.3
Novelty of Dislocated Crystals

The claimed dislocated crystals must be distinguishable from pre-existing crystal forms of the same substance. The demonstration of novelty is, in principle, relatively easy provided adequate parameters, particularly physicochemical ones, are used.

Patent examiners are instructed to be particularly suspicious whenever applicants use unusual characterization means for a new product. It is risky for an applicant to use an unusual parameter for characterizing a dislocated crystal, as, in any case, the prior art will – by force – not contain any data on which a comparison with the invention could be based. Provided there is another parameter characterizing the dislocated crystal, which exhibits the same value as in the prior art crystals, the examiner will systematically contest the novelty. The only way out is then to reproduce the crystals of the prior art and to determine the value of the unusual parameter for them. If it is significantly different, novelty can be recognized; if not, the patent application will be rejected. The same is likely to apply if the applicant refuses to make the effort of reproducing the prior art crystals or if he purportedly does not strictly follow the process for obtaining the prior art crystals.

A general line of defense in such a case can be to show that the prior art indications do not allow obtaining the crystals that the prior art alleges to have obtained. In such a case, the prior art is said to be "non-enabling" and can be dismissed. This may convince a patent examiner during his examination of the patent application (*ex parte* proceedings). However, most patent systems now provide for an easy challenge of such statements, and an adverse party, typically an opponent in opposition proceedings, might challenge the patentee's (or the applicant's, in pre-grant opposition systems) allegations in this respect and cast doubts on the "non-enablement" based on their own experimental findings or declaration of experts they have hired.

It can also genuinely happen that a new analytical method (and possibly new equipment) renders possible a distinction of the dislocated crystals over the prior art crystals. If a direct comparison is not possible because the prior art crystals cannot be submitted to that new method, the case can possibly still be saved by providing a plurality of hints that the dislocated crystals are truly different from them. This, however, will rarely be an easy battle.

16.1.4
Inventive Step of Dislocated Crystals

According to the problem – solution approach, the tiny difference exhibited by the dislocated crystals over prior art (non-dislocated) crystals must produce a technical effect different in intensity (much more rarely, in nature) from what the prior art crystals could achieve. This different technical effect underlies a technical problem that was not solved by the prior art, at least not to the same extent as with the dislocated crystals. The technical problem must have an industrial significance. Hence, it can reside in the properties of a more complex product, such as a composition or a galenic form, using the dislocated crystals.

Provided the patent applicant is able to establish that his or her own invention does solve the technical problem, it will still be necessary to show that the skilled person would not have solved the problem as proposed by the applicant, in view of the existing prior art, to which the skilled person can add his standard knowledge. This knowledge does not amount to the knowledge of an expert but only to what the average skilled person could be assumed to know in the field, as well as in related fields. Thus, using a dislocation technique that does not differ from one already used for preparing dislocated crystals of another API can be regarded as obvious for the skilled person, unless some technical prejudice existed against this use, in the case of the chemical entity at stake. The obviousness is, in general, based on a surprising effect: if the skilled person would have been surprised by the nature or the intensity of a technical effect caused by the dislocated crystals of a given substance, this is, in general, a positive hint in favor of an inventive step.

16.2
Patentability of Co-crystals

Co-crystal formation is a potential and very important source of innovation for improving APIs considering that about 30% of the latter marketed in the United States possess carboxylic groups capable of promoting co-crystal formation.

The idea is to associate certain APIs exhibiting a chemical function that is reactive, but is also capable of eliciting specific noncovalent interactions, with a complementary molecule. Typically, such APIs bear an acidic functionality such as a carboxylic group or a basic functionality such as an amine group. They are associated with at least one crystal former exhibiting a complementary functionality (i.e., a basic or an acidic functionality, respectively) in order to form an assembly that is solid at room temperature and in which the two components are connected by noncovalent interactions such as hydrogen bonding, van der Waals, and $\pi-\pi$ interactions.

The real nature of this assembly is often debatable because of the ambivalence of the functionality borne by the API. But means exist to distinguish them from solvates, where the pK_a between the two partners is higher, and two-component systems, which are a mere physical mixture of two ingredients. The interactions involved in the formation of co-crystals are based on far more subtle and weaker forces than those involved in covalent binding. They also distinguish co-crystals from the other forms dealt with in this chapter since the latter all share the same intermolecular bonds (but differ mainly – and considerably – by the range of disorder).

This is, however, enough to allow co-crystals to be patentable in spite of the scientific debate surrounding them. The first co-crystals were discovered in 1844, and the idea of using co-crystals for making drugs is said to have been reported in 1895. But it is only relatively recently that methods for the systematic screening of the possibility of forming co-crystals with candidate APIs have become available and directed inventions of pharmaceutical co-crystals were made. Accordingly, no specific case law in this field is available. The following analysis is therefore only deduced from a logical legal approach.

16.2.1 Clarity Problems in Claims for Co-crystals

The debate on the real nature and the characterization of co-crystals does not affect necessarily the clarity of a claim relating to co-crystals. What is important is that the ingredients forming the co-crystal – the API and the crystal former – are well and unambiguously identified and that the co-crystal obtained by associating them in certain (solvent) conditions can be distinguished from the mere physical mixture of these two ingredients and from a real salt (or solvate) if the formation of such a salt (or solvate) is also possible. This second condition implies, in general, a further characterization of the co-crystal by way of parameters.

Consequently, the claims for co-crystals are analyzed by patent specialists as claims for compositions complemented by a parametric definition.

The precautions already mentioned for dislocated crystals also apply to co-crystals. The parameters used must be reproducible and the conditions under which the parameters were measured should be defined (e.g., apparatus used, heating rate used in DSC (differential scanning calorimetry), etc.). Again, it must be stressed that the use of unusual parameters in the field should be avoided unless duly justified by the peculiarities of a co-crystal or a progress of analytical science, because it is a source of considerable difficulty for applicants using them. The choice of the parameter can be, in practice, influenced by the closest prior art. Analytical data used in co-crystal claims should always be carefully checked, as the correction of errors is, in general, impossible. Only obvious errors can be corrected (Rule 139 EPC; the regulation also exists in all patent laws.).

A claim such as "Co-crystal of API 'A' and compound "X"" might be criticized by a patent examiner if the experimental data disclosed in the patent application do not clarify whether the compound is a co-crystal or a salt. Such a claim would particularly stand exposed if other authors have shown that by following the patent's route only a salt is obtained.

What is important is that the public and the judge are presented with a definition of the product that is measurable and reliable and allows a distinction over prior art products.

Whenever the uncertainty about the structure (co-crystal or salt with low pK_a?) remains, a valid option may be to draft the claim as follows: "a co-crystal compound obtainable by the following process ... and characterized by the following parameters ...." Of course, the process must then be fully disclosed in an appropriate manner in the description of the patent application. As usual, if, however, the co-crystal happens to be already known with the same parameters but as the result of a different process, the claim – although it will be regarded as clear – will not be regarded as novel.

16.2.2
Sufficient Disclosure of Co-crystals

Article 83 of EPC states that "the European patent application must disclose the invention in a manner sufficiently clear and complete for it to be carried out by a person skilled in the art."

Accordingly, it is of paramount importance to provide a detailed disclosure of the critical steps for preparing the co-crystal.

The problem is seldom correctly perceived by inventors and researchers for whom the essential matter is to have found a process leading to the co-crystal formation. Nonspecialist patent attorneys are at risk of being unable to identify correctly what makes the process for preparing a co-crystal successful. The risk is aggravated by the fact that most processes for making co-crystals are standard ones. However, if a condition is essential in the process, it is very advisable to mention it in the disclosure. A lot of later discussions, necessitating almost always the involvement of costly experts, can then be avoided or maintained within reasonable limits. Hiding essential features that a skilled person cannot otherwise deduce from the application documents or from general knowledge has consistently led to the refusal of patent applications or revocation or nullification of the patent in related fields (salts, polymorphs, etc.).

Also, examples play an important role here. The more detailed the examples, the easier is it for the skilled person to carry out the invention over the claimed area. It should be borne in mind that disclosure issues can also be raised in *inter partes* proceedings, such as opposition proceedings, based on experimental reports by the parties having an adverse position to the patentee. When enough data is provided by the patent applicant upon filing, less leeway is left to the opponents later.

Every time an invention is dependent on data for the recognition of its patentability – and this is particularly the case for most solid-state inventions – it is appropriate to remember that flaws in the original disclosure can be extremely difficult to remedy after the filing date (Art. 123(2) EPC).

Later in this chapter, in the framework of amorphous forms, the above mentioned decision T 1066/03 will be discussed, which illustrates such a situation of an irredeemable flaw *ab initio*. The conclusions of this decision apply *mutatis mutandis* to co-crystal inventions.

16.2.3
Novelty of Co-crystals

The claimed co-crystals must be distinguishable not only from pre-existing co-crystal forms of the same substance but also from salts and solvates that involve the same co-crystal former, in another role.

The assessment of novelty is therefore, in principle, relatively easy for co-crystal inventions provided that adequate parameters, particularly the physicochemical parameters, can be used, this coming in addition to the definition of the constituents involved.

In this latter respect, co-crystal inventions can hardly be generalized since each co-crystal formation is practically "accidental." Therefore, most claims will, by force, mention one specific API and one specific co-crystal former. This greatly facilitates establishing whether the subject matter of such an invention is novel or not.

There are three standard situations with co-crystal applications:

The first one is when a co-crystal of a specific API "A" and a crystal former "X" is claimed and when the prior art discloses co-crystals of API "A" with defined compounds. If the co-crystal in question is neither mentioned not exemplified in the prior art document, the co-crystal is regarded as novel.

The second case is where the prior art discloses co-crystals of "A" with a co-crystal former "B, C, ..., X or Z" and where the co-crystal of "A" with "X" is not specifically exemplified. In such a case, the inventor is confronted with a typical case of a selection from one list, and an objection of lack of novelty is likely to be raised and maintained until rejection unless the prior art does not provide an enabling disclosure for the co-crystal of "A" with "X."

The third common case is where the co-crystal of API "A" with "X" is characterized as recommended in 2.1 above by experimental data, for example, a powder X-ray diffraction (PXRD) spectrum (indication of most relevant values of 2θ), and the prior art also discloses alleged co-crystals of API "X" with "A" but only exemplified and characterized, for example, by its NMR spectrum and its melting point. The patent examiner is likely to raise an objection of lack of novelty, *prima facie.* The applicant, nevertheless, has the possibility to overcome the objection by preparing the alleged prior art co-crystal, establishing its PXRD spectrum, and showing that it is different from the co-crystal of the invention.

16.2.4 Inventive Step of Co-crystals

According to the problem–solution approach, the difference exhibited by the claimed co-crystals over the prior art must produce a technical effect different in intensity (much more rarely, in nature) from what the prior art form of the API achieves.

In the case of the co-crystals, the question of what is the closest prior art must be posed, which is not always a straightforward one to answer.

It must be well understood that the principle governing the problem–solution approach is that the technical problem plays a central role. The choice of the closest prior art is therefore not primarily guided by the structural similarities with the invention. Thus, if the prior art discloses a co-crystal of the same API with another crystal former than the invention but refers to properties that do not exhibit any relationship with the pharmaceutical activity of the invention, the closest prior art may be another document that involves, for example, the API in mechanical admixture with an adjuvant having a certain similarity with the crystal former, provided the document in question relates to the same technical problem in the pharmaceutical field as the invention alleges to solve.

In general, no co-crystal with the API is known in the prior art, and the proposed co-crystal is the first one and has the same use as the API. But in future, it is likely that previous co-crystal(s) will sometimes constitute the closest relevant prior art for new co-crystals of the same API.

The difference between the prior art will therefore be, in general, that the API, instead of being in a crystalline or amorphous form, is proposed as a specific co-crystal.

The difference is then very well identifiable, and it will induce a technical effect and properties different from those of the previous forms. Typically, the solubility, dissolution rate, chemical stability, melting point, and the hygroscopicity will be influenced.

The technical problem to be solved will then typically be an increase of the bioavailability (solubility, dissolution rate), storage ability (stability properties: chemical or humidity uptake), and/or workability (melting point).

It is not infrequent that co-crystals exhibit polymorphism as well, with this exerting a further influence on the above properties.

Whether the invention itself solves the technical problem is a matter of proof in the hands of the patent applicant. The breadth of the claim being relatively narrow due to the specificity of this type of invention, whether the technical problem is solved for the whole ambit of the claim will, in general, not be a difficult issue. Still, the identification of a co-crystal with enhanced properties will with difficulty allow – in the absence of sufficient evidence – the extension of the claim, for instance, to other (non-exemplified) co-crystal formers. When an inventor argues that a co-crystal of an API is surprisingly formed, it is contradictory – in the absence of experimental evidence – to state that the same API could form also co-crystals with other crystal formers.

The discussion of whether solving the problem was obvious to the skilled person is a matter that is case-dependent, but there is no reason to think that specific standards for this discussion should be involved. Those normally applied to compositions or to compositions added with parametric conditions will apply.

Clearly, if a sophisticated, unusual co-crystal former is used, or if the chosen co-crystal former had been considered as inappropriate by the prior art, this should play in favor of the recognition of an inventive step. If the process for making the co-crystal is the core of the invention, the merits of the claim may be shifted, in part or in totality, to the selected process conditions.

16.3 Patentability of Amorphous Forms

APIs in amorphous – noncrystalline – form exhibit, in general, enhanced solubility due to their high thermodynamic energy level. Certain very lipophilic substances can thus be rendered more bioavailable than any crystalline form of the same. This high surface energy also causes the water absorption of the amorphous forms to be, in general, greater than in the corresponding crystalline forms, if any, where the water is more often adsorbed than absorbed. Amorphous forms can therefore be potentially more attractive for the first reason than crystalline forms or even co-crystals but more difficult to formulate or to store for the second reason.

In any case, several problems arise when considering the patentability of these forms.

First, a full definition and correct characterization of an amorphous form are not easy. This will cause a specific problem in terms of the claim's clarity. In addition, it is rarely possible to obtain a fully amorphous product because, in general, a certain proportion of the API is still in the (micro)crystalline form. It is never easy to describe clearly (and sufficiently) an amorphous form that, in general, presents large zones lacking long-range order but almost always other zones having local molecular assemblies. This will also cause specific problems with respect to sufficiency of disclosure directly linked to lack of reproducibility. In addition to this, the variability of the amorphous form may change from one production batch to another, and, because of thermodynamics, the forms are rarely stable over long periods. A consequence is that comparisons for establishing novelty and for assessing the inventive step are complicated by these variability and stability aspects and are delicate, as the prior art may itself suffer from the same uncertainties or lack of clear definition and/or characterization as the claimed invention.

Amorphous forms are, in general, more difficult to patent than any other solid-state form. They are, as a matter of fact, the forms under which substances are produced when no special care is taken at the final stage of a synthesis process and hence they can be regarded as being *prima facie* quite obvious. However, it is a quite different matter when one succeeds in controlling the achievement of the amorphous form more strictly and is able to master well the degree or the proportion of reasonably stable amorphization.

16.3.1 Clarity Problems in Claims for Amorphous Forms

16.3.1.1 Pure Amorphous Forms

It is rare that a fully amorphous form of any substance can be obtained. Unavoidably, some crystalline forms are included in the amorphous mass, which will anyway spontaneously tend to recrystallize in course of time. However, patent specialists can accommodate most aspects of this kind of situation. There is no doubt that transient products can be patentable, although they can obviously pose problems in the implementation of the claim in case of a patent suit. Compared to other substances, such as certain catalysts, amorphous forms have in any case (in general) a rather slow rate of recrystallization.

In general, when an amorphous form of an API is claimed, it is specified in the description of the application or even in the claim that it also includes a certain number of zones exhibiting one or several crystalline forms. The definition of the complex, pure product thus obtained can rely upon any standard parameter recognized in the field. XPRD, thermal analysis (e.g., DSC and thermogravimetric analysis (TGA), and solid-state nuclear magnetic resonance (SS-NMR), *inter alia*, can be used to determine and/or quantitate the amorphous form present in an API. Fourier transform (FT)-Raman/spectral measurements are often used, and patent examiners certainly welcome the introduction of these parameters in the claims to ensure their clarity. With respect to pair distribution functions (PDFs), it must be underlined that PDF analysis is essentially useful for the purpose of comparing one amorphous form with another. Even if these parameters are difficult to express using an absolute value, they are preferable to a – sole – negative definition such as the absence of individual peaks in X-ray powder diffraction (XPRD). However, the form and position of the diffuse halo obtained can help to distinguish one amorphous form from another, even when the presence of a halo – alone – might not be sufficient to distinguish the form. Obviously, enthalpy, volumetric relaxation, or water vapor absorption, which can also characterize an amorphous form, can be used as additional parameters for a proper definition of an amorphous form.

Ideally, a claim for a pure amorphous form should therefore indicate, in addition to the API's name and to the statement that it is in an amorphous state – some indication on its degree of amorphousness and how this degree was established. Alternatively, a reference to the process by which the form was obtained can be used, although the resulting protection is more difficult to enforce.

A claim referring to a compound in a "substantially" amorphous form can be accepted in the case of a new API as a sub-claim of the API itself, but this format is poorly appropriate when the API is already known and when there is a prior art relating to the API itself that does not specify in which form it was obtained. The amorphous form being the most probable, the examiner – even if he or she does not object to the clarity of the claim – will certainly exploit the breadth of such a claim in order to raise an objection for the lack of novelty.

One question that is often posed is whether an amorphous form can be claimed in the form of a plurality of independent claims (i.e., with claims that do not refer to another one): for instance, by claiming an amorphous form of an API "X" characterized by its DSC, then also and independently by claiming an – same – amorphous form of "X" characterized by its IR spectrum, and so on. This is clearly not advisable, as the examiner will almost certainly argue that, as these claims relate obviously to the same entity, the multiplicity of independent claims is not justified and even renders the claims, taken as a whole, unclear. A far better way to claim in such a case is to have only one independent claim 1 for the amorphous form of "X" characterized by a first parameter that is regarded by the applicant as one of the most relevant and reliable ones in the specific case, and then one or several claims *dependent* on claim 1 and characterized by another additional parameter. It is not necessary in most cases, *prima facie*, to have only one claim with all the different parameters at hand in it, unless the disclosure of the prior art requires it. It is also not advisable to associate in the same independent claim parameters the applicant is sure of with parameters the relevance of which is more disputable or whose accuracy is more uncertain. It is also not advisable – if this is not the case – to state in a claim that all such parameters must be present at the same time: the deletion of one of these parameters will be difficult at a later stage, if not impossible.

16.3.1.2 Mixtures of Amorphous (and/or Polymorphic) Forms

In recent years, patent applications have appeared in which the API was present as a combination of amorphous and/or polymorphic crystalline forms. Often, microphotographs are filed that illustrate a sample of the product. These forms are very difficult to determine and their definition needs in any case a clear indication of the approximate proportion of each form present. A reference to a figure cannot be used in a claim for the purpose of improving its clarity. A reference to two amorphous forms that are macroscopically distinguishable but both exhibiting an XPRD pattern that does not contain sharp peaks is not sufficient to impart clarity to the claim. Then, it seems unavoidable to include a reference to the process leading to the mixture of different forms.

This situation must be distinguished from that of a mechanical mixture of amorphous forms obtained separately by different processes that can be handled as an ordinary mixture of two solids. The clarity of a claim pertaining to such a mixture will mainly depend on the clarity of the definition of each amorphous component of it. The same applies to mixtures of amorphous form(s) and polymorphic form(s), be they obtained in a one-pot process or by mixing these forms, obtained separately by different processes. Invariably, whenever an amorphous form is involved, the reference to the process for obtaining it in a claim is an ultimate, but relatively good and safe, option provided the disclosure of the process is detailed and sufficient.

16.3.1.3 Mixtures Comprising Amorphous Forms and a Stabilizer

The spontaneous tendency of amorphous forms to crystallize can be countered by various methods. Mechanical means can be best protected by process claims. Chemical additives, for example, polyvinylpyrrolidone (PVP), can play a role as recrystallization inhibitors. In such a case, the compositions comprising the inhibitors can potentially be the object of patent protection. The problems are exactly the same as encountered above under points 3.1.1 and 3.1.2 as far as the amorphous form itself is concerned. The presence of a stabilizer merely reinforces the characterization of the nature of the invention (invention of a composition) and changes the technical problem to be solved from the provision of a mere amorphous form in general to the provision of a stabilized one. This kind of claim for compositions does not pose any unusual clarity problem.

16.3.2 Sufficient Disclosure of Amorphous Forms

Amorphous forms can be obtained by many different methods, which are part of the routine work of the skilled person. However, it is indispensable in the description of a patent application to pay sufficient attention to the fair recitation of the means and conditions that have to be deployed in order to obtain them.

A board of appeal of the European Patent Office has issued the already mentioned decision, referred to as T 1066/03, which is exemplary in this respect. The object of the patent was a process for preparing an amorphous form of the hemi calcium salt of atorvastatin. The starting material used in the process was a crystalline form I, which, unfortunately, was spelt out neither in the description nor in the prior art. All examples of the patent specification were rather detailed, and all used seed crystals of this crystalline form I, but no method for obtaining them was disclosed. In opposition proceedings, the opposition division decided to revoke the patent for lack of sufficiency of disclosure, and this decision was confirmed by the board of appeal.

This decision demonstrates the danger associated with an incomplete description of the process for obtaining an amorphous form. As already indicated before for dislocated crystals, all solid-state form inventions are based on the use of well-known techniques, and the steps that are critical for obtaining them are often overlooked and possibly not well identified by the inventors themselves. Failing to identify these steps and to disclose them will generally result in a difficult substantive examination and most often in a rejection of the patent application or a revocation or nullification of the patent if by chance it had been granted. This flaw is intrinsic to the patent application and does not depend on the prior art. Hiding certain critical steps of a process for obtaining an amorphous form has, in general, sooner or later, disastrous consequences.

It must again be stressed that solid-state inventions do not relate to new chemical entities, for which the approach could be more lenient. They relate to a very specific and, in general, small improvement, which incites the patent examiner to adopt strict approaches.

16.3.3
Novelty of Amorphous Forms

A relatively favorable situation is when an amorphous form is proposed for the first time because the API was always obtained in the prior art in a crystalline form. In such a case, the recognition of the novelty of a first amorphous form will be a straightforward matter.

In most cases, the prior art, however, either describes the API in an amorphous form or does not specify in which form it was obtained.

In both cases, novelty will be challenged systematically. The milestone decision T 296/87 (30/08/1988, "Enantiomers") of the board of appeal of the European Patent Office has stated that " … a chemical substance is held to be new if it **differs** from a known substance **in a reliable parameter** … ."

This is another reason why amorphous forms are often defined in a claim by the parameter values. In general, an amorphous material still contains some zones exhibiting long-range order (a certain degree of crystallinity), which can be measured.

A frequent difficulty in that case is identified in the Guidelines Relating to the Substantive Examination of the EPO G-VI, 6 (version of November 2015): " …

It may happen that in the relevant prior art a different parameter, or no parameter at all, is mentioned. If the known and the claimed products are identical in all other respects then in the first place an objection of lack of novelty arises ...."

The objection will be particularly strong if the process for making it is poorly disclosed in the description of the patent application and/or if the prior art teaches a process for making the API that is very similar to the one used and reported in the patent application. In such a case, appropriate comparative tests may be required in order to establish novelty. It is in any case possible to distinguish – by using additional parameters – two amorphous forms exhibiting by chance the same halo in XRPD when they result from different sources, such as milling of the crystalline API and extruding a melt mass of the same API.

An even more favorable situation is when the prior art document can be shown to be not enabling. According to the Guidelines G-VI, 4 of the EPO (version of November 2015), "... a chemical compound, the name or formula of which is mentioned in a prior art document, is not thereby considered as known, unless the information in the document, together, where appropriate, with the knowledge generally available on the relevant date of the document, enables it to be prepared and separated ...." It is possible to refer to decision T 605/02, from which it results that a prior art document that discloses the analytical data of an amorphous form but does not contain sufficient information for its preparation is NOT enabling.

If a claimed amorphous form is defined by parameters, the situation may arise where the prior art discloses an amorphous form defined without parameters or using other parameters than in the claimed invention. The comparative tests necessary for showing that the products are different, to be convincing, unavoidably would involve the preparation of the amorphous form of the prior art, applying the enabling process disclosed in the prior art as faithfully as possible and exhibiting any property announced by the prior art. The best strategy in such a case is certainly to ask an independent entity or expert to carry out the necessary work. If the parameters exhibited by the prior art amorphous form are clearly different from those of the invention's claim, novelty is established.

The problem is more difficult when the difference between the two products is not clear-cut. For instance, some parameters can be identical for both forms and some others different. Depending on the distinctiveness of the parameters that are different, novelty may or may not be acknowledged by the examiner or the judge. In this respect, decision T 885/02 (decision of 15/12/2004, "Paroxetine methanesulfonate/SMITHKLINE BEECHAM") (see point 3.4.10), which was issued in a case involving polymorphs, is certainly applicable to amorphous forms. Using certain – secondary – IR peaks or secondary properties, which are not the ones that the expert would consider as decisive and important for the characterization of the amorphous form, can fail to prove novelty. Similarly, decision T 226/98 (decision of 07/02/2001, "Famotidine/RICHTER GEDEON") has clearly shown that the parameter used for distinguishing the invention from the prior art solid form must be a reliable one. In that case, relating to a polymorph but applicable to the amorphous forms, the melting points were very close to each other. The fact that the processes for obtaining the solid forms were different was not sufficient to convince the Board of Appeals that the two forms were different. However, this should not discourage innovators since it is always possible by using (an) additional appropriate parameter(s) to show the difference between two amorphous forms exhibiting some identical parameters if one is a true solid dispersion and the other is phase-separated.

A recent decision of a board of appeal of the EPO has, however, brought to light the possible definition of broad peaks characterizing an amorphous form. In T 1723/10 (01/10/2013, "Amorphous Forms/ZENTIVA"), the Board stated that a broad peak defined in an X-ray diffractogram by its extension between 15° and 25° for 2θ and its center (at 17°) was to be considered as anticipated by a disclosure exhibiting a peak also centered at 17° and defined by an interval of 12.5°–22°. However, a claim restricted to another amorphous form (which was defined as "semi-crystalline" by the Board) characterized by a broad band between 17.4° and 20.2° and two sharp peaks that could not be attributed to impurities (but rather to the probable presence of a certain proportion of crystalline material) was novel over the same prior art.

16.3.4
Inventive Step of Amorphous Forms per se

Inventions of amorphous forms, like other inventions of solid forms, are submitted to methods usually employed for assessing the inventive step.

In patent systems where this assessment relies upon an approach that is as objective as possible, as in the European Patent Office, new amorphous forms will be first confronted with the problem of the choice of the closest prior art.

If the API concerned crystallizes easily, the candidate for being the closest prior art is likely to be a crystalline form characterized by its general description (typically, by the form or habit of the crystals, their color, their melting point, etc.) or by parameters such as spectroscopic data in the case of a polymorphic form.

If the API concerned does not crystallize easily, the candidate for being the closest prior art is likely to be another amorphous form. If no form is indicated, implicitly, it will generally be assumed that the prior art is an amorphous form.

However, the final choice of the closest prior art depends on whether the technical problem allegedly solved by the new amorphous form according to the inventor's own declarations is also solved or tentatively solved by the prior art.

The technical problem to be solved results from an objective chain of deductions.

First, the difference(s) that exist(s) between the closest prior art and the invention must be ascertained. In the case of a polymorphic form, the difference will be that the API exhibits the feature characterizing an amorphous form. In the case of another amorphous form, the difference will be the different values of the parameters characterizing an amorphous state.

These differences – which invariably must exist, otherwise the invention would not even be novel – produce a technical effect. This effect may affect directly the intrinsic property of the form or become visible only when the form is used, for instance, in a given galenic form.

As a consequence of this technical effect, a technical problem is solved by the amorphous form, which is significant for the envisaged field of use. That is the reason why the closest prior art to be considered is not necessarily another amorphous form but possibly a crystalline form if it is unknown whether the previous amorphous form solved the problem (or any other technical problem) whereas the crystalline form did. Obviously, it will often be the case that the relevant closest prior art will relate to another amorphous form if both a prior art disclosing such an amorphous form and a crystalline form exist.

The technical problems that can be solved are potentially too numerous to be all cited here. The limited solubility of an API is a general problem that can be solved by providing an amorphous form of that API, especially if the API was only known in a poorly soluble crystalline form (e.g., $<1\,\mu g\,ml^{-1}$). The technical problem to be solved will then be presented as increasing either the solubility of the API or its bioavailability, which is directly influenced by the former. Also, the stability of an amorphous form is of paramount importance for the storage of any composition containing the API. Increasing the storage stability of a form can constitute a typical technical problem to be solved. Amorphous forms have also frequently the ability to increase the facility with which the API can be formulated, for instance, as a compressed tablet. The control of the water uptake or wettability (hygroscopicity), the thermal stability, and the shifting of the glass transition temperature are also properties that can be influenced or improved and can be translated into technical problems to be solved with high significance in galenics.

All these problems, however, are based on technical effects, which are basically foreseeable as being inherent to any amorphous form coming after a crystalline form. Therefore, the main difficulty with which the inventor of a new amorphous form will be confronted is that the skilled person would have expected that any amorphous form would solve the corresponding technical problem. A solubility differential of amorphous over crystalline of up to 1000 is reported to be expected. It is consequently to be expected by inventors that the patent examiner will consider an amorphous form coming after a crystalline form as *prima facie* obvious. However, for instance, if the amorphous form was really not easy to obtain as documented by long and fruitless efforts by others or requiring a special treatment to become stable, the involvement of an inventive step can be acknowledged. Also, it has been reported that the expected bioavailability increase is not necessarily obtained if the amorphous form does not stay amorphous after the solubilization process has started. The desirable supersaturated solution that was initially present can then vanish, thereby leading to disappointing results. The existence of such more subtle phenomena should lead the patent examiner (and the judge) to refrain from too hasty conclusions as to obviousness or non-obviousness of amorphous state inventions.

In the decision T 1723/10, already mentioned above, the semicrystalline product that could overcome the novelty hurdle over the prior art was also recognized as involving an inventive step based on its superior dissolution properties.

16.3.4.1 **Other Inventions Involving Amorphous Forms**

Although numerous patent applications relate to amorphous forms *per se*, other inventions can be claimed that involve amorphous forms: two examples are processes for making an amorphous form, and pharmaceutical compositions comprising such forms.

Processes for making amorphous forms are in general of three types. One consists of obtaining an amorphous phase using solid- and solution-based methods from an API, generally in the crystalline form. These processes may take quite different technological forms: for example, milling (comprising also cryogrinding), lyophilization, and quench-melting. Another type consists of stabilizing amorphous forms (e.g., via amorphous phase dispersions). The last type consists of transforming one amorphous form into another, in general, more stable or showing an extended proportion of the amorphous form.

All these processes require, besides a clear recitation of the essential features of the process, a solid and exhaustive disclosure of the latter in the description of the patent. Also the features of the final amorphous product must be given in a very precise, in principle parametrically defined, manner. The assessment of the novelty and the inventive step is carried out as for any other process claim. Consequently, it will often be difficult to obtain a (broad) patent protection for such processes, which are in general all based on well-known technologies.

Pharmaceutical compositions comprising amorphous forms are treated like any other compositions. However, for the inventive step, the technical problem is more distanced from the amorphous form itself, and is that solved by the composition as a whole. In the case of compositions comprising different amorphous forms or amorphous form(s) and crystalline form(s), both must be carefully defined and the merits (such as a synergy, a different, better liberation profile, or an improved storage stability) of the composition must be carefully documented: basically, the composition is otherwise considered as obvious as *prima facie* exhibiting the added properties of each of the forms mentioned.

16.3.4.2 **Glass Materials**

These materials are amorphous forms exhibiting a glass transition phase, that is, a reversible transition from a hard and relatively brittle state into a molten or rubber-like state.

Glassy materials are mainly obtained by processes involving passing the API in general through a molten state, whereas other amorphous forms result from freeze-dried systems. However, some glassy materials have been obtained using quench-cooling and spray-drying.

Accordingly, all that has been said above regarding amorphous forms applies to glassy materials as well.

Obviously, the glass transition temperature and all parameters that can reflect the glassy state of such forms can be used in addition to the parameters that can be used for other amorphous forms.

As explained throughout the preceding paragraphs, the conditions for measuring the characterizing parameters should preferably be specified in the description. Thus when measuring T_g, the experimental conditions should be specified, in particular, the heating rate and the method (DSC in general).

Glass matrixes where the glassy material is dispersed in an also glassy matrix, such as a sugar, are treated like standard compositions.

16.4 Patenting (Disordered) Nanocrystals

Nanocrystals have in common with amorphous forms that they have lost their long-range crystalline order and have become disordered. However, they are not amorphous, which renders their characterization slightly easier than that of amorphous forms. They are, in general, the result of a long and thorough grinding process whereby the crystals may also undergo disordering due to high local mechanical forces in the grinder. By this typical process, the properties of the substance are continuously influenced: the peaks observed in XRPD continuously broaden upon grinding.

Nanocrystals being highly comminuted, the (average) particle size is also a characterizing parameter for them.

16.4.1 Clarity Problems in Claims for (Disordered) Nanocrystals

There is no general agreement on a definition of nanocrystals. Depending on the technical fields and their use, the upper limit for their particle size may vary between 500 and 200 nm. For pharmaceutical uses, it would appear that 200 nm is generally regarded as a maximum.

Accordingly, using the expression "nanocrystal" without further precision may be regarded as unclear in a patent claim. The lack of clarity will become acute when, in practice, particles of less than 200 nm are meant where the prior art discloses particles with an average particle size around 500 nm. It is therefore advisable to define the particle size in connection with the use of the expression "nanoparticle" to be on the safe side.

Disordered nanocrystals might be relatively more difficult to define than the corresponding crystals since some of the physicochemical methods used for defining "crystals" will not be usable in their case (e.g., XPRD). Characterization of the "disorder" is, however, possible using some of the methods applied to "dislocated crystals."

It is important to tackle the problem seriously in the description (at least, and preferably also in the claim) of a patent application. Characterizing features cannot be added after the filing date, as this would be beyond any doubt regarded as added subject matter.

The option of defining the nanocrystals as the result of a grinding process is not a very good one since it is probable that the claim drafted on the basis of this indirect definition could be easily circumvented by a competitor.

16.4.2
Sufficient Disclosure of (Disordered) Nanocrystals

Although the above-mentioned problems can impact the criterion of sufficiency of disclosure as well, the issue here is more that of reproducibility. Nanocrystals are in general less prone to instability than their amorphous forms. However, the difference between nanocrystals and normal, very fine crystals is not an obvious one. It is therefore important to disclose in the description of a patent application how they are obtained. The conditions under which the grinding operation is carried out (type of grinder, duration, temperature, etc.) are features that should be disclosed with accuracy, as they are essential for reproducing the object of the invention.

Clearly, by defining some of the conditions only, it will become often obvious that the process does not differ in an apparent manner from other comminuting methods. By giving all due care to the working conditions, inventors of nanoparticles will increase their chance to fulfill not only the criterion of sufficiency but also to pass the hurdles of the tests of novelty and inventive step.

When the invention relates to the use of nanoparticles in a pharmaceutical composition, such as a tablet, the applicable standard is the one applied to compositions. Beside the indication of all components and their proportion in the composition, no particular recommendation is necessary. If the formulation process implies a wet process, it seems obvious that some indication is necessary in the description of the formulation process if a solvent is used that could cause a recrystallization of the nanoparticles.

16.4.3
Novelty of (Disordered) Nanoparticles

As already outlined above, the main concern of inventors of nanoparticles or compositions containing them will be to show that they are distinct from previous particles or compositions having a very small size.

Showing that the prior art particles are not enabling because the process for making them cannot lead to nanometer sizes is an option that can be successfully exercised only by proper argumentation or convincing comparative data.

Basically, nanoparticles can be recognized as being novel compared to particles exhibiting a larger size. Thus the particle size should preferably mentioned in the claim, and a precise method for measuring it should be recited in the description, possibly by reference to a publication recognized in the field.

16.4.4
Inventive Step of (Disordered) Nanoparticles

This patentability criterion may be the major obstacle to the patentability of nanocrystals: it is rare that there are major technical obstacles to the grinding of an API down to the nanometer sizes. If so, there exist standard solutions in order to avoid, for instance, thermal degradation of the product upon prolonged grinding. Therefore, in general, the inventive step will be difficult to demonstrate. A nano-product can be distinguishable from another nano-product that has been ground for a shorter time by exhibiting a broader halo, but being quite expected, this is of no help in the assessment of the inventive step.

16.5
Conclusions

Patents in the field of disordered solid forms of APIs can be obtained provided certain strict precautions are taken. A prerequisite is that the invention constitutes a significant advance, in particular, with respect to the bioavailability of the API and/or its industrial formulation. Because of their *prima facie* relatively small input to technology, particular attention must be paid to the claims relating to these inventions. The parametric definition of the forms must be exact and refer to well-known parameters. Also, the patent application must contain a clear and complete description of the processes for obtaining these forms.

Taking these precautions into account will reduce the difficulties applicants would otherwise encounter with patent examiners and, later, with judges. It will also increase the acceptance of patent applications relating to these inventions in countries that object to their patenting.

Index

Disordered Pharmaceutical Materials, First Edition. Edited by Marc Descamps.

d

p